BASIC BIOMECHANICS

BASIC BIOMECHANICS

Susan J. Hall, Ph.D.
Department of Kinesiology
California State University, Northridge
Northridge, California

with 478 illustrations

St. Louis Baltimore Boston Chicago London Philadelphia Sydney Toronto

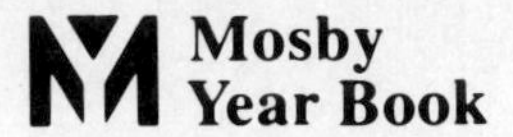

Dedicated to Publishing Excellence

Editor Vicki Van Ry
Senior Developmental Editor Michelle Turenne
Project Manager Carol Sullivan Wiseman
Book Designer Susan Lane
Illustrator Donald O'Connor Graphic Studio
Cover Image *Video digitized dancers*
© Thomas Porett/Photo Researchers, Inc.

Printed in the United States of America

Mosby–Year Book, Inc.
11830 Westline Industrial Drive
St. Louis, MO 63146

Library of Congress Cataloging in Publication Data
Hall, Susan J. (Susan Jean)
Basic biomechanics.

Includes bibliographical references and index.
1. Biomechanics. 2. Biomechanics—Problems, exercises, etc. I. Title.
QP303.H35 1991 612.7'6 90-19975
ISBN 0-8016-2087-2

CL/DC 9 8 7 6 5 4 3 2 1

PREFACE

I often say that when you can measure what you are speaking about
and express it in numbers, you know something about it; but when you cannot
measure it, when you cannot express it in numbers, your knowledge is
of a meagre and unsatisfactory kind.

Kelvin, 1891

Biomechanics is a quantitative field of study. Unlike many of the existing textbooks, *Basic Biomechanics* was written to provide an introduction to biomechanics without deemphasizing the quantitative nature of the topics presented. This approach includes the use of both quantitative and qualitative problems and applications that are designed to illustrate biomechanical principles. However, this book is also written with sensitivity to those students entering undergraduate courses in biomechanics who may possess limited backgrounds in mathematics. Consequently, the quantitative aspects are presented in a manageable and progressive fashion throughout the text. A portion of the first chapter is devoted specifically to problem-solving skills. A review of relevant mathematical procedures is also provided in Appendix A.

There are two advantages to using this approach. First, the concepts presented become more concrete and meaningful to students when numerical problems are given *and* when students are able to calculate the answers based on those concepts. Second, the review and use of basic mathematics in structured applications is of inherent educational value.

CONTENT HIGHLIGHTS

Several content features included in *Basic Biomechanics* will be effective in both learning and applying the concepts.

Balanced Coverage

The Kinesiology Academy of AAHPERD has recommended that preparation for undergraduate students in the area of biomechanics should be devoted to approximately one-third anatomical considerations, approximately one-third to mechanical considerations, and the remainder to applications. All three topical areas are important for a comprehensive understanding of the biomechanics of human movement. However, to treat these as though they are mutually exclusive of each other is inappropriate. Therefore, a second approach used in *Basic Biomechanics* is the integration of all three of these considerations throughout the text. Although some chapter titles may suggest that the topics addressed therein are largely either anatomical or mechanical, mechanically based applications are included in the anatomical chapters and vice versa.

Applications Oriented

All of the chapters contain discussions of a broad range of human movement applications, many of which are taken from the recent biomechanics research literature. These include examples derived from activities of daily living, as well as from exercise, sport, and clinical settings. The applications chosen also span the human age range from a child first learning to walk to the modified gait of an elderly individual using a cane.

Sample problems and solutions: To further integrate the applications coverage throughout the text, I have included sample problems with solutions from a variety of situations. For example, sports-related problems can be found on pp. 14 to 15 and p. 279, and samples from everyday activities can be found on pp. 54 to 55. These clearly identify the correct problem-solving progressions, and they are consistently presented to follow the format of the problem-solving skills described in Chapter 1. Helpful illustrations are also included where appropriate to assist in developing problem-solving skills.

Qualitative and quantitative balance: Each chapter concludes with introductory and additional problems for students to complete. These consist of both qualitative and quantitative applications designed to further illustrate the chapter's content. Here again, helpful illustrations are included where appropriate to assist in developing problem-solving skills.

Organization

Chapter 1, "Problems to Solve: Biomechanics and the Problem-Solving Approach," provides an introduction to the text. It overviews the field of biomechanics, includes a description of the types of research conducted by biomechanists, and explains the strategy for problem-solving that is used throughout the text.

Chapter 2, "The Basics: Terminology and Simple Mechanical Concepts," introduces important basic mechanical concepts used throughout the remainder of the text.

Biomechanical aspects of the skeletal and neuromuscular systems and of joints are discussed in Chapter 3 ("Bones: The Biomechanics of Bone Growth and Development"), Chapter 4 ("Muscles: Biomechanical Aspects of Neuromuscular Function"), and Chapter 5 ("Joints: The Biomechanics of Human Skeletal Articulations"), while the biomechanics of the upper and lower extremities and the spine are addressed in Chapter 6 ("Movement: The Biomechanics of the Upper Extremity"), Chapter 7 ("Movement: The Biomechanics of the Lower Extremity"), and Chapter 8 ("Movement: The Biomechanics of the Spine and Pelvis"). Coverage of bone and the spine as separate chapters (in Chapters 3 and 8 respectively) allows for the inclusion of topics of recent concern such as osteoporosis and low back pain.

Chapters 9 and 10 present the topics of linear and angular kinematics. Linear kinetic concepts, including use of the equations for static equilibrium, are covered in Chapter 11, and Chapter 12 addresses equilibrium. Chapter 13 is devoted to angular kinetics, while Chapter 14, "The Importance of Fluids: Introduction to Fluid Mechanics," covers fluids concepts including theories of propulsion in human swimming.

Chapter 15, "Assessment: Movement Analysis," presents strategies for analyzing human movement, along with a description of the tools commonly used for the collection of human movement data.

Although the order of chapters in the book is designed for the presentation of concepts in a logically progressive fashion, an instructor may choose to deemphasize or omit selected chapters depending on the orientation of the course and the prerequisite background expected of students. For example, the content of Chapter 2 constitutes review material for students who have had an introductory physics course. Chapters 3 through 8 may serve largely as review for students who have had coursework in anatomy and physiology. However, the content of these chapters includes biomechanical considerations such as common joint injury mechanisms and the development of common pathological conditions such as osteoporosis and low back pain. Although Chapter 15 could be incorporated at an earlier point in the course, it is expected that students will understand the process of human movement analysis more thoroughly after covering Chapters 9 through 14.

PEDAGOGICAL FEATURES

Basic Biomechanics contains a variety of pedagogical features designed to reinforce the content of the text. These include the following:

Chapter objectives: Listed at the beginning of each chapter, these introduce students to the points that will be highlighted. Accomplishing the objectives indicates fulfilling the chapter's intent.

Marginal definitions and notes: Additional information including definitions, notes, and tips are contained in the margins and are designed to assist student comprehension, to help apply the content learned, and to enhance overall visual appeal.

Chapter summaries: Each chapter has a summary outlining and reinforcing the major points covered.

Sample problems with solutions: Sample problems are presented from a variety of situations, including exercise, sports, clinical settings, and activities of daily living, and provide the correct problem-solving progressions to reinforce students' understanding.

References: Lists of the most up-to-date documentation are provided at the end of each chapter for the student who wishes to read further on the subject being discussed.

Annotated readings: Provided with annotations, these present additional resources for further information.

Appendixes: These include a basic mathematics review, trigonometric functions, common units of measurement, and anthropometric parameters.

Glossary: As a convenient reference, a comprehensive glossary of all terms defined in the text has been included at the end of the text.

ANCILLARY

The instructor's manual that accompanies *Basic Biomechanics* provides additional pedagogical materials, including solutions to selected problems from the text, additional activities, and resources. A test bank containing multiple choice and essay questions corresponds to each chapter of the text. Transparency masters conclude the manual. These include selected illustrations and sample problems to help explain more difficult concepts and to facilitate classroom instruction.

ACKNOWLEDGMENTS

The publisher's reviewers made excellent suggestions and criticisms that were carried out through three drafts of the manuscript. Their contributions are present in every chapter. I would like to express my sincere appreciation for both their critical and comparative readings of the early drafts:

Lawrence D. Abraham
University of Texas-Austin

Joseph Hamill
University of Massachusetts-Amherst

Phillip P. Watts
Northern Michigan University

Gail G. Evans
San Jose State University

Kathryn Lewis
Brigham Young University

Phillip E. Martin
Arizona State University

Maurita Robarge
University of Wisconsin-LaCrosse

William F. Straub
Ithaca College

Patricia Ann Sherman
East Carolina University

Ross E. Vaughn
Boise State University

Carole J. Zebas
University of Kansas

Jerry Wilkerson
Texas Woman's University

Sincere appreciation is extended to Ann Stutts, who very capably shot most of the photographs for the book, to Don O'Connor who did the line drawings with exceptional creativity, and to Michelle Turenne, Senior Developmental Editor with Mosby-Year Book, Inc., who has remained helpful and cheerful throughout the development of the book.

Susan J. Hall

CONTENTS IN BRIEF

CONTENTS

1 PROBLEMS TO SOLVE

Biomechanics and the Problem-Solving Approach

After reading this chapter, the student will be able to:

Define the terms biomechanics, statics, dynamics, kinematics, and kinetics and explain the ways in which they are related.

Explain the particular role played by biomechanics in the study of human movement.

Describe the scope of the research topics addressed by biomechanists from different academic fields.

Distinguish between qualitative and quantitative approaches for describing human movement.

Use the 11 steps identified in the chapter in solving formal problems.

Learning to walk is a momentous task for a young child. Different physical and neurological factors are involved in enabling a child to abandon crawling and assume an upright mode of locomotion.

We have all admired the fluid, graceful movements of highly skilled performers in various sports. We have also observed the awkward movements associated with the first steps taken by a young child, the laborious process of maneuvering a casted leg, and the hesitant, uneven gait of an elderly person using a cane. Virtually every physical education activity class includes a student who excels during the performance of every skill and a student who trips during a long jump takeoff or misses the volleyball when attempting to serve. What enables some individuals to execute complex movements so easily, while others appear to have difficulty with relatively simple movement skills?

Questions relating to human movement have been approached from different perspectives. Exercise physiologists, for example, would tell us that graceful movement requires adequate levels of strength, flexibility, and, for certain activities, aerobic capacity. Motor development specialists would relate movement patterns to the individual's maturational status, and motor control and learning specialists would question the appropriateness of motor planning. Sport psychologists would key in on the individual's state of arousal or relaxation or perhaps on the level of the individual's motivation. Another aspect of this question can be addressed from the perspective of biomechanics.

BIOMECHANICS: DEFINITION AND PERSPECTIVE

The term **biomechanics** became accepted during the early 1970s as the internationally recognized descriptor of the field of study concerned with the mechanical analysis of living organisms (14). Biomechanics has been aptly defined by Hatze as "the study of the structure and function of biological systems by means of the methods of mechanics" (5). Biomechanics involves the use of the tools of **mechanics,** the branch of physics involving analysis of the actions of forces, in the study of anatomical and functional aspects of living organisms. This relationship is shown in Figure 1-1.

The field of mechanics is composed of two major sub-branches known as **statics** and **dynamics.** Statics is the study of systems that are in a state of constant motion, that is, either at rest (with no motion) or moving with a constant velocity. Dynamics is the study of systems in motion in which acceleration is present. The study of a moving body may involve **kinematics** and/or **kinetics.** Kinematics is the study of the motion of bodies with respect to time or the precise pattern and speed of movement sequencing by the body segments that often translates to the degree of coordination an individual displays. Whereas kinematics describes the appearance of motion, kinetics is the study of the forces associated with motion, including forces causing motion and forces resulting from motion. The relationships among these subcategories of mechanics are displayed in Figure 1-2.

In addressing human movement, biomechanists examine the kinematics of the movement or the technique or form displayed by the

biomechanics
application of mechanical principles in the study of living organisms

mechanics
branch of physics that analyzes the actions of forces on particles and on mechanical systems

statics
the branch of mechanics that deals with systems not subject to acceleration

dynamics
the branch of mechanics that deals with systems subject to acceleration

kinematics
the study of the description of motion including considerations of space and time

kinetics
the study of the forces causing or resulting from motion

Figure 1-1
Biomechanics uses the principles of mechanics in approaching problems related to the structure and function of living organisms.

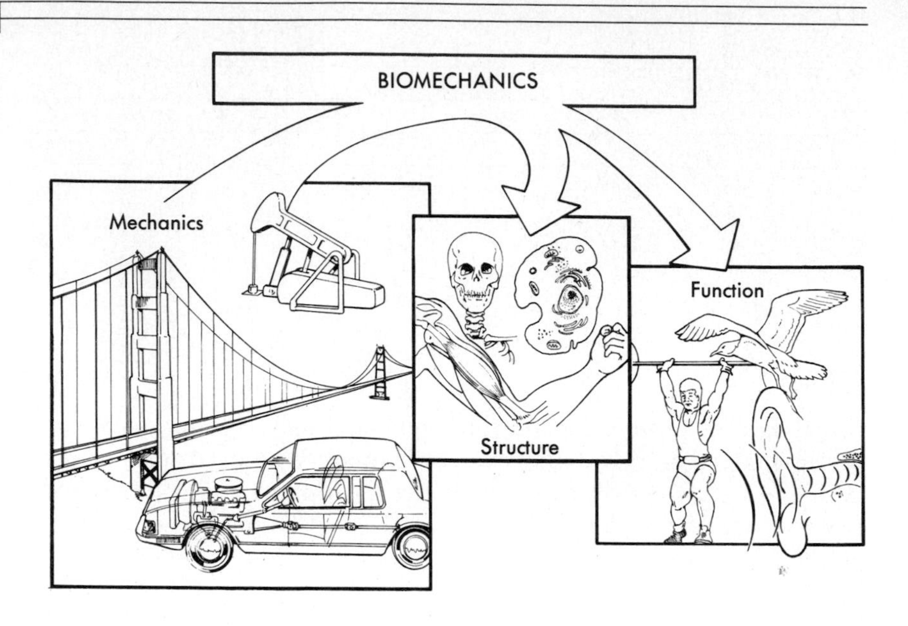

Figure 1-2
The sub-branches of mechanics.

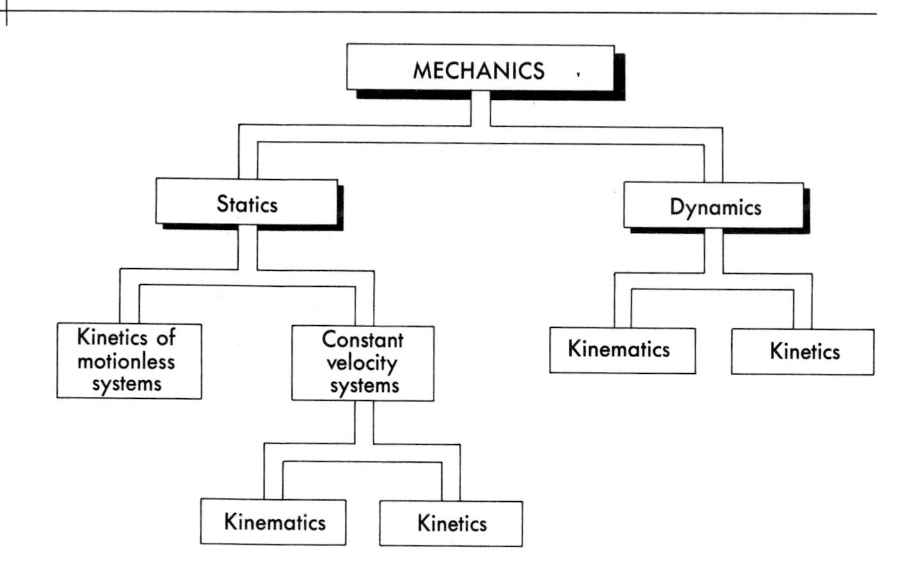

anthropometric
related to the dimensions and weights of body segments

performer. A biomechanical approach also involves questions pertaining to kinetics, such as whether the muscles produce amounts of force that are optimal for the intended purpose of the movement. **Anthropometric** factors including the size, shape, and weight of the body segments are other important considerations in this type of analysis.

Anthropometric characteristics may predispose an athlete to success in one sport and yet be disadvantageous for participation in another.

In reality, a combination of factors, such as inadequate strength, low motivation level, or improper sequencing of movement by the body segments, may limit the effectiveness and/or gracefulness of a given movement. What is clear, however, is that the study of human movement is multifaceted, with specialists from different areas of expertise (including biomechanists) often examining the same problem.

■ Courses in anatomy, physiology, mathematics, physics, and engineering are generally involved in the academic training of biomechanists.

Traditionally, courses in **kinesiology** (the study of human movement) have addressed topics pertaining to the structure and function of the human musculoskeletal system and have served as precursors for modern courses in biomechanics. However, the more literal semantic interpretation of kinesiology also produced courses entitled Psychological Kinesiology, Physiological Kinesiology, and Mechanical Kinesiology. In keeping with this broader usage of the term, many university departments of physical education changed their names to include the word *kinesiology*. In 1989 the American Academy of Physical Education voted to endorse the name *kinesiology* as the most appropriate descriptor of the entire field of the traditionally known area of physical education. As shown in Figure 1-3, biomechanics may be considered as a subdiscipline of kinesiology. In parts of the

kinesiology
the study of human movement from the perspectives of art and science

United States, however, *kinesiology* is still the title used for physical education courses that deal with the anatomical and mechanical bases of human movement.

sports medicine
clinical and scientific aspects of sports and exercise

As shown in Figure 1-4, biomechanics is also a scientific branch of **sports medicine.** Sports medicine has been defined by Lamb as "an umbrella term that encompasses both clinical and scientific aspects of exercise and sport" (9). The American College of Sports Medicine is an example of an organization that promotes interaction between scientists and clinicians with interests in sports medicine-related topics.

■ Biomechanics, exercise physiology, motor behavior, and sport psychology are considered to be subdisciplines of kinesiology.

Although biomechanics is relatively young as a recognized field of scientific inquiry, biomechanical considerations are a topic of interest in several different disciplines. Biomechanists may have academic backgrounds in zoology; orthopedic, cardiac, or sports medicine;

Figure 1-3
The subdisciplines of kinesiology.

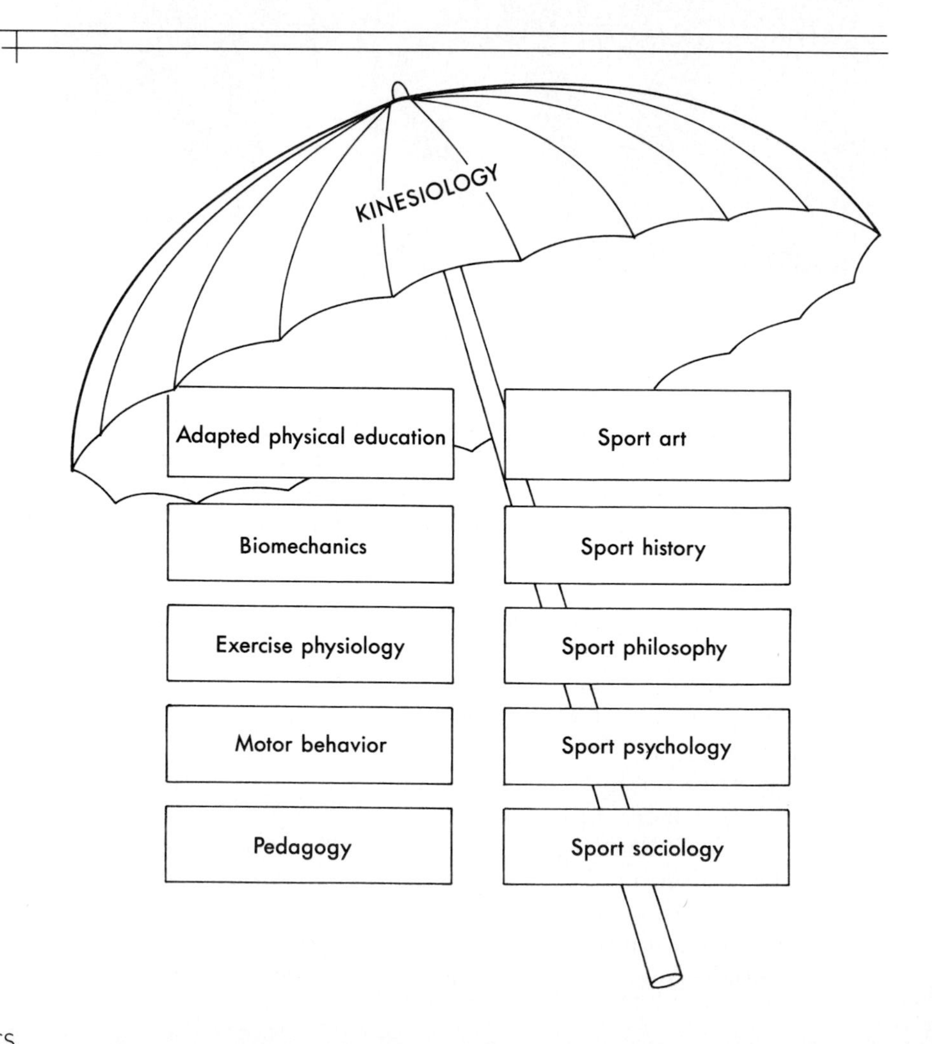

biomedical or biomechanical engineering; physical therapy; or kinesiology (physical education), with the commonality being an interest in the biomechanical aspects of the structure and function of living things.

Problems Addressed Through Biomechanics Research

As expected with the different professional and scientific fields represented by biomechanists, the questions or problems addressed in biomechanics research are diverse. In a biomechanical investigation conducted by zoologists (17), the locomotion patterns of 20 different species of animals were examined at controlled speeds on treadmills to identify the reasons for changes in gait pattern at particular speeds (for example, from walking to running or trotting to galloping). The researchers concluded that most vertebrates, including hu-

A single research study, investigation, or experiment is typically designed to address a single, often narrow, question.

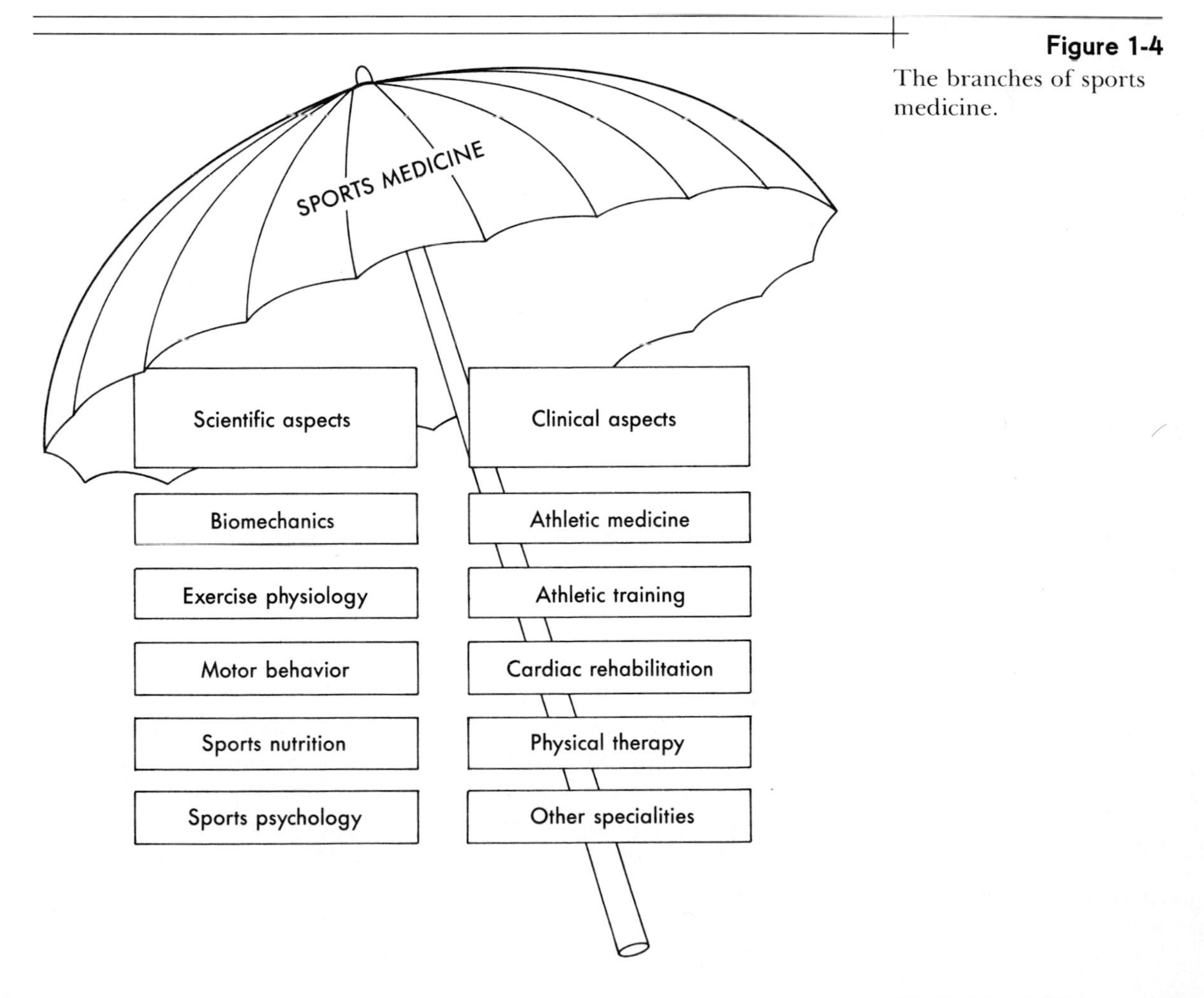

Figure 1-4
The branches of sports medicine.

Figure 1-5
Are cats easier or harder to motivate than humans?

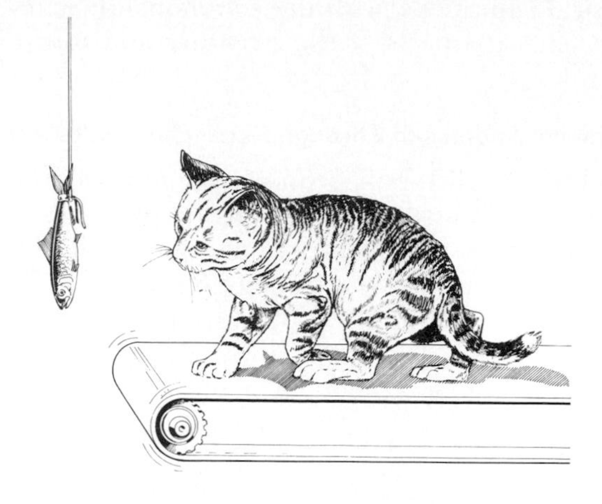

Race walking requires a greater caloric expenditure than running at the same speed.

■ The design of an experiment includes ways in which the experiment is planned, organized, and carried out and the procedures used for analyzing the data.

mans, select the gait associated with the most efficient use of body size and structure at a given speed. The results of the study suggested that Olympic race walkers expend considerably more caloric energy than runners performing at the same speed because running is a more economical gait pattern at fast speeds. Researchers must have been challenged by an experimental design that included persuading a cat, a pig, or a lion to run on a treadmill (Figure 1-5).

Other biomechanical issues addressed by researchers from subfields of medicine and engineering include the following: 1. How much weight can be safely lifted on a repetitive basis by an individual of a given strength in an industrial setting? 2. How can an artificial hand be designed to provide some volitional control by the person wearing it? 3. What manmade materials are sufficiently durable and compatible with biological tissues for use in artificial hearts and other implants? 4. How much and what type(s) of exercise does an astronaut need to perform to maintain bone and muscle strength in the absence of a gravitational field?

Engineers have also contributed to performance improvements in selected sports. The recent innovations in wheel and helmet designs for cycle racing have resulted from findings of experiments conducted in wind tunnels that involved controlled simulation of the air resistance actually encountered during cycling (8). Wind tunnel experiments have been conducted to identify optimal body configuration during ski jumping (15) and the effects of wind resistance and altitude on track events such as sprinting (1,7). Figure 1-6 demonstrates one way equipment can be tested in a wind tunnel.

Other efforts at improving athletic performance have been generated by sport biomechanists. It has been documented, for example, that the best world-class sprinters have longer strides, faster stride rates, and shorter support times than their less skilled counterparts

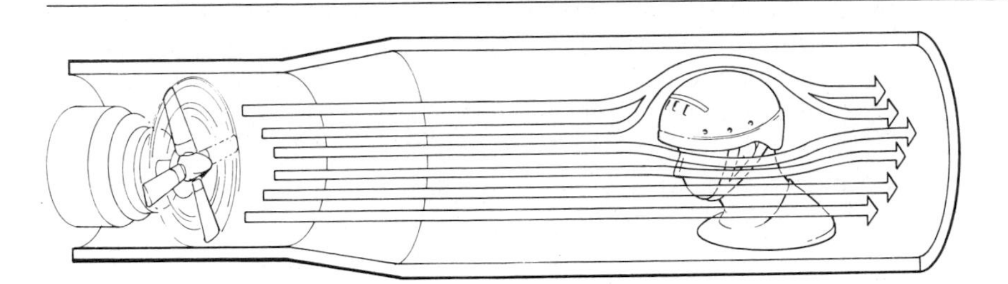

Figure 1-6

A wind tunnel is a large chamber designed to measure the drag forces created on objects at different wind speeds. New designs in cycling helmets have resulted from wind tunnel tests.

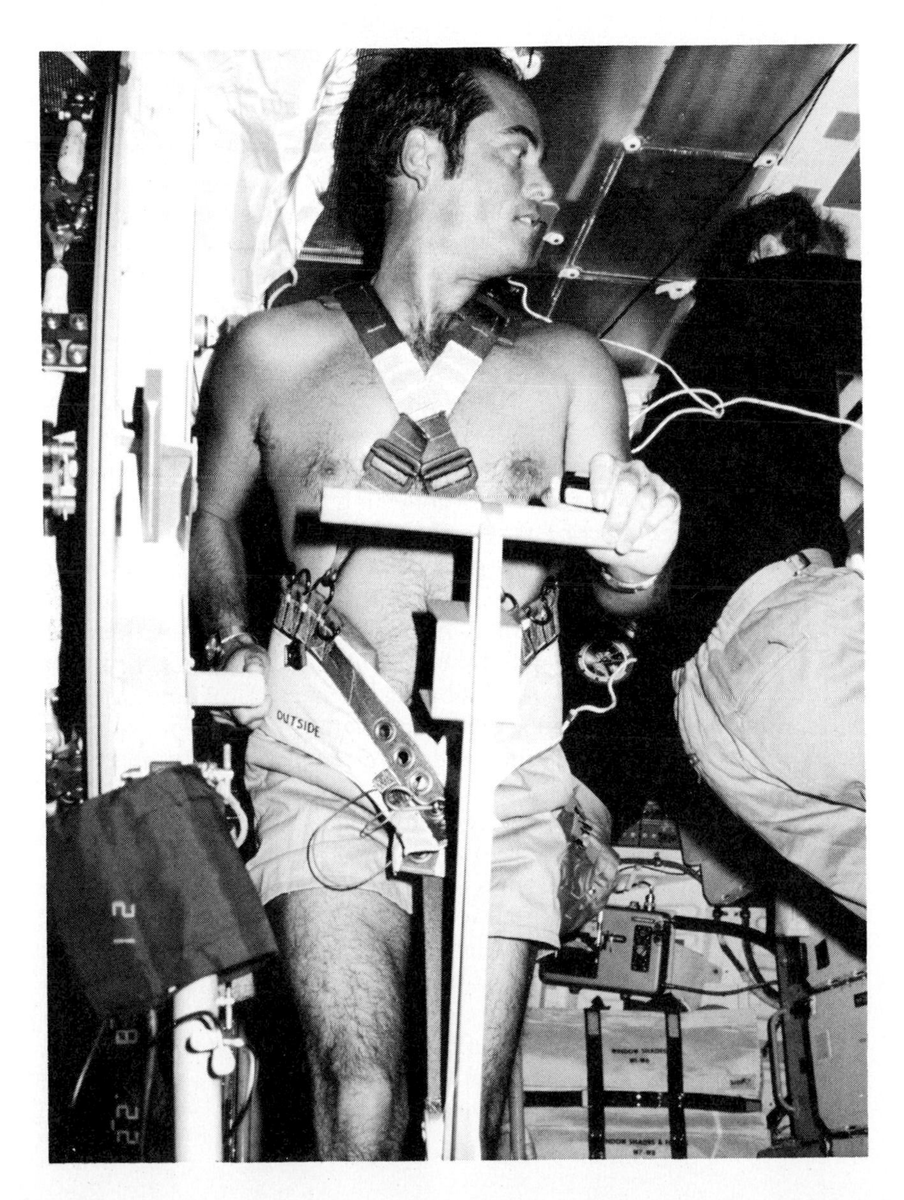

Exercise in space is critically important for the prevention of bone loss among astronauts.

The aerodynamic cycling equipment introduced in the 1984 Olympic Games contributed to the new world records set.

(12). Factors found to contribute to superior performance in the long jump are the athlete's horizontal velocity at the beginning of the fourth-last stride of the approach and the change in horizontal velocity during the next support phase (6). More predictable contributors are the athlete's horizontal and total velocities at takeoff and the flight distance (6). Analysis of the best world-class table tennis players has shown that they usually hit defensive shots earlier and on the rise more often than less-skilled players but often hit offensive shots when the ball is moving downward to generate topspin (4). It has also been documented that although some elite swimmers excel because they maintain a relatively constant forward velocity, other elite swimmers' stroke cycles are characterized by extremely high velocities with moderate velocity fluctuations (11). The role of sport biomechanists in assisting elite athletes in different sports will be increasingly important in the future.

One rather dramatic example of performance improvement partially attributable to biomechanical analysis is the case of four-time Olympic discus champion, Al Oerter. Mechanical analysis of the discus throw requires precise evaluation of the major mechanical factors affecting the flight of the discus. These factors are 1) the speed of the discus when it is released by the thrower, 2) the projection angle at which the discus is released, 3) the height above the ground at which the discus is released, and 4) the angle of attack (the orientation of the discus relative to the prevailing air current). By using computer simulation techniques, researchers can predict the needed combination of values for these four variables that will result in a throw of maximum distance for a given athlete. High-speed cinematography, a commonly used data-collection technique among sport biomechanists, can record performances in great detail, and when the film is analyzed, the actual projection height, velocity, and angle of attack can be compared to the computer-generated values required for optimal performance. At the age of 43, Oerter bettered his best Olympic performance by 8.2 meters, partially as a result of improvement of his technique following biomechanical analysis (16). Most adjustments to skilled athletes' techniques produce more modest results because their performances are characterized by above-average technique initially.

The high-speed camera is one of the data collection tools commonly used by sport biomechanists.

■ The high-speed movie cameras commonly used by sport biomechanists can be operated at as many as 500 frames (or pictures) per second.

Some of the research produced during the past decade by sport biomechanists has been done in conjunction with the Sports Medicine Division of the United States Olympic Committee (USOC). The general goal of USOC research is to examine the ways in which mechanical factors limit the performances of elite American athletes training for Olympic and other international competition (2). Typically, this work is done in direct cooperation with the national coach of the sport to ensure the practicality of results. USOC-sponsored research has yielded much new information about the mechanical characteristics of elite performance in various sports.

Other concerns of sport biomechanists relate to minimizing sport injuries through both the identification of dangerous practices and the design of safe equipment and apparel. For example, running or aerobic exercise on a hard surface such as cement probably increases the risk of developing stress fractures in the lower extremities. Protective helmets are evaluated to ensure that their impact characteristics offer reliable protection and that they do not overly restrict wearers' peripheral vision. Several models of knee braces are available to provide extra lateral stability for athletes' knees—particularly those of football players. A microcomputer-controlled snowski binding system now enables automatic release of skis during potentially injurious situations (10).

■ The USOC began to fund sports medicine research in 1981. Research aimed at assisting elite athletes has been occurring much longer in several other countries.

■ Impact testing of protective sport helmets is carried out scientifically in engineering laboratories.

Biomechanists recognize that the ground or playing surface, the shoe, and the human body compose an interactive system. The shoe plays a role not only in cushioning the body from impact forces but in influencing the kinematics of body movement (3). Consequently, athletic shoes are specifically designed for particular sports, surfaces, and anatomical considerations. Aerobic dance shoes are constructed to cushion the metatarsal arch. Football shoes to be used on artificial turf are designed to minimize the risk of knee injury. Running shoes are available for training, racing, and running on snow and ice and for individuals whose feet *pronate,* that is, roll inward at the point of contact with the ground. At Harvard University a running track was biomechanically designed to provide a surface that promotes running speed (13).

These examples illustrate the diversity of biomechanics research and the different contributions biomechanists can make to the knowledge of human movement and performance. Although different, all research and design topics are based on applications of mechanical principles to particular problems of living organisms. This book is designed to provide an introduction to many of those principles and to focus on some of the ways in which biomechanical principles may be applied in the analysis of human movement.

PROBLEM-SOLVING APPROACH

Scientific research is usually aimed at providing a solution for a particular problem or answering a specific question. Even for the nonresearcher, however, the ability to solve problems is a practical necessity for functioning in modern society. The use of specific problems is also an effective approach for illustrating basic biomechanical concepts.

Formal Versus Informal Problems

Different problems are commonly encountered in daily activities. Questions such as what clothes to wear, whether to major in botany or English, and whether to study or watch television are all problems in the sense that they are uncertainties that may create difficulties. Thus, a large portion of our daily lives is devoted to the solution of problems.

When confronted with a stated problem taken from an area of mathematics or science, many students believe they are inept at finding a solution. Clearly a stated math problem is different than a problem such as what to wear to a particular social gathering. In some ways, however, the informal type of problem is the more difficult one to solve. According to Wickelgren (18) a formal problem (such as a stated math problem), is characterized by three discrete components: 1) a set of given information; 2) a particular goal, answer, or desired finding; and 3) a set of operations or processes that can be used to arrive at the answer from the given information. In dealing with informal problems, however, individuals may find the given information, the processes to be used, and even the goal itself to be unclear or not readily identifiable.

Quantitative Versus Qualitative Problems

quantitative
the aspect involving the use of numbers

qualitative
the aspect pertaining to a nonnumerical quality

Analysis of human movement may be either **quantitative** or **qualitative.** The word *quantitative* implies that numbers are involved, and *qualitative* refers to a description of quality without the use of numbers. After watching the performance of a standing long jump, an observer might state qualitatively, "That was a very good jump." Another observer might announce quantitatively that the same jump was 2.1 meters in length. Other examples of qualitative and quantitative descriptors are displayed in Figures 1-7 and 1-8.

Figure 1-7

It is important to recognize that the term *qualitative* does not mean *general.* Qualitative descriptions may be general, but they may also be extremely detailed. It can be stated qualitatively and generally, for example, that a man is walking on the street. It might also be stated that the same man is walking very slowly, appears to be leaning to the left, and is bearing weight on his right leg for as short a time as possible. The second description is entirely qualitative but provides a more detailed picture of the movement.

Both qualitative and quantitative descriptions play important roles in the biomechanical analysis of human movement. Biomechanic researchers rely heavily on quantitative techniques to analyze specific problems related to the mechanics of living organisms. Clinicians, coaches, and teachers of physical activities regularly employ qualitative observations of their patients, athletes, or students to formulate opinions or give advice. (Procedures for qualitative and quantitative analyses of human movement are discussed in Chapter 15.)

Coaches rely heavily on qualitative observations of athletes' performances in formulating advice about technique.

Solving Formal Quantitative Problems

Formal problems are effective vehicles for translating nebulous concepts into well-defined, specific principles that can be readily understood and applied in the analysis of human motion.

People who believe themselves incapable of solving formal stated problems do not recognize that, to a large extent, problem-solving skills can be learned. Entire books on problem-solving approaches

Figure 1-8
Quantitatively, the robot missed the coffee pot by 5 centimeters. Qualitatively, he malfunctioned.

and techniques have been written. However, most students are not exposed to course work involving general strategies of the problem-solving process. A simple procedure for approaching and solving problems involves 11 sequential steps, which are as follows:

1. Read the problem *carefully*. It may be necessary to read the problem several times before proceeding to the next step. Only when the information given and the question(s) to be answered are clearly understood should step 2 be undertaken.
2. Write down the given information in list form. It is acceptable to use symbols (such as v for velocity) to represent physical quantities if the symbols are meaningful.
3. Write down what is wanted or what is to be determined, using list form if more than one quantity is required.
4. Draw a diagram representing the problem situation, clearly indicating all known quantities and representing those to be identified with question marks. (Although certain types of problems may not easily be diagrammatically represented, it is critically important to carry out this step whenever possible to accurately visualize the problem situation.)
5. Identify and write down the relationships or formulas that might be useful in solving the problem. (More than one formula may be useful and/or necessary.)
6. From the formulas that were written down in step 5, select the formula(s) containing both given variables (from step 2) and the variables that are desired unknowns (from step 3). If a formula contains only one unknown variable that is the variable to be determined, skip step 7 and proceed directly to step 8.
7. If a workable formula cannot be identified (in more difficult problems), certain essential information was probably not specifically stated but can be determined by using **inference** and further thought and analysis of the given information. If this occurs, it may be necessary to repeat step 1 and review the pertinent information relating to the problem presented in the text.
8. Once the appropriate formula(s) have been identified, write the formula(s) and carefully substitute the known quantities given in the problem for the variable symbols.
9. Using the simple algebraic techniques reviewed in Appendix A, solve for the unknown variable by a) rewriting the equation so that the unknown variable is isolated on one side of the equal sign and b) reducing the numbers on the other side of the equation to a single quantity.

inference
the process of forming deductions from given information

10. Do a common sense check of the answer derived. Does it seem too small or too large? If so, recheck the calculations. Also check to ensure that *all* information originally requested in the statement of the problem has been found.
11. Clearly box in the answer and include the correct units of measurement.

> SUMMARY OF STEPS FOR SOLVING FORMAL PROBLEMS
>
> 1. Read the problem carefully.
> 2. List the given information.
> 3. List the desired information to be determined.
> 4. Draw a diagram showing known and unknown information.
> 5. Write down the formulas that may be useful.
> 6. Identify the appropriate formula(s) to use.
> 7. If necessary, reread the problem statement to determine whether additional information may be inferred.
> 8. Carefully substitute the given information into the appropriate formula.
> 9. Solve for the unknown variable through algebraic manipulation of the known quantities.
> 10. Check that the answer is both reasonable and complete.
> 11. Clearly box in the answer.

This procedure for solving problems is summarized in the box. These steps should be carefully studied, referred to, and used in working the quantitative problems included at the end of each chapter. Figures 1-9 and 1-10 illustrate the use of this procedure.

Units of Measurement

Providing the correct units of measurement associated with the answer to a quantitative problem is important. Clearly an answer of 2 centimeters is quite different than an answer of 2 kilometers. It is also important to recognize the units of measurement associated with particular physical quantities. Communication would be impaired if someone ordered 10 kilometers of gasoline for a car when traveling in a foreign country.

The predominant system of measurement still used in the United States is the **English system.** The English system of weights and measures was developed over several centuries primarily for purposes of commerce and land parceling. Specific units were derived largely from royal decrees. For example, a yard was originally defined as the distance from the end of the nose of King Henry I to the thumb of his extended arm. The English system of measurement displays little logic. There are 12 inches to the foot, 3 feet to the

English system
the system of weights and measures originally developed in England and used in the United States today

Figure 1-9

SAMPLE PROBLEM 1

A baseball player hits a triple to deep left field. As he is approaching third base, he notices that the incoming throw to the catcher is wild and he decides to run for home plate. The catcher retrieves the ball 10 m from the plate and runs back toward the plate at a speed of 5 m/s. As the catcher starts running, the base runner, who is traveling at a speed of 9 m/s, is 15 m from the plate. Given that time = distance/speed, who will reach the plate first?

Solution

STEP 1 Read the problem carefully.

STEP 2 Write down the given information:

Base runner's speed	=	9 m/s
Catcher's speed	=	5 m/s
Distance of base runner from plate	=	15 m
Distance of catcher from plate	=	10 m

STEP 3 Write down the variable to be identified:
Find which player reaches home plate in the shortest time.

STEP 4 Draw a diagram of the problem situation.

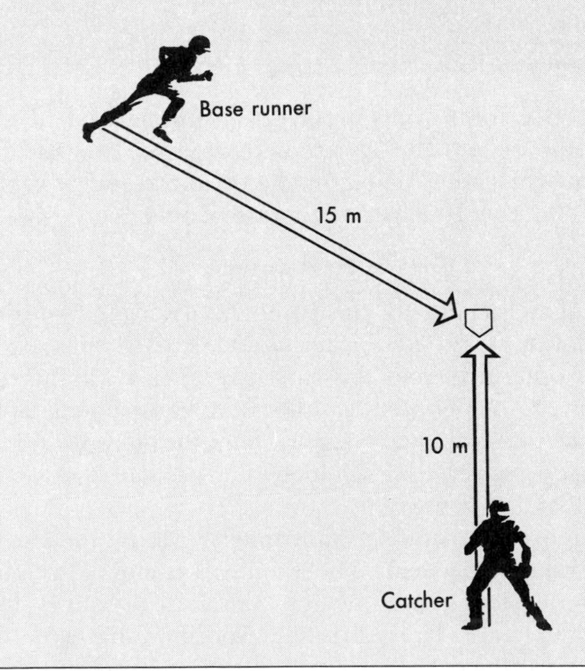

STEP 5 Write down formulas of use:
Time = distance/speed

STEP 6 Identify the formula to be used:
It may be assumed that the formula provided is appropriate because no other information relevant to the solution has been presented.

STEP 7 Reread the problem if all necessary information is not available:
It may be determined that all information appears to be available.

STEP 8 Substitute the given information into the formula:

$$\text{time} = \frac{\text{distance}}{\text{speed}}$$

Catcher:

$$\text{time} = \frac{10\text{ m}}{5\text{ m/s}}$$

Base runner:

$$\text{time} = \frac{15\text{ m}}{9\text{ m/s}}$$

STEP 9 Solve the equations.
Catcher:

$$\text{time} = \frac{10\text{ m}}{5\text{ m/s}}$$

$$\text{time} = 2\text{ s}$$

Base runner:

$$\text{time} = \frac{15\text{ m}}{9\text{ m/s}}$$

$$\text{time} = 1.67\text{ s}$$

STEP 10 Check that the answer is both reasonable and complete.

STEP 11 Box in the answer:

The base runner arrives at home plate first, by 0.33 seconds.

Figure 1-10

SAMPLE PROBLEM 2

A man sits in a 20 kg wheelchair at the top of a short ramp. When the brakes on the wheelchair are suddenly released, the wheelchair begins to roll down the ramp, accelerating at a rate of 0.5 meters per second^2. If the net force causing the wheelchair to roll down the ramp is 45 Newtons, what is the mass of the man in the wheelchair? (*Hint:* The relationship to be used in solving this problem is force = mass × acceleration. For more information on the metric units of kilograms, meters, and Newtons, refer to Appendix C.) *This time, work through the 11 steps on your own.*

STEP 2

Net force = 45 Newtons
Wheelchair mass = 20 kg
Acceleration = 0.5 m/s^2

STEP 3 Find: mass of man

STEP 4 $mass_{total} = mass_{chair} + mass_{man}$

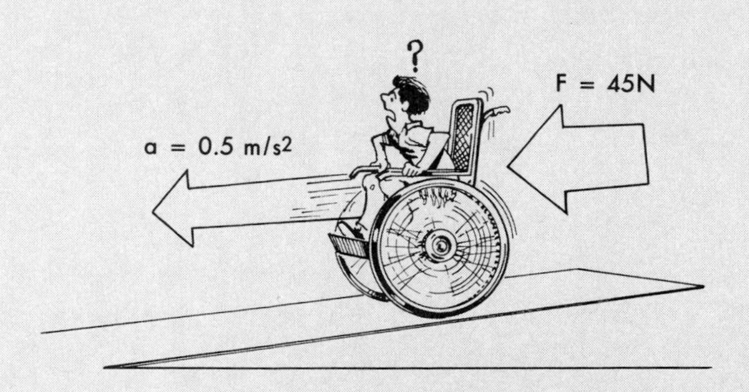

STEPS 5 AND 6 Force = mass × acceleration

STEPS 8 AND 9 Force = mass × acceleration

$$45 \text{ N} = (mass_{chair} + mass_{man}) \times 0.5 \text{ m/s}^2$$

$$45 \text{ N} = (20 \text{ kg} + mass_{man}) \times 0.5 \text{ m/s}^2$$

$$\frac{45 \text{ N}}{0.5 \text{ m/s}^2} = (20 \text{ kg} + mass_{man})$$

$$\frac{45 \text{ N}}{0.5 \text{ m/s}^2} - 20 \text{ kg} = mass_{man}$$

$$mass_{man} = 70 \text{ kg}$$

STEP 11 70 kg

yard, 5280 feet to the mile, 16 ounces to the pound, and 2000 pounds to the ton.

The system of measurement that is presently used by every major country in the world except the United States is Le Systeme International d'Unites, (the International System of Units or SI), which is commonly known as the **metric system.** The metric system was originated at the request of King Louis XVI by the French Academy of Sciences during the early 1790s. Although the system fell briefly from favor in France, it was readopted in 1837. In 1875 the Treaty of the Meter was signed by 17 countries agreeing to adopt the use of the metric system.

metric system
the system of weights and measures used in scientific applications and adopted by every major country except the United States

Since that time the metric system has enjoyed worldwide popularity for several reasons. First, it is derived from only four base units—the meter of length, the kilogram of mass, the second of time, and the degree Kelvin of temperature. Second, the base units are precisely defined, reproducible quantities that are independent of factors such as gravitational force. Third, all units are related by factors of 10, in contrast to the numerous conversion factors necessary in converting English units of measurement. Last, it is an international system.

Because the metric system is the system of units used almost exclusively by the scientific community, it is the system that is used in this book. The relevant units of measurement in both systems and common English-metric conversion factors are presented in Appendix C.

SUMMARY

The study of human movement is multifaceted, with researchers from subdisciplines such as exercise physiology, motor behavior, and sport psychology often studying the same questions from different perspectives. One of the areas of inquiry regarding human movement is biomechanics—the application of mechanical principles in the study of the structure and function of living things.

Biomechanics is a multidisciplinary science with applications in many professional fields. Because biomechanists come from different academic backgrounds, biomechanical research addresses a spectrum of problems and questions.

INTRODUCTORY PROBLEMS

1. Locate and read three articles from the scientific literature that report the results of biomechanical investigations. (The *Journal of Biomechanics, International Journal of Sport Biomechanics,* and *Medicine and Science in Sports and Exercise* are possible sources.) Write a half-page summary of each article and iden-

tify whether the investigation focused on static or dynamic aspects and on kinetic or kinematic considerations.

2. List 8 to 10 scientific journals that regularly or frequently include articles relating to biomechanics.
3. Discuss how knowledge of biomechanics may be useful in your intended profession or career.
4. Choose five jobs or professions and discuss the ways in which each is quantitative and qualitative.
5. Write a description of one informal problem and one formal problem.
6. Write a summary list of the problem-solving steps identified in the chapter using your own words.
7. Step by step, show how to arrive at a solution to one of the problems you described in Problem 5.
8. Solve for x in each of the equations below. Refer to Appendix A for help if necessary.
   a. $x = 5^3$
   b. $7 + 8 = x/3$
   c. $4 \times 3^2 = x \times 8$
   d. $-15/3 = x + 1$
   e. $x^2 = 27 + 35$
   f. $x = \sqrt{79}$
   g. $x + 3 = \sqrt{38}$
   h. $7 \times 5 = -40 + x$
   i. $3^3 = x/2$
   j. $15 - 28 = x \times 2$

   (Answer: a. 125; b. 45; c. 4.5; d. −6; e. 7.9; f. 8.9; g. 3.2; h. 75; i. 54; j. −6.5)
9. Two school children race across a playground for a ball. Tim starts running at a distance of 15 meters from the ball, and Jan starts running at a distance of 12 meters from the ball. If Tim's average speed is 4.2 m/s and Jan's average speed is 4.0 m/s, which child will reach the ball first? Show how you arrived at your answer. (See Figure 1-9.) (Answer: Jan reaches the ball first.)
10. A 0.5 kg ball is kicked with a force of 40 Newtons. What is the resulting acceleration of the ball? (See Figure 1-10.) (Answer: 80 m/s^2)

ADDITIONAL PROBLEMS

1. Select a specific movement or sport's skill of interest and read two or three articles from the scientific literature that report the results of biomechanical investigations related to the topic. Write a short paper that integrates the information from your sources into a scientifically based description of your chosen movement.

2. When attempting to balance your checkbook, you discover that your figures show a different balance in your account than was calculated by the bank. List an ordered, logical set of procedures that may be used to discover the error. You may use list, outline, or block diagram format.
3. Sarah goes to the grocery store and spends half of her money. On the way home, she stops for an ice cream cone that costs $0.78. Then she stops and spends a quarter of the remaining money on a $5.50 bill at the dry cleaners. How much money did Sarah have originally? (Answer: $45.56)
4. Wendell invests $10,000 in a stock portfolio made up of Petroleum Special at $30 per share, Newshoe at $12 per share, and Beans & Sprouts at $2.50 per share. He places 60% of the money in P.S., 30% in N, and 10% in B & S. With market values changing (P.S. down $3.12, N up 80%, and B & S up $.20), what is his portfolio worth 6 months later? (Answer: $12,087)
5. The hypotenuse of right triangle ABC (shown below) is 4 cm long. What are the lengths of the other two sides? (Answer: A = 2 cm; B = 3.5 cm)

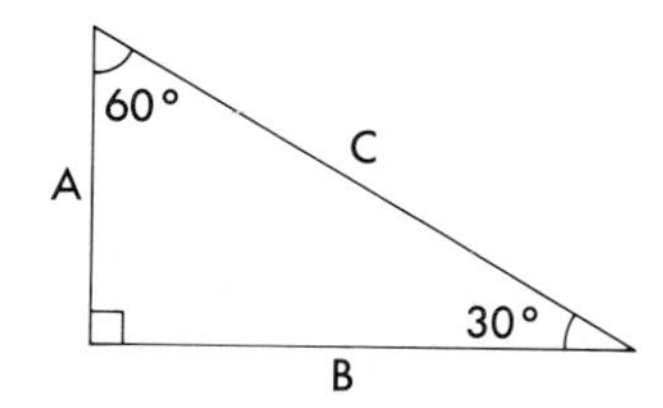

6. In triangle ABC, side B is 4 cm long and side C is 7 cm long. If the angle between sides B and C is 50 degrees, how long is side A? (Answer: 5.4 cm)
7. An orienteer runs 300 m north and then 400 m to the southeast (at a 45 degree angle to north). If he has run at a constant speed, how far away is he from the starting position? (Answer: 283.4 m)
8. John is out for his daily noontime run. He runs 2 km west, then 2 km south, and then runs on a path that takes him directly back to the place he started. a. How far did John run? b. If he has run at an average speed of 4 m/s, how long did the entire run take? (Answer: a. 6.83 km; b. 28.5 minutes)
9. John and Al are in a 15 km race. John averages 4.4 m/s during the first half of the race and then runs at a speed of 4.2 m/s until the last 200 m, which he covers at 4.5 m/s. At what average speed must Al run to beat John? (Answer: >4.3 m/s)
10. A sailboat heads north at 3 m/s for 1 hour and then tacks back to the southeast (at 45 degrees to north) at 2 m/s for 45 minutes. a. How far has the boat sailed? b. How far is it from its starting location? (Answer: a. 16.2 km; b. 8.0 km)

REFERENCES

1. Davies CTM: Effects of wind resistance and assistance on the forward motion of a runner, J Appl Physiol 48:702, 1980.
2. Dillman CJ: Overview of the United States Olympic Committee sports medicine biomechanics program. In Butts NK, Gushiken TT, and Zarins B, eds: The elite athlete, New York, 1985, Spectrum Publications.
3. Frederick EC: Biomechanical consequences of sport shoe design, Exerc Sport Sci Rev 14:375, 1986.
4. Groppel JL: A mechanical analysis of stroking techniques of world class table tennis players. In Dillman CJ and Bauer SJ, eds: Abstracts of biomechanical research, Colorado Springs, Colo, 1983, United States Olympic Committee.
5. Hatze H: The meaning of the term "biomechanics," J Biomech 7:189, 1974.
6. Hay JG, Miller JA, and Canterna RW: The techniques of elite male long jumpers, J Biomech 19:85, 1986.
7. Kyle CR: Reduction of wind resistance and power output of racing cyclists and runners traveling in groups, Ergonomics 22:387, 1979.
8. Kyle CR: Athletic clothing, Sci Am 254:104, 1986.
9. Lamb DR: The sports medicine umbrella, Sports Med Bull 5:8, 1984.
10. MacGregor D, Hull ML, and Dorius LK: A microcomputer controlled snow ski binding system—I. instrumentation and field evaluation, J Biomech 18:255, 1985.
11. Maglischo EW and Maglischo CW: Three dimensional cinematographic analyses of world class swimmers. In Dillman CJ and Bauer SJ, eds: Abstracts of biomechanical research, Colorado Springs, Colo, 1983, United States Olympic Committee.
12. Mann R et al: Kinematic trends in elite sprinters, In Terauds J et al, eds: Sports biomechanics, Del Mar, Calif, 1984 Research Center for Sports.
13. McMahon TA and Greene PR: Fast running tracks, Sci Am 239:148, 1978.
14. Nelson RC: Biomechanics: past and present. In Cooper JM and Haven B, eds: Proceedings of the Biomechanics Symposium, Bloomington, Ind, 1980.
15. Remizov LP: Biomechanics of optimal flight in ski jumping, J Biomech 17:167, 1984.
16. Ruby D: Biomechanics—how computers extend athletic performance to the body's far limits, Popular Science p 58, Jan 1982.
17. Taylor CR, Heglund NC, and Maloiy MO: Energetics and mechanics of terrestrial locomotion: metabolic energy consumption as a function of speed and body size in birds and mammals, J Exp Biol 97:1, 1982.
18. Wickelgren WA: How to solve problems, San Francisco, 1974, WH Freeman & Co.

ANNOTATED READINGS

Asmussen E: Movement of man and study of man in motion: a scanning review of the development of biomechanics. In Komi PV, ed: Biomechanics V-A, vol 1A, Baltimore, 1976, University Park Press.

Provides a historical overview of the development of biomechanics as a discrete field of study.

Attwater AE: Kinesiology/biomechanics: perspectives and trends, Res Q Exerc Sport 51:193, 1980.

Includes a historical perspective on the development of the biomechanics of human movement.

Bransford JD and Stein BS: The ideal problem solver, New York, 1984, WH Freeman & Co.

Discusses techniques for improving problem-solving skills.

Gardner M: Aha! insight, New York, 1978, Scientific American, Inc.

Creatively presents different types of problems and discusses their solutions.

Jacobs HR: Mathematics: a human endeavor, ed 2, New York, 1982, WH Freeman & Co.

Introduces basic mathematical concepts in an entertaining fashion.

Kluger J: The human machine, Science Digest p 64, Jun 1982.

Describes parallels between the human body and machines and discusses the research work of selected biomechanists.

Kyle CR: Athletic clothing, Sci Am 254:104, 1986.

Provides insights on the types of work done in designing modern cycling equipment and clothing, running shoes, and helmets.

Muybridge E: Muybridge's complete human and animal locomotion, New York, 1979, Dover Publications.

Includes a collection of photographic plates done by early cinematographer Eadweard Muybridge during the late 1800s, including the running stills used by Muybridge to prove the then controversial claim that all four hooves of a horse are simultaneously elevated from the ground during part of the gallop stride.

Nelson W: Application of biomechanical principles: optimization of sport technique. In Butts NK, Gushiken TT, and Zarins B, eds: The elite athlete, New York, 1985, Spectrum Publications.

Describes a general procedure for application of biomechanical analysis techniques toward performance optimization in a sport, with rowing presented as an example.

Norman RW: Biomechanical evaluations of sports protective equipment: Exerc Sport Sci Rev 11:232, 1983.

Provides a comprehensive review of the processes and results of biomechanical testing of protective equipment designed for regions of the body during participation in different sports.

Vaughan CL: Computer simulation of human motion in sports biomechanics, Exerc Sport Sci Rev 12:373, 1984.

Discusses the advantages and limitations of computer simulation techniques and reviews the research involving computer simulation on several basic movement patterns and sport skills.

2 THE BASICS

Terminology and Simple Mechanical Concepts

After reading this chapter, the student will be able to:

Provide examples of linear, rotary, and general forms of motion.

Identify and describe the reference positions, planes, and axes associated with the human body.

Define and appropriately use directional terms and joint movement terminology.

Define and identify common units of measurement for the quantities mass, force, weight, pressure, volume, density, weight density, and torque.

Construct free body diagrams for elementary force analyses.

Distinguish between vector and scalar quantities.

Solve quantitative problems involving vector quantities using both graphic and trigonometric procedures.

Mastering the terminology associated with an unfamiliar body of knowledge is an essential first step in learning about that subject. In this chapter many of the specific terms and concepts that are used in both kinematic and kinetic approaches to the study of human movement are introduced.

FORMS OF MOTION

Most human movement is a complex combination of the basic forms of motion. In analyzing movement from a mechanical perspective, it is often practical to separate complex movements into their **linear** and **angular** components.

Linear Motion

Pure linear motion is characterized by uniform motion of the body, with all body parts moving in the same direction at the same speed. (The term *body* will be used throughout the book in the generic sense, that is, to refer to objects other than just the human body.) Linear motion is also referred to as translatory motion, or **translation.** When a body is translated, the body as a unit is in motion but portions of the body do not move relative to each other. For example, a sleeping passenger who is motionless on a smooth airplane flight is being translated through the air. If the passenger lifts an arm to reach for a magazine, however, pure translation is no longer occurring because the position of the arm respective to the body has changed.

Linear motion is also motion along a line. If the line is straight, the motion is **rectilinear.** If the line is curved, the motion is **curvilinear.** A motorcyclist maintaining a motionless posture as the bike moves along a straight path is in rectilinear motion. If the motorcy-

linear
motion following a path along a line that may be straight or curved, with all parts of the body moving in the same direction at the same speed

angular
motion involving rotation around a central line or point

translation
term synonymous with linear motion

rectilinear
motion along a straight line

curvilinear
motion along a curved line

clist jumps the bike and the frame of the bike does not rotate, both rider and bike (with the exception of the spinning wheels) are in curvilinear motion while airborne. Likewise, a Nordic skier coasting in a locked static position down a short hill is in rectilinear motion. If the skier jumps over a gulley with all body parts moving in the same direction at the same speed along a curved path, the motion is curvilinear. When a motorcyclist or skier move over the crest of a hill, the motion is *not* linear because the top portions of the body are moving at a greater speed than lower body parts. Figure 2-1 displays a gymnast in rectilinear, curvilinear, and rotational motion.

Angular Motion

axis of rotation
imaginary line oriented perpendicular to the plane of rotation and passing through the center of rotation

Angular motion is rotation around a central imaginary line known as the **axis of rotation,** which is oriented perpendicular to the plane in which the rotation occurs. When a gymnast performs a giant circle on a bar, the entire body is in rotation, with the axis of rotation passing through the center of the bar. When a springboard diver executes a somersault in mid-air, the entire body is again in rotation, this time around an imaginary axis of rotation that moves along with the body. Almost all volitional human movement involves rotation of a body segment around an imaginary axis of rotation that passes

Rotation of a body segment at a joint occurs around an imaginary line known as the axis of rotation that passes through the joint center.

Figure 2-1

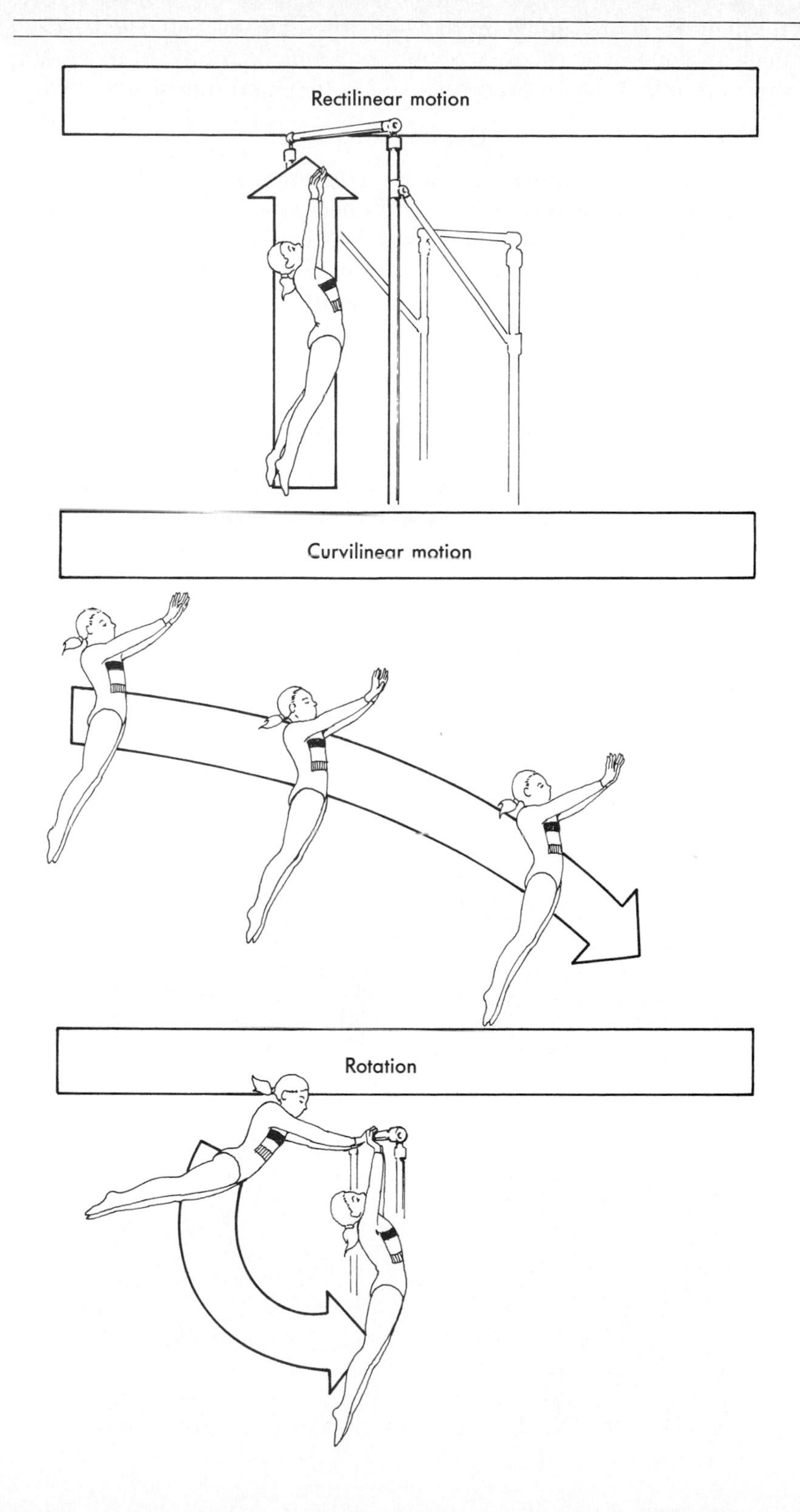

through the center of a joint to which the segment is attached. When angular motion or rotation occurs, portions of the body in motion are constantly being moved relative to other portions of the body.

General Motion

general motion
motion involving translation and rotation simultaneously

■ Most human movement activities are categorized as general motion.

When translation and rotation are combined the resulting movement is **general motion.** A football kicked end over end is translated through the air as it simultaneously rotates around a central axis (Figure 2-2). A runner is translated along by the angular movements occurring at the hip, knee, and ankle joints. Human movement usually consists of general motion rather than pure linear or angular motion.

STANDARD REFERENCE TERMINOLOGY

Identifying whether a movement is linear, angular, or general is often the first step in describing the movement. Communicating more specific information about human movement requires knowledge

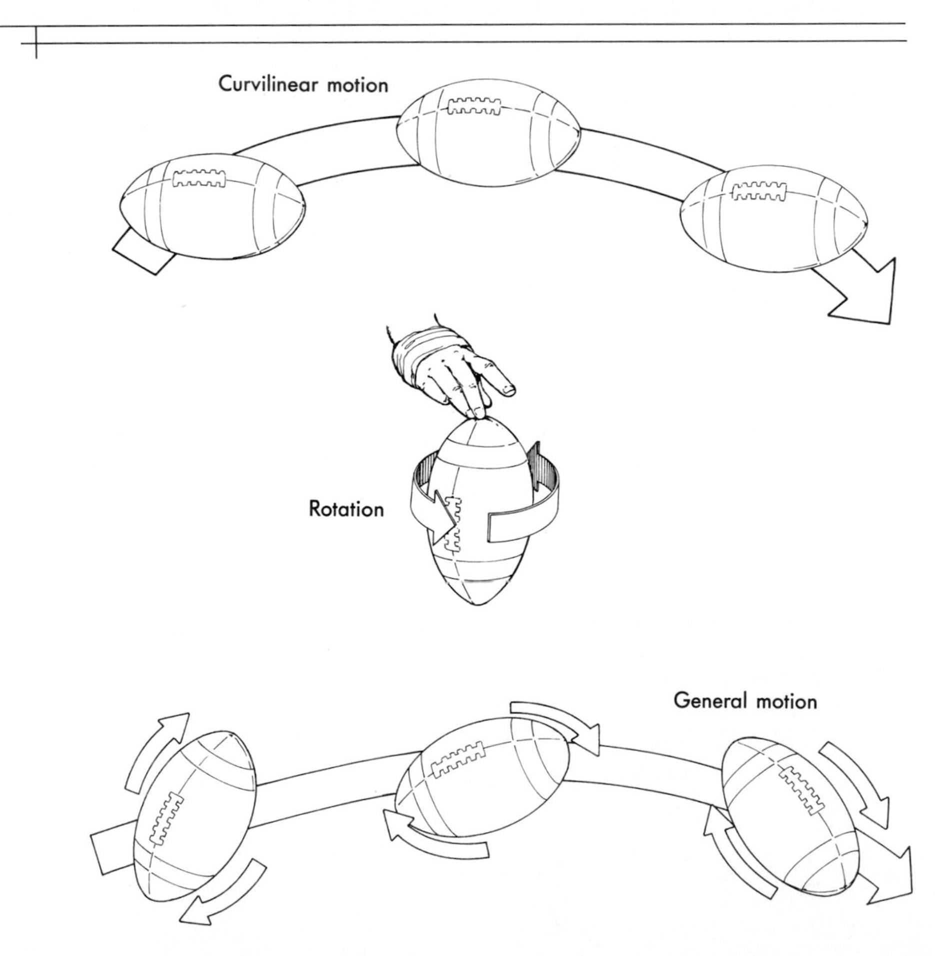

Figure 2-2
General motion is a combination of linear and angular motion.

and specialized terminology that precisely identifies the different actions occurring at the joints of the human body.

Anatomical reference position.

Anatomical Reference Position

Anatomical reference position is an erect standing position with the feet just slightly separated and the arms hanging relaxed at the sides, with the palms of the hands facing forward. It is not a natural standing position, but is the body orientation conventionally used as the reference position or starting place when movement terms are defined.

anatomical reference position
erect standing position with all body parts, including the palms of the hands, facing forward; used as the starting position for body segment movements

Directional Terms

In describing the relationship of body parts or the location of an external object with respect to the body, the use of directional terms is necessary. The following are commonly used directional terms:

Superior: Closer to the head (In zoology the synonymous term is *cranial.*)
Inferior: Farther away from the head (In zoology the synonymous term is *caudal.*)
Anterior: Toward the front of the body (In zoology the synonymous term is *ventral.*)
Posterior: Toward the back of the body (In zoology, the synonymous term is *dorsal.*)
Medial: Toward the midline of the body
Lateral: Away from the midline of the body
Proximal: Closer to the trunk (For example, the knee is proximal to the ankle.)
Distal: Away from the trunk (For example, the wrist is distal to the elbow.)
Superficial: Toward the surface of the body
Deep: Inside the body and away from the body surface

All of these directional terms can be paired as antonyms—words having opposite meanings. Saying that the elbow is proximal to the wrist is as correct as saying that the wrist is distal to the elbow. Similarly, the nose is superior to the mouth and the mouth is inferior to the nose.

Anatomical Reference Planes

The three imaginary **cardinal planes** bisect the body in three dimensions. A *plane* is a two-dimensional surface with an orientation defined by the spatial coordinates of three discrete points within the plane that are not all contained in the same line. It may be thought of as an imaginary flat surface. The **sagittal plane,** also known as the anteroposterior (AP) cardinal plane or median cardinal plane, divides the body vertically into left and right halves, with each half weighing the same. The **frontal plane,** also referred to as the coronal cardinal plane or lateral cardinal plane, splits the body vertically

cardinal planes
the three imaginary reference planes that divide the body in half by mass or weight

sagittal plane
a plane in which forward and backward movements of the body and of the body segments occur

frontal plane
a plane in which lateral movements toward and away from the midline of the body occur

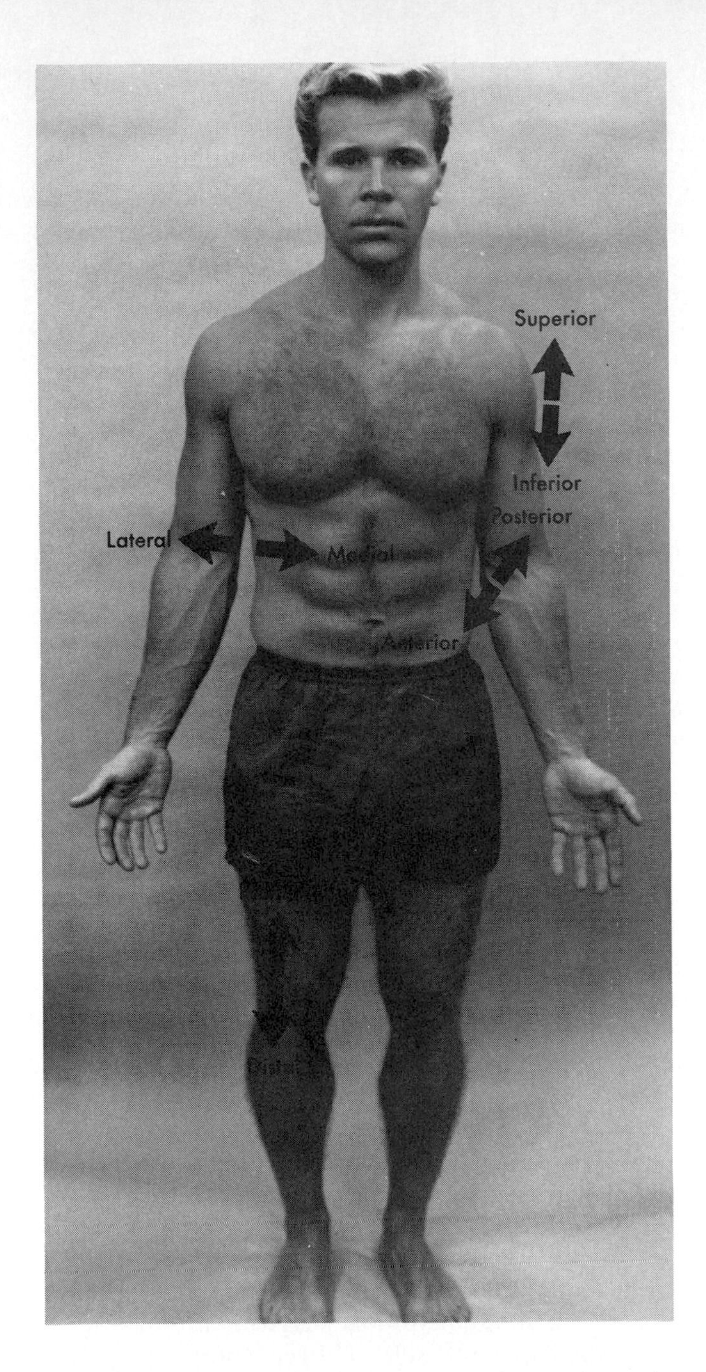

Directional terms.

■ Directional terms are used to relate the positions of body parts to other body parts and to external objects.

into front and back halves of equal weight. The horizontal or **transverse plane** separates the body into top and bottom halves of equal weight. For an individual standing in anatomical reference position, the three cardinal planes all intersect at a single point known as the body's center of gravity (Figure 2-3). These imaginary reference planes exist only with respect to the human body. If a person turns

transverse plane
a plane in which body movements parallel to the ground occur when the body is in an erect standing position

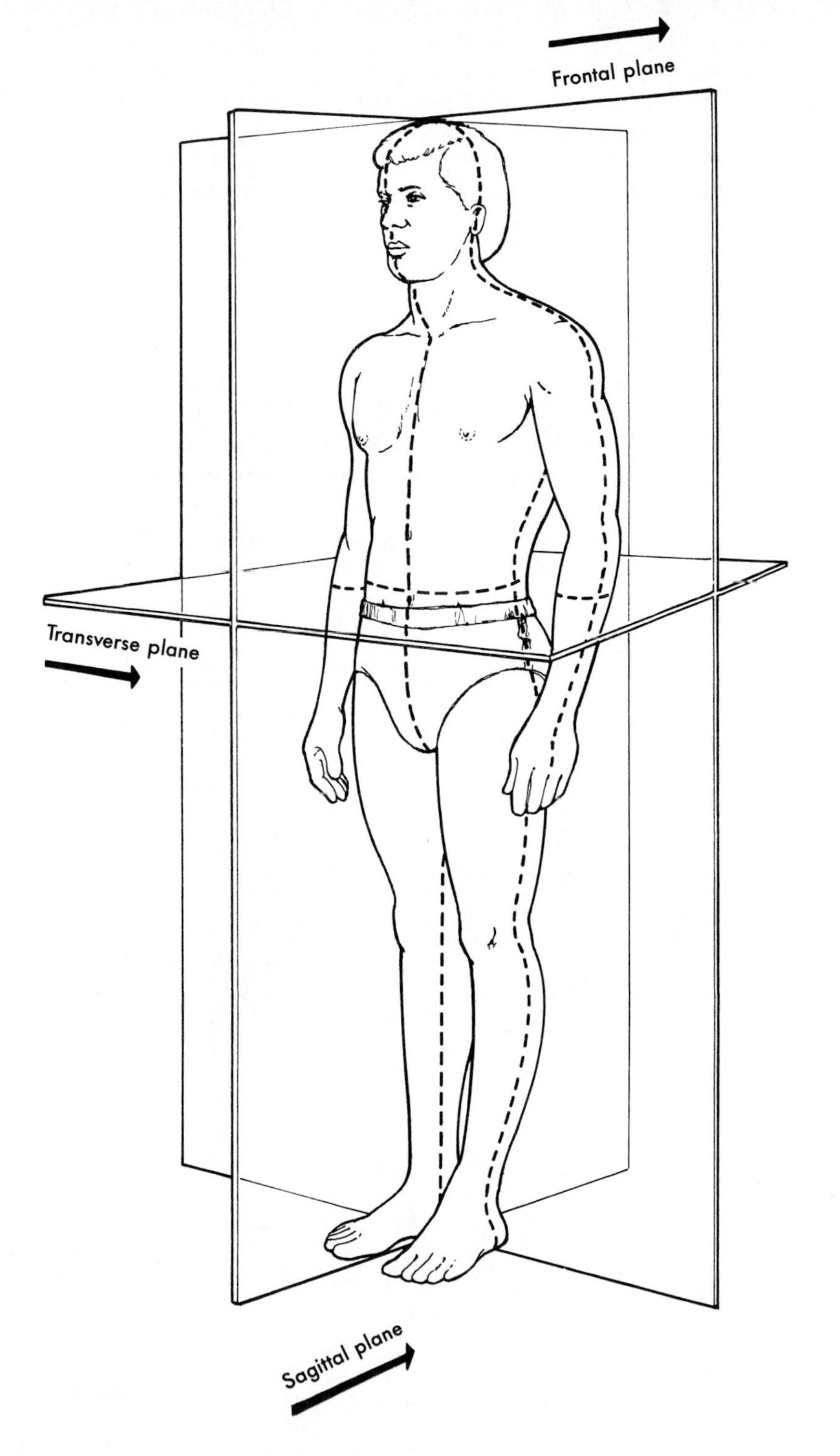

Figure 2-3
The three cardinal reference planes.

at a 45 degree angle to the right, the reference planes also turn at a 45 degree angle to the right.

Although the entire body may move in the orientation of one cardinal plane, the movements of individual body segments may also be described as sagittal plane movements, frontal plane movements, and transverse plane movements. When this occurs, the movements being described are usually in a plane that is parallel to one of the cardinal planes. For example, movements that involve forward and backward motion are referred to as sagittal plane movements. When a forward roll is executed, the entire body moves in the direction of the sagittal plane. During running in place, the motion of the arms and legs is generally forward and backward, although the planes of motion pass through the shoulder and hip joints rather than the center of the body. Marching, bowling, and cycling are all largely sagittal plane movements (Figure 2-4). Frontal plane movement is lateral (side-to-side) movement, and an example of total body frontal plane movement is the cartwheel. Jumping jacks, side stepping, and side kicks in soccer require frontal plane movement at certain body joints. Examples of total body transverse plane movement include a

Figure 2-4
Cycling requires sagittal plane movement of the legs.

twist executed by a diver, trampolinist, or airborne gymnast or a dancer's pirouette.

Although many of the movements conducted by the human body are not oriented sagittally, frontally, or transversely or are not planar at all, the three major reference planes are still useful. Gross body movements and specifically named movements that occur at joints are often described as primarily frontal, sagittal, or transverse plane.

Reference planes and axes are useful in describing gross body movements and in defining more specific movement terminology.

Anatomical Reference Axes

When a segment of the human body moves, it undergoes angular motion around an imaginary axis of rotation that passes through a joint to which it is attached. There are three reference axes for describing human motion, and each is oriented perpendicular to one of the three planes of motion. The **transverse axis,** also known as the frontal or mediolateral axis, is perpendicular to the sagittal plane. Rotation in the frontal plane occurs around the sagittal or **anteroposterior (AP) axis.** Transverse plane rotation is around the vertical or **longitudinal axis.** It is important to recognize that each of these three axes is always associated with the same single plane—the one to which the axis is perpendicular. Figures 2-5, 2-6, and 2-7 depict the anatomical reference axes for different movements.

transverse axis
an imaginary line around which sagittal plane rotations occur

anteroposterior axis
an imaginary line around which frontal plane rotations occur

longitudinal axis
an imaginary line around which transverse plane rotations occur

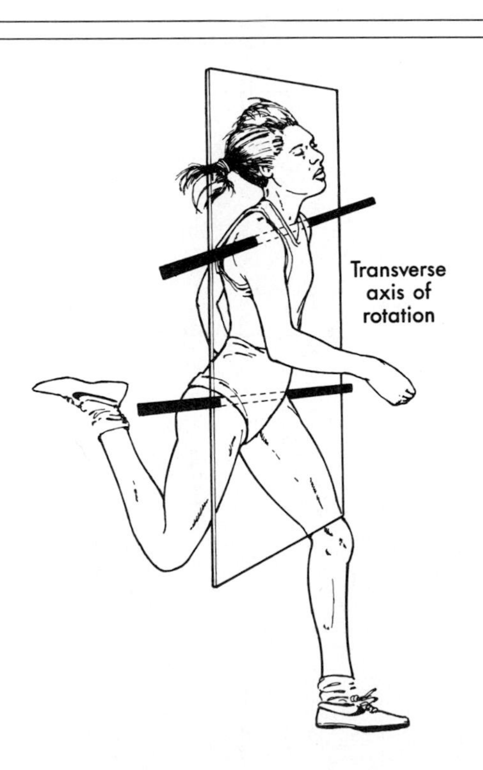

Figure 2-5
For running, the major axes of rotation are frontal axes passing through the shoulders and the hips.

Figure 2-6
For a jumping jack, the major axes of rotation are sagittal axes passing through the shoulders and the hips.

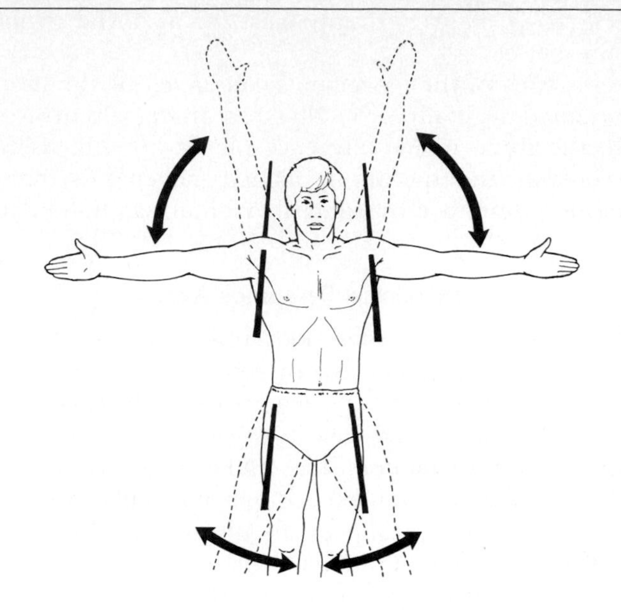

Figure 2-7
Rotation in the transverse plane occurs around a vertical axis.

JOINT MOVEMENT TERMINOLOGY

Sagittal Plane Movements

From anatomical position, the three primary movements occurring in the sagittal plane are *flexion, extension,* and *hyperextension.* Flexion is generally defined as sagittal plane movement that results in a lessening of the angle present at a joint. In anatomical position, the angles at the shoulder, elbow, wrist, hip, and knee are all 180 degrees. As the upper arm, forearm, thigh, or shank are elevated, reduction in the angle at the proximal joint occurs. When the arm is extended vertically overhead, the shoulder is in full flexion, or at 0 degrees. As shown in Figure 2-8, extension is defined as the movement that returns a body segment to anatomical position from a flexed position, and hyperextension is the movement beyond anatomical position in the direction opposite the direction of flexion. If the body segment is aligned in a plane other than the sagittal plane, flexion, extension, and hyperextension at the knee and elbow also occur in a different plane.

▪ Sagittal plane movements include flexion, extension, hyperextension, dorsiflexion, and plantar flexion.

When the body is in anatomical reference position with the toes pointing straight ahead, upward and downward rotation at the ankle is also sagittal plane movement. Elevation of the top of the foot toward the lower leg is known as *dorsiflexion,* and depression of the ball of the foot, is termed *plantar flexion* (Figure 2-9). The latter term is misleading because the motion increases rather than decreases the angle between the foot and lower leg.

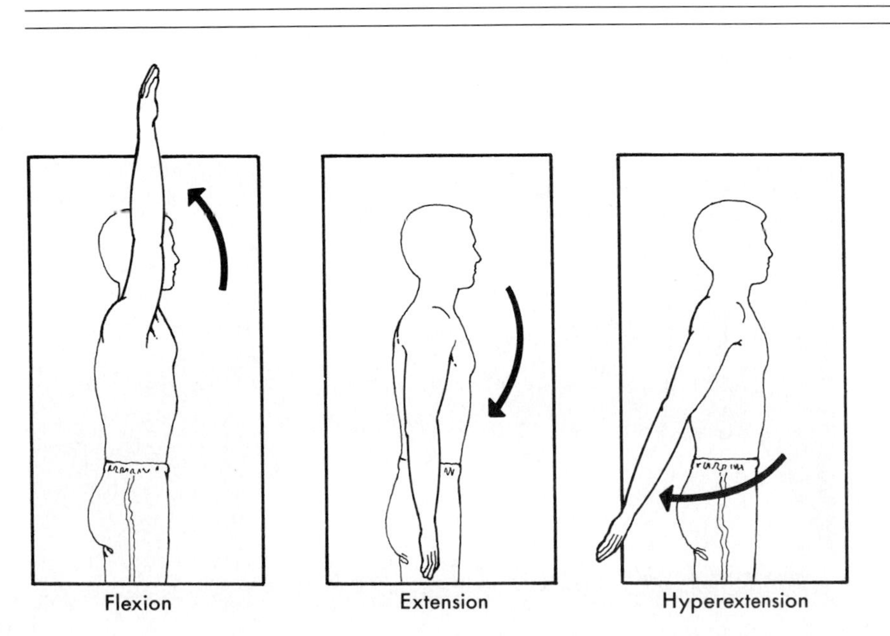

Figure 2-8
Sagittal plane movements at the shoulder.

Figure 2-9
Sagittal plane movements of the foot.

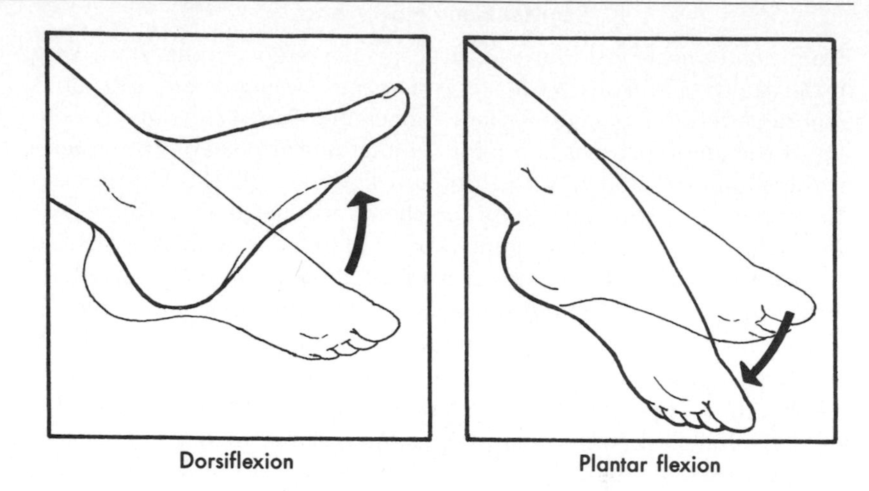

Frontal Plane Movements

■ Frontal plane movements include abduction, adduction, lateral flexion, elevation, depression, radial flexion, and ulnar flexion.

The major frontal plane movements are termed *abduction* and *adduction*. Abduction consists of motion conducted in the frontal plane away from the midline of the body. Adduction is the frontal plane movement in the opposite direction that brings a body segment closer to the midline of the body. Both movements are shown in Figure 2-10.

Other frontal plane movements include sideways movement of the trunk, which is termed *lateral flexion* or lateral bending (Figure 2-11). *Elevation* and *depression* of the shoulder girdle refer to movement of the shoulder girdle in superior and inferior directions re-

A cartwheel is frontal plane movement.

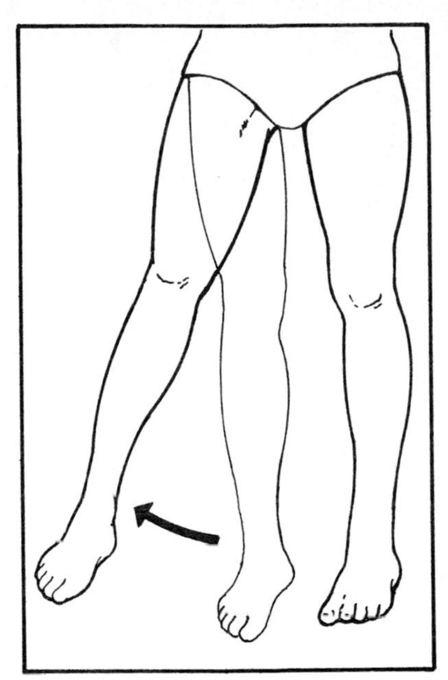

Adduction

Figure 2-10

Frontal plane movements at the hip.

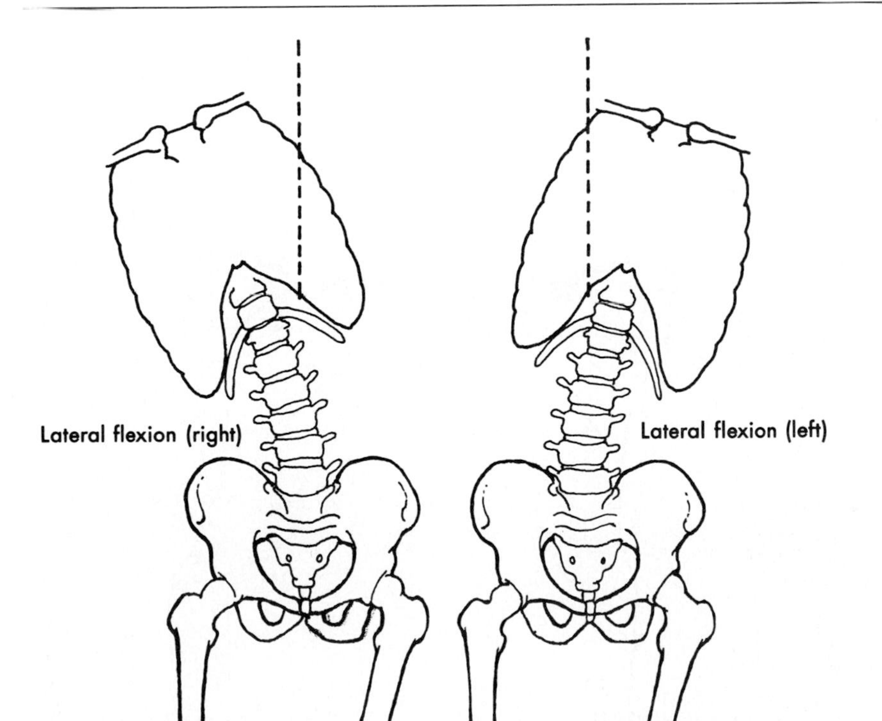

Figure 2-11

Frontal plane movements of the spinal column.

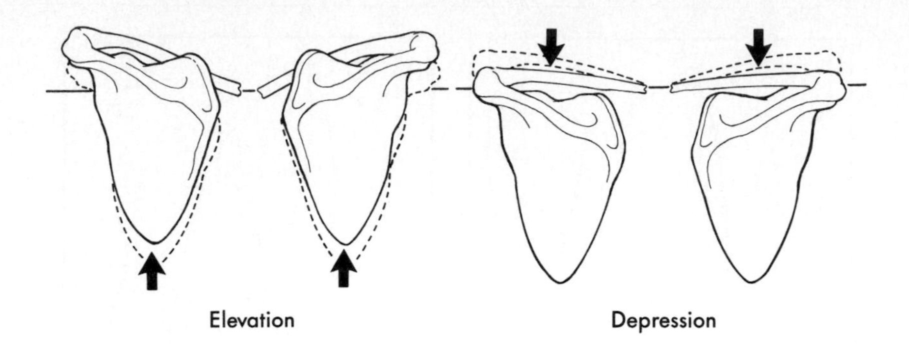

Figure 2-12
Frontal plane movements of the shoulder girdle.

spectively (Figure 2-12). Movement of the hand at the wrist in the frontal plane toward the radius (thumb side) is referred to as *radial flexion,* and *ulnar flexion* is hand deviation in the opposite direction (little finger side). These movements are shown in Figure 2-13.

Transverse Plane Movements

▪ Transverse plane movements include left and right rotation, inward and outward rotation, supination and pronation, inversion and eversion, and horizontal abduction and adduction.

Body movements in the transverse plane are rotational movements about a longitudinal or vertical axis. *Left rotation* and *right rotation* are used to describe transverse plane movements of the head, neck, and trunk. Movement of an arm or leg as a unit in the transverse plane is called medial or *inward rotation* when rotation is toward the midline of the body and lateral or *outward rotation* when the rotation is away from the midline of the body. Figure 2-14 displays these movements.

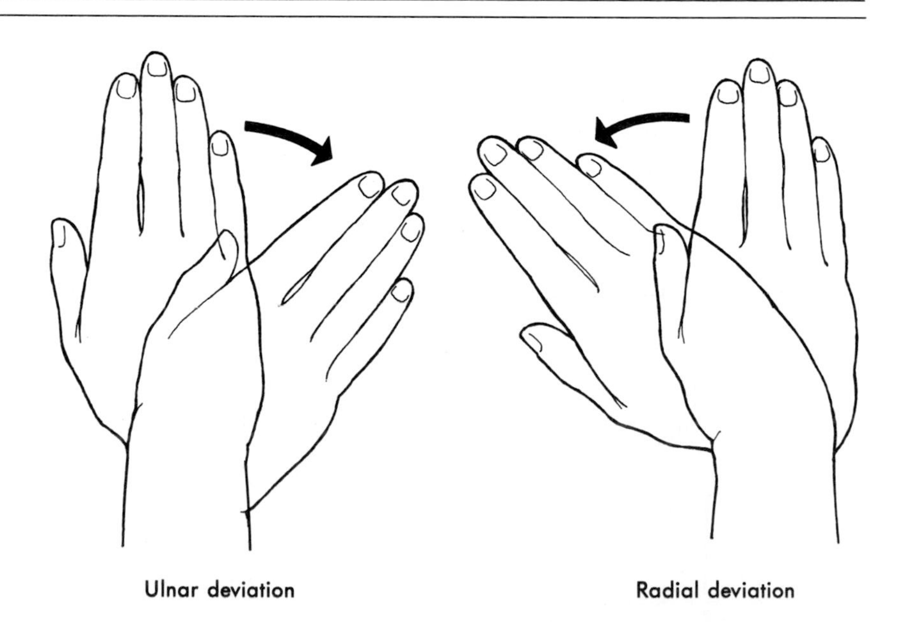

Figure 2-13
Frontal plane movements of the hand.

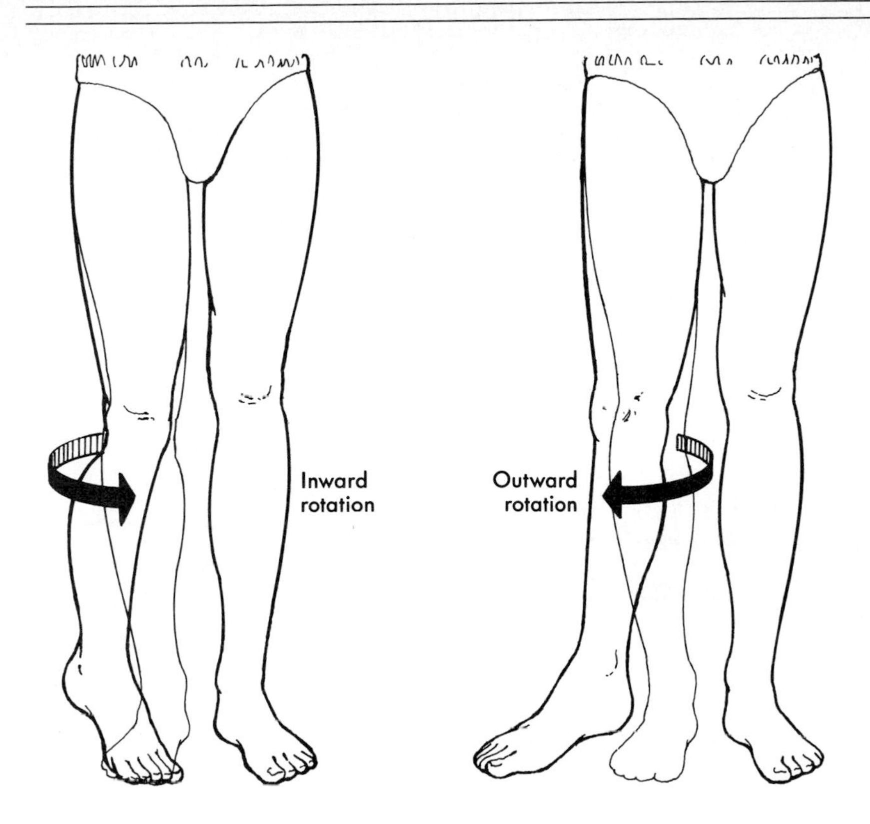

Figure 2-14
Transverse plane movements of the leg.

Specific terms are used for rotational movements of the forearm and of the foot. Outward rotation of the forearm is known as *supination.* In anatomical position the forearm is in a supinated position. *Pronation* is the term for inward rotation of the forearm (Figure 2-15). Outward rotation of the sole of the foot is termed *eversion,* and inward rotation of the sole of the foot is called *inversion.* These movements are illustrated in Figure 2-16. Abduction and adduction are also used to describe outward and inward rotation, of the entire foot. Pronation and supination are often used to describe motion occurring at the subtalar joint. Pronation at the subtalar joint consists of a combination of eversion, abduction, and dorsiflexion, and supination involves inversion, adduction, and plantar flexion (1).

Although abduction and adduction are frontal plane movements, movement of the arm or thigh in the transverse plane away from the midline of the body is termed *horizontal abduction* or horizontal flexion when either is flexed to a 90 degree position. Movement toward the midline of the body in the transverse plane is called *horizontal adduction* or horizontal extension. These movements are shown in Figure 2-17.

Figure 2-15
Transverse plane movements of the forearm.

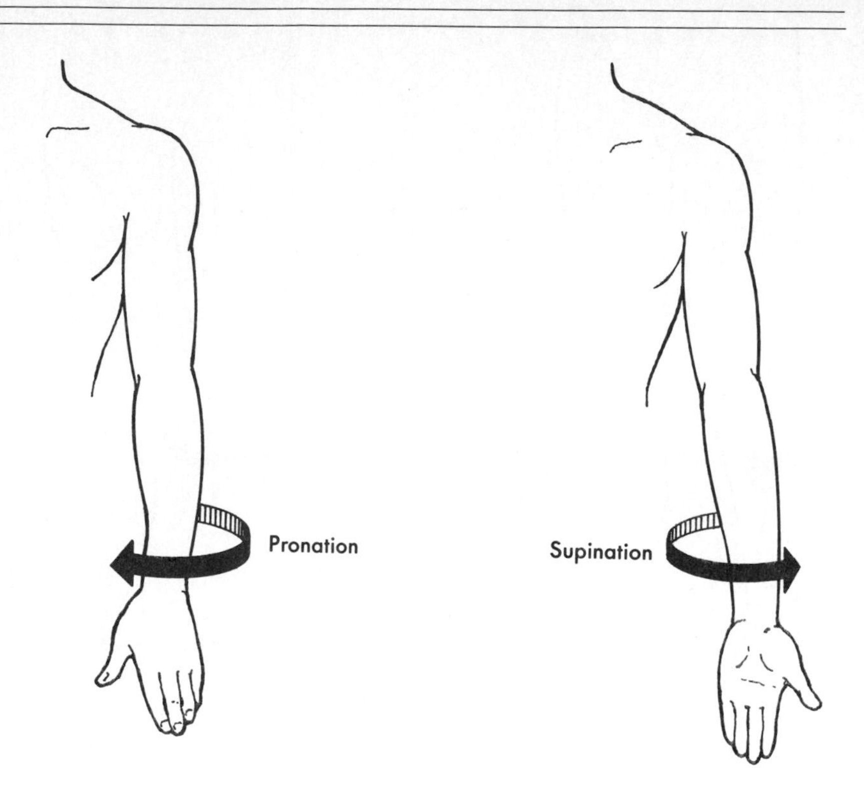

Figure 2-16
Rotational movements of the foot.

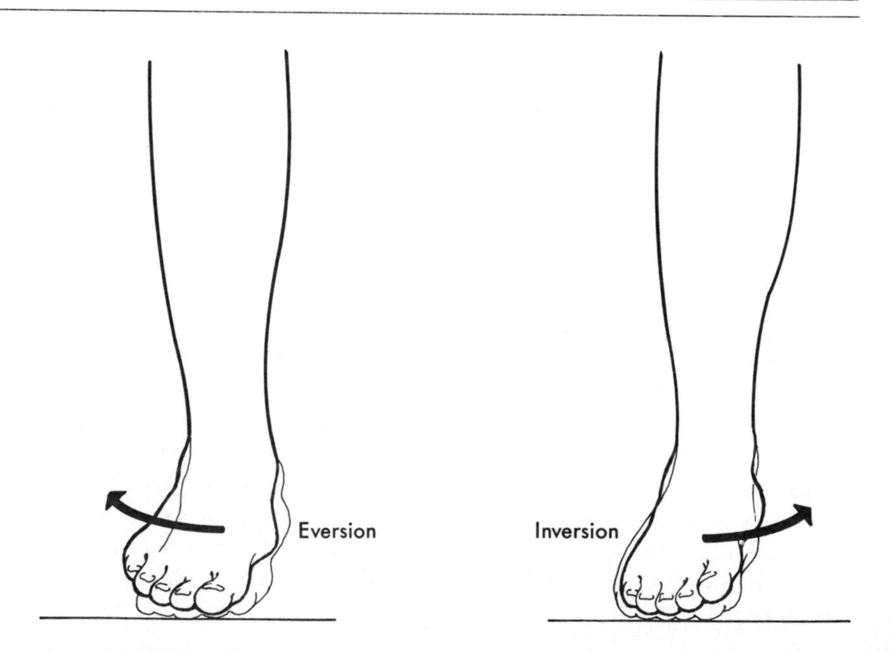

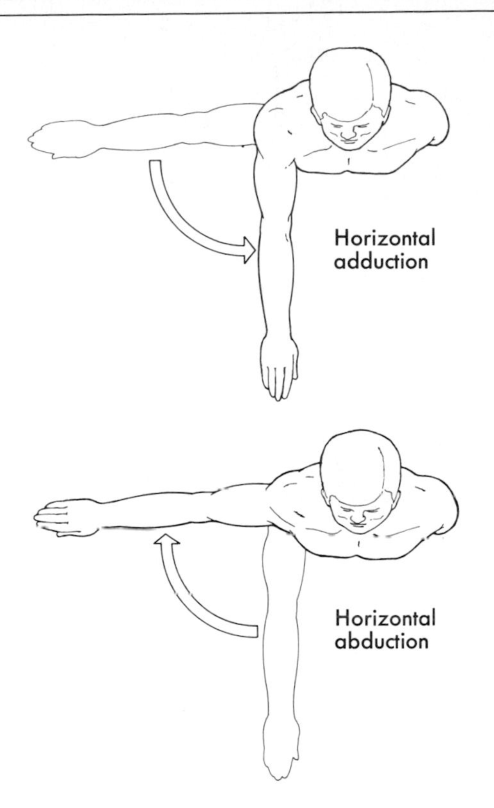

Figure 2-17
Transverse plane movements at the shoulder.

A tennis serve requires arm movement in a diagonal plane.

Other Movements

Many movements of the body limbs take place in planes that are oriented diagonally to the three traditionally recognized cardinal planes. Some authors (3) have assigned names to these **diagonal planes.** Because of the complex variations present in most human movements, nominal identification of every plane of movement would be difficult or impossible.

Movements that transect more than one plane are included in the category of general motion. Although most general motion is not specifically termed *nonplanar,* circular movement of a body segment is designated as *circumduction.* Tracing an imaginary circle in the air with a fingertip while the rest of the hand is stationary requires circumduction at the metacarpophalangeal joint. As shown in Figure 2-18, the path of the finger as a whole during the movement shapes a cone. Circumduction combines flexion, extension, abduction, and adduction at a joint.

diagonal planes
planes of movement oriented diagonally to the traditionally recognized planes of movement

Figure 2-18
Circumduction of the index finger at the metacarpophalangeal joint.

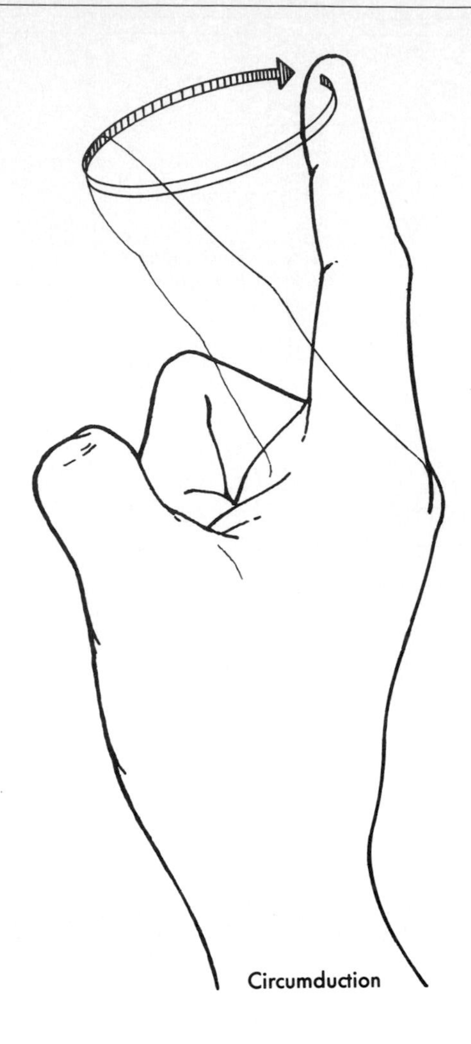

BASIC KINETIC CONCEPTS

Mass

mass
the quantity of matter contained in an object

Mass is the quantity of matter composing a body. The conventional symbol for mass is m. The common unit of mass in the metric system is the kilogram (kg), with the English unit of mass being the *slug*, which is much larger than a kg.

Inertia

inertia
the tendency of a body to maintain a motionless state or a state of constant velocity

In common usage, **inertia** means resistance to action or to change (Figure 2-19). Similarly, the mechanical definition is resistance to acceleration. Inertia is the tendency of a body to maintain its current state of motion, whether motionless or moving with a constant veloc-

Figure 2-19
A static object tends to maintain its motionless state because of its inertia.

A skater has a tendency to continue gliding with constant speed and direction due to the skater's inertia.

ity. For example, a 150 kg weight bar lying motionless on the floor has a tendency to remain motionless. A skater gliding on a smooth surface of ice has a tendency to continue gliding in a straight line with a constant speed.

Although inertia has no units of measurement, the amount of inertia a body possesses is directly proportional to its mass. The more massive an object the more it tends to maintain its current state of motion and the more difficult it is to disrupt that state.

Force

Force is a push or a pull acting on a body. If the push or pull is strong enough, it alters the body's state of motion, regardless of the body having no velocity, a constant positive or negative velocity, or a changing velocity. If the push or pull is relatively weak, it may not change the body's state of motion but produces a *tendency* for that state of motion to be altered. For example, although the force exerted by one person attempting to move a heavy refrigerator may not cause motion, an additional force supplied by a second individual may do so. When an applied force does not result in motion of the body, other forces such as friction or the body's weight have counteracted the applied force. When a force that is not counteracted acts on a body, motion occurs, regardless of the size of the force.

force
a push or pull that causes or tends to cause motion and the product of a body's mass and its acceleration

Force may also be defined as the product of a body's mass and the acceleration of that body resulting from the application of the force:

$$F = ma$$

F is the conventional symbol for force. Units of force are units of mass multiplied by units of acceleration. In the metric system, the most common unit of force is the Newton (N), which is defined as the product of 1 kg of mass and 1 m/s^2 of acceleration:

$$1 \text{ N} = (1 \text{ kg})\ (1 \text{ m/s}^2)$$

In the English system, the most common unit of force is the pound (lb). A lb of force is the amount of force necessary to accelerate a mass of 1 slug at 1 ft/s^2, and 1 lb is equal to 4.45 N:

$$1 \text{ lb} = (1 \text{ slug})\ (1 \text{ ft/s}^2)$$

free body diagram
a sketch that shows a given body in isolation with vector representations of all the forces acting on the body

A force is characterized by its size or magnitude, its direction, and the location of the point at which it is applied to a given body. Because a number of forces act simultaneously in most situations, constructing a **free body diagram** when analyzing the effects of forces on a body of interest is commonly the first step. A free body diagram consists of a sketch of the body or bodies being analyzed and vector representations of the forces present. Figure 2-20 shows several free body diagrams.

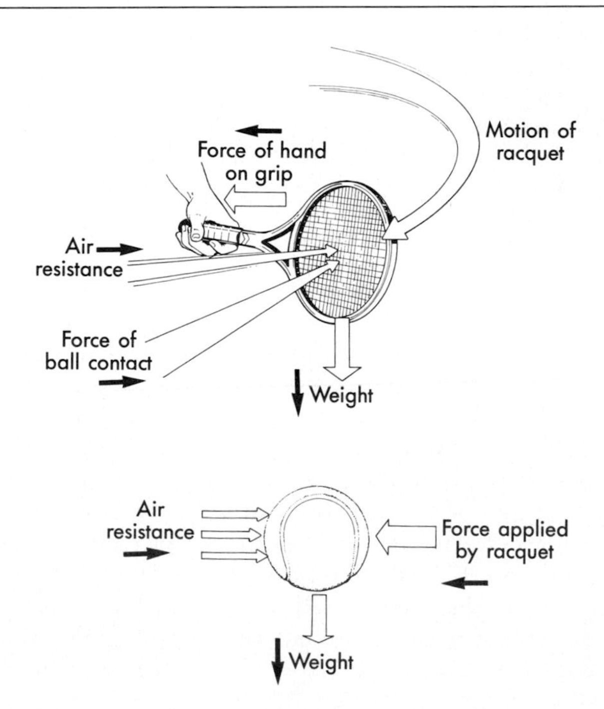

Figure 2-20
Free body diagrams. The black arrows are vectors.

Weight

Weight is defined as the amount of gravitational force exerted on a body. Algebraically, its definition is a modification of the general definition of a force, with weight (wt) being equal to mass multiplied by the acceleration of gravity (a_g).

weight
the attractive force that the earth exerts on a body

$$wt = ma_g$$

Since weight is a force, units of weight are units of force—either N or lb.

As the mass of a body increases, its weight increases proportionally. The factor of proportionality is the acceleration of gravity, which is $-9.81\ m/s^2$. The negative sign indicates that the acceleration of gravity is directed downward or toward the center of the earth. On the moon or another planet with a different gravitational acceleration, a body's weight would be different, although its mass would remain the same.

■ The acceleration of gravity is an unchanging constant, equal to approximately $-9.81\ m/s^2$.

Because weight is a force, it is also characterized by magnitude, direction, and point of application. The direction in which the force of weight acts is always toward the center of the earth. The point at which the weight force is assumed to act on a body is the body's center of gravity.

Although a body's mass remains unchanged on the moon, its weight is less due to a smaller gravitational acceleration.

Figure 2-21

SAMPLE PROBLEM 1

1. If a scale shows that an individual has a mass of 68 kg, what is that individual's weight?

Known

m = 68 kg

Solution

Wanted: weight

Formulas: $wt = ma_g$

$1\ kg = 2.2\ lb$

(Mass may be multiplied by the acceleration of gravity to convert to weight within either the English or the metric system.)

$$wt = ma_g$$
$$wt = (68\ kg)\ (9.81\ m/s^2)$$
$$wt = 667\ N$$

Mass in kg may be multiplied by the conversion factor 2.2 lb/kg to convert to weight in pounds:

$$(68\ kg)\ (2.2\ lb/kg) = \boxed{150\ lb}$$

2. What is the mass of an object weighing 1200 N?

Known

wt = 1200 N

Solution

Wanted: mass

Formula: $wt = ma_g$

$$1200\ N = m\ (9.81\ m/s^2)$$
$$\frac{1200\ N}{9.81\ m/s^2} = m$$
$$m = \boxed{122.32\ kg}$$

(Weight may be divided by the acceleration of gravity within a given system of measurement to convert to mass.)

3. What is the mass of an object weighing 100 lb?

Solution

$$\frac{100\ lb}{32.2\ ft/s^2} = \boxed{3.11\ slugs}$$

Although body weights are often reported in kilograms, the kilogram is actually a unit of mass. Weights should be identified in Newtons and masses reported in kilograms. The sample problem presented in Figure 2-21 illustrates the relationship between mass and weight.

Pressure

Would you prefer to have your foot stepped on by a woman wearing a spike-heeled shoe or by the same woman wearing a flat, smooth-soled shoe? The reason that the spike-heeled shoe would cause more pain relates to the difference in the amount of pressure exerted by the two different shoes.

Pressure (p) is defined as the amount of force acting over a given area (A).

pressure
force per unit of area

$$p = \frac{F}{A}$$

Units of pressure are units of force divided by units of area. Common units of pressure in the metric system are N per square centimeter (N/cm^2) and Pascals (Pa).

One Pascal represents one Newton per square meter ($Pa = N/m^2$). In the English system, the most common unit of pressure is pounds per square inch (psi or lb/in^2).

■ When forces are sustained by the human body, the smaller the area over which the force is distributed, the greater the likelihood of injury.

In the described situation the amount of force exerted on someone's foot is the woman's body weight. The area over which the force is distributed is the contact area between the sole of her shoe and the foot. As illustrated in the sample problem in Figure 2-22, the smaller amount of surface area on the bottom of a spike heel as compared to a flat sole results in a much greater amount of pressure being exerted.

■ Units of measurement for physical quantities often parallel the definitions of the quantities themselves.

Volume

A body's **volume** is the amount of space that it occupies. Because we describe space as being three dimensional, that is, having width, depth, and breadth, a unit of volume is a unit of length multiplied by a unit of length multiplied by a unit of length. In mathematical shorthand, this is a unit of length raised to the exponential power of three, or a unit of length *cubed*. In the metric system, common units of volume are cubic centimeters (cm^3) and cubic meters (m^3), and 1000 cubic centimeters is a liter (l):

volume
space occupied by a body

$$1 \text{ l} = 1000 \text{ cm}^3$$

In the English system of measurement, common units of volume are cubic inches (in^3) and cubic feet (ft^3). Another unit of volume in the English system is the quart (qt):

$$1 \text{ qt} = 57.75 \text{ in}^3$$

Figure 2-22

SAMPLE PROBLEM 2

Is it better to be stepped on by a woman wearing a spike heel or by a woman wearing a smooth-soled court shoe? If a woman's weight is 556 N, the surface area of the spike heel is 4 cm^2, and the surface area of the court shoe is 175 cm^2, how much pressure is exerted by each shoe?

Known

wt = 556 N

$A_s = 4\ cm^2$

$A_c = 175\ cm^2$

Solution

Wanted: Pressure exerted by the spike heel

Pressure exerted by the court shoe

Formula: $p = F/A$

Deduction: It is necessary to recall that weight is a force.

For the spike heel: $p = \frac{556\ N}{4\ cm^2}$

$$\boxed{p = 139\ N/cm^2}$$

For the court shoe: $p = \frac{556\ N}{175\ cm^2}$

$$\boxed{p = 3.18\ N/cm^2}$$

Comparison of the amounts of pressure exerted by the two shoes:

$$\frac{p_{\text{spike heel}}}{p_{\text{court shoe}}} = \frac{139}{3.18} = 43.75$$

Therefore, 43.75 times more pressure is exerted by the spike heel than by the court shoe worn by the same woman.

Pairs of balls that are similar in volume but markedly different in weight.

Volume is not the same as weight or mass. An 8 kg shot and a softball occupy approximately the same volume of space, but the weight of the shot is much greater than that of the softball. If a lean, muscular individual and an obese person have identical body weights, the obese person's body volume will be greater.

Density

The concept of **density** combines the mass or weight of a body with the body volume. Density is defined as mass per unit of volume. The conventional symbol for density is the small Greek letter rho (ρ).

density
mass per unit of volume

$$\text{density } (\rho) = \frac{\text{mass}}{\text{volume}}$$

Units of density are units of mass divided by units of volume. In the metric system, a common unit of density is the kilogram per cubic meter (kg/m^3). In the English system of measurement, units of density are not commonly used. Instead, units of specific weight (weight density) are employed.

specific weight weight per unit of volume

Specific weight (weight density) is defined as weight per unit volume. Because weight is proportional to mass, weight density is proportional to density. Units of weight density are units of weight divided by units of volume. The metric unit for specific weight is Newtons per cubic meter (N/m^3), and the English system uses pounds per cubic foot (lb/ft^3).

Although an 8 kg shot and a softball occupy approximately the same volume, the shot has a greater density and specific weight than the softball because the shot is heavier. Similarly, a lean person with the same body weight as an obese person has a higher body density because muscle is denser than fat.

Torque

When a force is applied to an object such as a pencil lying on a desk, either translation or general motion may result. If the applied force is directed parallel to the desk top and through the center of the pencil (a *centric force*), the pencil will be translated in the direction of the applied force. If the force is applied parallel to the desk top but directed through a point other than the center of the pencil (an *eccentric force*), the pencil will undergo both translation and rotation (Figure 2-23).

torque the rotary effect of a force

The rotary effect created by an eccentric force is known as **torque** (T) or moment of force. Torque, which may be thought of as *rotary force,* is the angular equivalent of linear force. Algebraically, torque

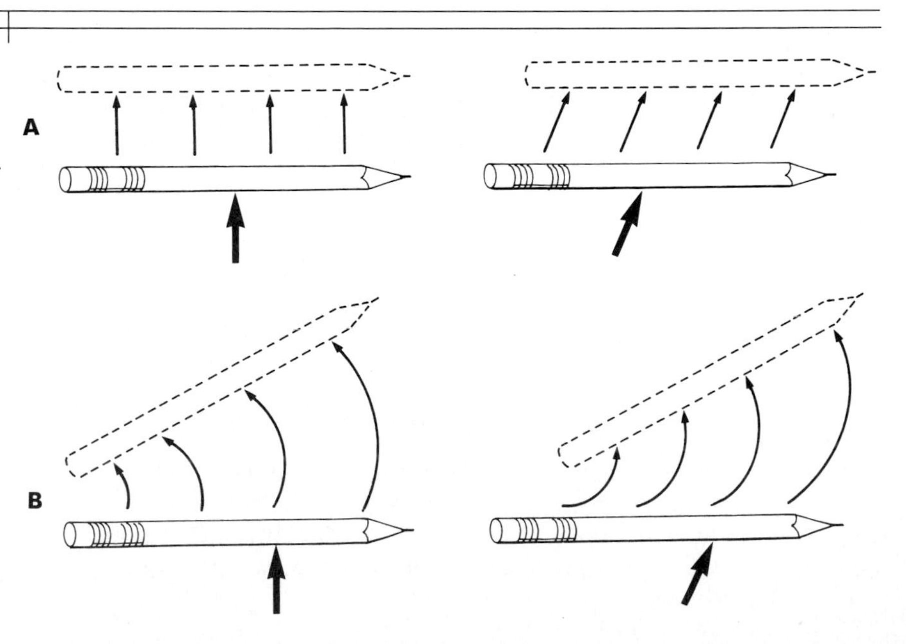

Figure 2-23
A, Centric forces produce translation. **B,** Eccentric forces produce translation and rotation.

Table 2-1
COMMON UNITS FOR KINETIC QUANTITIES

QUANTITY	SYMBOL	METRIC UNIT	ENGLISH UNIT
Mass	m	kg	slug
Force	F	N	lb
Pressure	p	Pa	psi (lb/ft^2)
Volume	V	m^3 liter	ft^3 quart gallon
Density	ρ	kg/m^3	
Specific weight		N/m^3	lb/ft^3
Torque	T	N-m	ft-lb

is the product of force (F) and the perpendicular distance (d) from the force's line of action to the axis of rotation.

$$T = Fd$$

The greater the amount of torque acting at the axis of rotation, the greater the tendency for rotation to occur. Units of torque in both the metric and English systems follow the algebraic definition. They are units of force multiplied by units of distance—Newton-meters (N-m) or foot-pounds (ft-lb). A summary of formulae and common units for the basic kinetic quantities discussed in the chapter is presented in Table 2-1.

VECTOR ALGEBRA

A **vector** is a symbol shaped like an arrow used to represent quantities that have both magnitude and direction. The magnitude of a quantity is its size; for example, the number 12 is of greater magnitude than the number 10. A vector's orientation on paper represents direction, and its length represents magnitude. Force, weight, pressure, weight density, and torque are **vector quantities.** None is fully defined without the identification of both its magnitude and direction, although with weight and weight density the direction is always downward. **Scalar quantities** possess magnitude but have no particular direction associated with them. Mass, volume, and density are examples of scalar quantities.

vector
an arrow symbol used to represent vector quantities

vector quantity
a physical quantity that possesses both magnitude and direction

scalar quantity
a physical quantity that is completely described when its magnitude is known

Vector Composition

When vectors are added together, the operation is called **vector composition.** The composition of two or more vectors that have exactly the same direction results in a single vector that has a magni-

vector composition
the process of determining a single vector from two or more vectors by vector addition

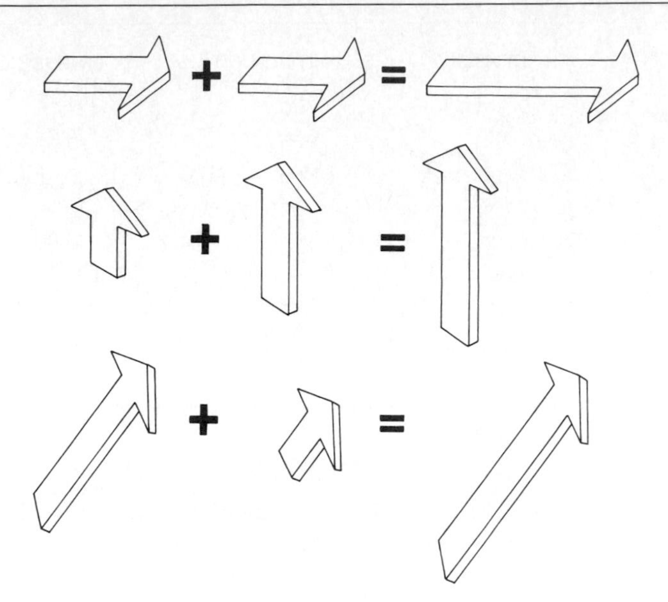

Figure 2-24
The composition of vectors with the same direction requires adding their magnitudes.

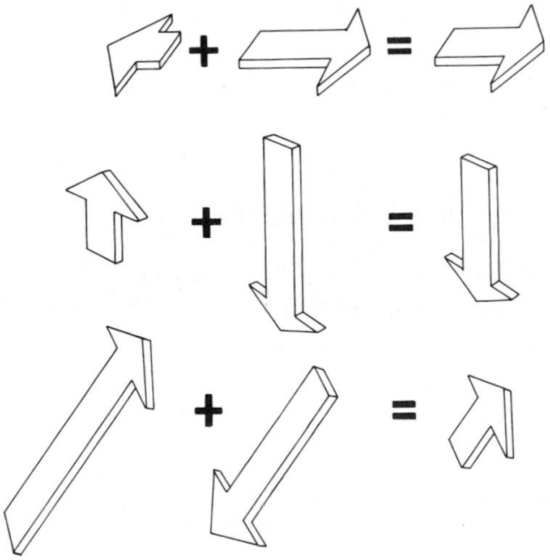

Figure 2-25
Composition of vectors with opposite directions requires subtracting their magnitudes.

resultant
the single vector that results from vector composition

tude equal to the sum of the magnitudes of the vectors being added (Figure 2-24). The single vector resulting from a composition of two or more vectors is known as the resultant vector or the **resultant.** If two vectors that are oriented in exactly opposite directions are composed, the resultant has the direction of the longer vector and a

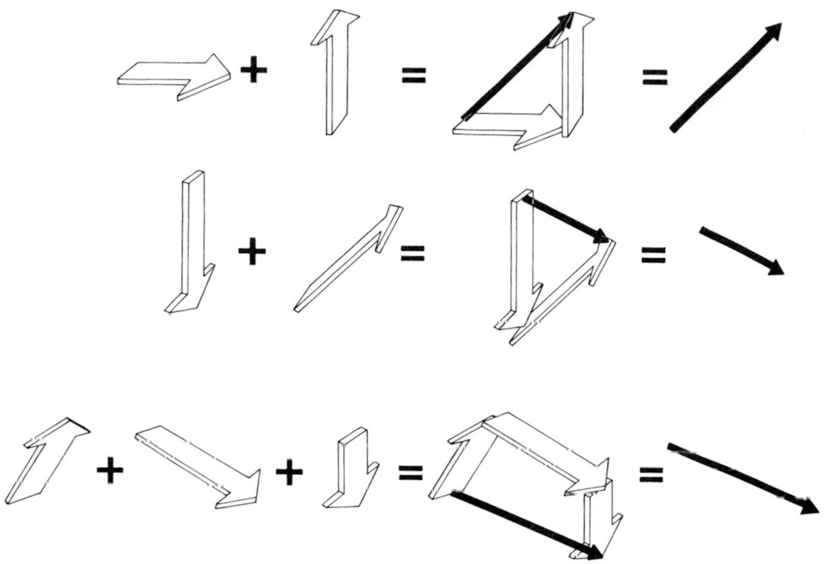

Figure 2-26
The "tip to tail" method of vector composition.

magnitude that is equal to the difference in the magnitudes of the two original vectors (Figure 2-25).

It is also possible to add vectors that are not oriented in the same or opposite directions. When the vectors are coplanar, that is, contained in the same plane, a procedure that may be used is the *tip to tail* method in which the tail of the second vector is placed on the tip of the first vector, and the resultant is then drawn with its tail on the tail of the first vector and its tip on the tip of the second vector. This procedure may be used for combining any number of vectors if each successive vector is positioned with its tail on the tip of the immediately preceding vector and the resultant connects the tail of the first vector to the tip of the last vector (Figure 2-26).

■ There are mathematical operations for adding, subtracting, and multiplying vectors.

Vectors are symbolic representations of actual physical quantities. Through the laws of vector combination, we often can calculate or better visualize the resultant effect of combined vector quantities. For example, a canoe floating down a river is subject both to the force of the current and the force of the wind. If the magnitudes and directions of these two forces are known, the single resultant or **net force** can be derived through the process of vector composition (Figure 2-27). The canoe would be expected to travel in the direction of the net force if no additional force or forces affect it.

net force
the resultant force derived from the composition of two or more forces

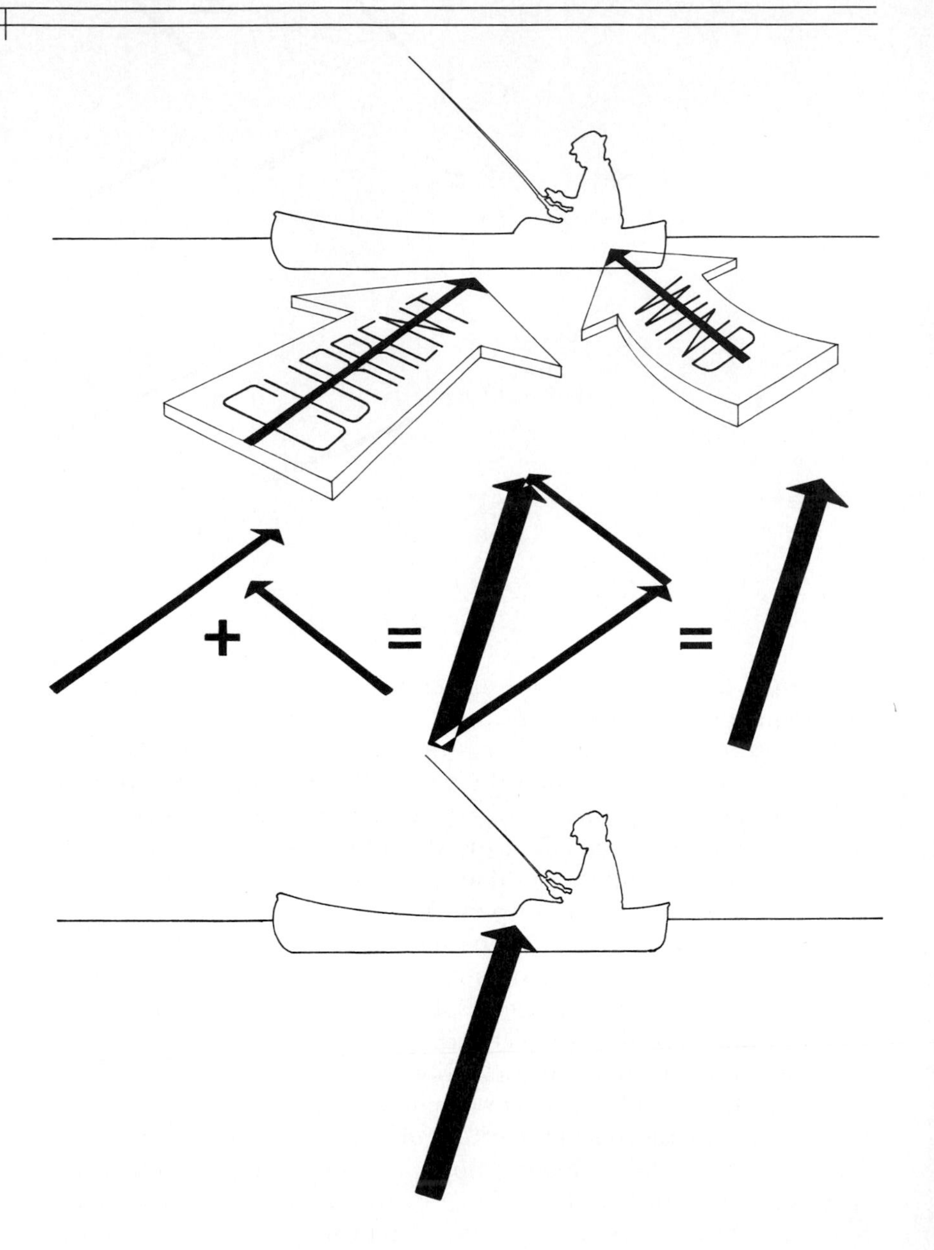

Figure 2-27
The net force is the resultant of all acting forces.

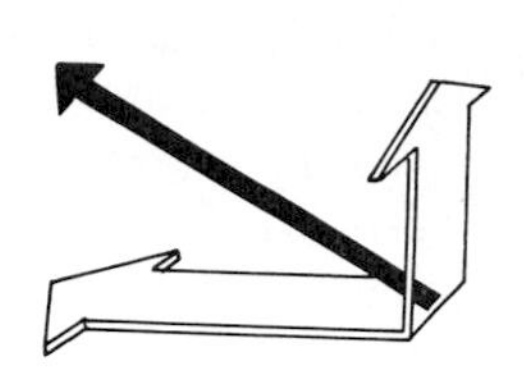

Figure 2-28
Vectors may be resolved into perpendicular components. The vector composition of each perpendicular pair of components yields the original vector.

Vector Resolution

Determining the perpendicular components of a vector quantity relative to a particular plane or structure is often important. For example, when a ball is thrown into the air, the horizontal component of its velocity determines the distance it travels, and the vertical component of its velocity determines the height it reaches. When a vector is resolved into perpendicular components—a process known as **vector resolution**—the vector sum of the components always yields a resultant that is equal to the original vector (Figure 2-28). The two perpendicular components therefore are a different but equal representation of the original vector.

vector resolution
an operation that replaces a single vector with two perpendicular vectors so that the vector composition of the two perpendicular vectors yields the original vector

Graphic Solution of Vector Problems

When vector quantities are uniplanar (contained in a single plane), vector manipulations may be done graphically to yield approximate results. Graphic solution of vector problems requires the careful measurement of vector orientations and lengths to minimize error. Vector lengths, which represent the magnitudes of vector quantities, must be drawn to scale. For example, 1 cm of vector length could represent 10 N of force. A force of 30 N would then be represented by a vector 3 cm in length, and a force of 45 N would be represented by a vector of 4.5 cm length.

Trigonometric Solution of Vector Problems

A more accurate procedure for quantitatively dealing with vector problems involves the application of trigonometric principles. Through the use of trigonometric relationships, the tedious process of measuring and drawing vectors to scale can be eliminated (see Appendix B). Figure 2-29 provides an example of the processes of both graphic and trigonometric solutions using vector quantities.

■ The mathematical (trigonometric) approach to solving vector problems offers the advantage of greater accuracy than the graphical approach.

Figure 2-29

SAMPLE PROBLEM 3

Terry and Charlie must move a refrigerator to a new location. They both push parallel to the floor, Terry with a force of 350 N and Charlie with a force of 400 N, as shown in the diagram below. a. What is the magnitude of the resultant of the forces produced by Terry and Charlie? b. If the amount of friction force that directly opposes the direction of motion of the refrigerator is 700 N, will they be able to move the refrigerator?

Graphic Solution

a. Use the scale 1 cm = 100 N to measure the length of the resultant.

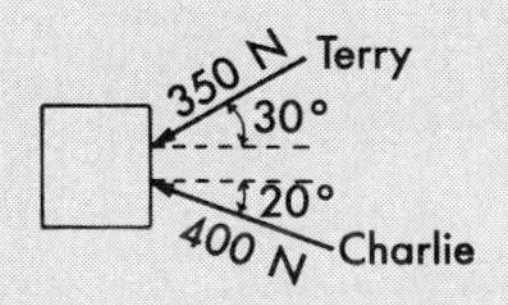

The length of the resultant is approximately 6.75 cm or 675 N.

b. Since 675 N $<$ 700 N, they will not be able to move the refrigerator.

Trigonometric Solution

Given: $F_T = 350$ N

$F_C = 400$ N

Wanted: The magnitude of the resultant force

Horizontal plane free body diagram:

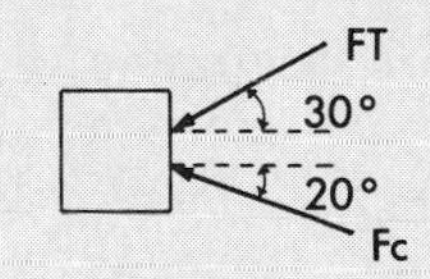

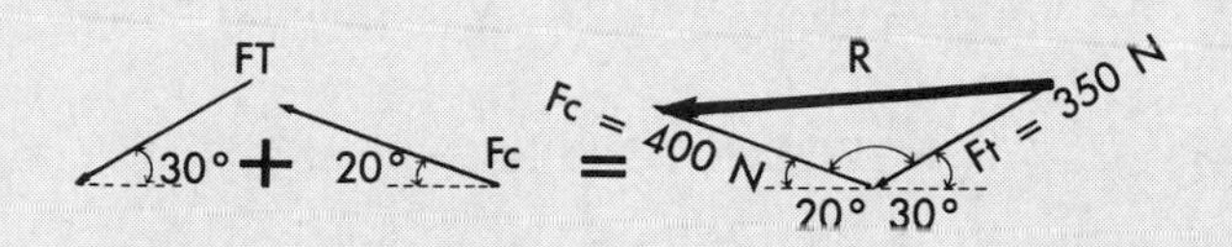

Formula:

$C^2 = A^2 + B^2 - 2(A)(B) \cos \theta$ (the Law of Cosines)

$R^2 = 400^2 + 350^2 - 2(400)(350)\cos 130$

R = 680 N

c. Since 680 N $<$ 700 N, they will not be able to move the refrigerator unless they exert more collective force while pushing at these particular angles. (If both Terry and Charlie pushed at a 90 degree angle to the refrigerator, their combined force would be sufficient to move it.)

SUMMARY

This chapter presents terminology that is used in kinematic and kinetic descriptions of human movement. The three general categories of movement are linear motion along a line, angular motion around an axis, and general motion, which is a combination of linear and angular motion. The anatomical reference position is used as the starting place for describing more specific movements of the human body. The sagittal, frontal, and transverse planes, with their respectively associated transverse, anteroposterior, and longitudinal axes, also provide frames of reference for the description of body movements. Terminology is presented that enables the description of directions with respect to the human body and the actions that occur at joints for communicating the particulars of human movement.

Basic concepts relating to kinetics are also introduced. Included are mass, the quantity of matter composing an object; inertia, the tendency of a body to maintain its current state of motion; force, a push or pull that alters or tends to alter a body's state of motion; weight, the gravitational force exerted on a body; pressure, the amount of force distributed over a given area; volume, the space occupied by a body; density, the mass or weight per unit of body volume; and torque.

Vector quantities have magnitude and direction; scalar quantities possess magnitude only. A vector is a symbol that can be used to represent a vector quantity. Vectors are effective aids in solving quantitative problems relating to vector quantities whether a graphic or a trigonometric approach is used. Of the two procedures, the use of trigonometric relationships is more accurate and less tedious.

INTRODUCTORY PROBLEMS

1. Using appropriate movement terminology, write a qualitative description of the performance of a single jumping jack.
2. Select a movement that occurs primarily in one of the three major reference planes. Qualitatively describe this movement in enough detail that the reader of your description can visualize the movement.
3. Select a familiar animal. Does the animal move in the same major reference planes in which humans move? What are the major differences in the movement patterns of this animal and the movement patterns of humans?
4. Identify five movements that occur primarily in each of the three cardinal planes. The movements may be either sport skills or activities of daily living.
5. William Perry, defensive tackle and part-time running back better known as *The Refrigerator,* weighed in at 1352 N during his 1985 rookie season with the Chicago Bears. What was Perry's mass? (Answer: 138 kg)

6. What is your own body mass?
7. Gravitational force on planet X is 40% of that found on the earth. If a person weighs 667.5 N on earth, what is the person's weight on planet X? What is the person's mass on the earth and on planet X? (Answer: Weight on planet X = 267 N; mass = 68 kg on either planet.)
8. How much force must be applied by a kicker to give a stationary 2.5 kg ball an acceleration of 40 m/s^2? (Answer: 100 N)
9. How much force must be applied to a 0.5 kg ice hockey puck to give it an acceleration of 30 m/s^2? (Answer: 15 N)
10. A rugby player is contacted simultaneously by three opponents who exert forces of the magnitudes and directions shown in the diagram at right. Using a graphical solution, show the magnitude and direction of the resultant force.

11. Using a graphical solution, compose the muscle force vectors to find the net force acting on the scapula.

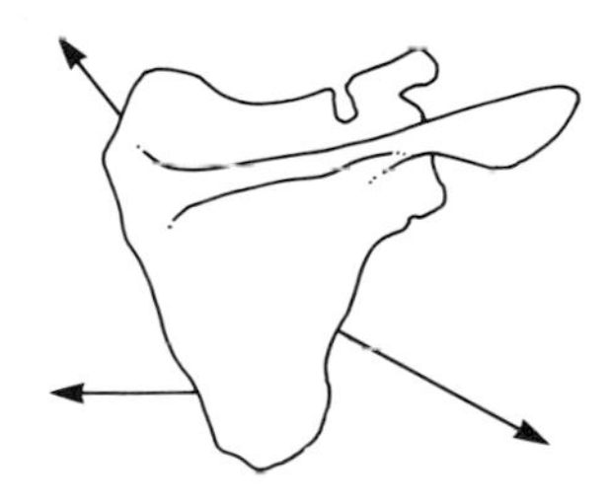

12. Draw the horizontal and vertical components of the vectors shown below.

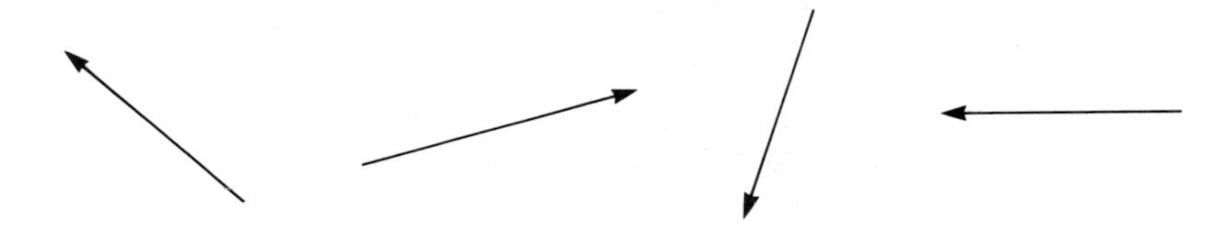

13. A gymnastics floor mat weighing 220 N has dimensions of 3 m × 4 m × 0.04 m. How much pressure is exerted by the mat against the floor? (Answer: 18.33 Pa)
14. If 37% of body weight is distributed above the superior surface of the L5 intervertebral disc and the area of the superior surface of the disc is 25 cm^2, how much pressure exerted on the disc is attributable to body weight for a 930 N man? (Answer: 13.8 N/cm^2)
15. What is the volume of a match box with sides of 4 cm, 3 cm, and 1.5 cm? (Answer: 18 cm^3)
16. What is the volume of a box whose sides are 2 m × 0.5 m × 0.7 m? (Answer: 0.7 m^3)

17. Choose three objects that are within your field of view and estimate the volume of each. List the approximate dimensions you used in formulating your estimates.
18. If the contents of the box described in Problem 16 weigh 100 N, what is the average density and specific weight of the box and contents? (Answer: 14.6 kg/m^3; 142.9 N/m^3)
19. Two children sit on opposite sides of a playground seesaw. Joey, who weighs 220 N, sits 1.5 m from the axis of the seesaw, and Suzy, who weighs 200 N, sits 1.7 m from the axis of the seesaw. How much torque is created at the axis by each child? In which direction will the seesaw tip? (Answer: Joey: 330 N-m; Suzy: 340 N-m; Suzy's end)
20. Two muscles develop tension simultaneously on opposite sides of a joint. Muscle A, attaching 3 cm from the axis of rotation at the joint, exerts 250 N of force. Muscle B, attaching 2.5 cm from the joint axis, exerts 260 N of force. How much torque is created at the joint by each muscle? What is the net torque created at the joint? In which direction will motion at the joint occur? (Answer: A: 7.5 N-m; B: 6.5 N-m; net torque equals 1 N-m in the direction of A)

ADDITIONAL PROBLEMS

1. Observe an individual with an injured joint attempting to perform a particular movement skill. Qualitatively identify the differences between the movement observed and a normal version of the movement.
2. Observe a single sport skill as performed by a highly skilled individual, a moderately skilled individual, and an unskilled individual. Qualitatively describe the differences observed.
3. Select a movement that is nonplanar and write a qualitative description of that movement sufficiently detailed to enable the reader of your description to picture the movement.
4. A football player is contacted by two tacklers simultaneously. Tackler A exerts a force of 400 N, and Tackler B exerts a force of 375 N. If the forces are coplanar and directed perpendicular to one another, what is the magnitude and direction of the resultant force acting on the player? (Answer: 548 N at an angle of 43 degrees to the line of action of Tackler A.)
5. A 75 kg skydiver in free fall is subjected to a crosswind exerting a force of 60 N and to a vertical air resistance force of 100 N. Describe the resultant force acting on the skydiver. (Answer: 638.6 N at an angle of 5.4 degrees to vertical)
6. Use a trigonometric solution to find the magnitude of the resultant of the following coplanar forces: 60 N at 90 degrees,

80 N at 120 degrees, and 100 N at 270 degrees. (Answer: 49.57 N)

7. In the nucleus pulposus of an intervertebral disc, the compressive load is 1.5 times the externally applied load. In the annulus fibrosus the compressive force is 0.5 times the external load (2). What are the compressive loads on the nucleus pulposus and annulus fibrosus of the L_5S_1 intervertebral disc of a 930 N man holding a 445 N weight bar across his shoulders, given that 37% of body weight is distributed above the disc? (Answer: 1183.7 N acts on the nucleus pulposus; 394.5 N acts on the annulus fibrosus.)
8. Estimate the volume of your own body. Construct a table that shows the approximate body dimensions you used in formulating your estimate.
9. Given the mass or weight and the volume of each of the following objects, rank them in the order of their densities.

OBJECT	WEIGHT OR MASS	VOLUME
A	50 kg	15.00 in^3
B	90 lb	12.00 cm^3
C	3 slugs	1.50 ft^3
D	450 N	0.14 m^3
E	45 kg	30.00 cm^3

10. An 890 N rock climber is supported in a horizontal position by a line that is oriented at an angle of 75 degrees to his body. Given that the vertical component of tension (force) in the line must be equal to his body weight, how much tension is present in the line? (Answer: 921 N)

REFERENCES

1. Frederick EC: Sport shoes and playing surfaces, Champaign, Ill, 1984, Human Kinetics Publishers, Inc.
2. Lindh M: Biomechanics of the lumbar spine. In Frankel VH and Nordin M: Basic biomechanics of the skeletal system, Philadelphia, 1980, Lea & Febiger.
3. Northrip JW, Logan GA, and McKinney WC: Introduction to biomechanic analysis of sport, ed 2, Dubuque, Ia, 1979, Wm C Brown Group.

ANNOTATED READINGS

Abbott EA: Flatland, New York, 1952, Dover Publications, Inc.

Uses geometrical figures as the characters and the setting for a delightful fantasy.

Barham JN: Introduction to kinetics. In Barham JN: Mechanical kinesiology, St Louis, 1978, The CV Mosby Co.

Introduces basic kinetic concepts from a mechanical perspective.

Wiktorin CV and Nordin M: Introduction to problem solving in biomechanics, Philadelphia, 1986, Lea & Febiger.

Provides an introduction to kinetic concepts in the context of clinical applications for physical therapists.

3 BONES

The Biomechanics of Bone Growth and Development

After reading this chapter, the student will be able to:

Identify and describe the different types of mechanical loads that act on the bones of the human body.

Explain how the material constituents and structural organization of bone affect its ability to withstand the mechanical loads to which it is subjected.

Describe the processes involved in the normal growth and maturation of bone.

Describe the effects of exercise and of weightlessness on bone mineralization.

Explain the significance of osteoporosis and speculate about ways to prevent it.

The word *bone* typically conjures up a mental image of a dead bone—a dry, brittle chunk of mineral that a dog would enjoy chewing. Given this picture, it is difficult to realize that living bone is actually an extremely dynamic tissue continually shaped and remodeled by the forces to which it is subjected. Bone fulfills two important mechanical functions for human beings: First, it provides a rigid skeletal framework that supports and protects other body tissues. Second, it forms a system of rigid **levers** that can be moved by forces from the attaching muscles (see Chapter 12).

lever
a relatively rigid object that may be made to rotate about an axis by the application of force

In recent years the role of exercise in maintaining proper bone health and mineralization has become an important research topic. There is concern because with the increasing proportion of senior citizens in American society, there has been a concomitant increase in the prevalence of **osteoporosis,** the abnormally high porosity of bones that may threaten the integrity of the skeletal system. Another topic addressed is the loss of bone mass outside the influence of the earth's gravitational field experienced by astronauts that may lead to serious problems during extended space flights. This chapter discusses the types of mechanical loads to which bones are subjected, the biomechanical aspects of bone composition and structure, bone growth and development, and the ways in which bones respond to the presence or absence of different forces.

osteoporosis
a pathological condition of decreased bone mass and strength

MECHANICAL LOADS ON BONES

Forces act on bone in different ways. Muscle forces, gravitational force, and the bone-breaking force encountered in a skiing accident all affect bone differently. One major consideration is the direction in which the force is applied to the bone. Forces can be divided into three categories in accordance with the directions in which they act on objects.

Compression, Tension, and Shear

Compressive force or **compression** can be thought of as a squeezing force (Figure 3-1). An effective way to press wildflowers is to place them inside the pages of a book and to stack other books on top of

compression
a pressing or squeezing force directed axially through a body

Figure 3-1

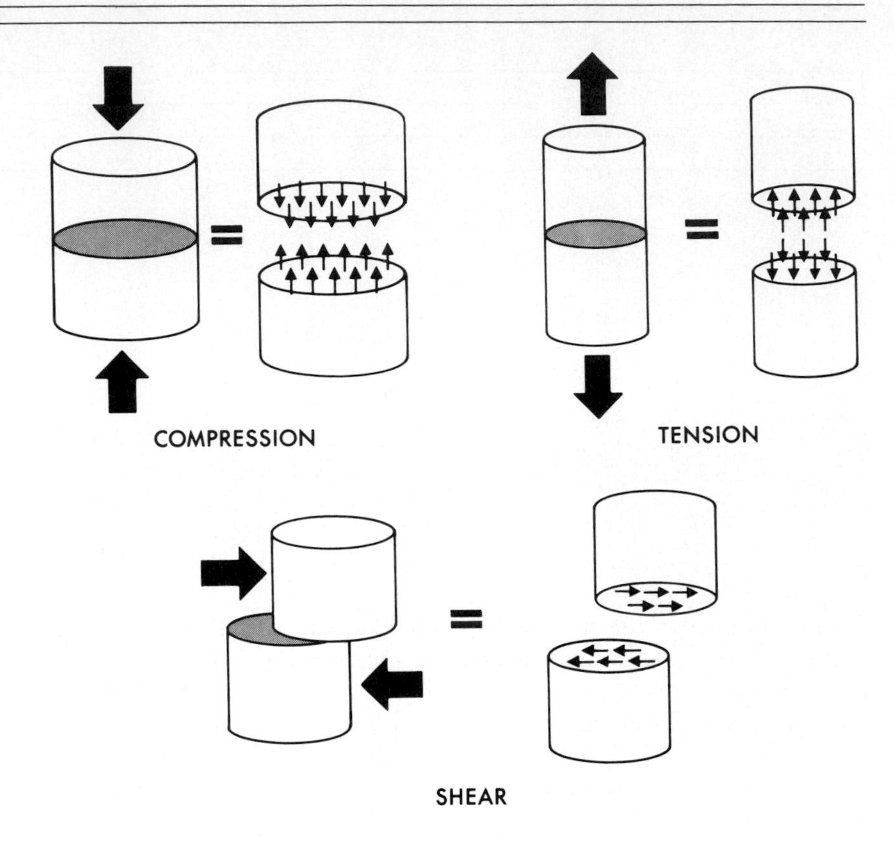

that book. The weight of the books creates a compressive force on the flowers. Similarly, the weight of the body acts as a compressive force on the bones that support it. Each vertebra in the spinal column must support the weight of all of the body above it.

tension
a pulling or stretching force directed axially through a body

The opposite of compressive force is tensile force or **tension** (Figure 3-1). Tensile force is a pulling force that creates tension in the object to which it is applied. When a child sits in a playground swing, the child's weight creates tension in the chains supporting the swing. A heavier child creates even more tension in the supports of the swing. Tensile force is exerted on a bone when the muscles attached to it contract or shorten.

shear
a force directed parallel to a surface

A third category of force is termed **shear.** Whereas compressive and tensile forces act along the long axis of a bone or other body to which they are applied, shear force acts parallel or tangent to a surface. Shear force tends to cause a portion of the object to slide, displace, or shear with respect to another portion of the object (Figure 3-1). For example, a force acting at the knee joint in a direction parallel to the tibial plateau is a shearing force at the knee (2). During

Loss of bone mass during periods of time spent outside of the earth's gravitational field is a problem for astronauts.

the performance of a squat exercise, joint shear at the knee is greatest at the full squat position (2). The largest amount of stress is placed on the ligaments and muscle tendons that are preventing the femur from sliding off of the tibial plateau (Figure 3-2).

Mechanical Stress

Another factor affecting the outcome of the action of forces on the bones of the human body is the way in which the force is distributed through the bone. Whereas pressure represents the distribution of force external to a body, **stress** represents the resulting force distribution inside of a body when an external force acts. Stress may be quantified in the same way as pressure—force per unit of area over

stress
the distribution of force within a body, which quantified as force divided by the area over which the force acts

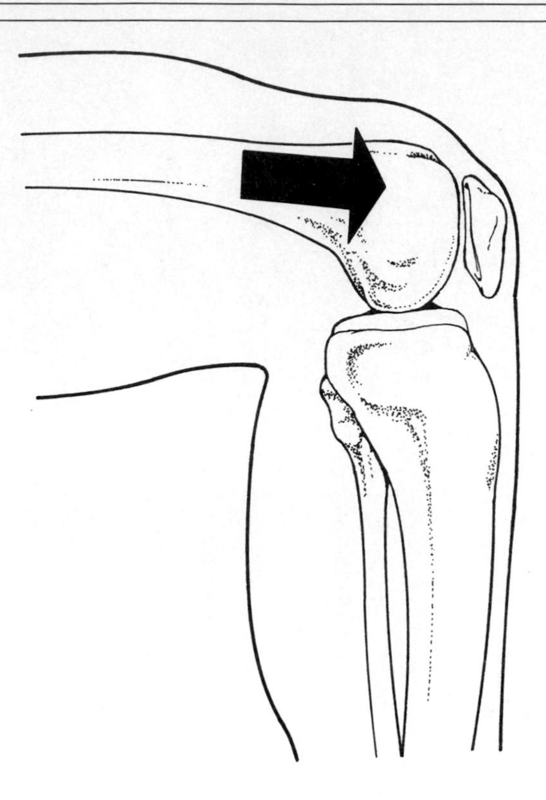

Figure 3-2
During performance of squat or knee bend exercises, shear force acting at the knee is maximal when flexion at the knee is maximal. The shear at the joint is produced by the axial force in the femur.

which the force acts. As shown in Figure 3-3, a given force acting on a small surface produces greater stress than the same force acting over a larger surface. When a blow is sustained by the human body, the likelihood of injury to body tissue is related to the magnitude and direction of the stress created by the blow. Compressive stress, tensile stress, and shear stress specifically indicate the direction of the stress present.

Because the lumbar vertebrae bear more of the weight of the body than the thoracic vertebrae when a person is in an upright position, the compressive stress in the lumbar region should logically be greater. However, the amount of stress present is not directly proportional to the amount of weight borne because the load-bearing surface areas of the lumbar vertebrae are greater than those of the vertebrae higher in the spinal column. Figure 3-4 illustrates this relationship. This increased surface area reduces the amount of compressive stress present. Nevertheless, the L5,S1 intervertebral disc (at the bottom of the lumbar spine) is the most common site of disc herniations. The amount of compressive stress sustained by the disc is probably not the only contributing factor (see Chapter 8). The quantification of mechanical stress is demonstrated in the sample problem shown in Figure 3-5.

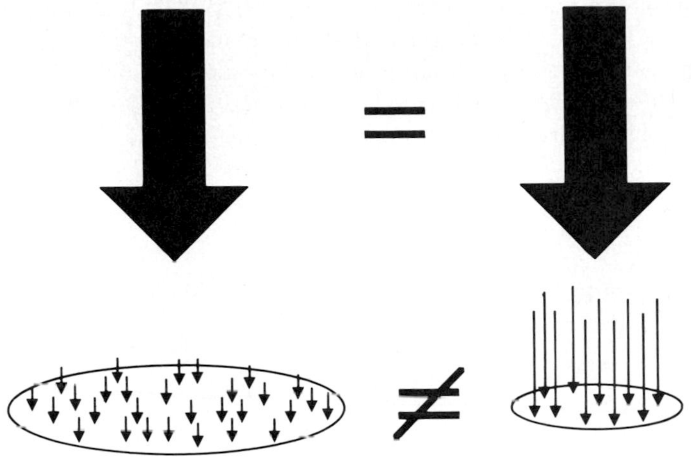

Figure 3-3
The amount of mechanical stress created by a force is dependent on the size of the area over which the force is spread.

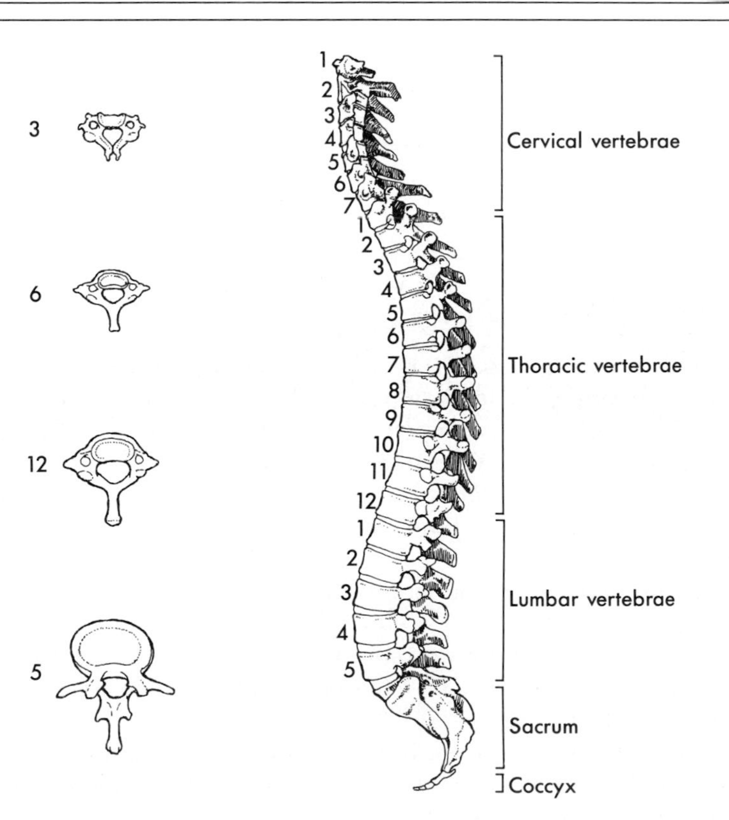

Figure 3-4
The surfaces of the vertebral bodies increase in surface area as more weight is supported.

Figure 3-5

SAMPLE PROBLEM 1

How much compressive stress is present on the L1, L2 vertebral disk of a 625 N woman, given that approximately 45% of body weight is supported by the disk a) when she stands in anatomical position? b) when she stands erect holding a 222 N suitcase? (Assume that the disk is oriented horizontally and that its surface area is 20 cm^2.)

Solution

a. Given: F = (625 N) (0.45)

A = 20 cm^2

Formula: Stress = F/A

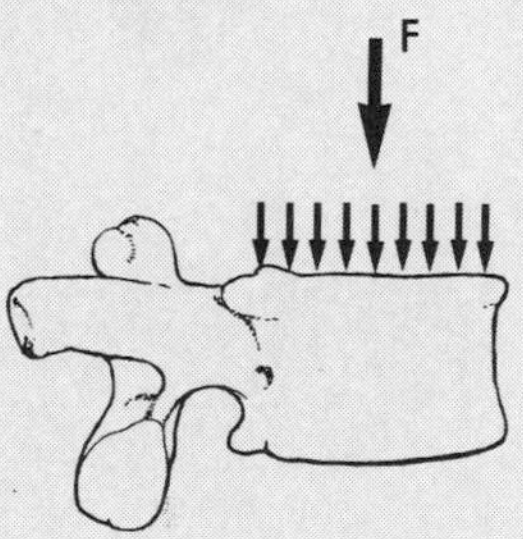

$$\text{Stress} = \frac{(625\ \text{N})\ (0.45)}{20\ \text{cm}^2}$$

Stress = 14 N/cm^2

b. Given: F = (625 N)(0.45) + 222 N

Formula: Stress = F/A

$$\text{Stress} = \frac{(625\ \text{N})\ (0.45) + 222\ \text{N}}{20\ \text{cm}^2}$$

Stress = 25.2 N/cm^2

bending
an asymmetric loading that produces tension on one side of a body's longitudinal axis and compression on the other side

axial
action directed along the longitudinal axis of a body

torsion
the twisting of a bone around a body's longitudinal axis

combined loading
the simultaneous action of more than one of the pure forms of loading

Torsion, Bending, and Combined Loads

A slightly more complicated type of loading that bones may undergo is called **bending.** Pure compression and tension are both **axial** forces; that is, they are directed along the longitudinal axis of the bone or other object. When an eccentric (or nonaxial) force is applied to the end of a bone, the bone bends, creating compressive stress on one side of the bone and tensile stress on the opposite side (Figure 3-6).

Torsion occurs when a bone is caused to twist about its longitudinal axis, typically when one end of the bone is fixed. Torsional fractures of the tibia are not uncommon in football and skiing accidents in which the foot is held in a fixed position while the rest of the body undergoes a twist.

Because the bones of the human body are subject to gravitational force, muscle forces, and often other forces, living bone is usually loaded in several different ways at once. The presence of more than one of the five pure forms of loading is known as **combined loading.**

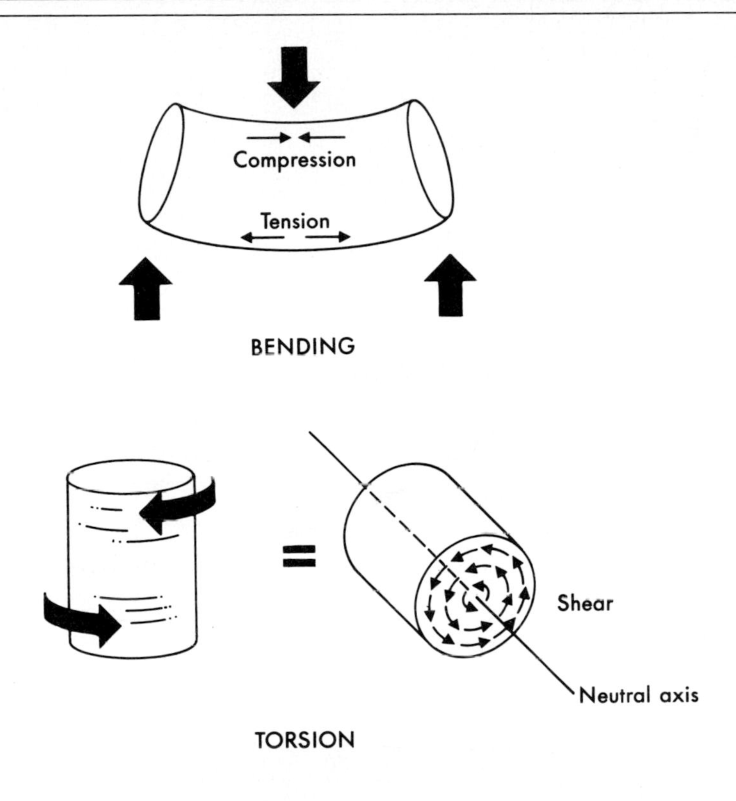

Figure 3-6
Bones loaded in bending are subject to compression on one side and tension on the other. Bones loaded in torsion develop internal shear stress, with maximal stress at the periphery and no stress at the neutral axis.

Irregular bone shapes and asymmetrical bone structures also contribute to the diversification of stresses present in response to a given external force.

Repetitive Versus Traumatic Loads

The distinction between **repetitive** and **traumatic loading** is also important. When a single force large enough to cause injury acts on biological tissues, the force is considered to be traumatic. The force produced by a fall, a rugby tackle, or an automobile accident may be sufficient to fracture a bone.

Injury can also result from the repeated sustenance of relatively small forces. For example, each time a foot hits the pavement during running a force of approximately 2 to 3 times body weight is sustained. Although a single force of this magnitude is not likely to result in a fracture of healthy bone, numerous repetitions of such a force may cause a fracture of an otherwise healthy bone somewhere in the lower extremity. Fractures resulting from repetitive loading are *fatigue fractures,* more commonly known as *stress fractures.* The relationship between the magnitude of the load sustained, the frequency of loading, and the likelihood of injury is shown in Figure 3-7.

repetitive loading
the repeated application of a nontraumatic load that is usually of relatively low magnitude

traumatic loading
the application of a single force of sufficient magnitude to cause injury to a biological tissue

Figure 3-7
The general pattern of injury likelihood as a function of load magnitude and repetition.

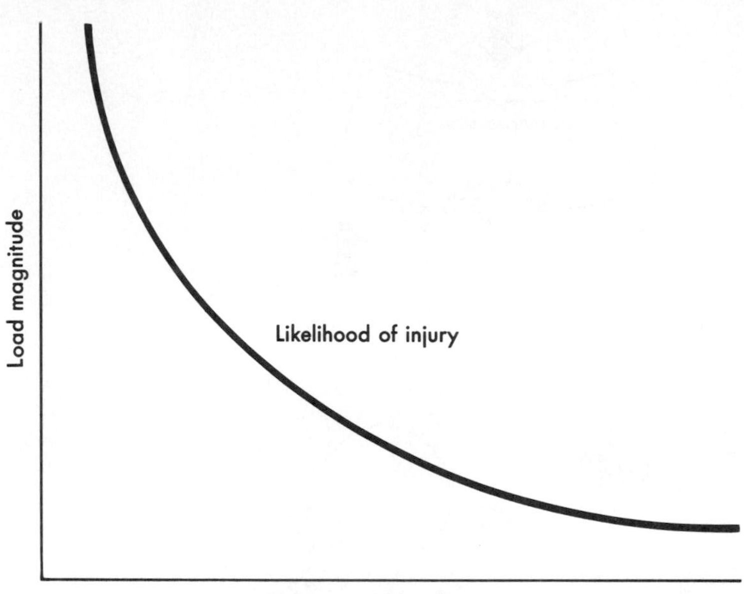

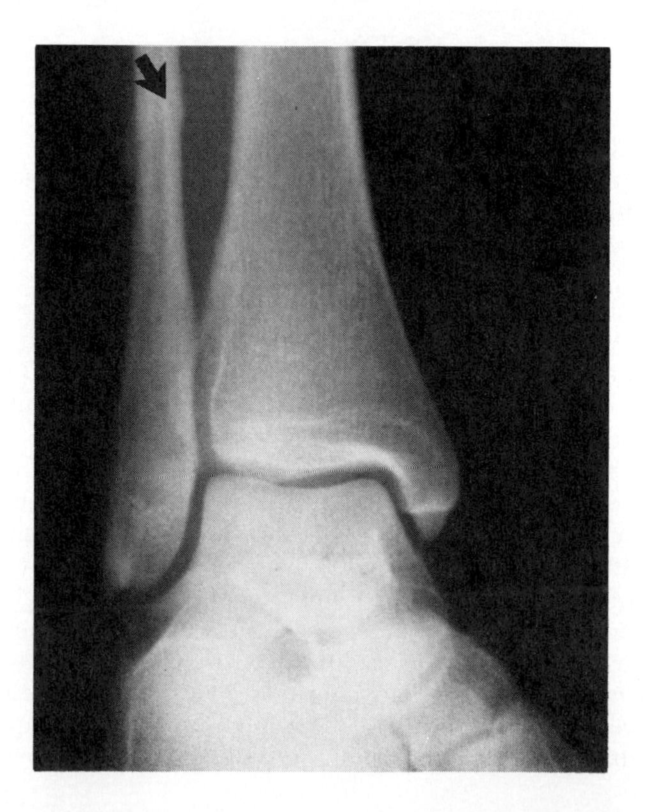
This x-ray shows a fatigue fracture, commonly known as a stress fracture, toward the distal end of the fibula. (Courtesy Lester Cohn, MD.)

PROPERTIES AND STRUCTURE OF BONE TISSUE

The ways in which bone behaves in response to the mechanical loads to which it is subjected relates to its material constituents and structural organization. The composition and structure of bone yield a material that is strong for its relatively light weight.

Material Constituents

The major building blocks of bone are calcium carbonate, calcium phosphate, collagen, and water. The relative percentages of these materials vary with the age and health of the bone. Calcium carbonate and calcium phosphate generally comprise approximately 60% to 70% of bone weight. These minerals give bone its **stiffness** and are the primary determiners of its **compressive strength.** Collagen is a protein that provides bone with **elasticity** and contributes to its **tensile strength.** Collagen is more prevalent in the bones of young children than in those of adults.

The water content of bone makes up approximately 25% to 30% of the total bone weight. The water present in bone tissue is an important contributor to bone strength (22). For this reason, scientists and engineers studying the material properties of different types of bone tissue must ensure that the bone specimens that they are testing do not become dehydrated.

Structural Organization

The relative percentage of bone mineralization varies not only with the age of the individual but also with the location of the bone in the body. Some bones within a given person are more **porous** than others. The more porous the bone, the smaller the proportion of calcium phosphate and calcium carbonate and the greater the proportion of nonmineralized bone tissue. Bone tissue has been classified into two categories based on porosity. If the porosity is low—5% to 30% of the bone volume occupied by nonmineralized tissue—the tissue is termed **cortical bone.** Bone tissue with a relatively high porosity—30% to greater than 90% of bone volume occupied by nonmineralized tissue—it is known as spongy or **cancellous bone.** Most of the bones of the human body have outer layers composed of cortical bone, with cancellous bone tissue underlying the cortical bone.

The porosity of bone tissue is of interest because it directly affects its mechanical characteristics. With its higher mineral content, cortical bone is stiffer so that it can withstand greater stress but less **strain** or relative **deformation** than cancellous bone. Cancellous bone is spongier than cortical bone. It can undergo more strain before fracturing. The shafts of the long bones contain a high proportion of strong cortical bone. The relatively high cancellous bone content of the vertebrae contributes to their shock-absorbing capability.

stiffness
the ratio of stress to strain in a loaded material; that is, the stress divided by the relative amount of change in the structure's shape

compressive strength
the resistance to compression

elasticity
the ability to regain original size and shape after a load is removed

tensile strength
the resistance to tension

porous
the property of containing pores or cavities

cortical bone
compact bone tissue with low porosity that is found in the shafts of long bones

cancellous bone
bone tissue with high porosity that is found in the ends of long bones and in the vertebrae

strain
the amount of deformation divided by the original length of the structure or by the original angular orientation of the structure

deformation
change in original shape

The structures of bones are largely determined by the nature of the forces to which they are subjected. Trabecular orientation, particularly in regions of high stress such as the femoral neck, most effectively resists the loads habitually encountered. Cancellous bone develops four types of structure depending on if it must withstand relatively high or relatively low forces and if the primary loading is axial (tension or compression) or asymmetric (bending) (6).

anisotropic
the property of exhibiting different mechanical properties in response to loads from different directions

Both cortical and cancellous bone are **anisotropic;** that is, they exhibit different strength and stiffness in response to forces applied from different directions. Bone is strongest in resisting compressive stress and weakest in resisting shear stress.

Types of Bones

axial skeleton
structure including the skull, vertebrae, sternum, and ribs

appendicular skeleton
bones composing the body appendages

The 206 bones of the human body are structured and shaped to fulfill specific functions. The skeletal system, displayed in Figure 3-8, is nominally subdivided into two categories—the **axial skeleton** and the **appendicular skeleton.** The axial skeleton includes the bones that form the axis of the body, which are the skull, the vertebrae, the sternum, and the ribs. The other bones form the body appendages or the appendicular skeleton. Bones are also categorized according to their general shapes and functions.

short bones
the small, cubical bones, including the carpals and tarsals

As shown in Figure 3-9, the **short bones** are approximately cubical. This category includes only the carpals and the tarsals. These bones provide limited gliding motions and serve as shock absorbers.

flat bones
bones that are largely flat in shape, for example, the scapula

Flat bones, shown in Figure 3-9, are also described by their name. These bones protect underlying organs and soft tissues and also provide large areas for muscle and ligament attachments. The flat bones include the scapulae, sternum, ribs, patellae, and some of the bones of the skull.

irregular bones
bones of irregular shapes, for example, the sacrum

The **irregular bones** have different shapes to fulfill special functions in the human body. For example, the vertebrae provide a bony, protective tunnel for the spinal cord, offer several processes for muscle and ligament attachments, and support the weight of the superior body parts while enabling movement of the trunk in all three cardinal planes (Figure 3-9). The sacrum, coccyx, and maxilla are other examples of irregular bones.

long bones
bones consisting of a long shaft with bulbous ends, for example, the femur

articular cartilage
a protective layer of cartilage over the ends of long bones where they articulate with other bones

Long bones form the framework of the appendicular skeleton (Figure 3-10). They consist of a long, roughly cylindrical shaft (also called the body or diaphysis) with bulbous ends known as condyles, tubercles, or tuberosities. The ends of long bones are covered with a self-lubricating **articular cartilage** that protects the bone from wear at points of contact with other bones. Long bones also contain a central hollow area known as the medullary cavity or canal.

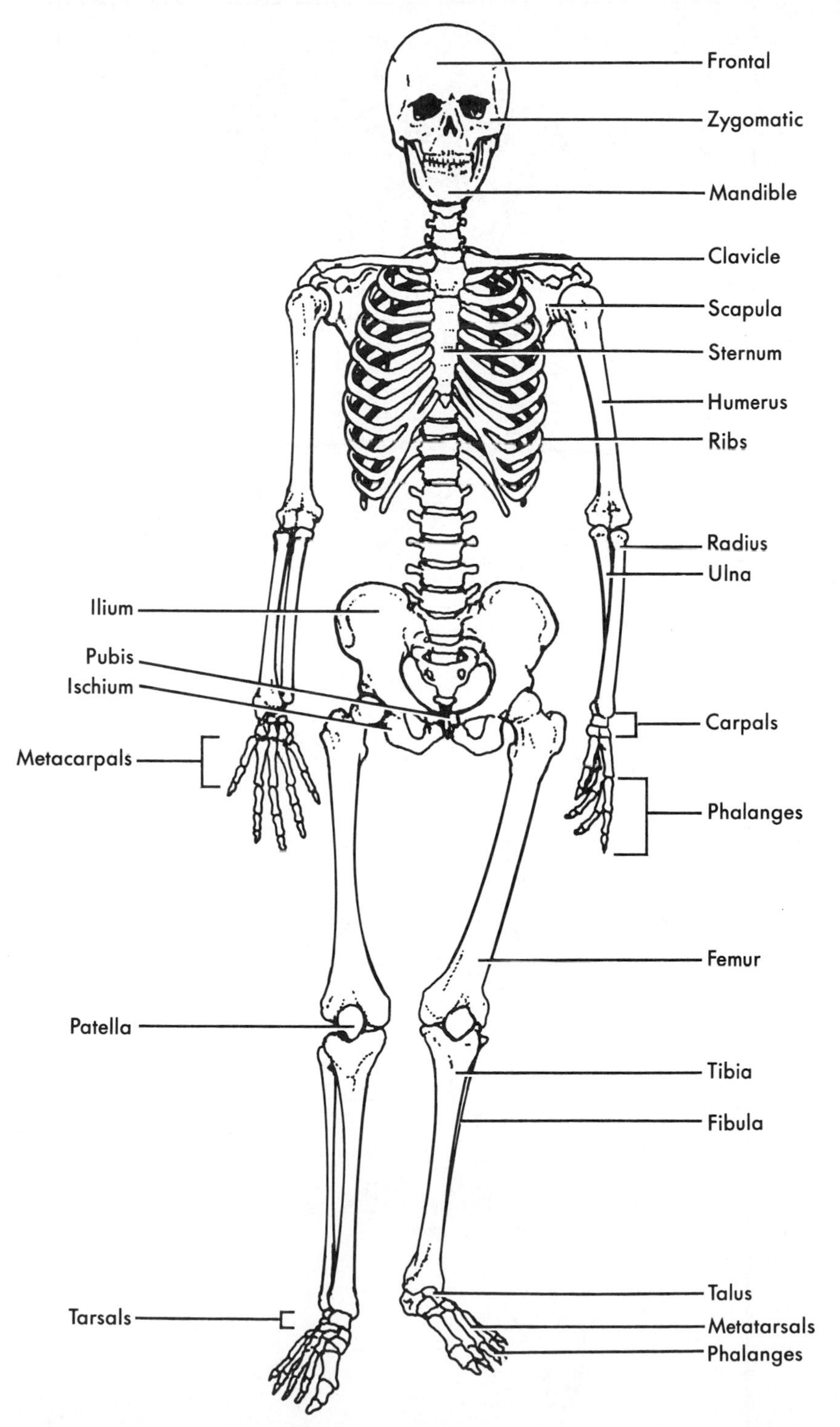

Figure 3-8
The human skeleton.

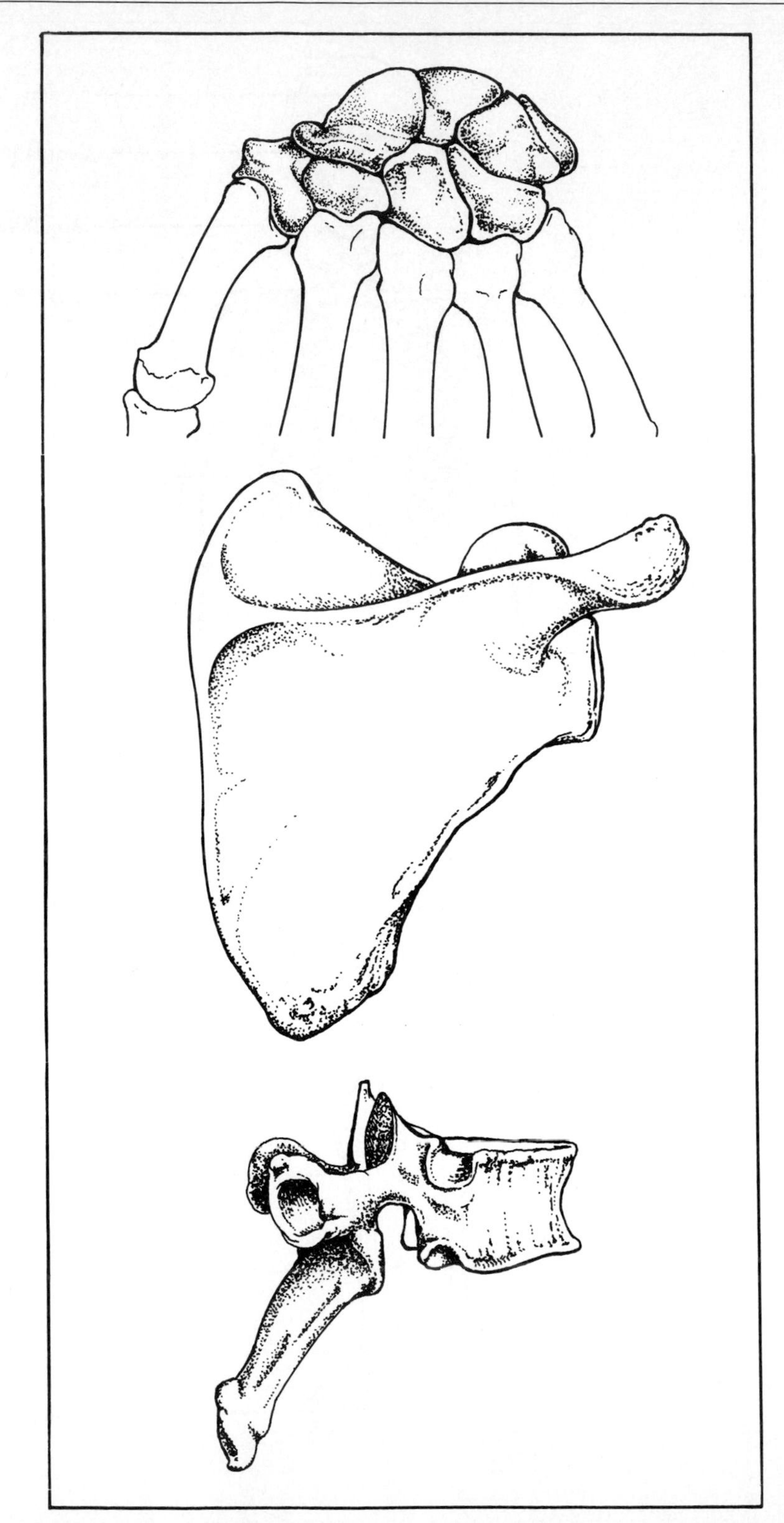

Figure 3-9

A, The carpals are categorized as short bones. **B,** The scapula is categorized as a flat bone. **C,** The vertebrae are examples of irregular bones.

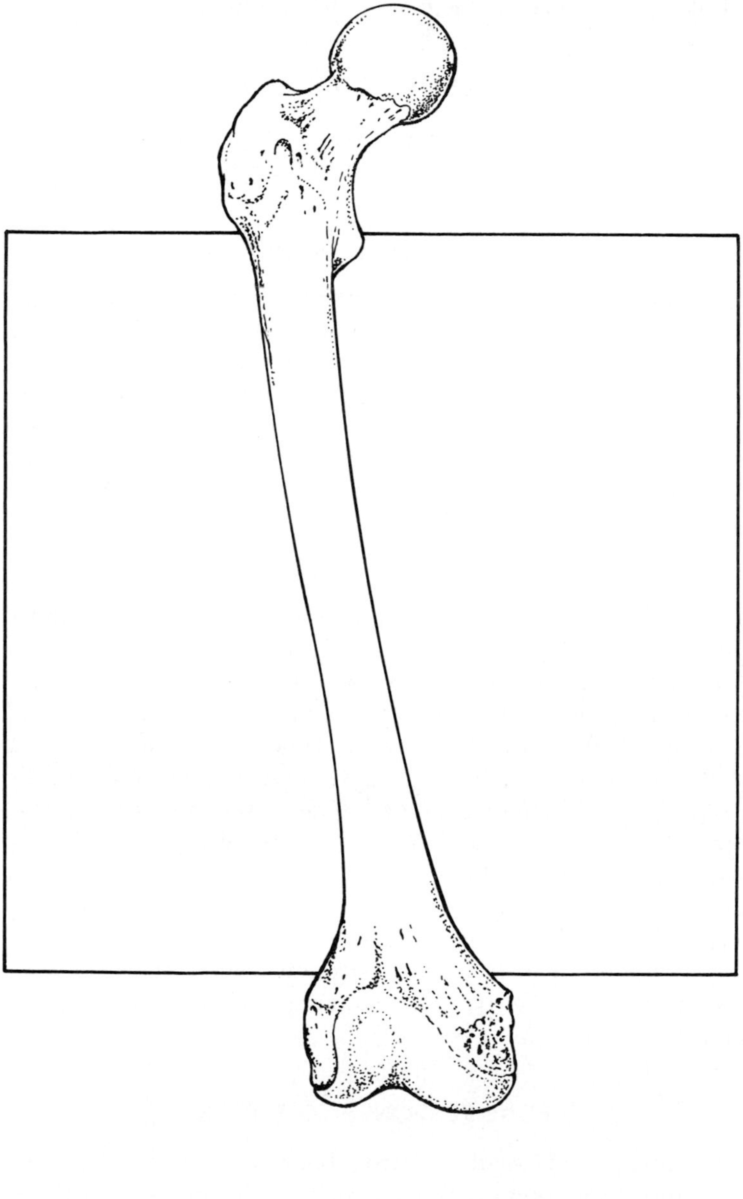

Figure 3-10
The femur represents the long bones.

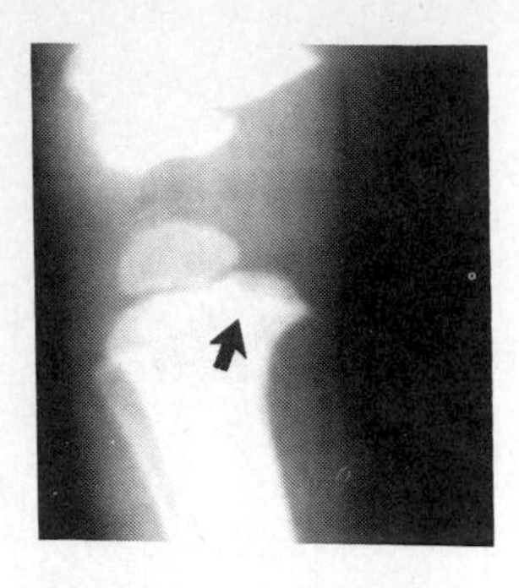

The epiphyseal plate is clearly observable in the femur of this young boy. (Courtesy Lester Cohn, MD.)

The long bones are adapted in size and weight for specific biomechanical functions. The tibia and femur are large and massive to support the weight of the body. The long bones of the upper extremity—the humerus, radius, and ulna—are smaller and lighter to promote ease of movement. Other long bones include the clavicle, the fibula, metatarsals, metacarpals, and the phalanges.

BONE GROWTH AND DEVELOPMENT

Living bone is continually changing during a person's lifespan. Many of these changes occur because of normal growth and maturation of bone.

Longitudinal Growth

Longitudinal growth of a bone occurs only when the bone's **epiphyses** are present. The epiphysis or epiphyseal plate continuously produces new bone cells on its diaphysis (central) side. When the epiphyseal plate stops producing cells, usually during the adolescent period, the plate disappears and the respective ends of the bone fuse to the shaft, terminating the longitudinal growth of the bone at that end. Most major epiphyses close at age 18, although some are present until approximately age 25.

epiphysis
the growth center of a bone that produces new bone tissue as part of the normal growth process until it closes during adolescence or early adulthood

Circumferential Growth

Bones grow in diameter throughout most of the lifespan, although the most rapid bone growth occurs before adulthood. The internal layer of the **periosteum** lays down bone tissue on top of the existing bone, forming concentric layers of bone tissue. At the same time, bone is **resorbed** around the circumference of the medullary cavity inside the bone, continually enlarging the cavity. The bone cells responsible for the formation of new bone tissue are **osteoblasts,** and **osteoclasts** resorb bone. The activity of osteoblasts and osteoclasts is largely balanced so that the total amount of bone present in an adult remains constant until approximately the fourth decade of life for women and the sixth decade of life for men. At this time, a decline in the total amount of bone mass present begins. Whether this loss of bone is an inevitable part of the aging process or a function of a sedentary lifestyle is not clear.

periosteum
double-layered membrane covering bone; muscle tendons attach to the outside layer, and the internal layer is a site of osteoblast activity

resorption
the process of breaking down and eliminating bone tissue

osteoblasts
bone cells that function to build new bone tissue during bone growth and in response to increased mechanical stress

osteoclasts
bone cells that resorb bone tissue during bone maturation and in response to reduced mechanical stress

BONE RESPONSE TO STRESS

Other changes that occur in living bone are unrelated to normal growth and development. Bone responds dynamically to the presence or absence of different forces with changes in size, shape, and density.

Bone Remodeling

The shapes and, to some extent, the sizes of the bones of a given human being are a function of the magnitude and direction of the mechanical stresses to which the bones are subjected. Through the actions of osteoblasts and osteoclasts, bone is continuously reshaped or remodeled.

■ The processes causing bone remodeling are not fully understood and continue to be researched by scientists.

The malleability of bone is dramatically exemplified by the case of an infant who was born in normal physical condition but missing one tibia, the major weight-bearing bone of the lower extremity. After the child was walking for a time, x rays revealed that the fibula in the abnormal leg had remodeled and could not be distinguished from a typical tibia (1).

Another interesting case is that of a construction worker who had lost all but the fifth finger of one hand in a war injury. After 32 years the metacarpal and phalanx of the remaining finger had been remodeled so that it was highly similar in length and thickness to the third finger of the other hand (16).

Research has shown that bone density is greater among female marathon runners than in sedentary women of the same age (10). Weight-bearing exercise such as running or walking can help prevent osteoporosis.

Bone Hypertrophy

Although cases of complete bone remodeling are unusual, bones often have hypertrophied in response to certain types of regular physical activity. The bones of physically active individuals are denser and therefore more mineralized than those of sedentary individuals of the same age and gender. Moreover, the results of several studies indicate that occupations and sports particularly stressing a certain limb or region of the body produce accentuated **bone hypertrophy** in the stressed area. For example, professional tennis players display not only muscular hypertrophy in the tennis arm but also hypertrophy of that arm's humerus (8) (Figure 3-11). Similar bone hypertrophy has been observed in the arms of baseball pitchers (20).

It also appears that the greater the forces or loads habitually encountered, the more dramatic the increased mineralization of the bone. In one interesting study, the density of the femur was measured among 64 nationally ranked athletes representing different sports (12). The femurs displaying the greatest density were those of the weight lifters, followed by throwers, runners, soccer players, and swimmers.

Regular exercise seems to increase bone density in both the limbs that are particularly stressed and the entire skeletal system. Investigations on both men and women runners have revealed bone densities higher than average in the upper and lower extremities (3, 4). These findings suggest that something other than mechanical stress to the bone tissue may contribute to bone hypertrophy. Factors such as improved circulation to the bone may also be significant.

bone hypertrophy
the increase in bone tissue resulting from a predominance of osteoblast activity

Figure 3-11
Bones display hypertrophy in response to repeated muscular stress.

Bone Atrophy

bone atrophy
the decrease in bone tissue resulting from a predominance of osteoclast activity

Whereas bone responds to increased mechanical stress by hypertrophying, it displays the opposite response to reduced stress. Some stress seems to be necessary for the maintenance of bone integrity. When the normal stresses exerted on bone by muscle contractions or weight bearing are removed, bone tissue atrophies. When **bone atrophy** occurs, the amount of calcium contained in the bone diminishes and both the weight and strength of the bone are decreased. Loss of bone mass has been found in bed-ridden patients, senior citizens, and astronauts.

Bone demineralization is a potentially serious problem. From a biomechanical standpoint, as bone mass diminishes, bone strength and thus resistance to fracture are progressively impaired. When calcium compounds are dissolved from bone tissue, they enter the blood stream and are filtered through the kidneys. Bone demineralization therefore contributes to the likelihood of developing kidney stones.

The results of calcium loss studies conducted during the Skylab flights indicate that urinary calcium loss is related to flight duration. The pattern of bone loss observed is highly similar to that documented among patients during periods of bed rest (19). Although space flights have resulted in only minimal amounts of bone demineralization among astronauts, a flight to Mars requiring approximately 1 year of space travel would result in a 25% loss in total body calcium (14). At this level, participation in normal activities of daily living would probably result in bone fractures.

It is not yet clear what specific mechanism or mechanisms are responsible for bone loss outside of the gravitational field. Simulating gravity inside space vehicles to combat this problem may be necessary. One technique that has been used to create artificial gravity is centrifuging. Centrifuging is a laboratory procedure that creates force on specimens spun at a high speed. Rats that were centrifuged aboard the Soviet Cosmos-936 biosatellite showed increased bone density and calcium content at the end of the experiment (13).

It remains to be seen if measures other than the artificial creation of gravity can effectively prevent bone loss during space travel. Astronauts' current exercise programs during flights in space are designed to prevent bone loss by increasing the mechanical stress placed on bones using muscular force. However, the muscles of the body exert mainly tensile forces on bone, whereas gravity provides a compressive force. Therefore, it may be that no amount of physical exercise alone can completely compensate for the absence of gravitational force.

COMMON BONE INJURIES

Considering the important mechanical functions performed by bone, bone health is an important part of general health. Bone health can be impaired by injuries and pathologies.

Fractures

A **fracture** is a disruption in the continuity of a bone. A fracture's characteristics are a function of the type and magnitude of mechanical load sustained and the health and maturational status of the bone at the time of injury. Different terms are used to describe general categories of fractures.

fracture
a disruption in the continuity of a bone

Traumatic fractures may result from several types of loading. **Avulsions** are caused by excessive tensile loading involving a tendon or ligament pulling a small chip of bone away from the rest of the bone (Figure 3-12). Extremely explosive throwing and jumping movements may result in avulsion fractures of the medial epicondyle of the humerus and the calcaneus. Fractures of the long bones are often produced by excessive torsional loads. Tibial fractures resulting from skiing accidents are an example. If the loading is rapid, the fracture is more likely to be **comminuted,** that is, to contain multiple fragments. Because bone is stronger in resisting compression than tension and shear, traumatic compression fractures of bone are rare.

avulsion
a fracture induced by tensile loading in which a chip of bone is pulled away by an attaching tendon or ligament

comminuted
characterized by numerous small fragments

Figure 3-12
An avulsion is a fracture produced by tensile loading.

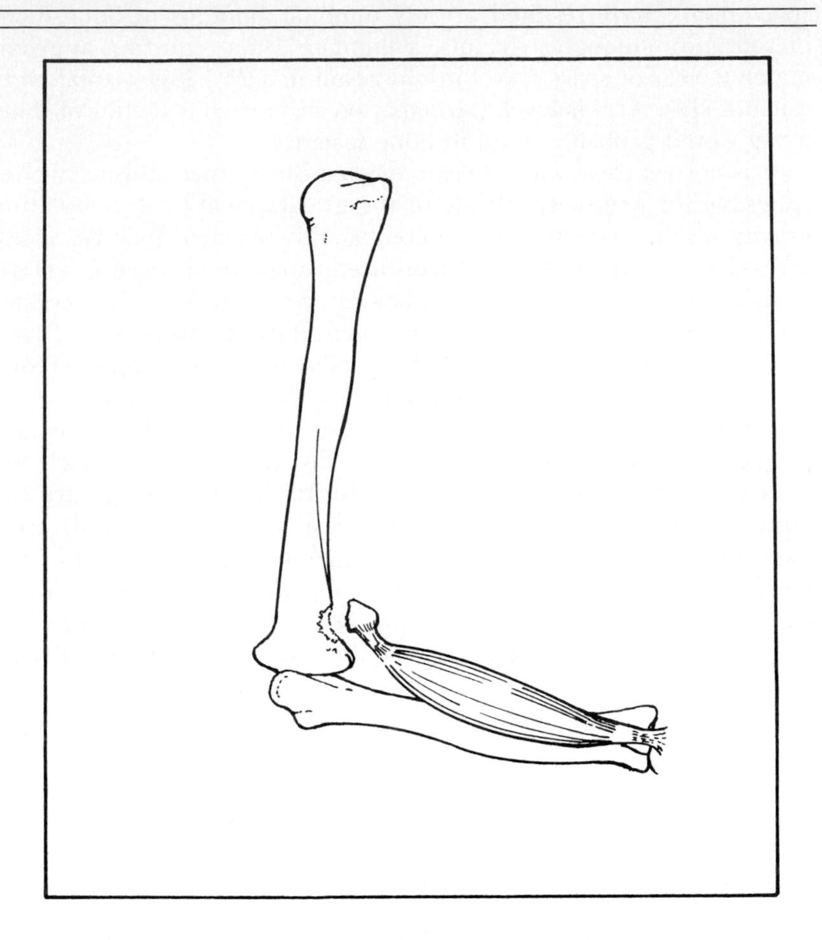

impacted
pressed together by a compressive load

fatigue fracture
a fracture resulting from a repeated loading of relatively low magnitude that is commonly referred to as a stress fracture

greenstick fracture
a fracture resulting from a bending or torsional load in which one side of the bone is fractured and the other side remains intact

In a combined loading situation, however, a fracture resulting from a torsional load may be **impacted** by the presence of a compressive load. An impacted fracture is one in which the opposite sides of the fracture are compressed together.

Fatigue fractures, also known as stress fractures, are derived from the repeated sustenance of nontraumatic forces. Fatigue fractures of the metatarsals, the femoral neck, and the pubis have been reported in runners who had overtrained. Fatigue fractures of the vertebrae have also been reported to occur in higher than normal frequencies in female gymnasts and football linemen (7).

Fractures in children. Because children's bones contain more collagen than adults' bones, they are more pliable. This pliability aids in resisting fractures. For this reason, **greenstick fractures** are found more often in children than in adults. As shown in Figure 3-13, a greenstick fracture is an incomplete fracture usually caused by excessive bending or torsional loads.

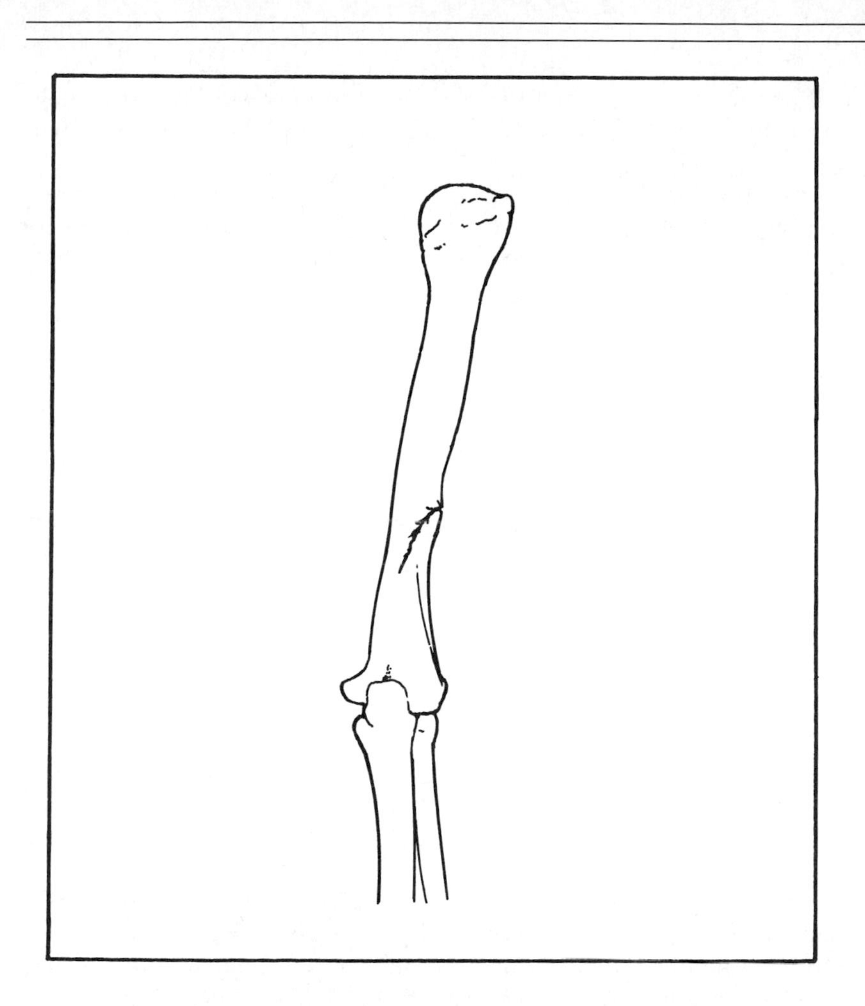

Figure 3-13
Greenstick fractures are incomplete fractures.

The bones of children are particularly vulnerable at the cartilaginous epiphyseal plate. Both traumatic and repetitive loading can injure the epiphysis, and certain sport activities are associated with an increased risk of epiphyseal injuries. For example, *Little Leaguer's elbow* is a stress injury to the medial epicondylar epiphysis of the humerus. Contact sports such as football increase the risk of traumatic injuries. Severe injury to an epiphysis may result in premature closure of the epiphyseal junction and termination of bone growth.

Osteoporosis

Bone atrophy is a problem not only for astronauts and bed-ridden patients but for a growing number of senior citizens—particularly women. Osteoporosis, a condition characterized by excessive loss of bone mineral mass and strength, is the most common metabolic dis-

■ Osteoporotic bone is normal in size, shape, and structural organization but deficient in mineralization.

ease in the United States (10). It is found in most elderly individuals and is becoming increasingly prevalent with the rise in the number of elderly.

Osteoporosis manifests during the fourth decade of life in women and the sixth decade of life in men when bone resorption begins to exceed bone production. Although some bone loss may be associated with aging of the skeleton, osteoporotic bone loss exceeds expected changes.

The most common symptom of osteoporosis is back pain derived from vertebral fractures. Crush fractures of the lumbar vertebrae resulting from compressive loads created by weight bearing during activities of daily living frequently cause reduction of body height. Wedging fractures of the thoracic vertebrae, in which the height of the anterior aspect of the vertebra is lessened while the posterior aspect remains unaffected, frequently result in a disabling deformity known as *dowager's hump* (21) Estimates are that 90% of all fractures occurring past age 60 are osteoporosis related (15).

The cause of osteoporosis is unknown. Possible contributing factors include hyperactivity of osteoclasts, hypoactivity of osteoblasts, hormonal factors, and insufficiencies of dietary calcium or other minerals or nutrients (21). Studies have shown that a regular exercise program among individuals with osteoporosis can increase bone mineralization (9,17,18). Exercise programs for the osteoporotic individual must be carefully designed to minimize the risk of bone fractures during exercise.

To lessen the likelihood of acquiring osteoporosis, women should maximize their bone mineral densities before age-related losses begin (5). Adequate dietary calcium intake and regular weight-bearing exercise are suggested as appropriate strategies (11).

SUMMARY

Bone is an important and interesting biological tissue. Its mechanical functions are to support and protect other body tissues and to act as a system of rigid levers that can be manipulated by the attaching muscles.

Within a living organism, bones are subjected to several types of mechanical loads. These include compression, tension, shear, bending, and torsion. Generally, some combination of these loading modes is present. The distribution of force within a bone is termed mechanical stress. The nature and magnitude of stress determines the likelihood of injury to biological tissues.

The mechanical characteristics of bone are based on its material composition and organizational structure. Minerals contribute to a bone's hardness and compressive strength, and collagen provides its elasticity and tensile strength. Cortical bone is stiffer and stronger than cancellous bone, whereas cancellous bone has greater shock-absorbing capabilities.

Bone is an extremely dynamic tissue that is continually being remodeled. Although bones grow in length only until the epiphyseal plates close at adolescence, bones continue to grow in diameter throughout a person's lifespan. Through the actions of osteoblasts and osteoclasts, bones also hypertrophy in response to increased mechanical stress and atrophy in the absence of mechanical stress.

Osteoporosis, a condition characterized by excessive loss of bone mineral mass and strength, is extremely common among the elderly. It affects women at an earlier age than men. Although the cause of osteoporosis remains unknown, a program of regular, weight-bearing exercise can improve bone mineralization among both individuals who do and who do not have the condition.

INTRODUCTORY PROBLEMS

1. Explain why the bones of the human body are stronger in resisting compression than tension and shear.
2. Identify a practical way in which compression, tension, and shear are each used in activities of daily living.
3. Rank the following activities according to their effect on increasing bone density: running, backpacking, swimming, cycling, weight lifting, polo, tennis. Write a paragraph providing the rationale for your ranking.
4. Would you expect bone density to be related to an individual's body weight? Explain why or why not.
5. When a runner's foot strikes the ground, a force of 2 to 3 times body weight is generated. Given this information, what range of forces are generated by the foot strikes of an 800 N runner? (Answer: 1600–2400 N)
6. A 750 N runner lands on the left foot with a force that is 2.5 times body weight. If the force is evenly distributed over the 180 cm^2 sole of a running shoe, how much stress is generated? (Answer: 10.4 N/cm^2)
7. The tibia is the major weight-bearing bone in the lower extremity. If 88% of body weight is proximal to the knee joint, how much compressive force acts on each tibia when a 650 N person stands in a stationary position on both feet? (Answer: 286 N)
8. Approximately 56% of body weight is supported by the fifth lumbar vertebra. How much stress is present on the 22 cm^2

surface area of that vertebra in an erect 756 N man? (Assume that the vertebral surface is horizontal.) (Answer: 19.2 N/cm^2)

9. In Problem 8, how much total stress is present on the 5th lumbar vertebra if 10 N of tension is also present in the spinal ligaments? (Answer: 18.8 N/cm^2)
10. In Problem 8, how much total stress is present on the 5th lumbar vertebra if the individual holds a 222 N weight bar balanced across the shoulders? (Answer: 29.3 N/cm^2)

ADDITIONAL PROBLEMS

1. Hypothesize about the way or ways in which each of the following bones is loaded when a person stands in anatomical position. Be as specific as possible in identifying which parts of the bones are loaded.
   a. Femur
   b. Tibia
   c. Scapula
   d. Humerus
   e. Third lumbar vertebra
2. Outline a 6-week exercise program that might be used with a group of osteoporotic elderly who are ambulatory.
3. Speculate about what exercises or other strategies may be employed in outer space to prevent the loss of bone mineral density in humans.
4. How does the ability of bone to resist compression, tension, and shear compare to the same properties of wood, steel, and plastic?
5. How are the bones of birds and fish adapted for their methods of transportation?
6. Why is bone tissue organized differently (cortical versus cancellous bone) within a given bone?
7. In the human femur, bone tissue is strongest in resisting compressive force, approximately half as strong in resisting tensile force, and only about one fifth as strong in resisting shear force. If a tensile force of 8000 N is sufficient to produce a fracture, how much compressive force will produce a fracture? How much shear force will produce a fracture? (Answer: compressive force = 16,000 N; shear force = 3200 N)
8. When an impact force is absorbed by the foot, the soft tissues at the joints act to lessen the amount of force transmitted upward through the skeletal system. If the force absorbed in Problem 6 on p. 81 is reduced 15% by the tissues of the ankle joint and 45% by the tissues of the knee joint, how much force is transmitted to the femur? (Answer: 750 N)

9. How much compression is exerted on the radius at the elbow joint when the biceps brachii, oriented at a 30 degree angle to the radius exerts a tensile force of 200 N? (Answer: 173 N)
10. If the anterior and posterior deltoids both insert at an angle of 60 degrees on the humerus and each muscle produces a force of 100 N, how much force is acting perpendicular to the humerus? (Answer: 173.2 N)

REFERENCES

1. Adrian MJ and Cooper JM: Biomechanics of human movement, Indianapolis, 1989, Benchmark Press, Inc.
2. Andrews JG, Hay JG, and Vaughan CL: The concept of joint shear. In Cooper JM and Haven B, eds: Biomechanics Symposium Proceedings, Bloomington, Ind, 1980, Indiana State Board of Health.
3. Brewer V et al: Role of exercise in prevention of involutional bone loss, Med Sci Sports Exerc 15:445, 1983.
4. Dalen N and Olsson KE: Bone mineral content and physical activity, Acta Orthop Scand 45:170, 1974.
5. Drinkwater BL: Exercise and aging: the female masters athlete. In Puhl J, Brown CH, and Voy RL: Sport science perspectives for women, Champaign, Ill, 1987, Human Kinetics Publishers, Inc.
6. Gibson LJ: The mechanical behaviour of cancellous bone, Biomech 18:317, 1985.
7. Jackson DW, Wiltse LL, and Cirincione RJ: Spondylolysis in the female gymnast, Clin Orthop Vol. 117:68, 1976.
8. Jones HH et al: Humeral hypertrophy in response to exercise, J Bone Joint Surg 59A:204, 1977.
9. Krolner B et al: Physical exercise as a prophylaxis against involutional bone loss: a controlled trial, Clin Sci 64:541, 1983.
10. Mercier LR: Practical orthopedics, ed 2, Chicago, 1987, Year Book Medical Publishers, Inc.
11. National Institute of Health: Osteoporosis, NIH Consensus Development Statement No 5, 1984.
12. Nilsson B and Westline N: Bone density in athletes, Clin Orthop 77:179, 1971.
13. Prodhonchukov AA et al: Comparative study of effects of weightlessness and artificial gravity on density, ash, calcium, and phosphorous content of calcified tissues, STAR Pub N80-34064, 1980.
14. Rambaut PC, Leach CS, and Whedon GD: A study of metabolic balance in crewmembers of Skylab IV, Acta Astronautica 6:1313, 1979.
15. Recker RR: Osteoporosis, Contemp Nutr 8:1, 1983.
16. Ross JA: Hypertrophy of the little finger, Brit Med J 2:987, 1950.
17. Smith EL Jr: The effects of physical activity on bone in the aged. In Mazess RB, ed: International conference on bone mineral measurements, DHEW Pub No NIH 75-683, Washington, DC, 1973.
18. Smith EL Jr, Reddan W, and Smith PE: Physical activity and calcium modalities for bone mineral increase in aged women, Med Sci Sports Exerc 13:60, 1981.

19. Smith MC et al: Bone mineral measurement—experiment M078. In Johnston RS and Dietlein LF, eds: Biomedical results from Skylab, Washington, DC, National Aeronautics and Space Administration.
20. Watson RC: Bone growth and physical activity. In Mazess RB, ed: International conference on bone measurements, DHEW Pub No NIH 75-683, Washington, DC, 1973.
21. Wiesel SW, Bernini P, and Rothman RH: The aging lumbar spine, Philadelphia, 1982, WB Saunders Co.
22. Yamada H: Mechanical properties of locomotor organs and tissues. In Evans FG, ed: Strength of biological materials, Baltimore, 1970, Williams & Wilkins.

ANNOTATED READINGS

Aloia JF: Exercise and skeletal Health, J Am Geriatr Soc 29:104, 1981.

Reviews experimental evidence suggesting that the loss of bone mass with aging can be prevented by physical exercise.

Lukert BP: Osteoporosis—a review and update, Arch Phys Med Rehabil 3:480, 1982.

Reviews the medical perspective on osteoporosis, including pathogenesis, symptoms, diagnosis, and treatment.

Montoye HJ: Exercise and osteoporosis. in Eckert HM and Montoye HJ, eds: Exercise and health: American academy of physical education papers, No. 17, 1984.

Reviews the scientific studies conducted on osteoporosis and related considerations, with emphasis on experimental techniques used.

Nordin M and Frankel VH: Biomechanics of whole bones and bone tissue. In Frankel VH and Nordin M: Basic biomechanics of the skeletal system, Philadelphia, 1980, Lea & Febiger.

Concisely describes the mechanical properties of bones, the loading conditions to which bone is subjected, and some of the factors affecting bone strength and stiffness.

Recker RR: Osteoporosis, Contemp Nutr 8:1, 1983.

Provides an overview of osteoporosis reflecting scientific information about the disease written in lay terminology.

Smith EL: Bone changes in the exercising older adult. In Smith EL and Serfass RC: Exercise and aging: the scientific basis, Hillside, NJ, 1981, Enslow Publishers, Inc.

Summarizes scientific knowledge on age-related changes in bone, including discussion of the evidence that regular exercise may change the pattern of bone aging.

Smith EL, Sempos CT, and Purvis RW: Bone mass and strength decline with age. In Smith EL and Serfass RC: Exercise and aging: the scientific basis, Hillside, NJ, 1981, Enslow Publishers, Inc.

Provides information about proposed etiologies, treatments, and assessment techniques for osteoporosis.

4 MUSCLES

Biomechanical Aspects of Neuromuscular Function

After reading this chapter, the student will be able to:

Identify the basic behavioral properties of muscle tissue and explain why muscle may shorten, lengthen, or remain at the same length while producing tension.

Describe the characteristics of the three major categories of muscle fiber type and discuss their distribution in human muscles.

Define motor unit and describe the general pattern by which motor units are recruited.

Identify the categories of muscle fiber arrangement within a muscle and explain the advantages and disadvantages associated with each.

Describe and discuss the force-velocity, force-length, and force-time relationships for muscle tissue.

Discuss the concepts of strength, power, and endurance from a biomechanical perspective.

Explain the ways in which the sensory receptors in muscle and tendon contribute to the neuromuscular control of human movement.

Muscle is the only tissue capable of actively developing tension. This capability enables smooth, skeletal, and heart muscle to fulfill different types of functions in the human body. The functioning of skeletal muscle enables in the maintenance of upright posture and the movement of the body limbs. Because muscle can only perform its functions when appropriately stimulated, the human nervous system and muscular system are often referred to collectively as the neuromuscular system.

Effective functioning of the neuromuscular system is essential for participation in the activities of daily living. Body builders and athletes are often interested in increasing muscle size, muscular strength, or muscular endurance. This chapter discusses the behavioral properties of muscle tissue, the biomechanical aspects of muscle fiber types and organizations, and the mechanical aspects of muscle function.

The sport of body building focuses on the development of muscular size and definition.

BEHAVIORAL PROPERTIES OF MUSCLE TISSUE

The four behavioral properties of muscle tissue are extensibility, elasticity, irritability, and the ability to develop tension. These properties are common to all muscle, including the smooth, skeletal, and heart muscle of human beings, as well as the muscles of other mammals, reptiles, amphibians, birds, and insects.

Extensibility and Elasticity

The properties of extensibility and elasticity are common to many biological tissues. As shown in Figure 4-1, extensibility is the ability to be stretched or to increase in length, and elasticity is the ability to return to normal length after extension or contraction.

The elastic behavior of muscle has been described as consisting of two major conceptual components (32). The **parallel elastic component** (PEC) provides resistive tension when a muscle is passively stretched. The **series elastic component** (SEC) acts as a spring to store elastic energy when an active muscle is stretched. The elasticity of human skeletal muscle is believed to be due primarily to the SEC.

parallel elastic component
the ability of muscle to resist passive stretching

Figure 4-1
The characteristic properties of muscle tissue are extensibility, elasticity, contractility, and irritability *(not shown).*

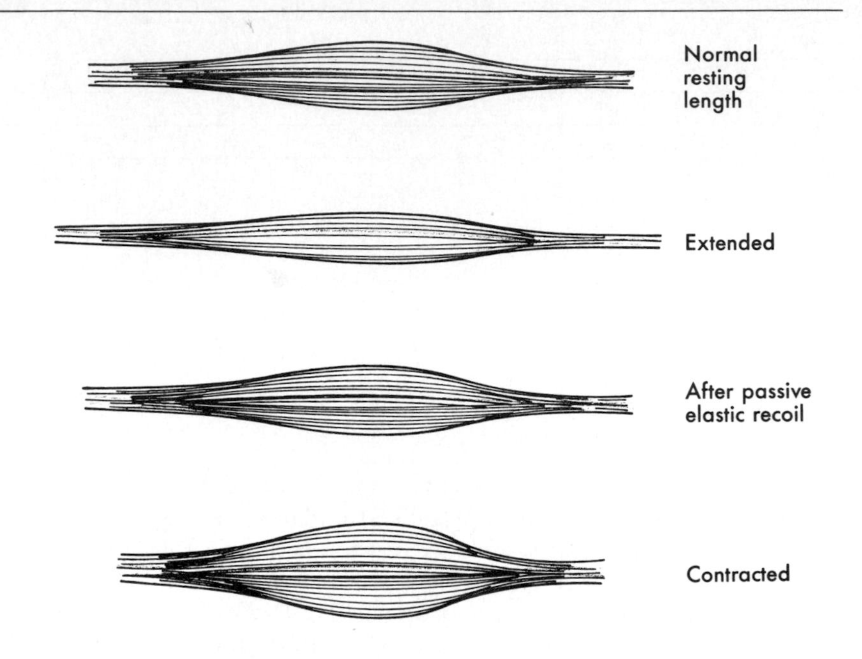

series elastic component
the ability of activated muscle to recoil elastically; acts in series with the contractile component of muscle

Although some researchers (1, 26) have associated the PEC and the SEC with the functioning of particular anatomical components of the muscle, others (11) believe it is erroneous to do so because these components describe the behavior of whole muscles.

The elastic recoil of tensed muscle tissue that is stretched (the SEC) contributes significantly to the force developed by the muscle. Muscle's elasticity also returns it to normal resting length following a stretch.

Irritability and the Ability to Develop Tension

Another of muscle's characteristic properties, irritability, is the ability to respond to a stimulus. Stimuli affecting muscles are either electrochemical, such as an action potential from the attaching nerve, or mechanical, such as an external blow to a portion of a muscle. If the stimulus is of sufficient magnitude, muscle responds by developing tension.

The ability to develop tension is the one behavioral characteristic unique to muscle tissue. Historically, the development of tension by muscle has been referred to as contraction or the **contractile component** of muscle function. Contractility is the ability to shorten in length. However, tension in a muscle may not result in the muscle's shortening.

contractile component
the ability of muscle to develop tension when stimulated

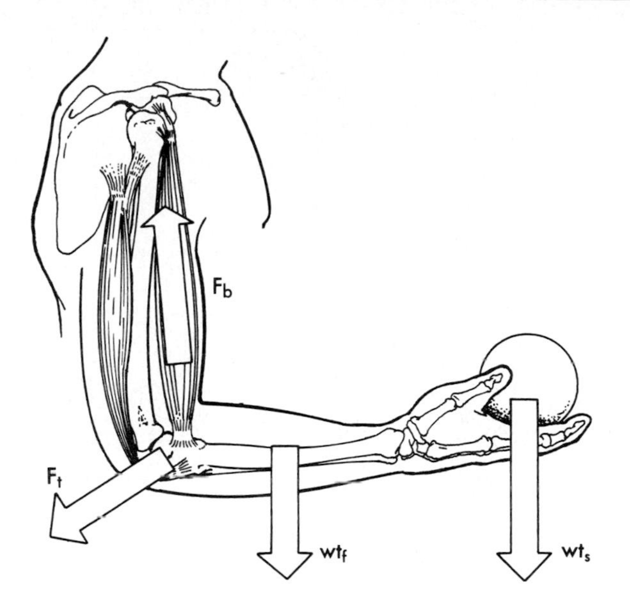

Figure 4-2
The torque exerted by the biceps brachii (F_b) must counteract the torques created by the force developed in the triceps brachii (F_t), the weight of the forearm and hand (wt_f), and the weight of the shot held in the hand (wt_s).

MUSCULAR TENSION

As discussed in Chapter 2, tension is a pulling force. When a stimulated (activated) muscle develops tension, the amount of tension present is constant throughout the length of the muscle and at the sites of the musculo-tendinous attachments to bone. The tensile force developed by the muscle pulls on the attached bones through the tendinous attachments and creates torque at the joint or joints crossed by the muscle. The magnitude of the torque generated is the product of the force developed by the muscle and the perpendicular distance of the line of action of that force from the center of rotation at the joint. In keeping with the laws of vector addition, the net torque present at that joint determines if movement results. The weight of a body segment, the simultaneous development of tension in a muscle on the opposite side of the bone, and external forces acting on the body may all generate torques at a joint (Figure 4-2).

Concentric Action

When muscular tension results in a torque in excess of the resistive torque at the joint crossed, the muscle shortens, thereby pulling the attached bones closer together and causing a change in the angle at the joint. When muscular tension results in the shortening of the muscle, the action is **concentric.** A single muscle fiber is capable of shortening to approximately one half of its normal resting length.

concentric
action involving shortening of a muscle

Figure 4-3
The execution of a soccer kick requires the development of concentric tension in the quadriceps group.

Muscular contraction (concentric tension development) is responsible for most volitional movement of the limbs of the human body. For example, concentric tension development in the quadriceps is used in kicking a ball (Figure 4-3).

Isometric Action

Muscles can also develop tension without shortening. If the opposing torque at the joint crossed by the muscle is equal to the torque produced by the muscle (with zero net torque present), muscle length remains unchanged and no movement occurs at the joint.

Figure 4-4
A static body-builder's pose involves the development of isometric tension in the opposing muscle groups.

When muscular tension develops but no change in muscle length occurs, the action is **isometric.** Because the development of tension increases the diameter of the muscle, body builders develop isometric tension to display their muscles when competing. Developing isometric tension simultaneously in muscles on opposite sides of a limb, such as in the triceps brachii and the biceps brachii, enlarges the tensed muscles although no movement occurs at either the shoulder or elbow joints (Figure 4-4).

isometric
action involving no change in muscle length

Eccentric Action

When opposing joint torque exceeds that produced by tension in a muscle, the muscle lengthens, pulling the attached bones farther apart and resulting in motion at the joint. This occurs only if the joint or joints crossed by the muscle are not at their full range of motion in that direction. When a muscle lengthens as it is being stimulated to develop tension, the action is **eccentric.** Eccentric tension occurs in the elbow flexors during the elbow extension or weight-lowering phase of a curl exercise (Figure 4-5). The eccentric tension acts as a braking mechanism. Without the presence of eccentric tension in the muscles, the weight would drop uncontrolled because of the force of gravity.

eccentric
action involving lengthening of a muscle

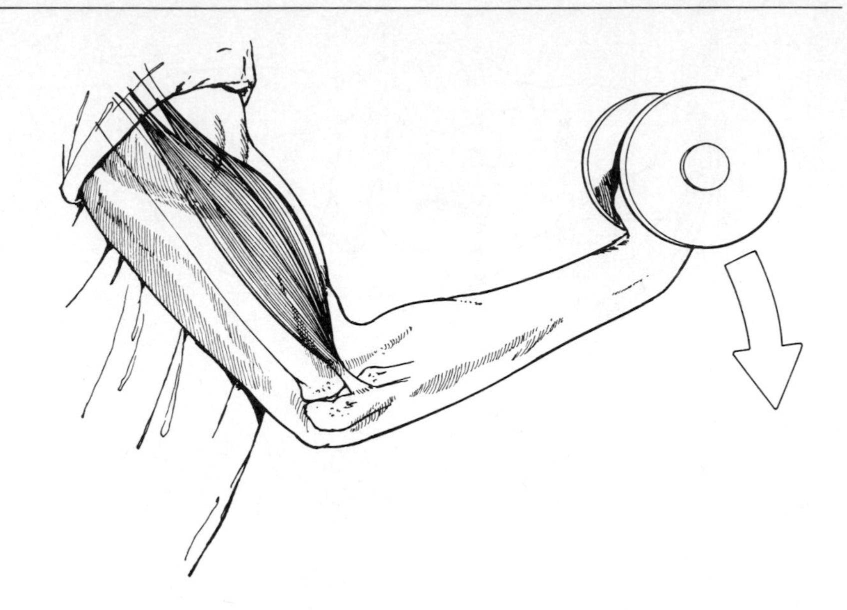

Figure 4-5
The lowering of a weight during a curl exercise involves the presence of eccentric tension in the biceps brachii.

FUNCTIONAL ORGANIZATION OF SKELETAL MUSCLE

When tension is developed in a muscle, biomechanical considerations such as the magnitude of the force generated, the speed with which the force is developed, and the length of time that the force may be maintained are affected by the particular anatomical and physiological characteristics of the muscle.

Muscle Fibers

A single muscle cell is termed a *muscle fiber* because of its threadlike shape. Considerable variation in the length and diameter of fibers is seen in adults. Some fibers may run the entire length of a muscle, whereas others are much shorter. Fibers over 30 cm in length have been identified in the sartorius (2). Skeletal muscle fibers grow in length and diameter from birth to adulthood, with a fivefold increase in fiber diameter during this period (33). Fiber diameter can also be increased by physical training.

In animals such as amphibians the number of muscle fibers present also increases with the age and size of the organism. However, this does not appear to occur in human beings. The number of muscle fibers present at birth is maintained throughout life, except for the occasional loss from injury (16). The increase in muscle size after resistance training represents an increase in fiber diameters rather than in the number of fibers (28). The number of muscle fibers present is genetically determined and varies from person to person.

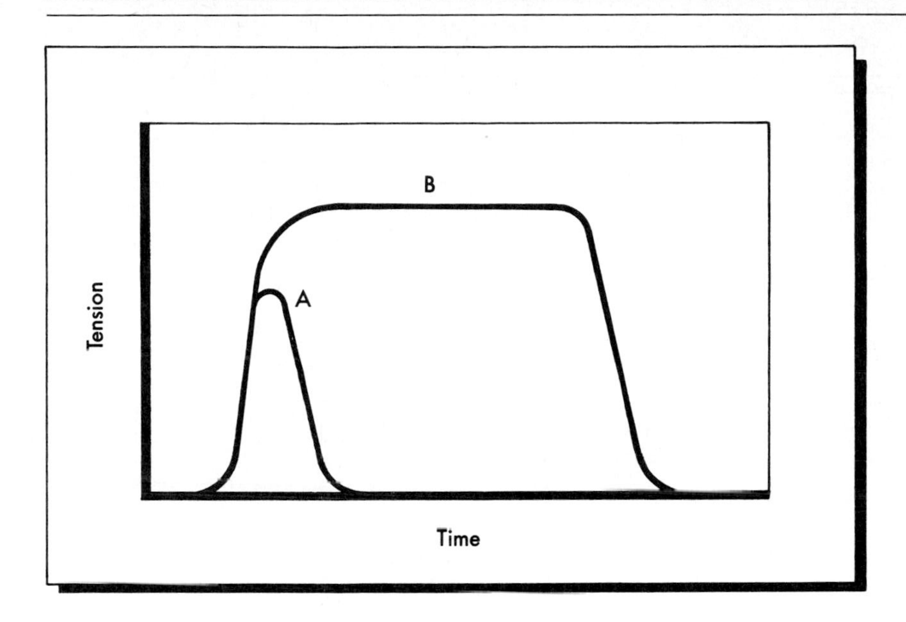

Figure 4-6
Tension developed in a muscle fiber in response to a single stimulus *(A)* and in response to repetitive stimulation (tetanus) *(B)*.

Most skeletal muscle fibers in mammals are *twitch type* cells that respond to a single stimulus by developing tension in a twitchlike fashion. The tension in a twitch fiber following the stimulus of a single nerve impulse rises to a peak value in less than 100 msec and then immediately declines. In the human body, however, fibers are generally activated by a volley of nerve impulses. When more than one impulse activates a fiber, the tension is progressively elevated until a maximum value for that fiber is reached (Figure 4-6). A fiber repetitively activated so that its maximum tension level is maintained for a time is in **tetanus.** As tetanus is prolonged, fatigue causes a gradual decline in the level of tension produced.

tetanus
state of muscle producing sustained maximal tension resulting from repetitive stimulation

Not all human skeletal muscle fibers are of the twitch type. Fibers of the *tonic type* are found in the oculomotor apparatus (37). These cells require more than a single stimulus before the initial development of tension.

Motor Units

Muscle fibers are organized into functional groups of different sizes. Composed of a single motor neuron and all fibers innervated by it, these groups are known as **motor units** (Figure 4-7). The axon of each motor neuron subdivides many times so that each individual fiber is supplied with a motor end plate. Typically, there is only one end plate per fiber, although multiple innervation of fibers has been reported in vertebrates other than humans (21). The fibers of a mo-

motor unit
a single motor neuron and all fibers it innervates

Figure 4-7

A motor unit consists of a single neuron and all muscle fibers innervated by that neuron.

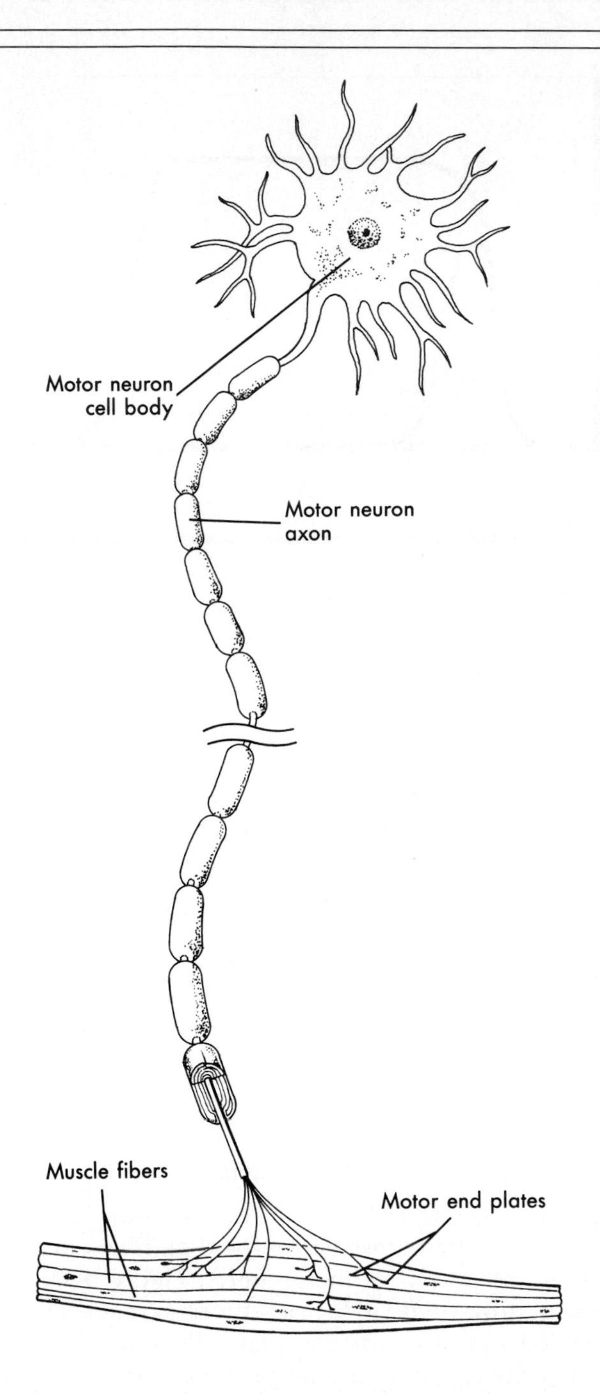

tor unit may be spread over a several centimeter area and be interspersed with the fibers of other motor units (8). With rare exceptions (13), motor units are confined to a single muscle and are localized within that muscle (10). A single mammalian motor unit may contain from less than 100 to 2000 fibers depending on the type of movements the muscle executes (7). Movements that are precisely controlled, such as those of the eyes or fingers, are usually produced by motor units with small numbers of fibers. Gross, forceful movements, such as those produced by the gastrocnemius, are more typically the result of the activity of large motor units.

Fiber Types

Skeletal muscle fibers exhibit many structural, histochemical, and behavioral characteristics. Because these differences have direct implications for muscle function, they are of particular interest to many biomechanists, exercise physiologists, and motor behaviorists. Consequently, much research has addressed the identification of discrete categories of fiber types. Based on histochemical differences, researchers have identified at least five categories of skeletal muscle fibers (18) and have proposed several classification schemes (17).

The fibers of some motor units contract to reach maximum tension more quickly than others after being stimulated. Based on this distinguishing characteristic, fibers have been generally categorized as **fast twitch** (FT) or **slow twitch** (ST). It takes FT fibers only about half the time required by ST fibers to reach peak tension and then to relax (Figure 4-8) (16). However, a wide range of twitch times to

fast twitch fiber
a fiber that reaches peak tension relatively quickly

slow twitch fiber
a fiber that reaches peak tension relatively slowly

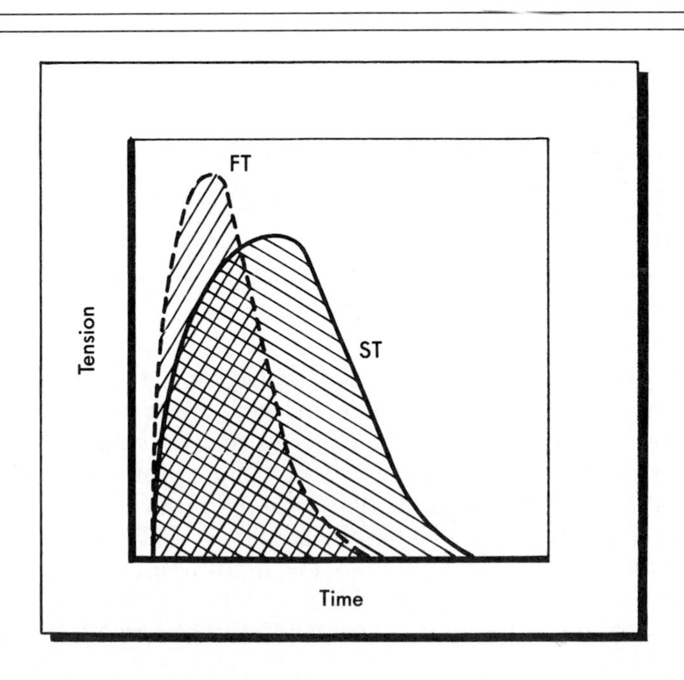

Figure 4-8
FT fibers respond to a super-threshold stimulus by developing tension more quickly than ST fibers.

achieve maximum tension exist within both categories (9). This difference in time to peak tension is attributed to higher concentrations of the enzyme myofibrillar ATPase in FT fibers. The FT fibers are also larger in diameter than ST fibers. Because of these and other differences, FT fibers usually produce higher tensions than ST fibers but also fatigue more quickly.

FT fibers are divided into two categories based on histochemical properties. The first type of FT fiber shares the resistance to fatigue that characterizes ST fibers, and the second type of FT fiber fatigues relatively rapidly.

Researchers have proposed several categorization schemes based on metabolic and contractile properties of these three general types of fiber. In one scheme, ST fibers are referred to as *Type I,* and the FT fibers are called *Type IIa* and *Type IIb* (6). Another system terms the ST fibers as *slow, oxidative* (SO), with FT fibers divided into *fast oxidative glycolytic* (FOG) and *fast glycolytic* (FG) fibers (36). An additional categorization includes ST fibers, and fast-twitch fatigue resistant (FFR) and fast-twitch fast fatigue (FF) fibers (10). These systems of classification are based on different fiber properties and are *not* interchangeable (17).

Although all fibers in a motor unit are the same type, most skeletal muscles contain both FT and ST fibers, with the relative amounts varying from muscle to muscle and individual to individual. For example, the soleus, which is generally used only for postural adjustments, primarily contains ST fibers. In contrast, the gastrocnemius may contain more FT than ST fibers. On average, males may have a slightly higher proportion of ST fibers than females (29), although this has not been adequately documented in the general population (26).

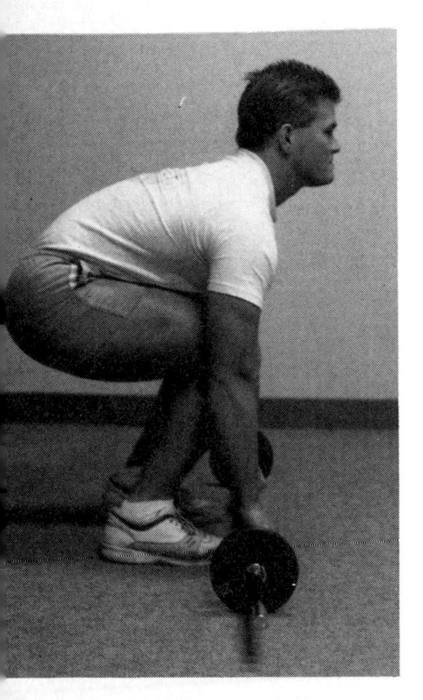

Elite power lifters are likely to have muscles composed of a high percentage of FT fibers.

The FT fibers are important contributors to a performer's success in events requiring fast, powerful muscular contraction such as sprinting or jumping. Endurance events such as distance running, cycling, or swimming require effective functioning of the more fatigue-resistant ST fibers. Using muscle biopsies, researchers have shown that world-class athletes in events requiring strength and power have unusually high proportions of FT fibers (4) and that elite endurance athletes have abnormally high proportions of ST fibers (30).

Although these findings suggest that athletic training may cause fibers to convert from one type to another, this has not been conclusively demonstrated (16). Gollnick states that the fiber type distributions of both elite strength-trained and elite endurance-trained athletes fall within the range of fiber type compositions found in untrained individuals (16). Within the general population a bell-shaped distribution of FT versus ST muscle composition exists; that is, most people have an approximate balance of FT and ST fibers, though a small percentage have a much greater number of FT fibers and a small percentage have an unusually high number of ST fibers. People genetically endowed with a high percentage of FT fibers may gravitate to sports requiring strength, and those with a high percentage of ST fibers may choose endurance sports.

Recruitment of Motor Units

The central nervous system exerts an elaborate system of control enabling muscle contraction according to the requirements of the movement so that smooth, delicate, or precise movements can be executed. The neurons that innervate ST motor units often have low thresholds and are relatively easy to activate, whereas FT motor units are supplied by nerves more difficult to activate. Consequently, the ST fibers are the first to be activated, even when the resulting limb movement is rapid (12). As the force requirement, speed requirement, and/or duration of the activity increases, motor units with higher thresholds are progressively activated, with Type IIa or FOG fibers added before the Type IIb or FG fibers (14). Within each fiber type, a continuum of ease of activation exists, and the central nervous system may selectively innervate more or fewer motor units.

During low intensity exercise, the central nervous system may recruit ST fibers almost exclusively. As activity continues and fatigue sets in, Type IIa and then Type IIb motor units are activated until all motor units are involved (19).

Fiber Architecture

Another variable influencing muscle function is the arrangement of fibers within a muscle. The orientations of fibers within a muscle and the arrangements by which fibers attach to muscle tendons vary considerably among the muscles of the human body. These structural considerations affect the strength of muscular contraction and the range of motion through which a muscle group can move a body segment. The two categories of muscle fiber arrangement are termed **parallel** and **pennate.** In a parallel fiber arrangement, all fibers are oriented parallel to the longitudinal axis of the muscle (Figure 4-9). The sartorius, rectus abdominis, and biceps brachii have parallel fiber orientations. A pennate fiber arrangement is one in which the fibers lie at an angle to the muscle's longitudinal axis. Each fiber in a pennate muscle attaches to one or more tendons, some of which extend the entire length of the muscle. The fibers of a muscle may exhibit more than one angle of pennation (angle of attachment) to a tendon. The tibialis posterior, rectus femoris, and deltoid muscles have pennate fiber arrangements.

Although numerous subcategories of parallel and pennate fiber arrangements have been proposed (14), the distinction between these two umbrella categories is sufficient for discussing biomechanical features.

parallel fiber arrangement
an arrangement of fibers within a muscle in which the fibers are roughly parallel to the longitudinal axis of the muscle

penniform fiber arrangement
an arrangement of fibers within a muscle with short fibers attaching to one or more tendons

Figure 4-9

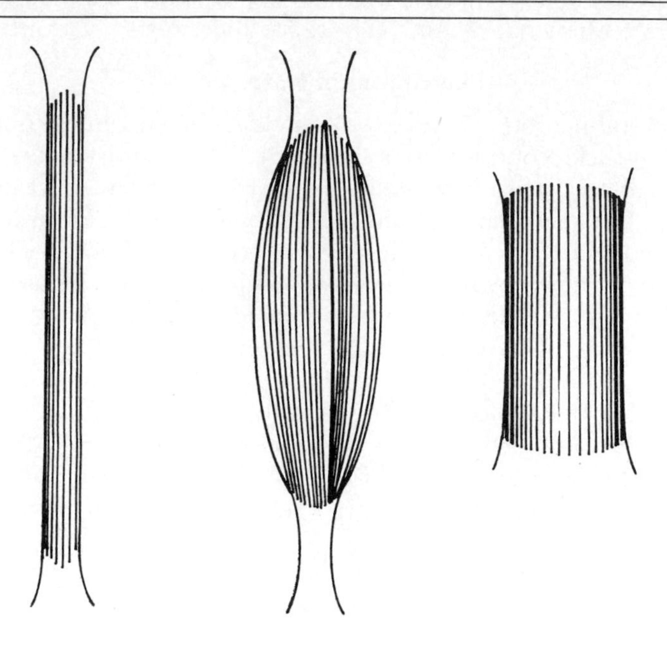

Parallel fiber arrangements

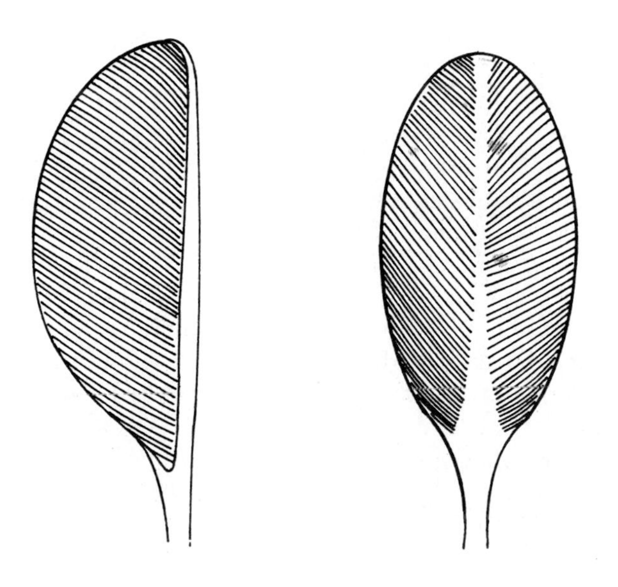

Pennate fiber arrangements

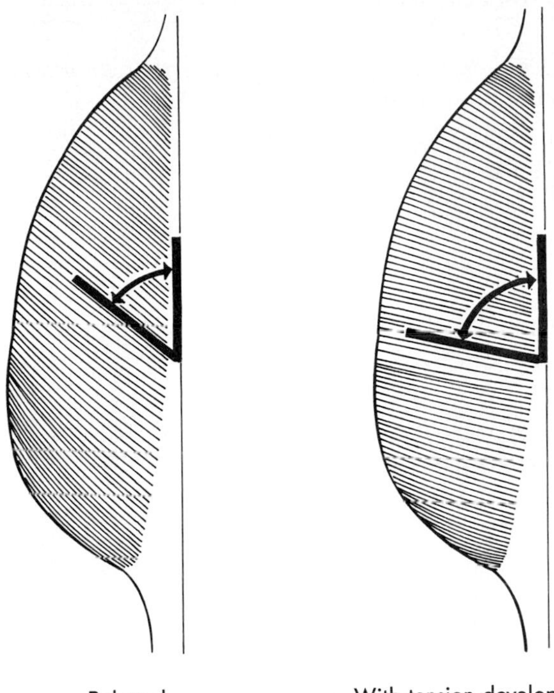

Figure 4-10
The angle of pennation increases as tension progressively increases in the muscle fibers.

When tension is developed in a parallel-fibered muscle, any shortening of the muscle is primarily the result of the shortening of its fibers. When the fibers of a pennate muscle shorten, they rotate about their tendon attachment or attachments, progressively increasing the angle of pennation (Figure 4-10). As demonstrated in the problem shown in Figure 4-11, the greater the angle of pennation, the less the amount of effective force actually transmitted to the tendon or tendons to move the attached bones. Once the angle of pennation exceeds 60 degrees, the amount of effective force transferred to the tendon is less than one half of the force actually produced by the muscle fibers.

Although pennation reduces the effective force generated at a given level of fiber tension, this arrangement allows the packing of more fibers than the amount that can be packed into a longitudinal muscle occupying equal space. Because pennate muscles contain more fibers per unit of muscle volume, they can generate more force than parallel-fibered muscles of the same size. However, the parallel fiber arrangement enables greater shortening of the entire muscle than is possible with a pennate arrangement. Parallel-fibered muscles can move body segments through larger ranges of motion than comparably sized pennate-fibered muscles.

Figure 4-11

SAMPLE PROBLEM 1

How much force is exerted by the tendon of a pennate muscle when the tension in the fibers is 100 N, given that the angle of pennation is

a. 40 degrees?
b. 60 degrees?
c. 80 degrees?

Known:

$F_{fibers} = 100$ N
Angle of pennation = 40°, 60°, 80°

Solution

Wanted: F_{tendon}

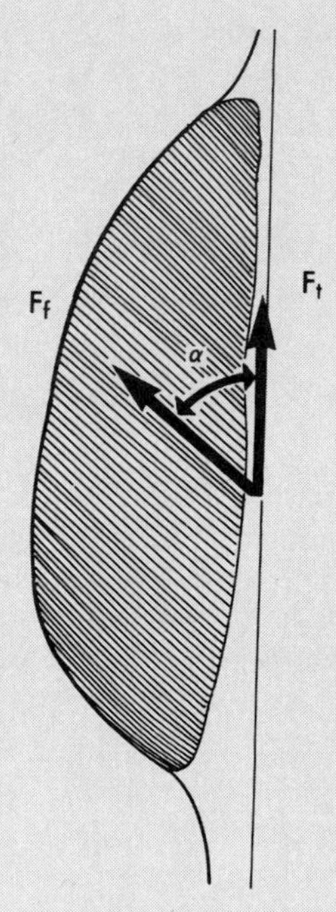

The relationship between the tension in the fibers and the tension in the tendon is

$$F_{tendon} = F_{fibers} \cos\alpha$$

a. For $\alpha = 40°$, $F_{tendon} = (100 \text{ N})(\cos 40)$

$$\boxed{F_{tendon} = 76.6 \text{ N}}$$

b. For $\alpha = 60°$, $F_{tendon} = (100 \text{ N})(\cos 60)$

$$\boxed{F_{tendon} = 50 \text{ N}}$$

c. For $\alpha = 80°$, $F_{tendon} = (100 \text{ N})(\cos 80)$

$$\boxed{F_{tendon} = 17.4 \text{ N}}$$

MECHANICAL FACTORS AFFECTING MUSCULAR FORCE

The magnitude of the force generated by muscle is also related to the velocity of muscle shortening, the length of the muscle at activation, and the length of time for which the muscle has been activated. Because these factors are significant determiners of muscle force, they have been extensively studied by scientists.

Force-Velocity Relationship

The classical force-velocity relationship for concentric tension development in muscle tissue was first documented by Hill in 1938 (23). The relationship between the concentric force exerted by a muscle and the velocity at which the muscle is capable of shortening is inverse, as shown in the portion of the curve above the line of isometric contraction in Figure 4-12. When a muscle develops concentric tension against a high load, the velocity (speed) of muscle shortening must be relatively slow. When the resistance is low, the velocity of shortening can be relatively fast.

The force-velocity relationship can be translated to body movements involving the action of a group of similarly functioning muscles such as the hip flexors or elbow extensors. For example, forearm curls are accomplished more quickly with a 45 N weight than with a 445 N weight because the higher the magnitude of the muscle force generated, the farther to the right the activity occurs on the force-velocity curve and the lower the velocity of muscle shortening. A young child can swing a hollow plastic baseball bat faster than a solid metal bat because low resistance activities occur farther to the

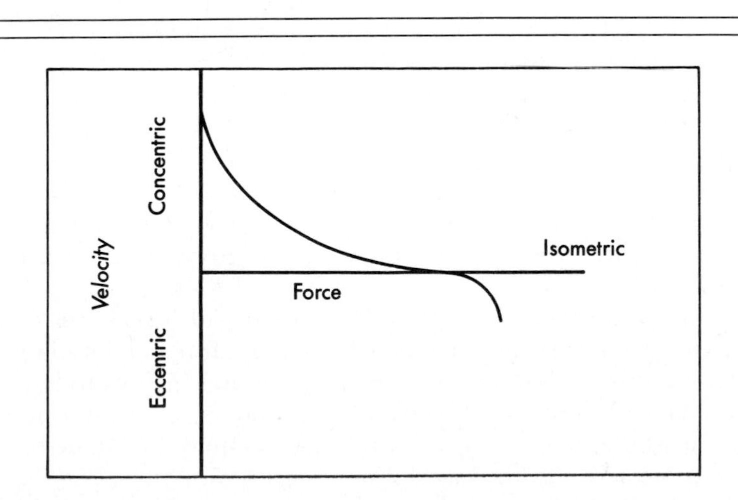

Figure 4-12
The force-velocity relationship for muscle tissue.

left on the force-velocity curve than activities involving higher resistances.

The force-velocity relationship does *not* imply that it is impossible to move a heavy resistance at a fast speed. The stronger the muscle is, the higher the magnitude of maximum isometric tension. This is the maximum amount of force that a muscle can generate before actually lengthening as the resistance is increased. However, the shape of the force-velocity curve remains the same, regardless of the magnitude of maximum isometric tension.

The force-velocity relationship also does not imply that it is impossible to move a light load at a slow speed. Most activities of daily living require slow, controlled movements of submaximum loads. The relationship does indicate that for a given load or muscular force requirement, the maximum velocity of muscle shortening is fixed in accordance with the pattern shown in the graph of Figure 4-12. At submaximum movement speeds the velocity of muscle shortening is subject to volitional control. Only the number of motor units required are activated. For example, a pencil can be picked up from a desk top quickly or slowly depending on the controlled pattern of motor unit recruitment in the muscle group or groups involved.

The force-velocity relationship has been tested for skeletal, smooth, and cardiac muscle in humans and other muscle tissues in different animals (24). It generally holds true for all types of muscle, even the tiny muscles responsible for the rapid fluttering of insect wings. Maximum values of force at zero velocity and maximum values of velocity at a minimal load vary with the size and type of muscle. Although the physiological basis for the force-velocity relationship is not completely understood, the shape of the concentric portion of the curve corresponds to the rate of energy production in a muscle.

The force-velocity relationship for muscle undergoing eccentric tension differs from that described by Hill for concentric tension (27). At loads less than the isometric maximum, the velocity of muscle lengthening is subject to volitional control. At loads greater than the isometric maximum, the muscle is forced to lengthen. The relationship between eccentric muscle tension and the velocity of muscle lengthening at loads greater than the isometric maximum is shown in the portion of the curve below the line of isometric contraction in Figure 4-12.

Eccentric strength training involves the use of resistances that are greater than the athlete's maximum isometric force generation capability. As soon as the load is assumed, the muscle begins to lengthen. This type of training is effective in increasing strength but no more so than isometric or concentric training techniques (3). It has also been associated with increased muscle soreness.

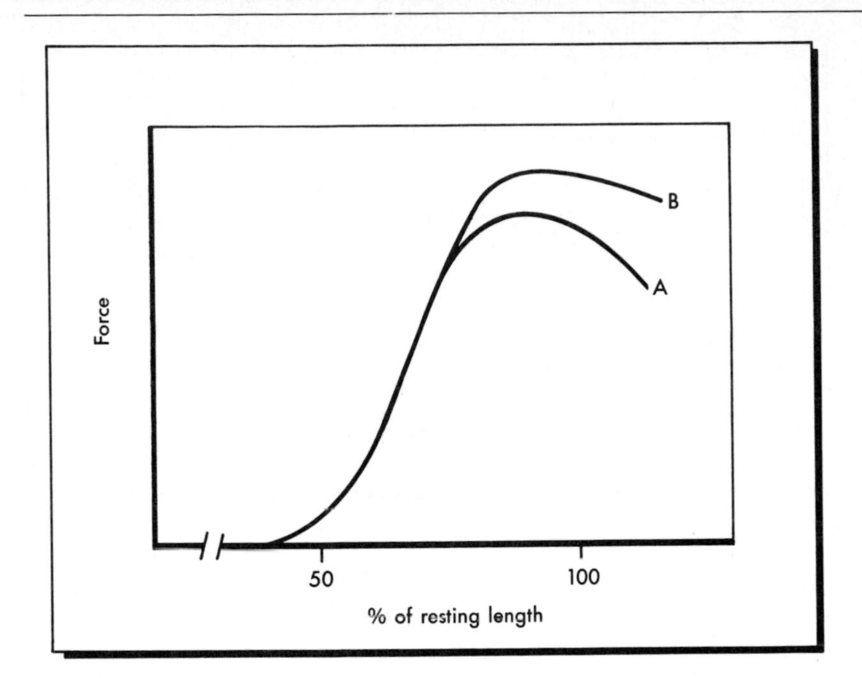

Figure 4-13
The force-length relationship for muscle tissue, showing the contribution of the contractile component alone *(A)* and the added contribution of the elastic component *(B)*.

Force-Length Relationship

The amount of maximum isometric force a muscle is capable of producing is partly dependent on the muscle's length. In single muscle fibers and isolated muscle preparations, force generation is at its peak when the muscle is at normal resting length (neither stretched nor contracted). When the length of the muscle increases or decreases beyond resting length, the maximum force the muscle can produce decreases, following the form of a bell-shaped curve (11).

Within the human body, however, force generation capability increases when the muscle is slightly stretched. Parallel-fibered muscles produce maximum tensions at just over resting length, and pennate-fibered muscles generate maximum tensions at between 120% and 130% of resting length (20). This phenomenon may be due to the contribution of the SEC, which adds to the tension present in the muscle when the muscle is stretched. Figure 4-13 shows the pattern of maximum tension development as a function of muscle length, with the contribution of the SEC indicated.

Force-Time Relationship

When a muscle is stimulated, a brief period of time elapses before the muscle begins to develop tension (Figure 4-14). Referred to as *electromechanical delay* (EMD), this period of time is believed to be

Figure 4-14
The brief period of time that elapses between the stimulation of a muscle and the initiation of tension development is known as electromechanical delay.

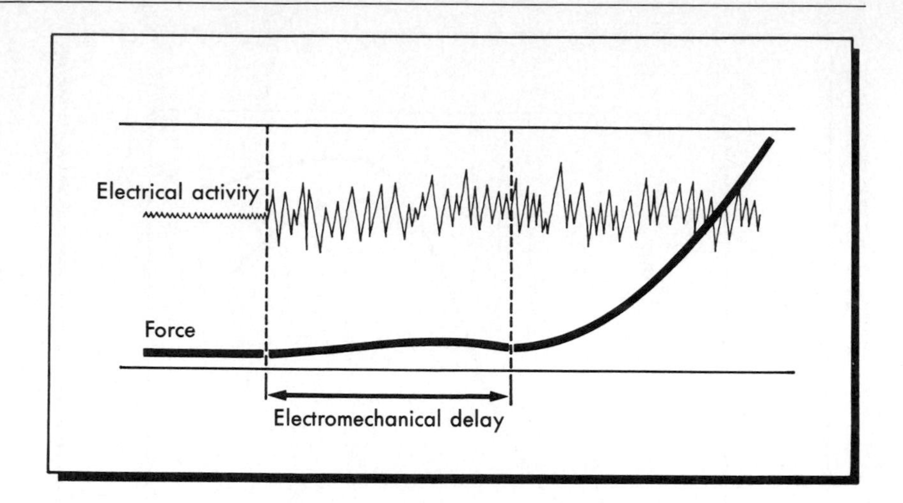

needed for the contractile component of the muscle to stretch the SEC. During this time, muscle laxity is eliminated. Once the SEC is sufficiently stretched, tension development proceeds. The length of EMD varies considerably among human muscles, with values of from 20 to 100 msec reported (28). Muscles with high percentages of ST fibers have longer EMDs than muscles with high percentages of FT fibers (34).

The time required for a muscle to develop maximum isometric tension may be a full second following EMD (28). Shorter force development times are associated with a high percentage of FT fibers in the muscle and with a trained state (40).

MUSCULAR STRENGTH, POWER, AND ENDURANCE

In practical evaluations of muscular function the force-generating characteristics of muscle are discussed within the concepts of muscular strength, power, and endurance. These characteristics of muscle function have significant implications for success in different forms of strenuous physical activity, such as splitting wood, throwing a javelin, or hiking up a mountain trail.

Muscular Strength

When scientists excise a muscle from an experimental animal and electrically stimulate it in a laboratory, they can directly measure the force generated by the muscle (Figure 4-15). It is largely from controlled experimental work of this kind that our understandings of the force-velocity and force-length relationships for muscle tissue are derived.

Figure 4-15
Understandings of the force-velocity and force-length relationships for muscle tissue are derived from controlled laboratory experimentation with excised muscle tissue and from research involving intact animals and intact humans.

Scientists working with isolated muscle preparations consider the muscle's strength as the maximum force it is capable of producing. In the human body, however, it is not possible to directly assess the force produced by a given muscle. The most direct assessment of muscular strength commonly practiced is a measurement of the maximum torque generated by an entire muscle group at a joint. Muscular strength may be considered as a function of the collective force-generating capability of a given functional muscle group. More specifically, muscular strength is the ability of a given muscle group to generate torque at a particular joint.

Torque is the product of force and the perpendicular distance at which the force acts from an axis of rotation. Because muscle force is a vector quantity, it can be resolved into its two perpendicular components. Accordingly, the torque generated by a single muscle is the product of the component of muscular force perpendicular to the bone and the distance from the muscle's attachment to the center of rotation at the joint (Figure 4-16). Only that component of muscle force directed perpendicular to the bone is effective in causing the bone to rotate about the joint. When the muscle pulls at a 90 degree angle to the bone, 100% of the force moves or tends to move the bone. As the joint angle changes and the angle of pull becomes pro-

■ The perpendicular distance from the line of action of the force produced by a muscle to the joint center is known as the muscle's moment arm.

Figure 4-16

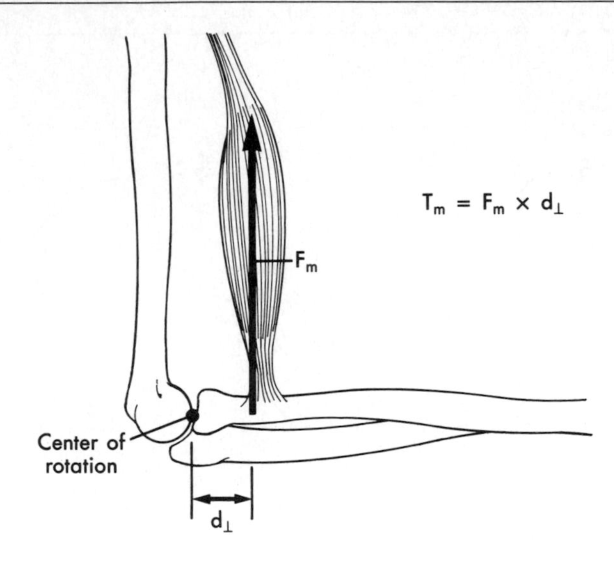

Figure 4-17
The component of muscular force that produces torque at the joint crossed (F_t) is directed perpendicular to the attached bone.

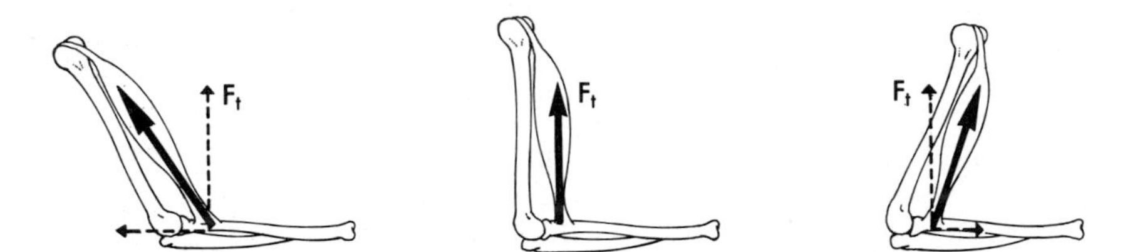

gressively larger or smaller than 90 degrees, less of the force pulls perpendicularly on the bone and a larger component of the force is directed parallel to the bone (Figure 4-17). The sample problem shown in Figure 4-18 demonstrates how the torque generated by a given muscle force changes as the angle of the muscle's attachment to the bone changes.

Therefore, muscular strength is derived from the maximum amount of tension that can be generated by the muscle tissue and from the amount of joint torque produced by the locations and orientations of the contributing muscles with respect to the joint center. Both sources are affected by several factors.

The tension-generating capability of a muscle is related to its cross-sectional area and, possibly to its training state. The force generation capability per cross-sectional area of muscle is 90 N/cm^2 (35).

SAMPLE PROBLEM 2

Figure 4-18

How much torque is produced at the elbow by the biceps brachii inserting at an angle of 60 degrees on the radius when the tension in the muscle is 400 N? (Assume that the muscle attachment to the radius is 3 cm from the center of rotation at the elbow joint.)

Known:

$F_m = 400$ N

$\alpha = 60°$

$d_\perp = 0.03$ m

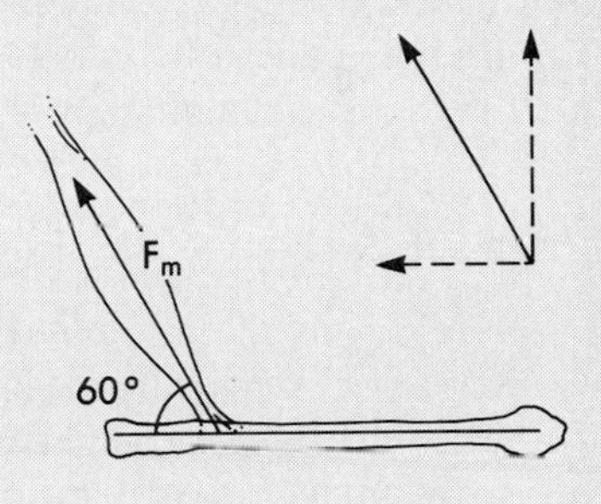

Solution

Wanted: T_m

Only the component of muscle force perpendicular to the bone generates torque at the joint. From the diagram, the perpendicular component of muscle force is

$$F_p = F_m \sin$$

Therefore

$$F_p = (400 \text{ N}) (\sin 60)$$

$$F_p = 346.4 \text{ N}$$

Because

$$T_m = F_p d_\perp$$

Thus

$$T_m = (346.4 \text{ N}) (0.03 \text{ m})$$

$$\boxed{T_m = 10.4 \text{ N-m}}$$

The sample problem shown in Figure 4-19 illustrates the way in which the cross-sectional area of a muscle relates to the approximate amount of force the muscle is capable of generating. When strength training with heavy loads occurs, initial gains in strength may be related to improved innervation of the trained muscle rather than to the increase in its cross-sectional area (22). According to Komi (8), however, the relationship of constant strength per cross-sectional area is generally true. The hypothesis that muscles composed primarily of FT motor units are capable of producing greater force per cross-sectional area than muscles composed largely of ST units has not been strongly documented (28).

Figure 4-19

SAMPLE PROBLEM 3

How much tension may be developed in muscles with the following cross-sectional areas:

a. 4 cm^2?

b. 10 cm^2?

c. 12 cm^2?

Known:

Muscle cross-sectional areas = 4 cm, 10 cm, and 12 cm

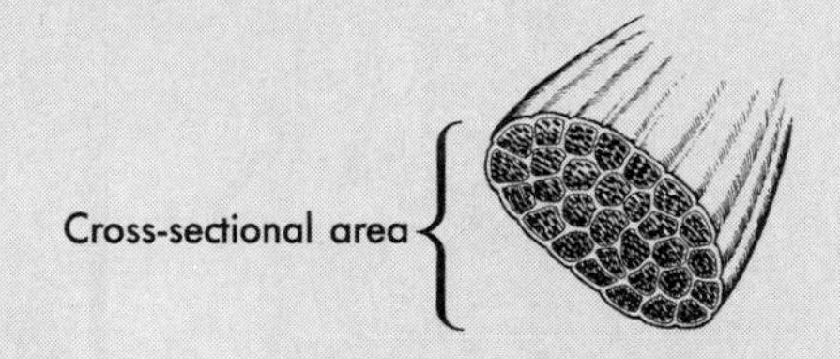

Solution

Wanted: Tension development capability

The tension-generating capability of muscle tissue is 90 N/cm^2. The force produced by a muscle is the product of 90 N/cm^2 and the muscle's cross-sectional area. So,

a. F = (90 N/cm^2) (4 cm)

F = 360 N

b. F = (90 N/cm^2) (10 cm)

F = 900 N

c. F = (90 N/cm^2) (12 cm)

F = 1080 N

The amount of torque generated by a given muscle at a joint depends on the amount of tension in the muscle, the angle of the muscle's attachment to bone, and the distance between the muscle's bone attachment and the axis of rotation at the joint center. The greatest amount of torque is produced by maximum tension in a muscle that is oriented at a 90 degree angle to the bone, as far from the joint center as possible.

Muscular Power

Mechanical power is the product of force and velocity. Muscular power is therefore the product of muscular force and the velocity of muscle shortening. Maximum power occurs at approximately one

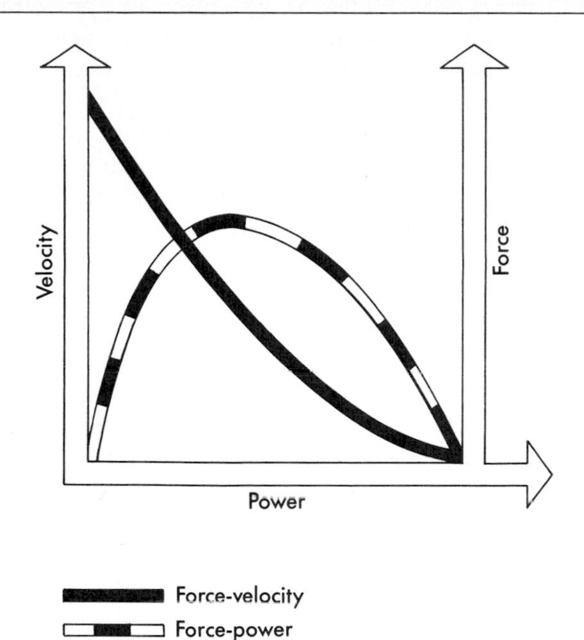

Figure 4-20
The relationship between muscular power and the concentric tension present in the muscle.

third of maximum velocity (23) and at approximately one third of maximum concentric force (28). Figure 4-20 shows muscular power plotted against concentric tension in the muscle.

Because neither muscular force nor the speed of muscle shortening can be directly measured in an intact human being, muscular power is more generally defined as the rate of torque production at a joint or the product of the net torque and the angular velocity at the joint. Accordingly, muscular power is affected by both muscular strength and movement speed.

Muscular power is an important contributor to activities requiring both strength and speed. The strongest shot putter on a team is not necessarily the best shot putter because the ability to accelerate the shot is a critical component of success in the event. Sports that require explosive movements, such as power lifting, throwing, jumping, and sprinting, are based on the ability to generate muscular power.

Since FT fibers develop tension more rapidly than ST fibers, a large percentage of FT fibers in a muscle are an asset for an individual training for a muscular power-based event. Individuals with a predominance of FT fibers generate more power at a given load than individuals with a high percentage of ST compositions. Those with primarily FT compositions also develop their maximum power at faster velocities of muscle shortening (39).

Sprinting requires muscular power, particularly in the hamstrings and the gastrocnemius.

Muscular Endurance

Long cycling races require muscular endurance, particularly in the quadriceps.

Muscular endurance is the ability of the muscle to exert tension over a period of time. The tension may be constant, as when a gymnast performs an iron cross, or varying, as during rowing, running, and cycling. The longer the time tension is exerted, the greater the endurance. Although maximum muscular strength and maximum muscular power are relatively specific concepts, muscular endurance is more nonspecific because the force and speed requirements of the activity dramatically affect the length of time it can be maintained.

The research literature typically addresses muscular endurance from the standpoint of muscle fatiguability. Fatiguability is the opposite of endurance. The more rapidly a muscle fatigues, the less endurance it has. A complex array of physiological and neurological factors affect the rate at which a muscle fatigues. A muscle fiber reaches absolute fatigue once it is unable to develop tension when stimulated by its motor axon. Fatigue may also occur in the motor neuron itself, rendering it unable to generate an action potential (5). FG fibers fatigue more rapidly than FOG fibers, and SO fibers are the most resistant to fatigue. The proportion of ST fibers in the vastus lateralis is directly related to the length of time that a level of 50% of maximum isometric tension can be maintained (31).

Effect of Muscle Temperature

As body temperature elevates, the speeds of nerve and muscle functions increase. This causes a shift in the force-velocity curve, with a higher value of maximum isometric tension and a higher maximum velocity of shortening possible at any given load (Figure 4-21). At an elevated temperature the activation of fewer motor units is needed to sustain a given load (37). The metabolic processes supplying oxygen and removing waste products for the working muscle also quicken with higher body temperatures. These benefits result in increased muscular strength, power, and endurance and provide the rationale for warming up before an athletic endeavor.

Muscle function is most efficient at 38.5° C (2). Elevation of body temperature beyond this point may occur during strenuous exercise under conditions of high ambient temperature and/or humidity and can be extremely dangerous, possibly resulting in heat exhaustion or heat stroke. Organizers of long distance events involving running or cycling should be particularly cognizant of the potential hazards associated with competition in such environments.

REGULATION OF MOVEMENT

Two-Joint and Multi-Joint Muscles

When a muscle develops tension concentrically, motion occurs at the joint crossed by it. However, many muscles in the human body cross two or more joints. These muscles affect motion at both or all of the

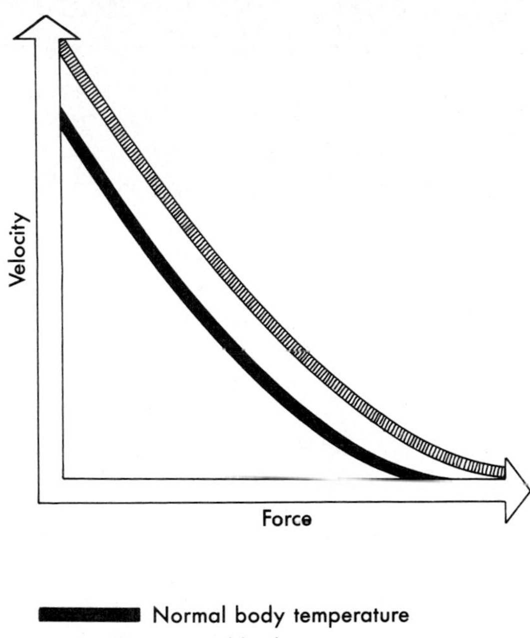

Figure 4-21
When muscle temperature is slightly elevated, the force-velocity curve is shifted. This is one benefit of warming up before an athletic endeavor.

joints crossed simultaneously, which is an efficient anatomical arrangement. The degree of effectiveness of a two-joint or multi-joint muscle in changing the angle at any joint crossed depends on the location and orientation of the muscle's attachment relative to the joint, the joint's relative degree of flexibility or resistance to motion, and the actions of other muscles that cross the joint.

Several examples of two-joint and multi-joint muscles are found in the human body. In the upper extremity the biceps brachii and the long head of the triceps brachii cross both the shoulder and elbow joints. A number of muscles cross the wrist and all finger joints. In the lower extremity the hamstring muscles and the rectus femoris of the quadriceps group cross both the hip and knee joints.

There are two disadvantages associated with the function of two-joint and multi-joint muscles. They are incapable of shortening to the extent required to produce a full range of motion at all joints crossed simultaneously, which is termed *active insufficiency*. For example, the finger flexors cannot produce as tight a fist when the wrist is in flexion as when it is in a neutral position (Figure 4-22). A second problem is that for most people, two-joint and multi-joint muscles cannot stretch to the extent required for full range of motion in the opposite direction at all joints crossed. This problem is referred to as

Figure 4-22
When the wrist is in full flexion, flexion of the fingers cause reduction of the amount of flexion present at the wrist. When the fingers are fully flexed, flexion capability at the wrist is restricted. This is active insufficiency of the multi-joint muscles crossing the palmar side of the fingers and wrist.

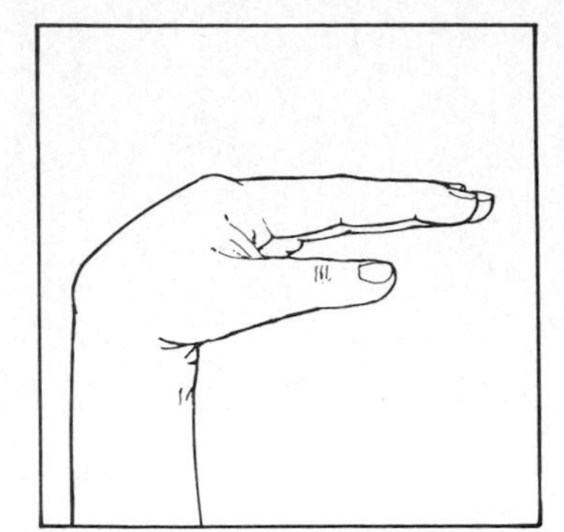

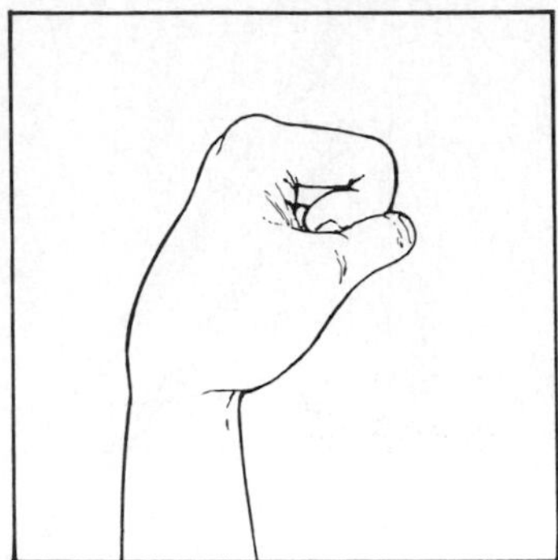

Figure 4-23
The wrist may be extended more fully when the finger flexor muscles are not on stretch. This restriction of the range of motion in extension by the stretched flexor muscles is called *passive insufficiency*.

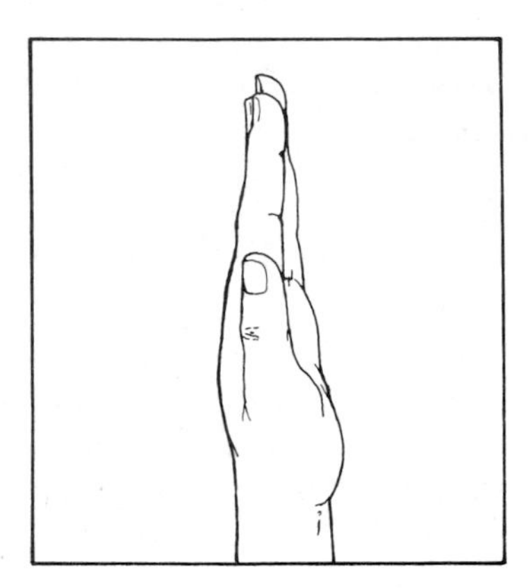

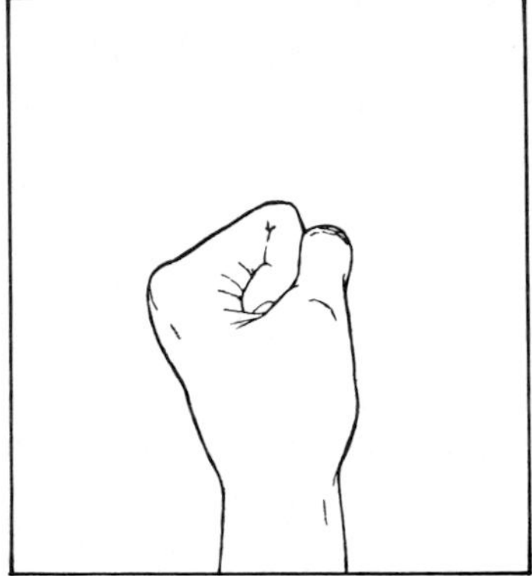

passive insufficiency (38). For example, a larger range of hyperextension is possible at the wrist when the fingers are not fully extended (Figure 4-23).

The action or actions occurring when two-joint and multi-joint muscles develop tension are greatly influenced by the presence of simultaneous tension in other muscles crossing the same joints. Coordinated movement typically requires the simultaneous activation of numerous muscles in a sequencing that is precisely controlled by the central nervous system. Because muscles may cause, slow down, or prevent a particular movement, muscles and groups of muscles are often referred to by the roles that they assume relative to a given movement.

During the elbow flexion phase of a forearm curl the brachialis and the biceps brachii act as the primary agonists, with the brachioradialis, extensor carpi radialis longus, and pronator teres serving as assistant agonists (38).

Roles Assumed by Muscles

When a muscle causes a movement at a joint through the concentric development of tension, it is acting as an **agonist.** Because several different muscles often contribute to a movement, the distinction between primary and assistant agonists is sometimes made. For example, during the elbow flexion phase of a forearm curl the brachialis and the biceps brachii act as the primary agonists, with the brachioradialis, extensor carpi radialis longus, and pronator teres serving as assistant agonists (15).

agonist
the role played by a muscle acting to cause a movement

When a muscle opposes a movement at a joint through the development of eccentric tension, it is acting as an **antagonist.** Antagonists often provide controlling or braking actions, particularly at the end of a fast, forceful movement. During the elbow extension phase of a forearm curl in which motion is caused by the action of gravity, the elbow flexors acting as antagonists control the speed of elbow extension.

antagonist
the role played by a muscle generating torque opposing that generated by the agonists at a joint

Another role assumed by muscles involves stabilizing a portion of the body against a particular force. The force may consist of tension in other muscles or be an external force, such as the weight of an object being lifted. The rhomboids act as **stabilizers** by developing tension to stabilize the scapulae against the pull of the tow rope during water skiing.

stabilizer
the role played by a muscle acting to stabilize a body part against some other force

A fourth role assumed by muscles is as a **neutralizer.** Neutralizers prevent unwanted accessory actions that normally occur when an agonist develops concentric tension. For example, if a muscle causes

neutralizer
the role played by a muscle acting to eliminate an unwanted action produced by an agonist

both flexion and abduction at a joint but only flexion is desired, the action of a neutralizer causing adduction can eliminate the abduction. When the biceps brachii develops concentric tension, it produces both flexion at the elbow and supination of the forearm. If only elbow flexion is desired, the pronator teres act as a neutralizer to counteract the supination of the forearm.

NEUROMUSCULAR RELATIONSHIPS

Golgi Tendon Organs

Sensory receptors known as *Golgi tendon organs* (GTOs) are located in tendons and aponeuroses and in the junctions between muscles and their tendons (Figure 4-24). These receptors are stimulated by the presence of active tension in the muscle. The receptors respond through their neural connections by inhibiting tension development in the muscle and by initiating the excitation and subsequent development of tension in the antagonist muscles. GTOs are encased in tendinous tissues that are relatively inelastic as compared to muscle. These receptors are unlikely to be stimulated by low magnitudes of tension that may result from gentle passive stretching.

The technique of muscle stretching known as *proprioceptive neuromuscular facilitation* (PNF) is based on the responses elicited by the GTOs. The basic PNF procedure consists of three steps. The individual places the muscle group to be stretched in a maximally stretched position and then develops forceful isometric tension in the same muscle group, which activates the GTOs, inhibiting any tension development in the muscle group to be stretched and exciting the antagonist muscle group. Immediately after tension subsides,

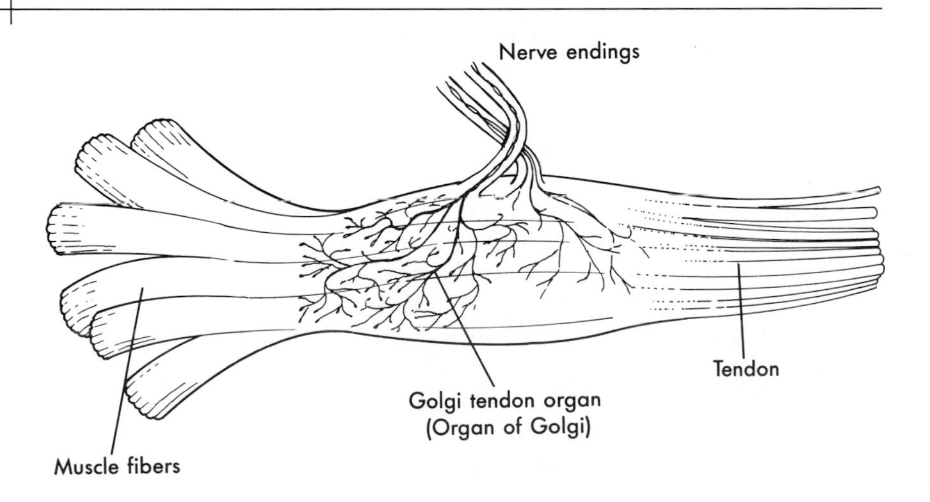

Figure 4-24
Schematic representation of a Golgi tendon organ.

a partner moves the involved limb farther in the direction causing muscle stretch, while the subject deliberately develops tension in the antagonist muscle group. Research indicates that PNF is superior to other forms of muscle stretching (25).

Muscle Spindles

Another type of sensory receptor is interspersed throughout the fibers of muscles. These receptors, which are oriented parallel to the fibers, are known as *muscle spindles* because of their shape (Figure 4-25). The muscle spindles also respond to stretch. Unlike the GTOs, however, muscle spindles respond to both passive and active stretching of the muscle, with a faster rate of stretching provoking a stronger response. Researchers have identified two general responses of stretched muscle spindles: First, they initiate the **stretch reflex.** Second, they inhibit the development of tension in the antagonist muscle group, a process known as **reciprocal inhibition.**

stretch reflex
a monosynaptic reflex initiated by stretching of muscle spindles and resulting in immediate development of muscle tension

reciprocal inhibition
the inhibition of the antagonist muscles resulting from activation of muscle spindles

The stretch reflex, also known as the *myotatic reflex,* is provoked by the activation of the spindles in a stretched muscle. This rapid response involves neural transmission across a single synapse, with afferent nerves carrying stimuli from the spindles to the spinal cord and efferent nerves returning an excitatory signal directly from the spinal cord to the muscle, resulting in the development of tension in the muscle housing the spindles. The knee-jerk test, a common neurological test of motor function, is an example of muscle spindle function producing a short-lived contraction in a stretched muscle. A tap on the patellar tendon initiates the stretch reflex, resulting in the jerk caused by the immediate development of tension in the quadriceps group.

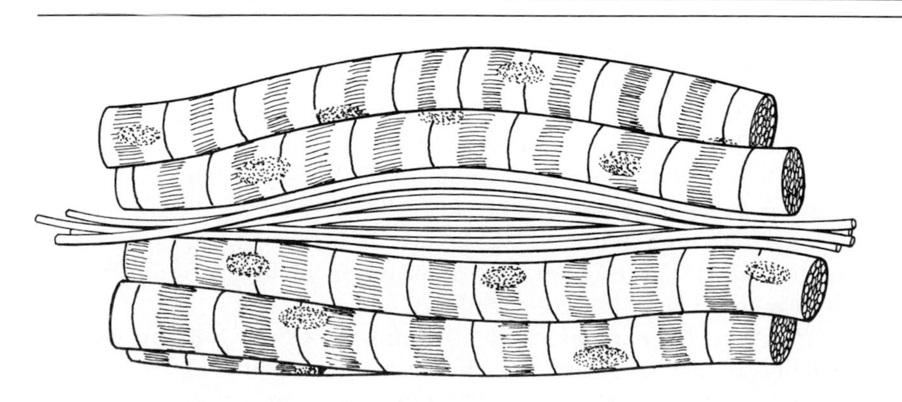

Figure 4-25
Schematic representation of a muscle spindle.

Baseball pitchers commonly initiate a forceful stretch of the shoulder muscles immediately before throwing the ball. The stretch reflex then contributes to the development of tension in these muscles.

The stretch reflex promotes the maintenance of a balanced standing position. When the body sways forward, the posterior muscles of the lower leg (especially the soleus) are stretched. Initiation of the stretch reflex causes these muscles to develop tension, arresting the forward sway. When the body sways backward, the muscles on the anterior aspect of the lower legs develop a counteractive tension. Because the stretch initiating the myotatic reflex is slow and of low magnitude compared to the response producing the knee-jerk, the development of tension produced is more controlled.

The stretch reflex also contributes to the *stretch-shortening cycle* (28). When a muscle is actively stretched, the SEC causes an elastic recoil effect, and the stretch reflex simultaneously initiates the development of tension in the muscle. Thus, a stretch promotes forceful shortening of the muscle. This phenomenon contributes to effective development of concentric muscular force in many sport activities. Football quarterbacks and baseball pitchers typically initiate a forceful stretch of the shoulder muscles immediately before throwing the ball. The same action occurs in muscle groups of the trunk and shoulders at the peak of the backswing in golf and just before swinging a baseball bat. The stretch-shortening cycle is also found during running, with alternating eccentric and concentric tension present in the gastrocnemius.

Reciprocal inhibition can be used to self-treat cramping of the leg muscles. Active contraction of the quadriceps alleviates cramping of the hamstrings, with the spindles of the quadriceps firing to produce reflex inhibition of the hamstrings. Forcefully contracting an antagonist muscle group can ease cramping in other muscle groups as well.

SUMMARY

Muscle is elastic and extensible and responds to stimulation. Most important, it is the only biological tissue capable of developing tension. The action resulting from the development of muscle tension is concentric, eccentric, or isometric, depending on if the muscle is shortening, lengthening, or remaining unchanged in length.

Muscle fibers are arranged in functional motor units that consist of a single motor neuron and all fibers attached to it. Twitch fibers of a given motor unit are either slow twitch, fast twitch fatigue resistant, or fast twitch fast fatigue. Both ST and FT fibers are typically found in all human muscles, although the proportional fiber composition varies. The number and distribution of fibers within muscles appear to be genetically determined.

Different fiber arrangements in muscles of the human body are usually parallel or penniform. Pennate fiber arrangements promote force production, and the parallel fiber arrangement enables greater shortening of the muscle.

Force production by a muscle is influenced by the velocity of muscle shortening, the length of the muscle at the time of stimulation, and the time since the onset of the stimulus.

Force production by a muscle is usually considered in terms of muscular strength, power, and endurance. From a biomechanical perspective, strength is the ability of a muscle group to generate torque at a joint, power is the rate of torque production at a joint, and endurance is resistance to fatigue.

The coordination of muscular force by the nervous system is a complex process. Muscles affect the force, speed, timing, and directional requirements of a movement. Golgi tendon organs and muscle spindles provide sensory feedback to assist the nervous system in its direction of movement and posture regulation.

INTRODUCTORY PROBLEMS

1. Write three general statements identifying the types of movement activities requiring concentric, eccentric, and isometric tension development in muscles.
2. List five movement skills for which a high percentage of fast twitch muscle fibers are an asset and list five movement skills for which a high percentage of slow twitch fibers are an asset. Provide brief statements of rationale for each of your lists.
3. Hypothesize about the pattern of recruitment of motor units in the major muscle group or groups involved during each of the following activities:
   a. Walking up a long flight of stairs
   b. Sprinting up a flight of stairs
   c. Throwing a javelin
   d. Cycling in a 100 km race
   e. Threading a needle
4. Identify three muscles that have parallel fiber arrangements and explain the ways in which the muscles' functions are enhanced by this arrangement.
5. Answer Problem 4 for pennate fiber arrangement.
6. How is the force-velocity curve affected by muscular strength training?
7. Write a paragraph describing the biomechanical factors determining muscular strength.
8. List five activities in which the production of muscular force is enhanced by the series elastic component and the stretch reflex.
9. Muscle can generate approximately 90 N of force per square cm of cross-sectional area. If a biceps brachii has a cross-sectional area of 10 square cm, how much force can it exert? (Answer: 900 N)

10. Using the same force/cross-sectional area estimate as in Problem 9, how much force can be exerted by a gastrocnemius with a cross-sectional area of 22 square cm? (Answer: 1980 N)
11. If a force of 500 N must be generated by the triceps to stabilize the elbow joint for a particular task, what must be the minimum cross-sectional area of the muscle? (Answer: 5.6 cm^2)
12. Estimate the cross-sectional area of your own biceps brachii. How much force should the muscle be able to produce?

ADDITIONAL PROBLEMS

1. Which of the structural components of the whole muscle would you hypothesize contribute to the parallel elastic component, the series elastic component, and the contractile component of muscle function? (You may wish to do some extra reading to learn what has been proposed by researchers.)
2. Considering both the force-length relationship and the concept of mechanical advantage, sketch what you would hypothesize to be the shape of a force versus joint angle curve for the biceps brachii at the elbow. Write a brief rationale in support of the shape of your graph.
3. Certain animals, such as the kangaroo and the cat, are well known for their jumping abilities. What would you hypothesize about the biomechanical characteristics of their muscles?
4. Identify the functional roles played by the muscle groups that contribute to each of the following activities:
   a. Carrying a suitcase
   b. Throwing a ball
   c. Rising from a seated position
5. If the fibers of a penniform muscle are oriented at a 45 degree angle to a central tendon, how much tension is produced in the tendon when the muscle fibers contract with 150 N of force? (Answer: 106 N)
6. How much force must be produced by the fibers of a penniform muscle aligned at a 60 degree angle to a central tendon to create a tensile force of 200 N in the tendon? (Answer: 400 N)
7. What must be the effective minimal cross-sectional areas of the muscles in Problems 5 and 6 above, given an estimated 90 N of force producing capacity per square cm of muscle cross-sectional area? (Answer: 1.2 cm^2; 4.4 cm^2)
8. If the biceps brachii, attaching to the radius 2.5 cm from the elbow joint, produces 250 N of tension perpendicular to the bone and the triceps brachii, attaching 3 cm away from the elbow joint, exerts 200 N of tension perpendicular to the bone, how much net torque is present at the joint? Will there be

flexion, extension, or no movement at the joint? (Answer: 0.25 N-m; flexion)

9. Calculate the amount of torque generated at a joint when a muscle attaching to a bone 3 cm from the joint center exerts 100 N of tension at the following angles of attachment:
   a. 30 degrees
   b. 60 degrees
   c. 90 degrees
   d. 120 degrees
   e. 150 degrees
   (Answer: a. 1.5 N-m; b. 2.6 N-m; c. 3 N-m; d. 2.6 N-m; e. 1.5 N-m)
10. Write a quantitative problem of your own involving the following variables: muscle tension, angle of muscle attachment to bone, distance of the attachment from the joint center, and torque at the joint. Provide a solution for your problem.

REFERENCES

1. Alexander RM and Bennet-Clark HC: Storage of elastic strain energy in muscle and other tissues, Nature 256:114, 1977.
2. Astrand PO and Rodahl K: Textbook of work physiology, ed 2, New York, 1977, McGraw-Hill, Inc.
3. Atha J: Strengthening muscle, Exerc Sport Sci Rev 9:1, 1981.
4. Bar-Or O et al: Anaerobic capacity and muscle fiber type distribution in man, Int J Sports Med 10:82, 1980.
5. Bigland-Ritchie B: EMG/force relations and fatigue of human voluntary contractions, Exerc Sport Sci Rev 9:75, 1981.
6. Brooke MH and Kaiser KK: The use and abuse of muscle histochemistry, Ann N Y Acad Sci 228:121, 1974.
7. Buchthal F and Schalbruch H: Motor unit of mammalian muscle, Physiol Rev 60:90, 1980.
8. Burke RE: Motor units: their physiological properties and neural connections, Am Zoologist 18:127, 1978.
9. Burke RE: Motor units: anatomy, physiology, and functional organization. In Brookhart JM et al, eds: Handbook of physiology. The nervous system, motor control, part 2, Baltimore, 1981, Williams & Wilkens.
10. Burke RE and Tsairis P: The correlation of physiological properties with histochemical characteristics in single motor units, Ann N Y Acad Sci 228:145, 1974.
11. Chapman AE: The mechanical Properties of human muscle, Exerc Sport Sci Rev 13:443, 1985.
12. Desmedt JE and Godaux E: Fast motor units are not preferentially activated in rapid voluntary contractions in man, Nature 267:717, 1977.
13. Emonet-Denand F, Laporte Y, and Proske V: Contraction of muscle fibers in two adjacent muscles innervated by branches of the same motor axon, J Neurophysiol 34:132, 1971.
14. Gans C: Fiber architecture and muscle function, Exerc Sport Sci Rev 10:160, 1982.
15. Gardner EB and O'Connell A: Understanding the scientific bases of human movement, Baltimore, 1972, Williams & Wilkins.

16. Gollnick PD: Muscle characteristics as a foundation of biomechanics. In Matsui H and Kobayashi K, (eds:) Biomechanics VIII-A, Champaign, Ill, 1983, Human Kinetics Publishers, Inc.
17. Gollnick PD and Hodgson DR: The identification of fiber types in skeletal muscle: a continual dilemma, Exerc Sport Sci Rev 14:81, 1986.
18. Gollnick PD, Parsons D, and Oakley CR: Differentiation of fiber types in skeletal muscle from the sequential inactivation of myofibrillar actomyosin ATPase during acid preincubation, Histochemistry 77:543, 1983.
19. Gollnick PD, Piehl K, and Saltin B: Selective glycogen depletion pattern in human muscle fibres after exercise of varying intensity and at varying pedalling rates, J Physiol 241:45, 1974.
20. Gowitzke BA and Milner M: Understanding the scientific bases of human movement, ed 2, Baltimore, 1980, Williams & Wilkins.
21. Guthe K: Reptilian muscle: fine structure and physiological parameters. In Gans C and Parsons TS, eds: Biology of the reptilia, vol 11, London, 1981, Academic Press, Inc.
22. Hakkinen K, Komi PV, and Tesch P: Electromyographic changes during strength training and detraining, Med Sci Sports Exerc 15:455, 1983.
23. Hill AV: The heat of shortening and the dynamic constants of muscle, Proc R Soc Lond B126:136, 1938.
24. Hill AV: First and last experiments in muscle mechanics, Cambridge, Mass, 1970, Cambridge University Press.
25. Holt LE and Smith RK: The effect of selected stretching programs on active and passive flexibility. In Terauds J, ed: Biomechanics in sports: proceedings of the international symposium of biomechanics in sports, Del Mar, Calif, 1982, Research Center for Sports.
26. Huxley AF and Simmons RM: Mechanical properties of the cross-bridges of frog striated muscle, J Physiol (Lond) 218:59, 1971.
27. Katz B: The relation between force and speed in muscular contraction, J Physiol (Lond) 96:45, 1939.
28. Komi PV: Physiological and biomechanical correlates of muscle function: effects of muscle structure and stretch—shortening cycle on force and speed, Exerc Sport Sci Rev 12:81, 1984.
29. Komi PV and Karlsson J: Skeletal muscle fiber types, enzyme activities and physical performance in young males and females, Acta Physiol Scand 103:210, 1978.
30. Komi PV et al: Anaerobic performance capacity in athletes, Acta Physiol Scand 100:107, 1977.
31. Komi PV et al: Effects of heavy resistance and explosive-type strength training methods on mechanical, functional, and metabolic aspects of performance. In Komi PV, ed: Exercise and sport biology, Champaigne, Ill, 1982, Human Kinetics Publishers, Inc.
32. Levin A and Wyman J: The viscous elastic properties of muscle, Proc R Soc Lond B101:218, 1927.
33. Lockhart RD: Anatomy of muscles and their relation to movement and posture. In Bourne GH, (ed:) The structure and function of muscle, vol 1, New York, 1973, Academic Press, Inc.

34. Nilsson J, Tesch P, and Thorstensson A: Fatigue and EMG of repeated fast and voluntary contractions in man, Acta Physiol Scand 101:194, 1977.
35. Norman RW: The use of electromyography in the calculation of dynamic joint torque, doctoral dissertation, University Park, Pa, 1977, Pennsylvania State University.
36. Peter JB et al: Metabolic profiles of three fiber types of skeletal muscle in guinea pigs and rabbits, Biochemistry 11:2627, 1972.
37. Rall JA: Energetic aspects of skeletal muscle contraction: implications of fiber types, Exerc Sport Sci Rev 13:33, 1985.
38. Rasch PJ and Burke RK: Kinesiology and applied anatomy, Philadelphia, 1974, Lea & Febiger.
39. Tihanyi J, Apor P, and Fekete GY: Force-velocity—power characteristics and fiber composition in human knee extensor muscles, Eur J Appl Physiol 48:331, 1982.
40. Viitasalo JT and Komi PV: Interrelationships between electromyographic, mechanical, muscle structure and reflex time measurements in man, Acta Physiol Scand 111:97, 1981.

ANNOTATED READINGS

Cavagna GA: Storage and utilization of elastic energy in skeletal muscle, Exerc Sport Sci Rev 5:82, 1977.

Reviews the research and models relating to the storage of elastic energy in muscle tissue.

Chapman AE: The mechanical properties of human muscle, Exerc Sport Sci Rev 13:443, 1985.

Reviews the mechanical models associated with human muscle functioning in a format written specifically for students of human movement.

Gans C: Fiber architecture and muscle function, Exerc Sport Sci Rev 10:160, 1982.

Reviews research pertaining to muscle fibers and motor units, arrangements of fibers, and arrangements of muscles, with examples drawn from a variety of animal species and from humans.

Gollnick PD: Muscle characteristics as a foundation of biomechanics. In Matsui H and Kobayashi K, eds: Biomechanics VIII-A, Champaign, Ill, 1983, Human Kinetics Publishers, Inc.

Discusses physiological and mechanical aspects of the properties of motor units, patterns of motor unit use, fatigue properties, fiber composition and performance, hereditary factors in muscle composition, adaptations in muscle with training, and conversion of fiber types.

Komi PV: Biomechanics and neuromuscular performance, Med Sci Sports Exerc 16:26, 1984.

Discusses the mechanical properties of muscle using a readable and informative format.

Komi PV: Physiological and biomechanical correlates of muscle function: effects of muscle structure and stretch—shortening cycle on force and speed, Exerc Sport Sci Rev 12:88, 1984.

Presents the research findings relating how the structural and physiological characteristics of muscle affect the muscle force and speed of tension development.

5 JOINTS

The Biomechanics of Human Skeletal Articulations

After reading this chapter, the student will be able to:

Describe the characteristics of the categories of joints based on structure and movement capabilities and identify a joint from each category.

Explain the functions of articular cartilage and fibrocartilage.

Describe the material properties of articular connective tissues.

Identify the factors contributing to joint stability and mobility.

Discuss the concept of joint flexibility.

Explain the advantages and disadvantages of the different approaches to increasing flexibility.

The joints of a body largely govern both the directions and the ranges of movements of body segments. The anatomical structure of a given joint, such as the uninjured knee, varies little from person to person, as do the directions in which the attached body segments, such as the thigh and lower leg, are permitted to move at the joint (Figure 5-1). However, differences primarily in the elasticity of the surrounding **collagenous tissues** result in differences in joint ranges of movement. This chapter discusses the biomechanical aspects of joint function, including the concept of joint flexibility.

collagenous tissue
tissue containing strong, slightly extensible protein collagen fibers

Gymnastics is a sport requiring a large amount of flexibility at the major joints of the body.

Figure 5-1
Mechanically, the human body may be considered as a series of rigid links or segments that are connected at joints.

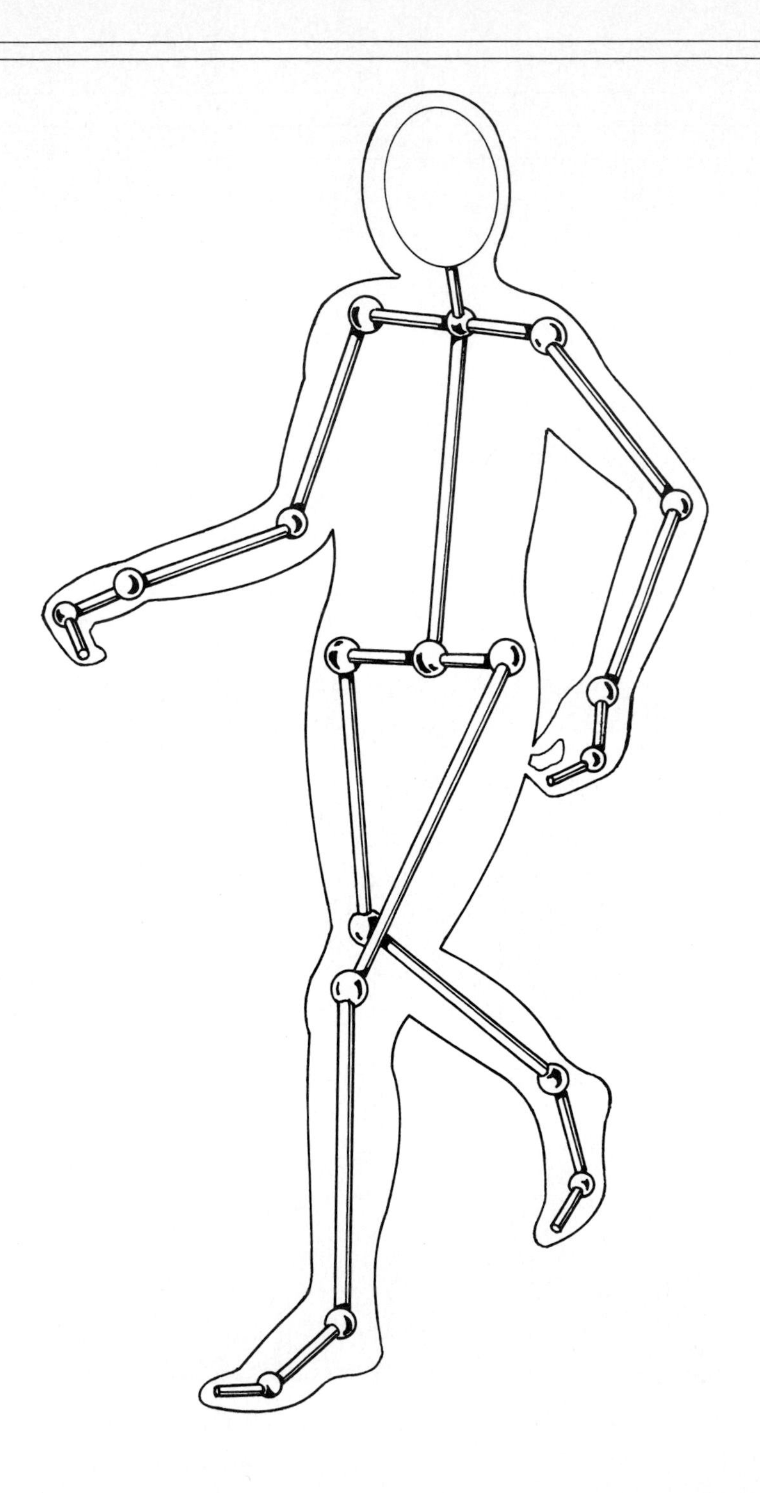

JOINT ARCHITECTURE
Classification of Joints

Various systems of joint classification are presented in courses dealing with anatomy, and they are typically based on either joint complexity, the number of axes present, joint geometry, or movement capabilities (14). A summary of the joint classification system based on the nature of the movement permitted follows, with examples of joints from each category shown in Figure 5-2.

1. *Synarthroses* (immovable): These fibrous joints can attenuate force (absorb shock) but permit little or no movement of the articulating bones.
   a. *Sutures:* These joints have only a slight separation of the adjacent bones, and the fibrous tissue of the joint is continuous with the periosteum. The only example in the human body is the sutures of the skull.
   b. *Syndesmoses:* In these joints, dense fibrous tissue binds the bones together, permitting extremely limited movement. Examples include the coracoacromial, mid radioulnar, mid tibiofibular, and inferior tibiofibular joints.
2. *Amphiarthroses* (slightly movable): These cartilaginous joints attenuate applied forces and permit more motion of the adjacent bones than synarthrodial joints.
   a. *Synchondroses:* In these joints the articulating bones are separated by a thin layer of fibrocartilage. Examples include the sternocostal joints and the epiphyseal plates (before ossification).
   b. *Symphyses:* In these joints, thin plates of hyaline cartilage separate a disc of fibrocartilage from the bones. Examples include the vertebral joints and the pubic symphysis.
3. *Diarthroses* (freely movable; synovial): These joints vary widely in structure and movement capabilities. At diarthrodial joints a separation or cavity is present between the articulating bones, a ligamentous capsule surrounds the joint, and a synovial membrane lining the interior of the joint capsule secretes a lubricant known as *synovial fluid* (Figure 5-3).
   a. *Gliding* (plane; arthrodial): In these joints the articulating bone surfaces are nearly flat, and the only movement permitted is nonaxial gliding. Examples include the intermetatarsal, intercarpal, and intertarsal joints and the facet joints of the vertebrae.
   b. *Hinge* (ginglymus): One articulating bone surface is convex and the other is concave in these joints. Strong collateral ligaments restrict movement to a planar, hingelike motion. Examples include the ulnohumeral and interphalangeal joints.
   c. *Pivot* (screw; trochoid): In these joints, rotation is permitted around one axis. Examples include the atlantoaxial joint and the proximal and distal radioulnar joints.

■ The joints most frequently of interest to biomechanists are the diarthrodial joints and the vertebral joints, which are symphyses.

Figure 5-2
Types of joints present in the human body.

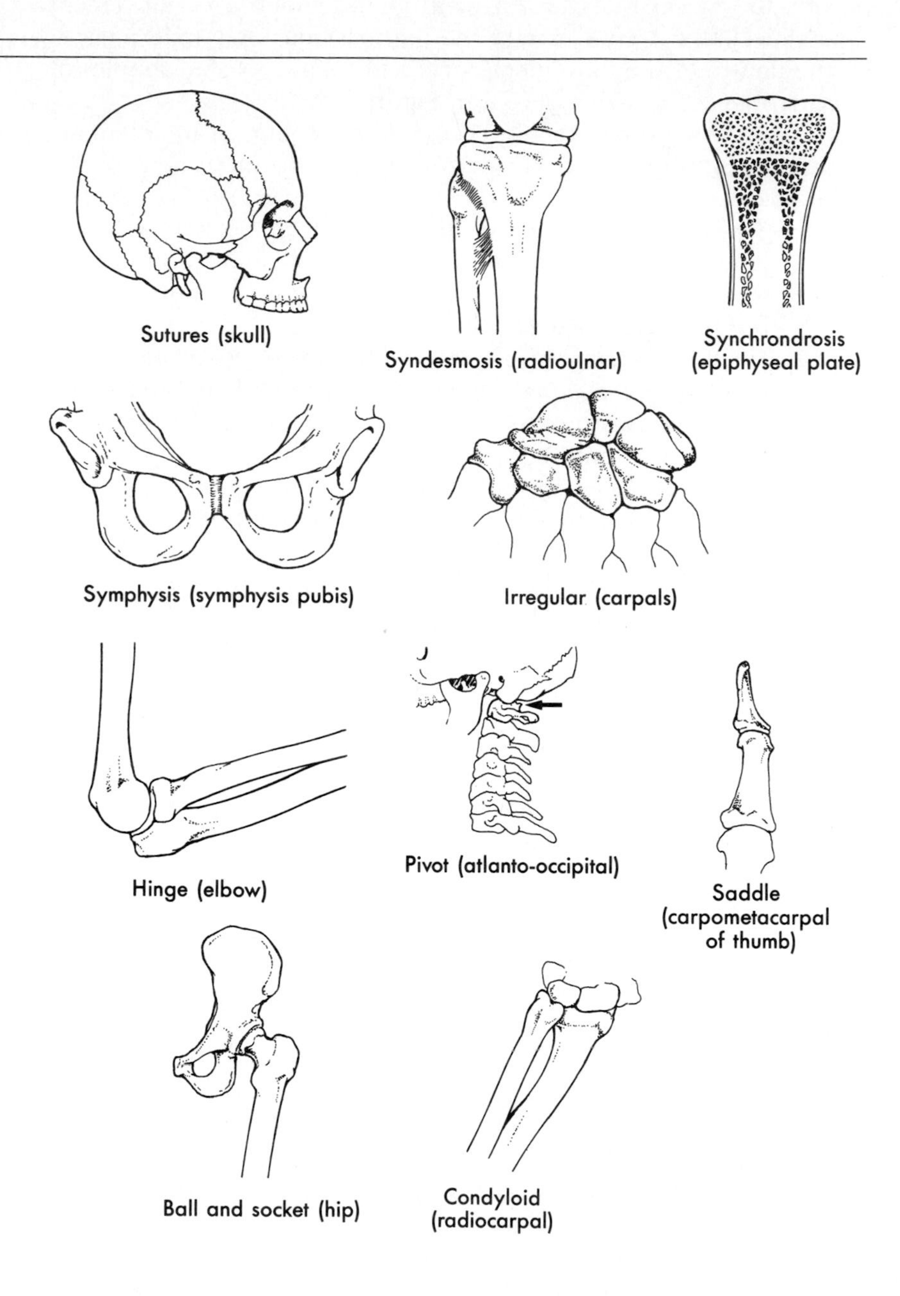

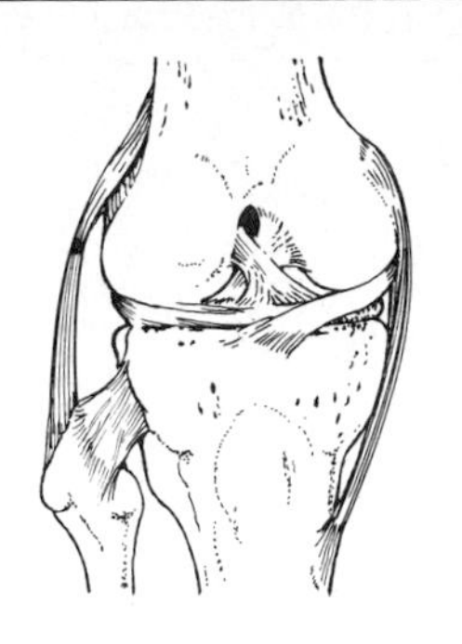

Exterior view

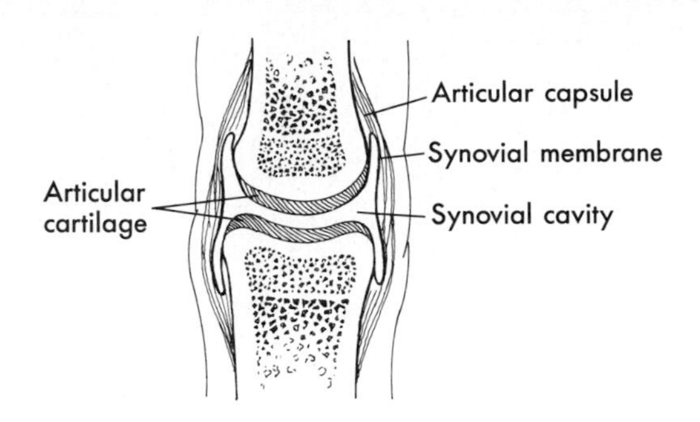

Cross-sectional view

Figure 5-3
Schematic representation of a diarthrodial joint (the knee), with ligamentous capsule, articular cavity, and articular cartilage from exterior and cross-sectional views.

d. *Condyloid* (ovoid; ellipsoidal): One articulating bone surface is an ovular convex shape, and the other is a reciprocally shaped concave surface in these joints. Flexion, extension, abduction, adduction, and circumduction are permitted. Examples include the second through fifth metacarpophalangeal joints and the radiocarpal joints.

e. *Saddle* (sellar): The articulating bone surfaces are both shaped like the seat of a riding saddle in these joints. Movement capability is the same as that of the condyloid joint but with greater range of movement allowed. An example is the carpometacarpal joint of the thumb.

f. *Ball and socket* (spheroidal): In these joints the surfaces of the articulating bones are reciprocally convex and concave. Rotation in all three planes of movement is permitted. Examples include the hip and shoulder joints.

■ The articulating bone surfaces at all joints are of approximately matching (reciprocal) shapes.

Articular Cartilage

The joints of a mechanical device must be properly lubricated if the movable parts of the machine are to move freely and not wear against each other. In the human body a special type of dense, white connective tissue known as **articular cartilage** provides a protective lubrication. A 1- to 7-mm thick protective layer of this material coats the ends of bones articulating at diarthrodial joints (19). Articular cartilage serves two important purposes: First, it spreads loads at the joint over a wide area so that the amount of stress at any contact point between the bones is reduced. Second, it allows movement of the articulating bones at the joint with minimal friction and wear (3).

articular cartilage
a protective layer of tissue that covers the ends of articulating bones at diarthrodial joints

Cartilage can reduce the maximum contact stress acting at a joint by 50% or more (20). The lubrication supplied by the articular cartilage is so effective that the friction present at a joint is approximately 17% to 33% of the friction of a skate on ice under the same load (1).

■ Excesses in the magnitude and/or frequency of loading at a joint can damage the articular cartilage, making it less effective in carrying out its functions.

Figure 5-4
The intervertebral discs function as hydrostatic cushions between the vertebrae.

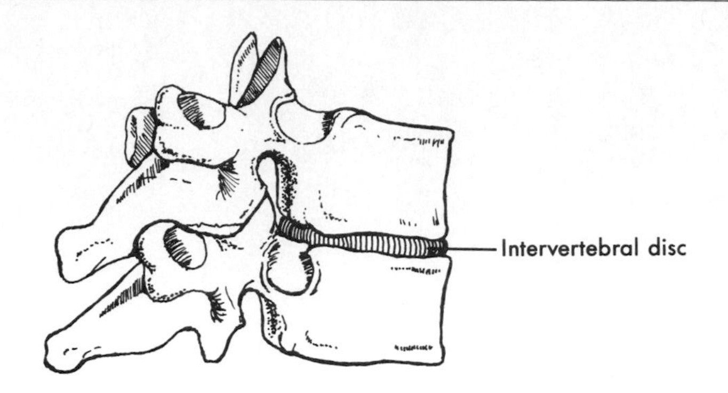

Articular Fibrocartilage

articular fibrocartilage discs or menisci composed of fibrocartilage that intervene between articulating bones at amphiarthrodial joints

At some joints, **articular fibrocartilage** in the form of either a fibrocartilaginous disc or partial discs known as menisci is also present between the articulating bones. The intervertebral discs (Figure 5-4) and the menisci of the knee (Figure 5-5) are examples. Although the function of discs and menisci is not clear, possible roles include the following (19):

1. Absorption and distribution of loads
2. Improvement of the fit of the articulating surfaces
3. Limitation of translation or slip of one bone with respect to another
4. Protection of the periphery of the articulation
5. Lubrication

▪ Intervertebral discs act as cushions between the vertebra, reducing stress levels by spreading loads.

Figure 5-5
The menisci at the knee joint help to distribute loads, lessening the stress transmitted across the joint.

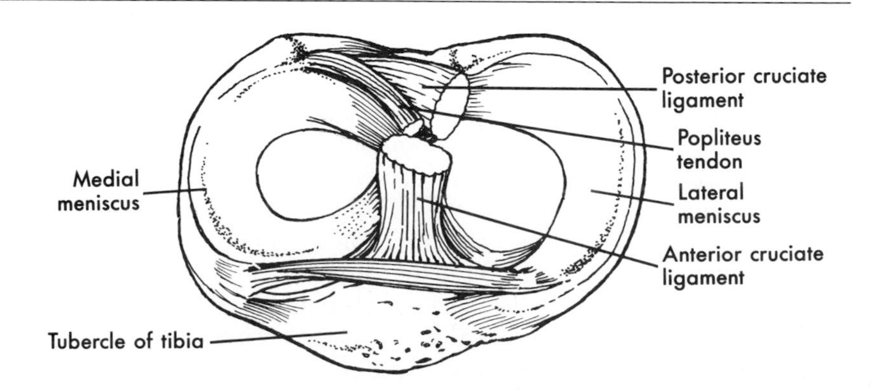

Articular Connective Tissue

Tendons, which connect muscles to bones, and ligaments, which connect bones to other bones, are passive tissues comprised primarily of collagen and elastic fibers. Tendons and ligaments do not have the ability to contract like muscle tissue, but they are slightly extensible. These tissues are elastic and will return to their original length after being stretched, unless they are stretched beyond their **elastic limits.** A tendon or ligament stretched beyond its elastic limit during an injury remains stretched and can be restored to its original length only through surgery.

elastic limit
a material stretched beyond its elastic limit remains lengthened beyond its original length after tension is released

■ Stretching or rupturing of the ligaments at a joint can result in abnormal motion of the articulating bone ends, with subsequent damage to the articular cartilage.

Tendons and ligaments respond to increased mechanical stress by hypertrophying. Research has shown that regular exercise results in increased size and strength of both tendons (16) and ligaments (22), as well as in increased strength of the junctions between tendons or ligaments and bone (17).

JOINT STABILITY

The stability of an articulation is its ability to resist dislocation. Specifically, it is the ability to resist the displacement of one bone end with respect to another while preventing injury to the ligaments, muscles, and muscle tendons surrounding the joint. Different factors influence **joint stability.**

joint stability
the ability of a joint to resist abnormal displacement of the articulating bones

Shape of the Articulating Bone Surfaces

In many mechanical joints, the articulating parts are exact opposites in shape so that they fit tightly together (Figure 5-6). In the human body, the articulating ends of bones are usually shaped as mating convex and concave surfaces.

Although most joints have reciprocally shaped articulating surfaces, these surfaces are not symmetrical, and there is typically one position of best fit in which the area of contact is maximum. This is known as the **close-packed position,** and it is in this position that joint stability is usually greatest (14). Any movement of the bones at the joint results in a **loose-packed position,** with reduction of the area of contact.

close-packed position
joint position in which the contact between the articulating bone surfaces is maximum

loose-packed position
any joint position other than the close packed position

■ The close-packed position occurs for the knee, wrist, and interphalangeal joints at full extension and for the ankle at full dorsiflexion (14).

Some articulating surfaces are shaped so that in both close- and loose-packed positions, there is either a large or a small amount of contact area and consequently more or less stability. For example, the acetabulum provides a relatively deep socket for the head of the femur, and there is always a relatively large amount of contact area between the two bones, which is one reason the hip is a stable joint. At the shoulder, however, the small glenoid fossa has a vertical diameter that is approximately 75% of the vertical diameter of the humeral head and a horizontal diameter that is 60% of the size of the humeral head (3). Therefore, the area of contact between these two

Figure 5-6
Mechanical joints are often composed of reciprocally shaped parts.

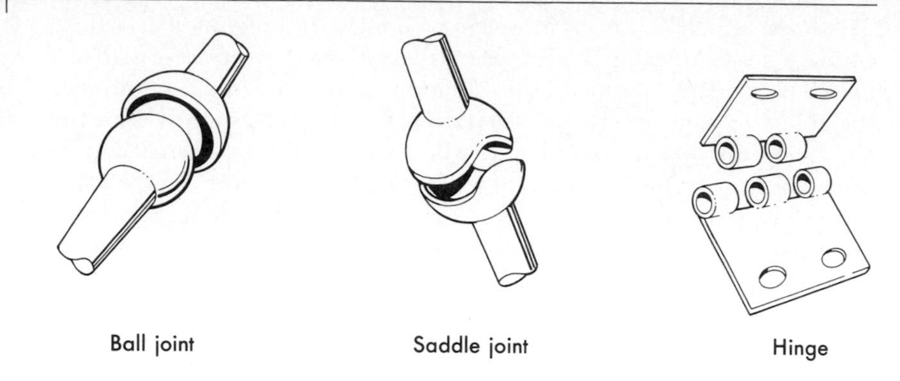

Figure 5-7
A larger bone-bone contact area results in the hip being a more stable joint than the shoulder.

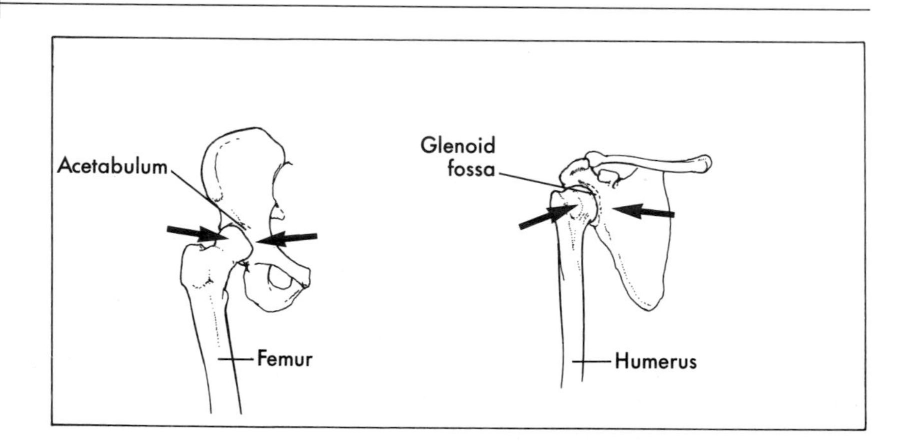

▪ One factor enhancing the stability of the glenohumeral joint is a posteriorly tilted glenoid fossa and humeral head. Individuals with anteriorly tilted glenoids and humeral heads are predisposed to shoulder dislocation.

bones is relatively small, contributing to the relative instability of the shoulder complex (Figure 5-7). Slight anatomical variations in shapes and sizes of the articulating bone surfaces at any given joint among individuals are found; therefore some people have joints that are more or less stable than average.

Arrangement of Ligaments and Muscles

Ligaments, muscles, and muscle tendons affect the relative stability of joints. At joints such as the knee and the shoulder in which the bone configuration is not particularly stable, the tension in ligaments and muscles contributes significantly to joint stability by helping to hold the articulating bone ends together. If these tissues are weak from disuse or lax from being overstretched, the stability of the joint is reduced. Strong ligaments and muscles often increase joint stability. For example, strengthening of the quadriceps and hamstring

Figure 5-8
At the knee joint, stability is primarily derived from the tension in the ligaments and muscles that cross the joint.

groups enhances the stability of the knee (12). The complex array of ligaments and tendons crossing the knee is illustrated in Figure 5-8.

The angle of attachment of most tendons to bones is arranged so that when the muscle exerts tension, the articulating ends of the bones at the joint crossed are pulled closer together, enhancing joint stability. This situation is usually found when the muscles on oppo-

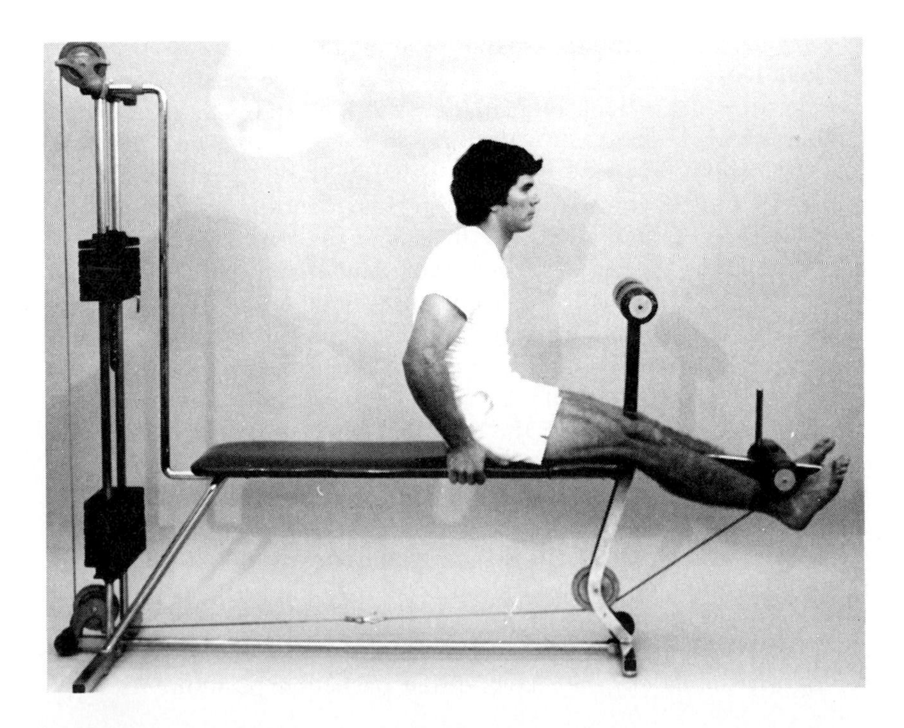

Strengthening the quadriceps and the hamstrings increases the stability of the knee joint.

Figure 5-9

Contraction of the biceps brachii produces a component of force at the elbow that may tend to be stabilizing or dislocating depending on the angle present at the elbow when contraction occurs.

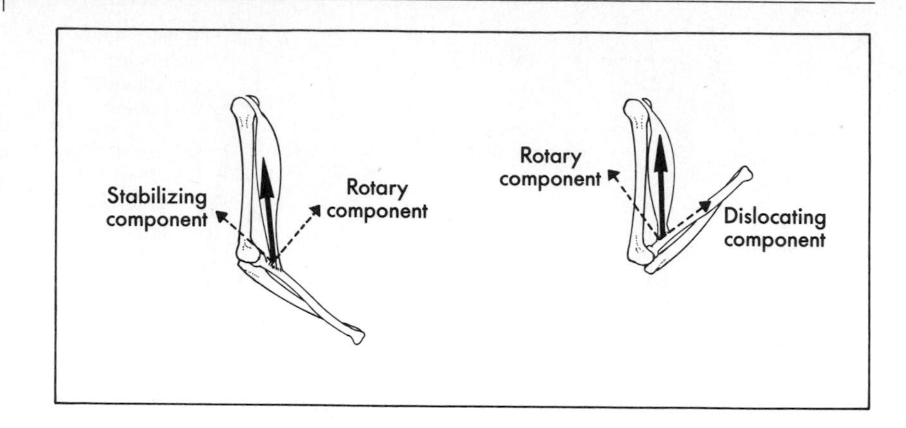

- Engaging in athletic participation with fatigued muscles increases the likelihood of injury.

- The joint angle associated with maximum torques is approximately 90 degrees. The exact angle varies slightly with the geometry of the joint and the location of the muscle attachment to the bone.

rotary component
the component of muscle force acting perpendicular to the attached bone

stabilizing component
the component of muscle force directed toward the center of the joint crossed

dislocating component
the component of muscle force directed away from the center of the joint crossed

site sides of a joint produce tension simultaneously. When muscles are fatigued, however, they are less able to contribute to joint stability. Rupture of the cruciate ligaments is most likely to occur when the tension in fatigued muscles surrounding the knee is inadequate to protect the cruciate ligaments from being stretched beyond their elastic limits (12). Gymnastics injuries also occur with greater frequency when the athletes are fatigued (11).

Tension in the muscles crossing a joint does not always promote the joint's stability. As discussed in Chapter 4, only the component of muscle force acting perpendicular to the long axis of the attached bone actually contributes to the rotation of a body segment around the joint crossed by the muscle. Accordingly, the perpendicular component of muscle force is termed the **rotary component.** When the angle between the line of muscle force and the long axis of the bone is greater than 90 degrees, the component of muscle force acting parallel to the bone pulls the articulating bone ends closer together at the joint crossed (Figure 5-9). For this reason, it is termed the **stabilizing component** of muscle force. When the angle between the muscle and bone is less than 90 degrees, the parallel component of muscle force tends to pull the attached bone away from the joint center and is therefore known as the **dislocating component** of muscle force.

Actual dislocation of a joint rarely occurs from the tension developed by a muscle, but if a dislocating component of muscle force is present, a tendency for dislocation occurs. If the angle at the elbow is less than 90 degrees, tension produced by the biceps tends to pull the radius away from its articulation with the humerus, thereby lessening the stability of the elbow in that particular position.

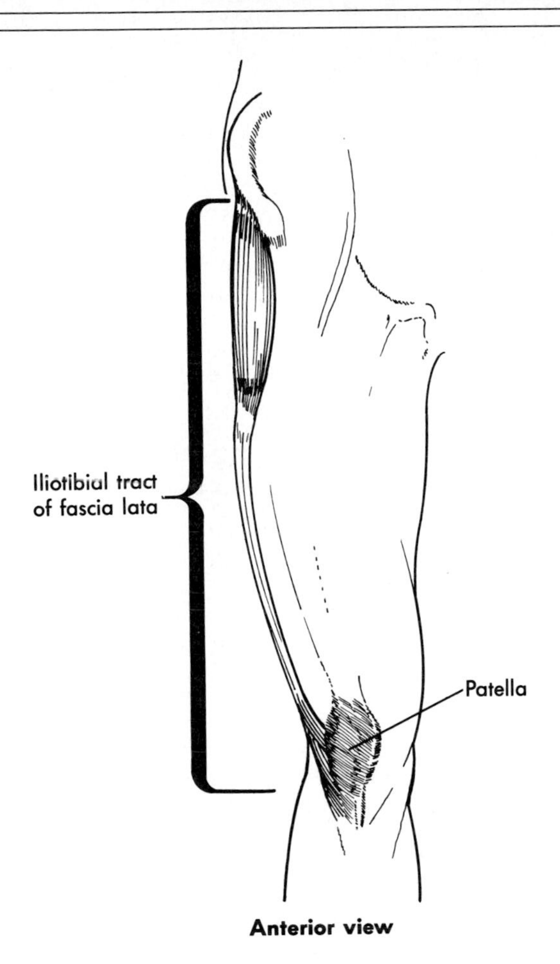

Figure 5-10
The iliotibial band is a strong thickened region of the fascia lata that crosses the knee, contributing to the knee's stability.

Connective Tissues

White fibrous connective tissue known as *fascia* surrounds muscles and the bundles of muscle fibers within muscles, providing protection and support. A particularly strong, prominent tract of fascia known as the *iliotibial band* crosses the lateral aspect of the knee, contributing to its stability (Figure 5-10). The fascia and the skin on the exterior of the body are other tissues that contribute to joint integrity.

Figure 5-11
The range of motion for flexion at the hip is typically measured with the individual supine.

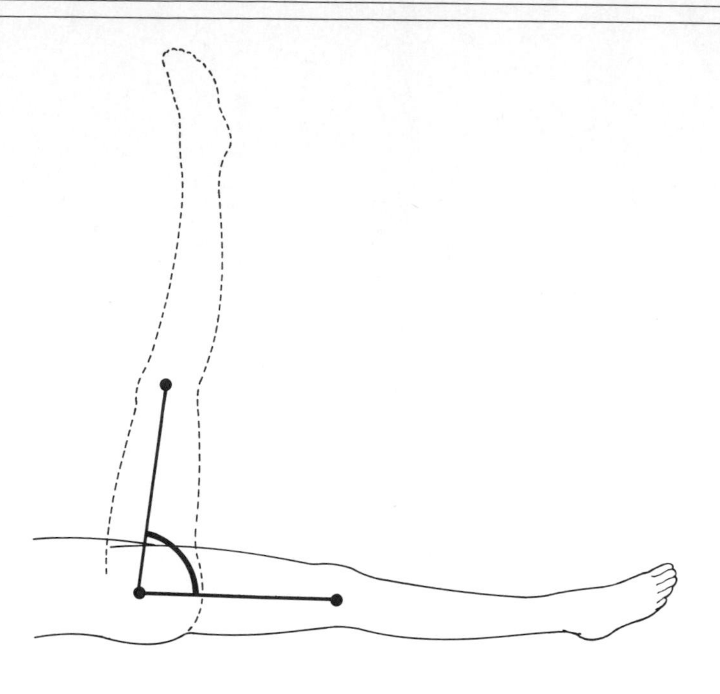

JOINT FLEXIBILITY
Factors Influencing Joint Mobility

joint mobility
a term indicating the relative degree of motion allowed at a joint

range of motion
the angle through which a joint moves from anatomical position to the extreme limit of segment motion in a particular direction

■ A joint with an unusually large range of motion is termed hypermobile.

Joint mobility is a qualitative term used to describe the **range of motion** (ROM) allowed in each of the planes of motion at a joint. Joint ROM is measured directionally. For example, the ROM for flexion at the hip is the difference in the angle at the hip when it is in anatomical position (fully extended) and the angle at the hip when it is in maximum flexion (Figure 5-11). The ROM for extension (return to anatomical position) is the same as that for flexion, with movement past anatomical position in the other direction quantified as the ROM for hyperextension. Several devices used for measuring joint range of motion are shown in Figure 5-12 and Figure 5-13.

Different factors influence joint mobility. The shapes of the articulating bone surfaces and intervening muscle or fatty tissue may terminate movement at the extreme of a ROM. When the elbow is in extreme hyperextension, for example, contact of the olecranon of the ulna with the olecranon fossa of the humerus restricts further motion in that direction. Muscle and/or fat on the anterior aspect of the arm may terminate elbow flexion.

For most individuals, joint mobility is primarily a function of the relative laxity and/or extensibility of the collagenous tissues crossing the joint. Tight ligaments and muscles with limited extensibility are

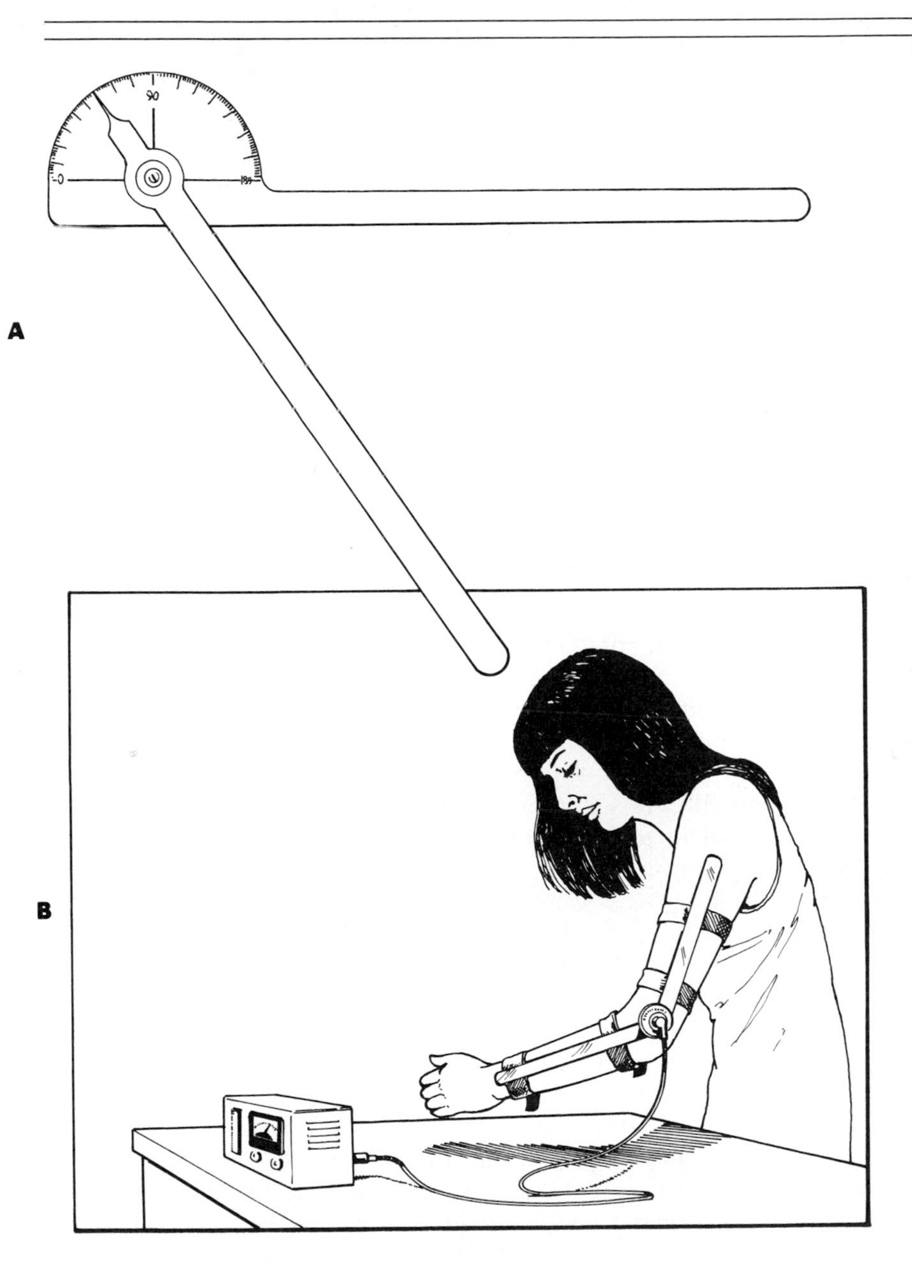

Figure 5-12

A, A goniometer is basically a protractor with two arms attached. The point where the arms intersect is aligned over the joint center while the arms are aligned with the longitudinal axes of the body segments to measure the angle present at a joint. **B,** An electrogoniometer is a goniometer with a potentiometer located at its axis of rotation so that as the angle between the goniometer arms changes, the amount of electrical current output to a recorder changes. Electrogoniometers can be fixed to a subject for the recording of changes in joint position during the performance of different activities.

Figure 5-13
A device for measuring changes in the position of a body segment is the Leighton flexometer, which consists of a weighted needle attached to a 360 degree scale.

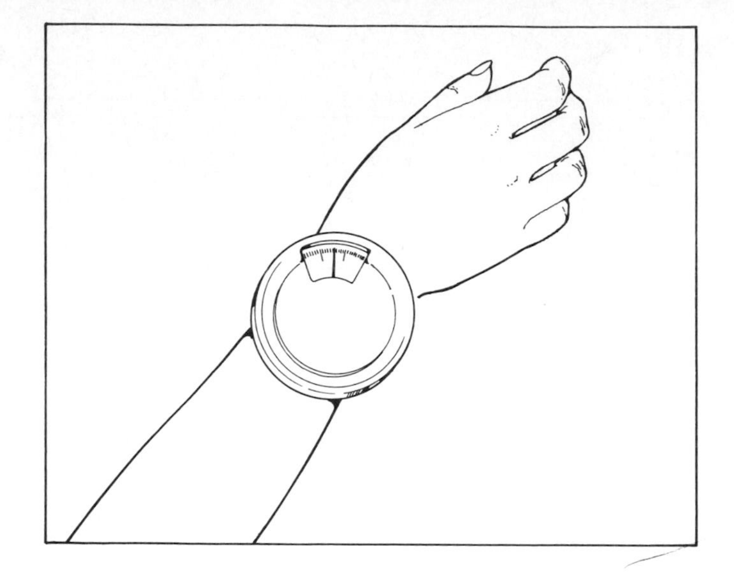

the most common inhibitors of a joint's ROM. When these tissues are not stretched, their extensibility usually diminishes. The fluid content of the cartilaginous discs present at some joints also influences the mobility at these joints.

Laboratory studies have shown that the extensibility of collagenous tissues increases slightly with temperature elevation (18). Although this finding suggests that a warm-up before an athletic endeavor would increase joint ROM, this has not been proven (21). More research is needed to identify the type of warm-up required to increase joint ROM.

Flexibility

flexibility
a qualitative term used to represent the ranges of motion present at a joint in different directions

The term **flexibility** is commonly used as an indicator of joint mobility. Although people's general flexibility is often compared, flexibility is actually joint-specific. That is, an extreme amount of flexibility at one joint does not guarantee the same degree of flexibility at all joints.

Extremely limited joint flexibility is undesirable because if the collagenous tissues crossing the joint are tight, the likelihood of their tearing or rupturing when the joint is forced beyond its normal ROM increases. In a study of competitive female gymnasts, those in a highly injury-prone category had less flexibility of the shoulder, elbow, wrist, hip, and knee joints than those in a low injury incidence category (15).

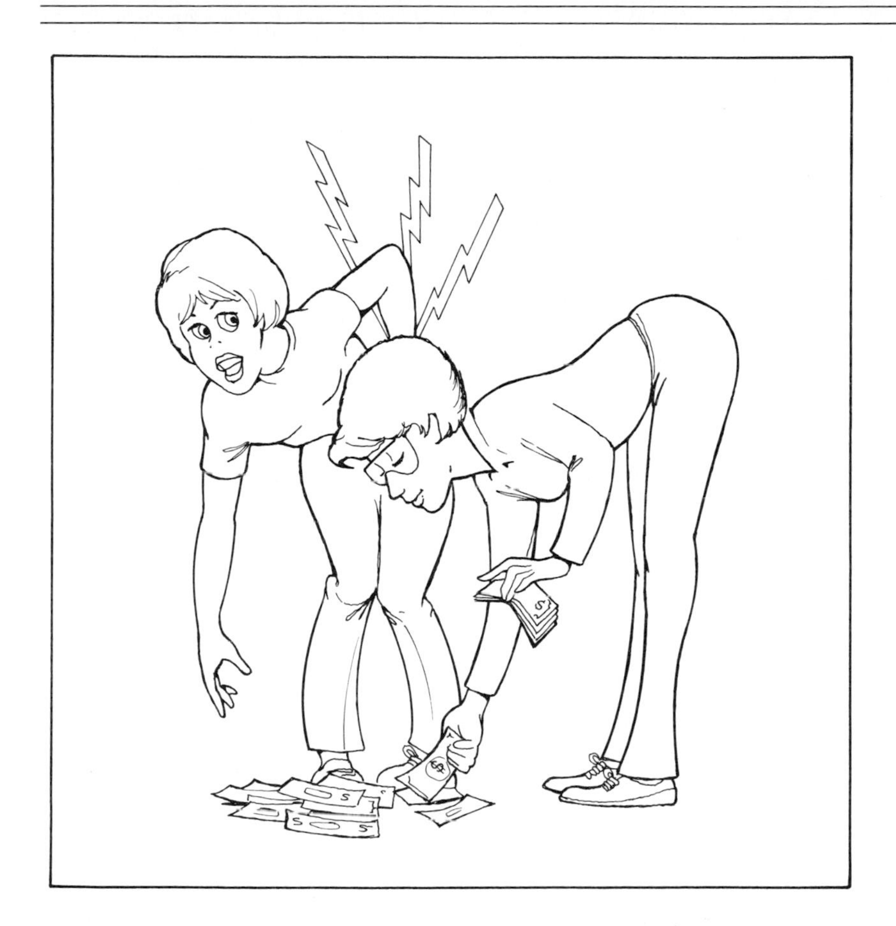

Figure 5-14
The ability to touch one's toes has often been used as a general test of hamstring flexibility.

In the general population, limited flexibility of the hips and lumbar spine may predispose individuals to developing low back pain, a problem that affects approximately 80% of people at some point during their lifespans (10) and is the leading cause of compensable absence from the workplace (14). One reason low back pain is so common is that normal daily activities do not stretch the hamstrings. Thus, this muscle group often tightens and loses its extensibility, with the ROM for hip flexion consequently reduced. When an individual with tight hamstrings leans over to make a bed or retrieve an item from the floor, the low back compensates for the lack of flexion capability at the hip, often resulting in injury. Because adequate hip flexion capability is believed to lessen the likelihood of the development of low back pain, toe-touching exercises are commonly included in health-related physical fitness tests (Figure 5-14).

■ Static toe touches that stretch both the hamstrings and the low back muscles are recommended for individuals who are prone to low back pain.

The desirable amount of joint flexibility is largely dependent on the activities in which an individual wishes to engage. Gymnasts and dancers obviously require greater joint flexibility than nonathletes. However, athletes such as gymnasts and dancers also require strong muscles, tendons, and ligaments to perform well and avoid injury. Although large, bulky muscles may inhibit joint ROM, an extremely strong, stable joint can also enable large ROMs.

Although people usually become less flexible as they age, this phenomenon is primarily related to decreased levels of physical activity rather than to changes inherent in the aging process. Regardless of the age of the individual, if the collagenous tissues crossing a joint are not stretched, they will shorten. Conversely, when these tissues are regularly stretched, they lengthen and flexibility is increased. The results of several studies indicate that flexibility can be significantly increased among elderly individuals who participate in a program of regular stretching and exercise (2, 7, 9).

Techniques for Increasing Joint Flexibility

Increasing joint flexibility is often an important component of therapeutic and rehabilitative programs and programs designed to train athletes for a particular sport. Increasing or maintaining flexibility involves stretching the collagenous tissues, particularly the ligaments and muscles, that limit the ROM at a joint. Several approaches for stretching these tissues can be used, with some being more effective than others.

active stretching
the stretching of muscles, tendons, and ligaments produced by active development of tension in the antagonist muscles

passive stretching
the stretching of muscles, tendons, and ligaments produced by a stretching force other than tension in the antagonist muscles

Stretching can be done either actively or passively. **Active stretching** is produced by tensile force generated by the muscles on the side of the joint opposite the muscles, tendons, and ligaments to be stretched. **Passive stretching** involves the use of gravitational force, force applied by another body segment, or force applied by another

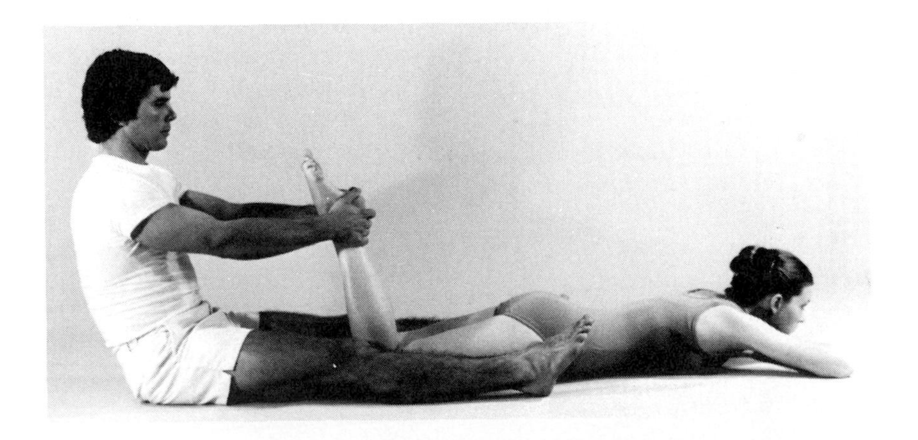

Passive stretching (without contraction of the antagonistic muscle groups) can be accomplished with the assistance of a partner.

person to stretch the collagenous tissues crossing a joint. Active stretching provides two advantages. First, the action of the stretched muscle spindles in the muscle groups supplying the force suppresses the active development of tension in the muscle groups being stretched (6). Second, the muscle groups used to develop force are exercised. With passive stretching, movement can be carried farther beyond the existing ROM than with active stretching, but the concomitant disadvantage of increased likelihood of injury is also present.

Both active and passive stretching can be done either statically or ballistically. In a static stretch the movement is extremely slow, and when the desired joint position is reached, it is maintained statically, usually from about 10 to 30 seconds. Ballistic or bouncing stretches make use of the momentum of body segments to extend joint position to or beyond the extremes of the ROM. Because a ballistic stretch activates the stretch reflex and results in the immediate development of tension in the muscle being stretched, injury to the collagenous tissues may occur.

■ Ballistic, bouncing types of stretches can be dangerous because they tend to promote contraction of the muscles being stretched and the momentum generated may carry the body segments far enough beyond the normal ROM to tear or rupture collagenous tissues.

An effective and increasingly popular procedure for increasing joint ROM is **proprioceptive neuromuscular facilitation** (PNF) (see Chapter 4). PNF is based on the responses elicited by the Golgi tendon organs in muscle tendons. Although the neuromuscular basis for the PNF procedure is not completely understood (8), it has been found to be more effective than other approaches for increasing joint ROM (5, 13).

proprioceptive neuromuscular facilitation
an effective stretching procedure

Active static stretching involves holding a position at the extreme of the range of motion for a period of approximately 10 to 30 seconds.

SUMMARY

The anatomical configurations of the joints of the human body govern the movement capabilities of the articulating body segments. From the perspective of movement permitted, there are three major categories of joints: synarthroses (immovable joints), amphiarthroses (slightly movable joints), and diarthroses (freely movable joints). Each major category is further subdivided into classes of joints with common anatomical characteristics.

The ends of bones articulating at diarthrodial joints are covered with articular cartilage, which reduces contact stress and regulates joint lubrication. Fibrocartilaginous discs or menisci present at some joints also may contribute to these functions.

Tendons and ligaments are strong collagenous tissues that are slightly extensible and elastic. These tissues are similar to muscle and bone in that they respond to levels of habitual mechanical stress by hypertrophying or atrophying.

Joint stability is the ability of the joint to resist displacement of the articulating bones that results in tissue injury. The major factors influencing joint stability are the size and shape of the articulating bone surfaces and the arrangement and strength of the surrounding muscles, tendons, and ligaments.

Joint mobility is primarily a function of the relative lengths of the muscles and ligaments that span the joint. If these tissues are not stretched, they tend to shorten. Approaches for increasing flexibility include active versus passive stretching and static versus dynamic stretching. PNF is a particularly effective procedure for stretching collagenous tissues.

INTRODUCTORY PROBLEMS

(Reference may be made to Chapters 6 to 8 for additional information on specific joints.)

1. Construct a table that identifies joint type and the plane or planes of allowed movement for the shoulder (glenohumeral joint), elbow, wrist, hip, knee, and ankle.
2. Describe the directions and approximate ranges of movement that occur at the joints of the human body during each of the following movements:
   a. Walking
   b. Running
   c. Performing a jumping jack
   d. Rising from a seated position
3. What factors contribute to joint stability?
4. Explain why athletes' joints are often taped before they participate in an activity. What are some possible advantages and disadvantages of taping?

5. What factors contribute to flexibility?
6. What degree of joint flexibility is desirable?
7. How is flexibility related to the likelihood of injury?
8. Discuss the relationship between joint stability and joint mobility.
9. Explain why grip strength diminishes as the wrist is hyperextended.
10. Why is ballistic stretching contraindicated?

ADDITIONAL PROBLEMS

1. Construct a table that identifies joint type and the plane or planes of movement for atlanto-occipital, the L5-S1 vertebral joint, the metacarpophalangeal joints, the interphalangeal joints, the carpometacarpal joint of the thumb, the radioulnar joint, and the talocrural joint.
2. Identify the position (for example, full extension, 90 degrees of flexion) for which each of the following joints is close packed:
   a. Shoulder
   b. Elbow
   c. Knee
   d. Ankle
3. How is articular cartilage similar to and different from ordinary sponge? (You may wish to consult the annotated readings).
4. Comparatively discuss the properties of muscle, tendon, and ligament. (You may wish to consult the annotated readings).
5. Discuss the relative importance of joint stability and joint mobility for athletes participating in each of the following sports:
   a. Gymnastics
   b. Football
   c. Swimming
6. What specific exercises would you recommend for increasing the stability of each of the following joints:
   a. Shoulder
   b. Knee
   c. Ankle
7. What specific exercises would you recommend for increasing the flexibility of each of the following joints:
   a. Hip
   b. Shoulder
   c. Ankle
8. In which sports are athletes more likely to incur injuries that are related to insufficient joint stability? Explain your answer.

9. In which sports are athletes likely to incur injuries related to insufficient joint flexibility? Explain your answer.
10. What exercises would you recommend for senior citizens interested in maintaining an appropriate level of joint flexibility?

REFERENCES

1. Brand RA: Joint lubrication. In Albright JA and Brand RA, eds: The scientific basis of orthopedics, New York, 1979, Appleton-Century-Crofts.
2. Chapman EA, deVries HA, and Sweezey R: Joint stiffness and effects of exercise on young and old men, J Gerontol 27:218, 1972.
3. Frankel VH and Nordin M: Basic biomechanics of the skeletal system, Philadelphia, 1980, Lea & Febiger.
4. Frymoyer JW and Mooney V: Occupational orthopaedics, J Bone Joint Surg 68-A:469, 1986.
5. Holt LE and Smith RK: The effect of selected stretching programs on active and passive flexibility. In Terauds J, ed: Biomechanics in sports: proceedings of the international symposium on biomechanics in sports, Del Mar, Calif, 1982, Research Center for Sports.
6. Jacobs M: Neurophysiological implications of slow active stretching, Am Corrective Ther J 30:151, 1976.
7. Lesser M: The effects of rhythmic exercise on the range of motion in older adults, Am Corrective Ther J 32:4, 1978.
8. Moore MA and Hutton RS: Electromyographic investigation of muscle stretching techniques, Med Sci Sports Exerc 12:322, 1980.
9. Munns K: Effects of exercise on the range of joint motion in elderly subjects. In Smith EL and Serfass RC, eds: Exercise and aging: the scientific basis, Hillside, NJ, 1981, Enslow Publishers.
10. Nachemson AL: Advances in low back pain, Clin Orthop 200:266, 1985.
11. Pettrone FA and Ricciardelli E: Gymnastic injuries: the Virginia experience, Am J Sports Med 15:59, 1987.
12. Radin EL: Role of muscles in protecting athletes from injury, Acta Med Scand Suppl 711:143, 1986.
13. Sady SP, Wortman M, and Blanke D: Flexibility training: ballistic, static or proprioceptive neuromuscular facilitation? Arch Phys Med Rehabil 63:261, 1982.
14. Soderberg GL: Kinesiology: application to pathological motion, Baltimore, 1986, Williams & Wilkins.

15. Steele VA and White JA: Injury prediction in female gymnasts, Br J Sports Med 20:31, 1986.
16. Tipton CM et al: Influence of exercise on strength of medial collateral ligaments of dogs, Am J Physiol 218:894, 1970.
17. Tipton CM et al: The influence of physical activity on ligaments and tendons, Med Sci Sports Exerc 7:165, 1975.
18. Warren CG, Lehman JF, and Koblanski JN: Heat and stretch procedures: an evaluation using rat tail tendon, Arch Phys Med Rehabil 57:122, 1976.
19. Warwick R and Williams PL: Gray's anatomy, ed 36, Philadelphia, 1980, WB Saunders Co.
20. Weightman BO and Kempson GE: Load carriage. In Freeman MAR, ed: Adult articular cartilage, London, 1979, Pitman.
21. Williford HN et al: Evaluation of warm-up for improvement in flexibility, Am J Sports Med 14:316, 1986.
22. Woo SLY et al: Long term exercise effects on the biomechanical and structural properties of swine tendons, Transactions of the twenty-fifth annual meeting of the Orthopaedic Research Society 4:3, 1979.

ANNOTATED READINGS

Gray H: The articulations. In Gray H, ed: Anatomy: descriptive and surgical, New York, 1977, Crown-Bounty Books.

The classical anatomical reference book commonly cited in anatomy and kinesiology texts.

Mow VC, Roth V, and Armstrong CG: Biomechanics of joint cartilage. In Frankel VH and Nordin M, eds: Basic biomechanics of the skeletal system, Philadelphia, 1980, Lea & Febiger.

Provides in depth information from the research literature on the structure and function of joint cartilage. Extensive reference list included.

Nordin M and Frankel VH: Biomechanics of collagenous tissues. In Frankel VH and Nordin M, eds: Basic biomechanics of the skeletal system, Philadelphia, 1980, Lea & Febiger.

Describes the material properties of tendons and ligaments, including excellent graphs and illustrations.

Soderberg GL: Articular mechanics and function. In Soderberg GL, ed: Kinesiology: application to pathological motion, Baltimore, 1986, Williams & Wilkins.

Provides an overview of joint function from a mechanical perspective.

6 MOVEMENT

The Biomechanics of the Upper Extremity

After reading this chapter, the student will be able to:

Explain how anatomical structure affects the movement capabilities of upper extremity articulations.

Identify the factors that determine the relative mobility and stability present at upper extremity articulations.

Identify the muscles that are active during specific upper extremity movements.

Describe the biomechanical contributors to common injuries of the upper extremity.

Pitching a ball requires the coordination of the muscles of the entire upper extremity.

The capabilities of the upper extremity are varied and impressive. With the same basic anatomical structure of the arm, forearm, hand, and fingers, major league baseball pitchers hurl fast balls at 40 m/s, swimmers cross the English Channel, gymnasts perform the iron cross, travelers carry suitcases, seamstresses thread needles, and students type on computer keyboards. In this chapter we will review the anatomical structures enabling these different types of movement and examine the ways in which the muscles cooperate to achieve the diversity of movement of which the upper extremity is capable.

STRUCTURE OF THE SHOULDER

Articulations

The shoulder is the most complex joint in the human body, largely because it includes four separate articulations—the glenohumeral joint, the sternoclavicular joint, the acromioclavicular joint, and the coracoclavicular joint. The glenohumeral joint is the articulation between the head of the humerus and the glenoid fossa of the scapula, which is the ball and socket joint typically considered as *the* major shoulder joint. The sternoclavicular and acromioclavicular joints provide mobility for the clavicle and the scapula—the bones of the shoulder girdle.

■ The glenohumeral joint is considered to be *the* shoulder joint.

■ The clavicle and the scapula make up the shoulder girdle.

Sternoclavicular Joint

The proximal end of the clavicle articulates with the clavicular notch of the manubrium of the sternum and with the cartilage of the first rib to form the **sternoclavicular joint.** This joint provides the major axis of rotation for movements of the clavicle and scapula (Figure 6-1). The sternoclavicular joint is a modified ball and socket, with frontal and transverse plane motion freely permitted and some forward and backward sagittal plane rotation allowed. Rotation occurs at the sternoclavicular joint during motions such as shrugging the shoulders, elevating the arms above the head, and swimming.

sternoclavicular joint
a modified ball and socket joint between the proximal clavicle and the manubrium of the sternum

■ Most of the motion of the shoulder girdle takes place at the sternoclavicular joint.

Figure 6-1
The sternoclavicular joint.

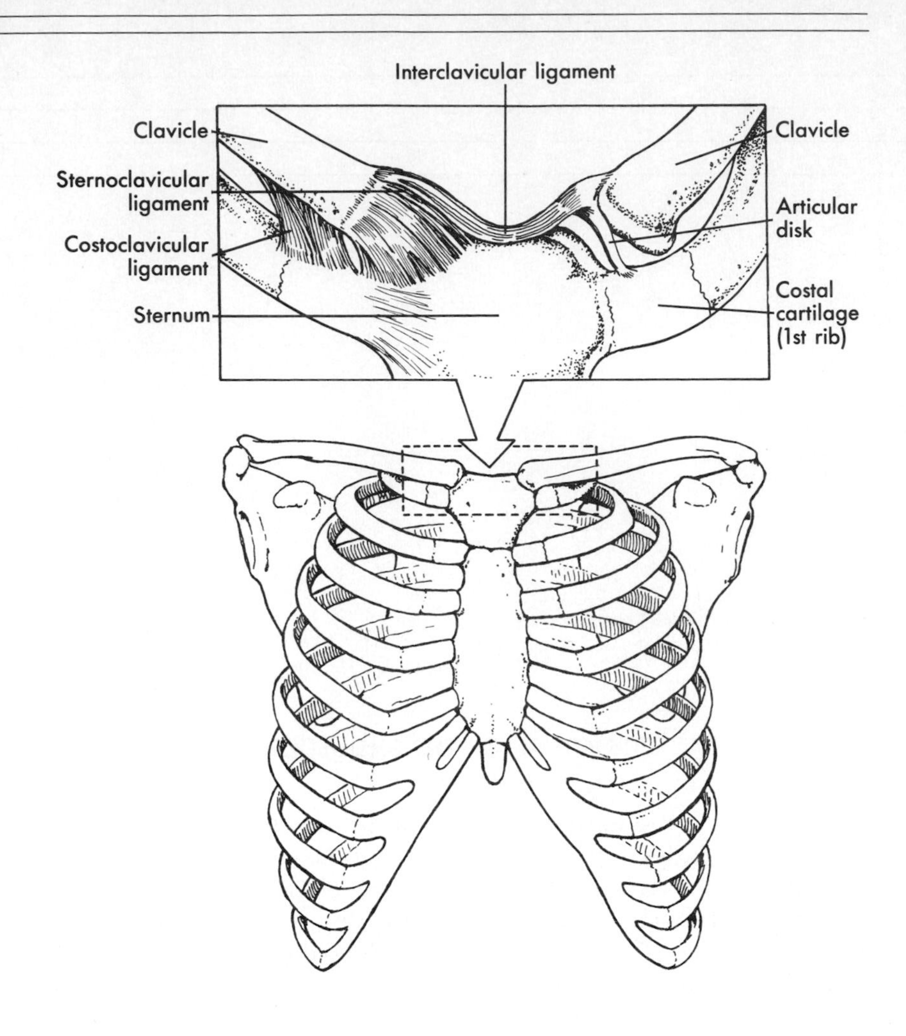

Acromioclavicular Joint

acromioclavicular joint
an irregular joint between the acromion process of the scapula and the distal clavicle

The articulation of the acromion process of the scapula with the distal end of the clavicle is known as the **acromioclavicular joint.** It is classified as an irregular diarthrodial joint, although the joint's structure allows limited motion in all three planes. Rotation occurs at the acromioclavicular joint during arm elevation.

Coracoclavicular Joint

coracoclavicular joint
a syndesmosis joint with the coracoid process of the scapula bound to the inferior clavicle by the coracoclavicular ligament

The **coracoclavicular joint** is a syndesmosis, formed where the coracoid process of the scapula and the inferior surface of the clavicle are bound together by the coracoclavicular ligament. This joint permits little movement. The coracoclavicular and acromioclavicular joints are shown in Figure 6-2.

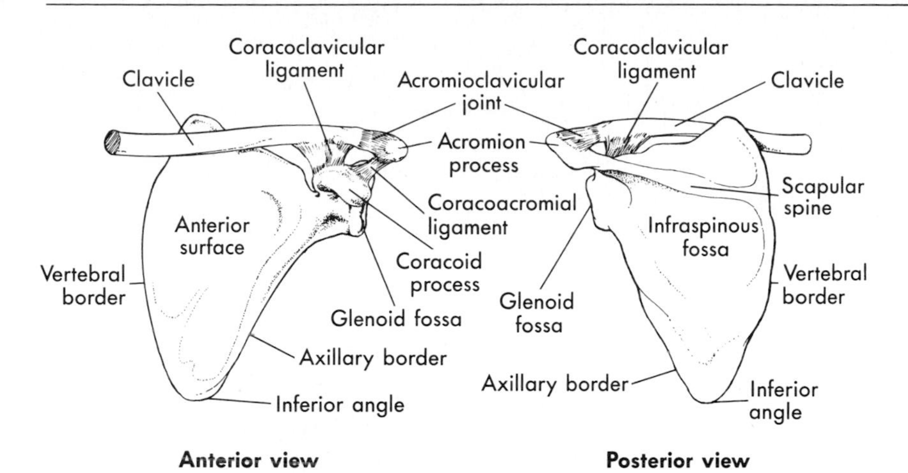

Figure 6-2
The acromioclavicular and coracoclavicular joints.

Glenohumeral Joint

The **glenohumeral joint** is the most freely moving joint in the human body, enabling flexion, extension, hyperextension, abduction, adduction, horizontal abduction and adduction, and medial and lateral rotation of the humerus (Figure 6-3). The almost hemispherical head of the humerus has three to four times the amount of surface area as the shallow glenoid fossa of the scapula with which it articulates. The glenoid fossa is also less curved than the surface of the humeral head, enabling the humerus to move linearly across the surface of the glenoid fossa in addition to its extensive rotational capability (12). At the perimeter of the glenoid fossa is a lip or **labrum** composed of part of the joint capsule, the tendon of the long head of the biceps brachii, and the glenohumeral ligaments. The labrum deepens the fossa and adds stability to the joint (22). The capsule surrounding the glenohumeral joint is shown in Figure 6-4.

Several ligaments merge with the glenohumeral joint capsule, including the superior, middle, and inferior glenohumeral ligaments on the anterior side of the joint and the coracohumeral ligament on the superior side. The tendons of four muscles—subscapularis, supraspinatus, infraspinatus, and teres minor—also join the joint capsule. These are known as the **rotator cuff** muscles because they contribute to rotation of the humerus and their tendons form a collagenous cuff around the glenohumeral joint. Tension in the rotator cuff muscles pulls the head of the humerus toward the glenoid fossa, contributing significantly to the joint's minimal stability.

glenohumeral joint
a ball and socket joint in which the head of the humerus articulates with the glenoid fossa of the scapula

■ The extreme mobility of the glenohumeral joint is achieved at the expense of joint stability.

glenoid labrum
the ring of soft tissue located on the periphery of the glenoid fossa that adds stability to the glenohumeral joint

rotator cuff
a structure consisting of subscapularis, supraspinatus, infraspinatus, and teres minor, which have tendinous attachments to the capsule of the glenohumeral joint

Figure 6-3
The glenohumeral joint.

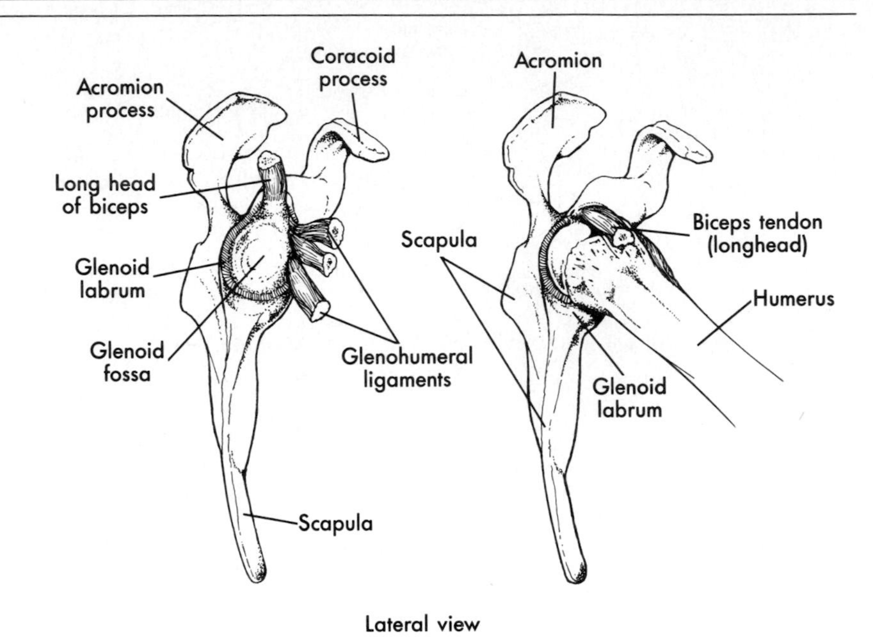

Figure 6-4
The capsule surrounding the glenohumeral joint contributes to joint stability.

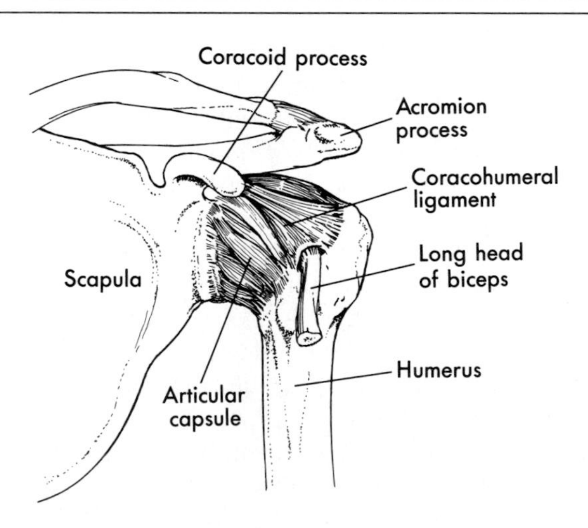

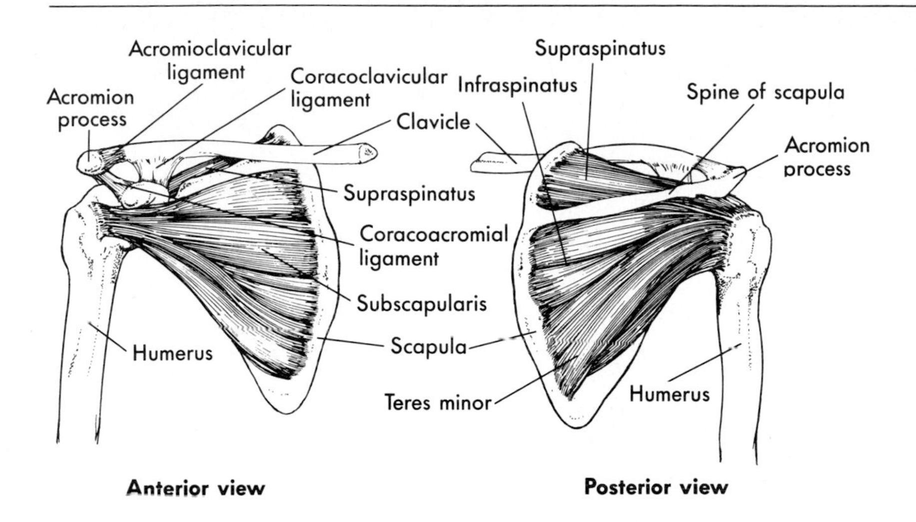

Figure 6-5
The four muscles of the rotator cuff and the associated subacromial bursa.

Bursae

Several fibrous sacs that secrete synovial fluid internally similar to a joint capsule are located in the shoulder region. These sacs, known as **bursae,** cushion and reduce friction between layers of collagenous tissues. The subacromial bursae are shown in Figure 6-5.

bursae
sacs secreting synovial fluid to lessen friction between soft tissues around joints

MOVEMENTS OF THE SHOULDER COMPLEX

Motion of the Humerus

Although a limited amount of glenohumeral motion may occur while the other shoulder articulations remain stabilized, movement of the humerus more commonly involves some movement at all three shoulder joints. As the arm is elevated beyond 30 degrees of abduction or 60 degrees of flexion, the scapula rotates approximately 1 degree for every 2 degrees of movement of the humerus (1). During the first 90 degrees of arm elevation (in sagittal, frontal, or diagonal planes), the clavicle is also elevated through approximately 35 to 45 degrees of motion at the sternoclavicular joint (16). Clavicular elevation and scapular rotation are depicted in Figure 6-6. Rotation at the acromioclavicular joint occurs during the first 30 degrees of humeral elevation and again as the arm is moved from 135 degrees to maximum elevation (10).

Figure 6-6
Abduction of the arm is accompanied by elevation of the clavicle and rotation of the scapula.

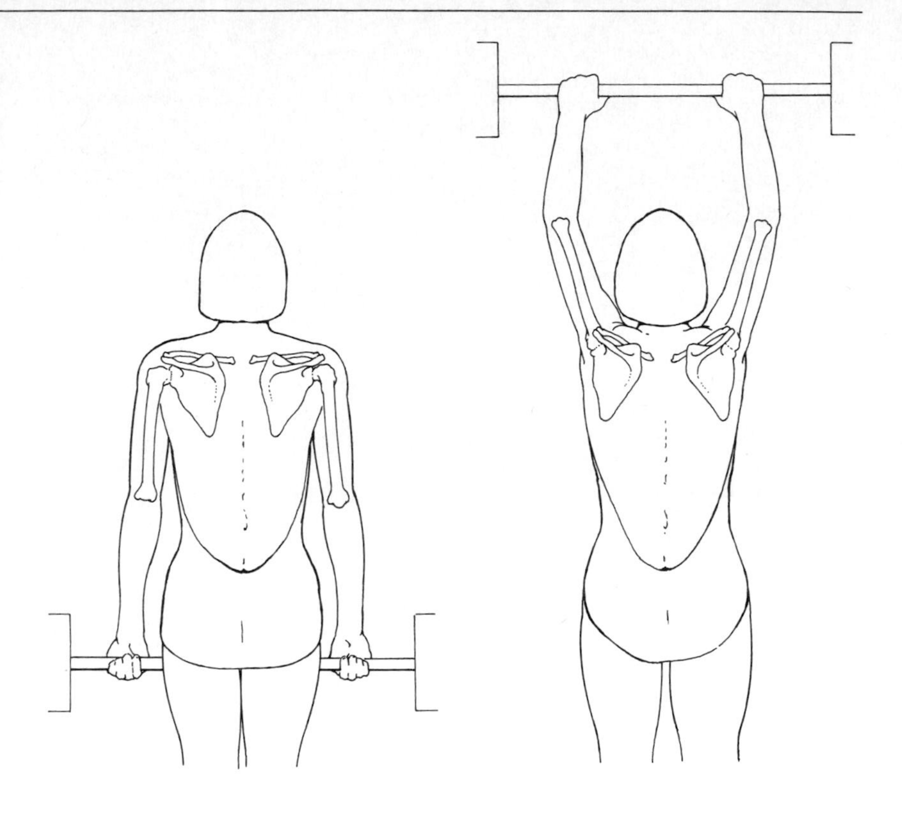

Muscles of the Scapula

The muscles that attach to the scapula are the levator scapula, rhomboids, serratus anterior, pectoralis minor, subclavius, and the four parts of the trapezius. Figures 6-7 and 6-8 show the directions in which each of these muscles exert force on the scapula when contracting. Scapular muscles have two general functions: First, they stabilize the scapula so that it forms a rigid base for muscles of the shoulder during the development of tension. For example, when a person carries a suitcase, the levator scapula, trapezius, and rhomboids stabilize the shoulder against the added weight. Second, scapular muscles facilitate movements of the upper extremity by positioning the glenohumeral joint appropriately. For example, during an overhand throw the rhomboids contract to move the entire shoulder posteriorly as the arm and hand move posteriorly during the preparatory phase. As the arm and hand move anteriorly to deliver the throw, tension in the rhomboids subsides to permit forward movement of the shoulder, facilitating outward rotation of the humerus.

■ The major functions of the scapular muscles are to stabilize and move and position the scapula to facilitate movement at the glenohumeral joint.

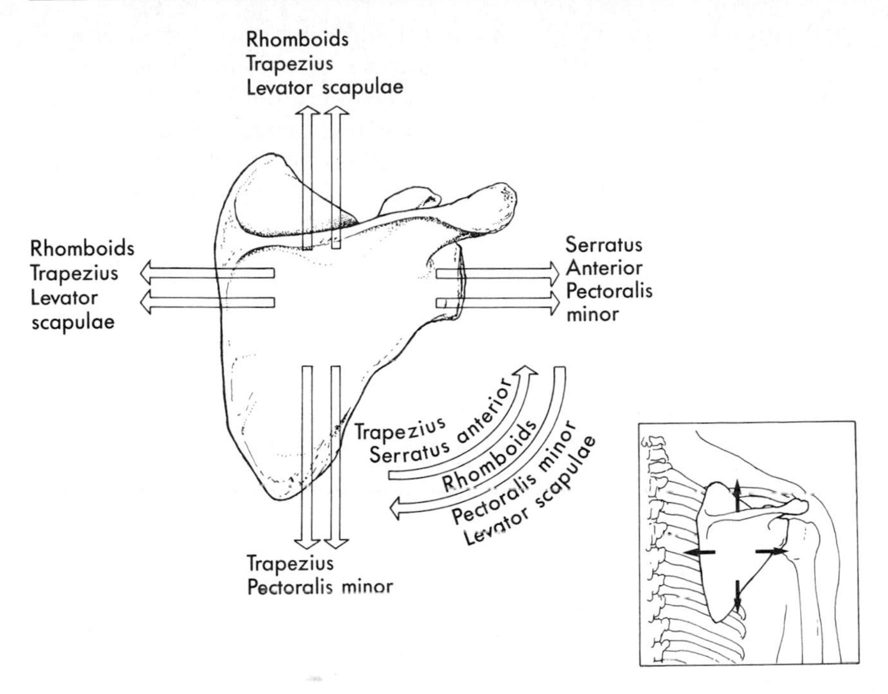

Figure 6-7

Actions of the scapular muscles.

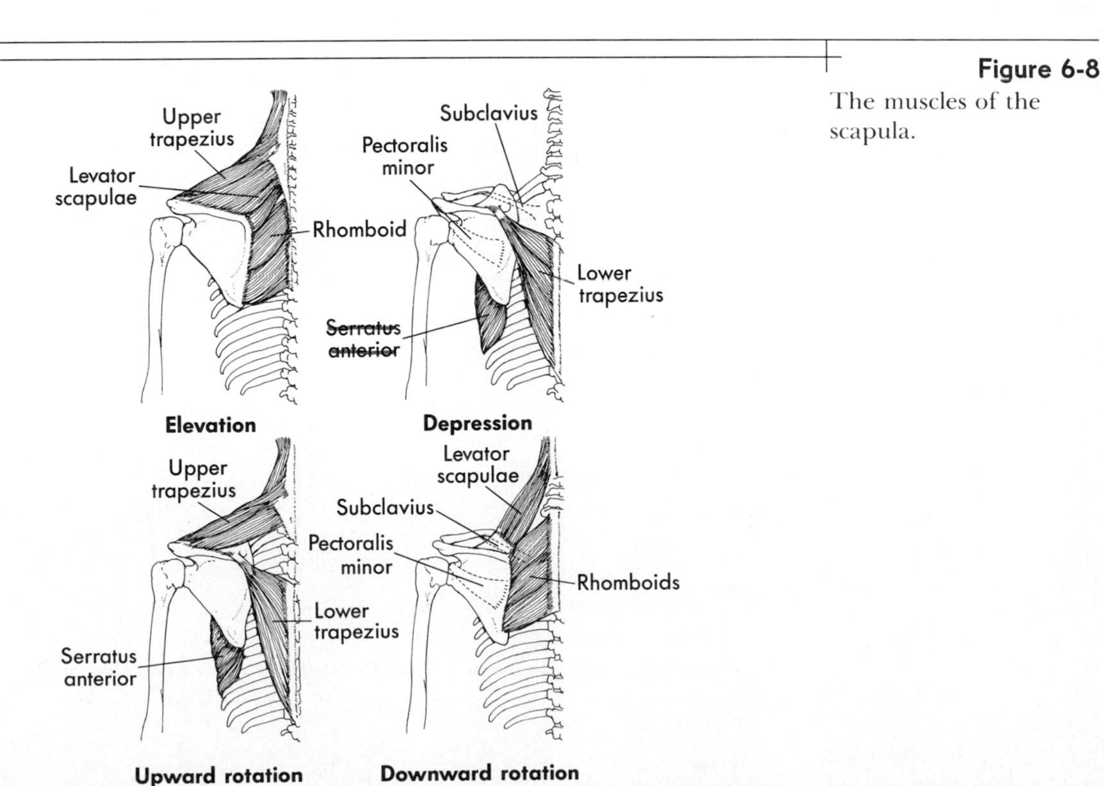

Figure 6-8

The muscles of the scapula.

Muscles of the Glenohumeral Joint

■ The development of tension in one shoulder muscle must frequently be accompanied by the development of tension in an antagonist to prevent dislocation of the humeral head.

Many muscles cross the glenohumeral joint. Because of their attachment sites and lines of pull, some muscles contribute to more than one action of the humerus. A further complication is that the action produced by the development of tension in a muscle may change with the orientation of the humerus because of the shoulder's large range of motion. With the basic instability of the structure of the glenohumeral joint, a significant portion of the joint's stability is derived from tension in the muscles and tendons crossing the joint. However, when one of these muscles develops tension, tension development in an antagonist may be required to prevent dislocation of the joint. A review of the muscles of the shoulder is presented in Table 6-1.

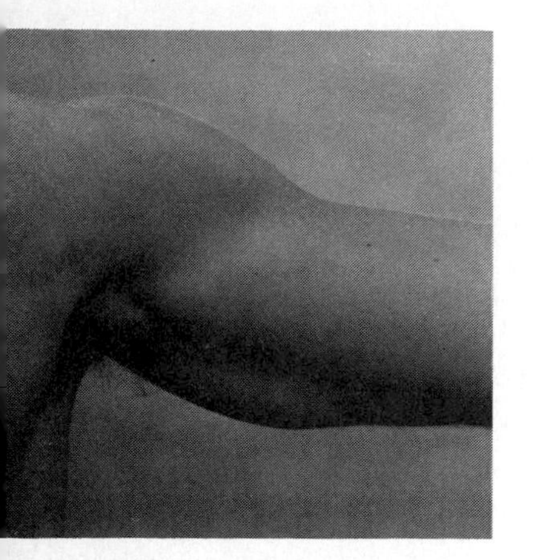

The contraction of muscles anterior to the glenohumeral joint produces humeral flexion.

Flexion at the Glenohumeral Joint

The muscles crossing the glenohumeral joint anteriorly participate in flexion at the shoulder (Figure 6-9). The prime flexors are the anterior deltoid and the clavicular portion of the pectoralis major. The small coracobrachialis assists with flexion, as does the short head of the biceps brachii. Because the biceps also crosses the elbow joint, it is more effective in its actions at the shoulder when the elbow is in full extension.

Extension at the Glenohumeral Joint

When shoulder extension is unresisted, gravitational force is the primary mover, with eccentric contraction of the flexor muscles controlling or braking the movement. When resistance is present, contraction of the muscles posterior to the glenohumeral joint, particularly the sternocostal pectoralis, latissimus dorsi, and teres major, extend the humerus. The posterior deltoid assists in extension, especially when the humerus is externally rotated. The long head of the tri-

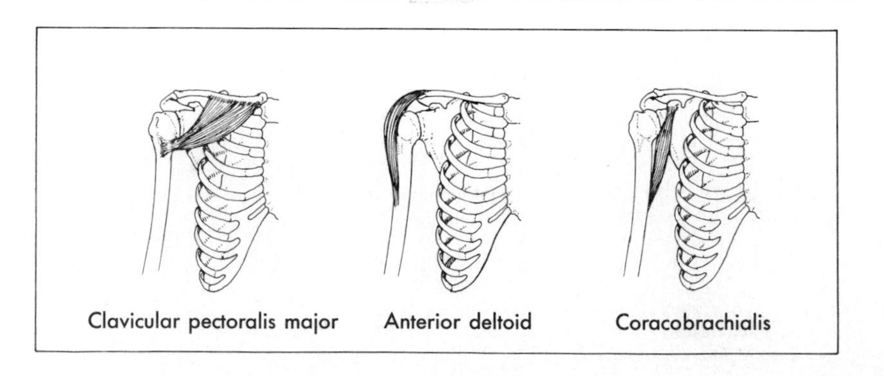

Figure 6-9
The major flexor muscles of the shoulder.

Table 6-1
MUSCLES OF THE SHOULDER

MUSCLE	PROXIMAL ATTACHMENT	DISTAL ATTACHMENT	PRIMARY ACTIONS
Deltoid (anterior)	Outer third of clavicle	Deltoid tuberosity of humerus	Flexion, horizontal adduction
Deltoid (middle)	Top of acromion	Deltoid tuberosity of humerus	Abduction, horizontal abduction
Deltoid (posterior)	Scapular spine	Deltoid tuberosity of humerus	Extension, horizontal abduction
Pectoralis major (clavicular)	Medial two thirds of clavicle	Lateral aspect of humerus just below head	Flexion, horizontal adduction
Pectoralis major (sternal)	Anterior sternum and cartilage of first six ribs	Lateral aspect of humerus just below head	Extension, adduction, horizontal adduction
Supraspinatus	Supraspinous fossa	Greater tuberosity of humerus	Abduction
Coracobrachialis	Coracoid process of scapula	Medial anterior humerus	Horizontal adduction
Latissimus dorsi	Lower six thoracic and all lumbar vertebrae, posterior sacrum, iliac crest, lower three ribs	Anterior humerus	Extension, adduction
Teres major	Lower, lateral, dorsal scapula	Anterior humerus	Extension, adduction, inward rotation
Infraspinatus	Infraspinous fossa	Greater tubercle of humerus	Outward rotation, horizontal abduction
Teres minor	Posterior, lateral border of scapula	Greater tubercle and adjacent shaft of humerus	Outward rotation, horizontal abduction
Subscapularis	Entire anterior surface of scapula	Lesser tubercle of humerus	Inward rotation
Biceps brachii (long head)	Upper rim of glenoid fossa	Tuberosity of radius	Assists with abduction
Biceps brachii (short head)	Coracoid process of scapula	Tuberosity of radius	Assists with flexion, adduction, inward rotation, horizontal adduction
Triceps brachii (long head)	Just inferior to glenoid fossa	Olecranon process of ulna	Assists with extension, adduction

Figure 6-10
The major extensor muscles of the shoulder.

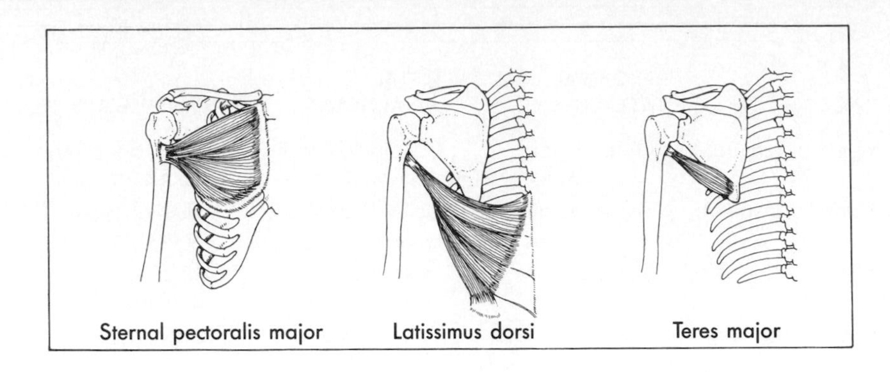

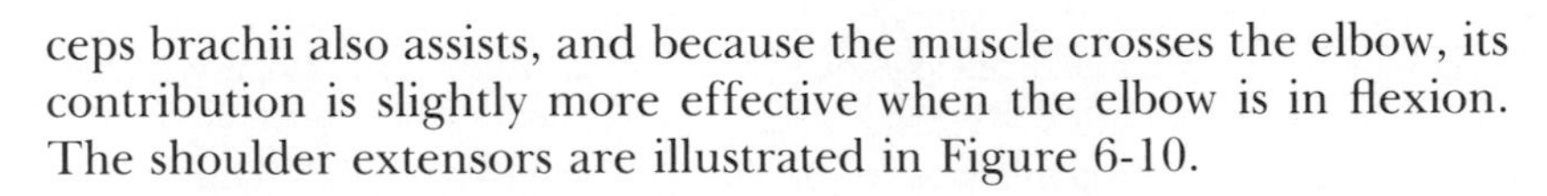

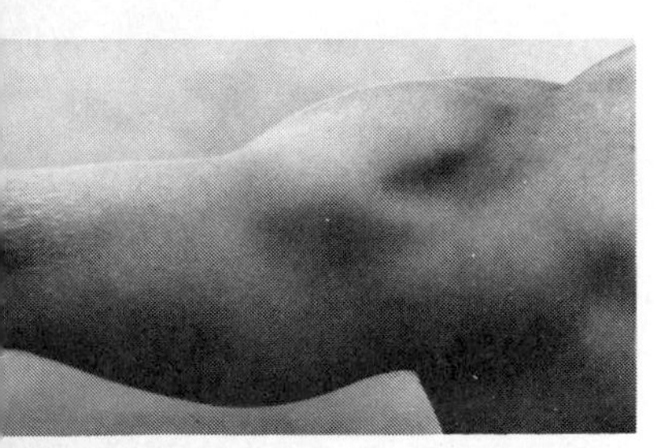

Muscles that cross the posterior aspect of the glenohumeral joint are responsible for resisted extension and hyperextension of the humerus.

ceps brachii also assists, and because the muscle crosses the elbow, its contribution is slightly more effective when the elbow is in flexion. The shoulder extensors are illustrated in Figure 6-10.

Abduction at the Glenohumeral Joint

The middle deltoid and supraspintaus are the major abductors of the humerus. Both muscles cross the shoulder superior to the glenohumeral joint (Figure 6-11). The supraspinatus, which is active through approximately the first 110 degrees of motion, initiates abduction (5). During the contribution of the middle deltoid (occurring from approximately 90 to 180 degrees of abduction) the infraspinatus, subscapularis, and teres minor neutralize the superiorly dislocating component of force produced by the middle deltoid.

Adduction at the Glenohumeral Joint

As with extension at the shoulder, adduction in the absence of resistance results from gravitational force, with the abductors controlling the speed of motion. With resistance added, the primary adductors are the latissimus dorsi, teres major, and the sternocostal pectoralis, which are located on the inferior side of the joint (Figure 6-12). The short head of the biceps and the long head of the triceps contribute minor assistance, and when the arm is elevated above 90 degrees, the coracobrachialis and subscapularis also assist.

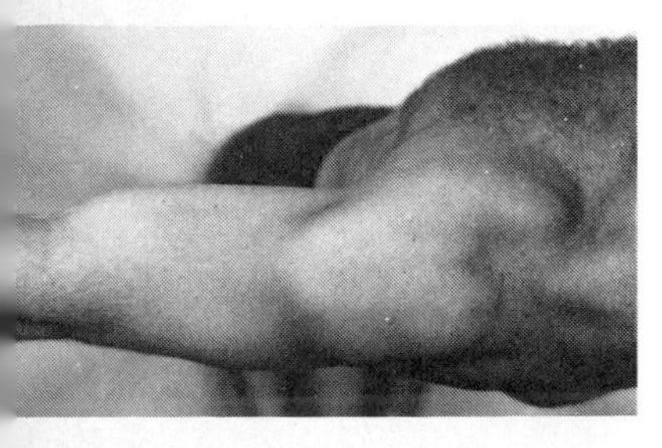

Because of its size and location, the middle deltoid is a powerful abductor of the humerus.

Medial and Lateral Rotation of the Humerus

Medial or inward rotation of the humerus results primarily from the action of the subscapularis and teres major, both attaching to the anterior side of the humerus. Both portions of the pectoralis major, the anterior deltoid, latissimus dorsi, and the short head of the biceps assist. Muscles attaching to the posterior aspect of the humerus, particularly infraspinatus and teres minor, produce lateral or outward rotation, with some assistance from the posterior deltoid.

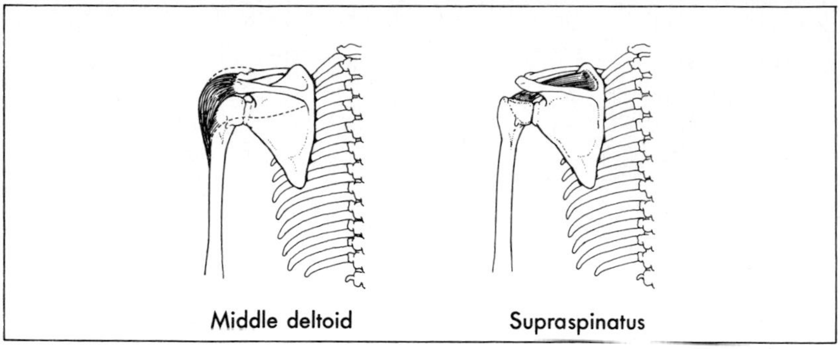

Middle deltoid

Supraspinatus

Figure 6-11
The major abductor muscles of the shoulder.

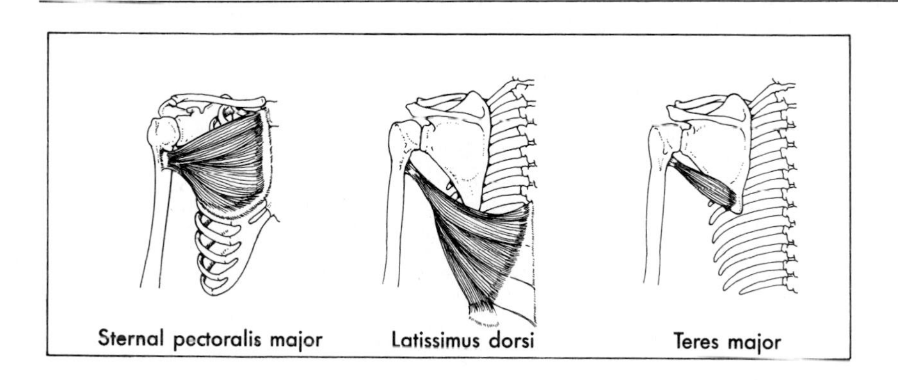

Figure 6-12
The major adductor muscles of the shoulder.

Horizontal Adduction and Abduction at the Glenohumeral Joint

The muscles anterior to the joint, including both heads of pectoralis major, the anterior deltoid, and coracobrachialis, produce horizontal adduction, with the short head of the biceps brachii assisting. Muscles posterior to the joint axis affect horizontal abduction. The major horizontal abductors are the middle and posterior portions of the deltoid, infraspinatus, and teres minor, with assistance provided by teres major and the latissimus dorsi. The major horizontal adductors and abductors are shown in Figures 6-13 and 6-14.

COMMON INJURIES OF THE SHOULDER

The shoulder is involved in 8% to 13% of all sport-related injuries (8). Although the weight of the arm is relatively low, the length of the arm creates large torques that must be countered by the shoulder muscles. The force acting on the articulating surfaces of the glen-

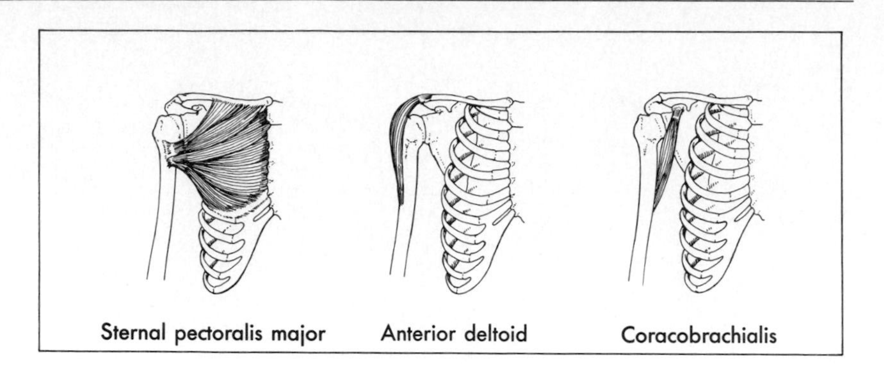

Figure 6-13
The major horizontaladductors of the shoulder.

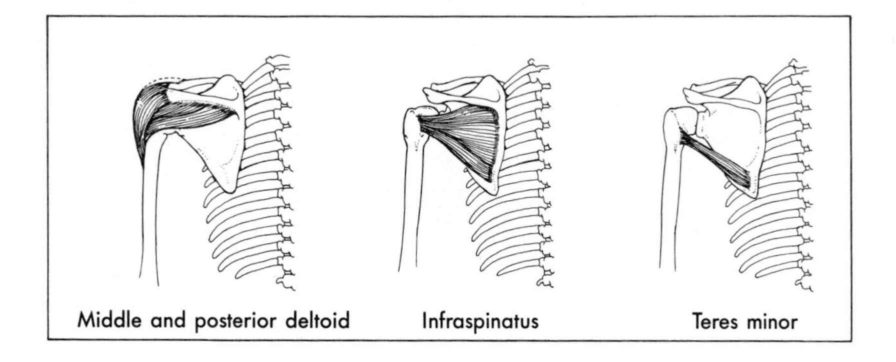

Figure 6-14
The major horizontal abductors of the shoulder.

ohumeral joint when the arm is abducted to 90 degrees reaches 90% of body weight (21). Consequently, the shoulder is susceptible to both traumatic and overuse types of injuries.

Dislocations

The loose structure of the glenohumeral joint enables extreme mobility but provides little stability, and dislocations may occur in anterior, posterior, and inferior directions. The strong coracohumeral ligament usually prevents displacement in the superior direction. Glenohumeral dislocations typically occur when the humerus is abducted and externally rotated, with anterior-inferior dislocations more common than those in other directions (20). Dislocation may result from an individual sustaining a large external force during an accident, such as in cycling, or during participation in a contact sport such as wrestling or football. Unfortunately, once the joint has been dislocated, the stretching of the surrounding collagenous tissues predisposes it to subsequent dislocations. Glenohumeral capsular laxity may also be present due to genetic factors. Individuals with this con-

Figure 6-15
Sustaining the force of a fall on an extended arm can result in injury to the bones and/or joints of the shoulder complex.

dition should strengthen their shoulder muscles before athletic participation (4).

Dislocations or separations of the acromioclavicular joint are also common among wrestlers and football players. When a rigidly outstretched arm sustains the force of a full-body fall, either acromioclavicular separation or fracture of the clavicle is likely to result (Figure 6-15).

Rotator Cuff Damage

A common injury among workers and athletes who engage in forceful movements involving abduction or flexion along with medial rotation is shoulder impingement syndrome or **rotator cuff impingement syndrome.** This condition involves pain and tenderness in the shoulder region and is exacerbated by rotatory movements of the humerus. Activities that may promote the development of shoulder impingement syndrome include throwing (particularly an implement like the javelin), serving in tennis, and swimming (especially the freestyle, backstroke, and butterfly). Among competitive swimmers, the syndrome is known as *swimmer's shoulder.* Reports indicate shoulder pain complaints in up to 50% of competitive swimmers (19).

rotator cuff impingement syndrome
pain and tenderness in the shoulder region probably caused by tendonitis of one or more of the rotator cuff muscles

Physicians have proposed two theories regarding the biomechanical cause of most rotator cuff problems (14). The impingement theory suggests that a genetic factor results in the formation of too narrow a space between the acromion process of the scapula and the head of the humerus. In this situation the rotator cuff and associated bursa are pinched between the acromion, the acromioclavicular ligament, and the humeral head each time the arm is elevated, with the resulting friction causing irritation and wear. An alternative theory proposes that the major factor is inflammation of the supraspinatus tendon caused by repeated overstretching of the muscle-tendon unit. When the rotator cuff tendons become stretched and weakened, they cannot perform their normal function of holding the humeral head in the glenoid fossa. Consequently, the deltoid muscles pull the humeral head up too high during abduction, resulting in impingement and subsequent wear and tear on the rotator cuff.

Another theory has been suggested regarding rotator cuff damage among swimmers. Research has shown that during the recovery phase of swimming, the serratus anterior rotates the scapula so that the supraspinatus, infraspinatus, and middle deltoid may freely abduct the humerus (15). The serratus develops nearly maximum tension to accomplish this task. It has been hypothesized that if the serratus becomes fatigued, the scapula may not be rotated sufficiently to abduct the humerus freely, and impingement may develop (15).

Rotational Injuries

Tears of the labrum, the rotator cuff muscles, and the biceps brachii tendon are among the injuries that may result from repeated, forceful rotation at the shoulder. Throwing, serving in tennis, and spiking in volleyball are examples of forceful rotational movements. If the attaching muscles do not sufficiently stabilize the humerus, it can articulate with the glenoid labrum rather than with the glenoid fossa, contributing to wear on the labrum. Most tears are located in the anterior-superior region of the labrum. Tears of the rotator cuff, primarily of the supraspinatus, have been attributed to the extreme tension requirements placed on the muscle group during the deceleration phase of a vigorous rotational activity. Tears of the biceps brachii tendon at the site of its attachment to the glenoid fossa may result from the forceful development of tension in the biceps when it negatively accelerates the rate of elbow extension during throwing (12).

Serving and hitting in volleyball may result in overuse injuries of the shoulder.

Other pathologies of the shoulder attributed to throwing movements are calcifications of the soft tissues of the joint and degenerative changes in the articular surfaces (17). Bursitis, the inflammation of one or more bursae, is another overuse syndrome, generally caused by friction within the bursa (18).

Subscapular Neuropathy

A shoulder injury that sometimes occurs among competitive volleyball players is subscapular neuropathy. This condition involves denervation of the infraspinatus, with accompanying loss of strength during external rotation of the humerus. It has been attributed to the repeated stretching of the nerve during the serving motion (3).

STRUCTURE OF THE ELBOW

The elbow region includes three articulations: the humeroulnar, humeroradial, and proximal radioulnar joints. All are enclosed in the same joint capsule, which is reinforced by the anterior and posterior radial collateral and ulnar collateral ligaments.

humeroulnar joint
a hinge joint in which the humeral trochlea articulates with the trochlear fossa of the ulna

Humeroulnar Joint

The joint considered to be the major elbow joint is the **humeroulnar.** In this joint the ovular trochlea of the humerus articulates with the closely reciprocally shaped trochlear fossa of the ulna. This is a classic hinge joint, permitting flexion, extension, and in some individuals a small amount of hyperextension.

▪ The humeroulnar hinge joint is considered to be *the* elbow joint.

Humeroradial Joint

The **humeroradial joint** is immediately adjacent to the humeroulnar joint and is formed between the spherical capitellum on the lateral side of the distal humerus and the proximal end of the radius. Although the humeroradial articulation is classified as a gliding joint, the immediately adjacent humeroulnar joint restricts motion to the sagittal plane. Medial and lateral views of the humeroulnar and humeroradial joints are shown in Figure 6-16.

humeroradial joint
a gliding joint in which the capitellum of the humerus articulates with the proximal end of the radius

▪ When pronation and supination of the forearm occur, the radius pivots around the ulna.

Proximal Radioulnar Joint

The articulation between the head of the radius and the radial notch of the ulna is the proximal **radioulnar joint.** This is a pivot joint in which the annular ligament binds the radius to the ulna. Forearm pronation and supination occur as the radius rolls around the ulna at this joint (Figure 6-17).

radioulnar joints
the proximal and distal radioulnar joints are pivot joints; the middle radioulnar joint is a syndesmosis

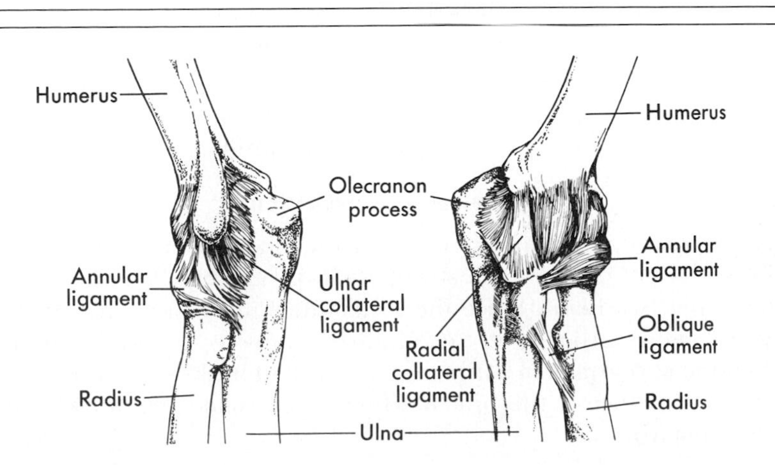

Figure 6-16
The major ligaments of the elbow.

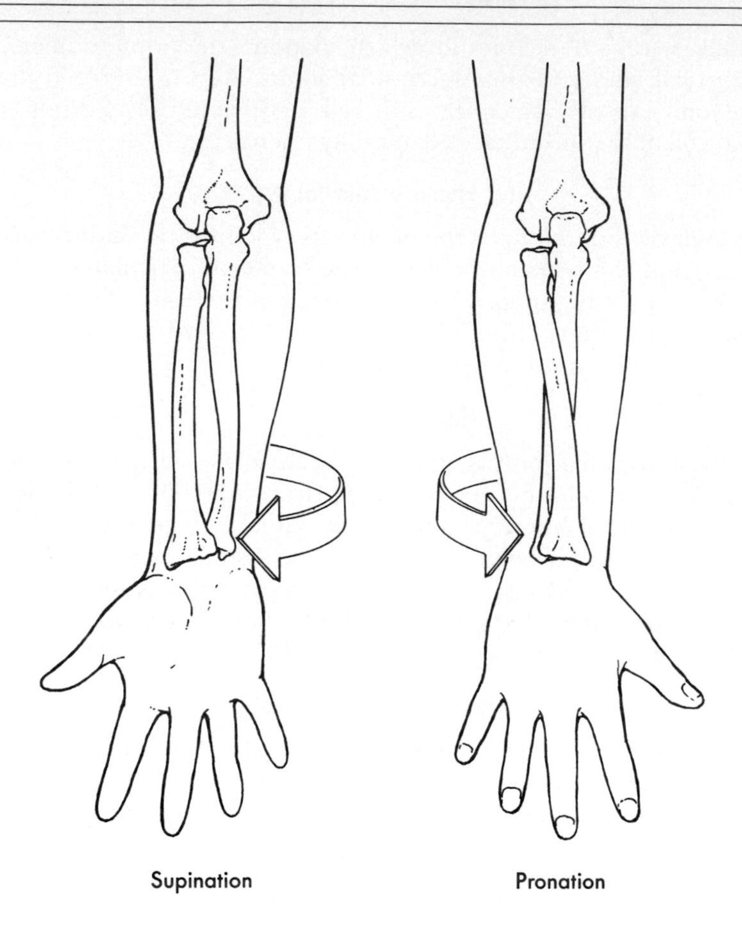

Figure 6-17
Pronation involves the rotation of the radius around the ulna.

MOVEMENTS AT THE ELBOW

Muscles Crossing the Elbow

Numerous muscles cross the elbow, including those that also cross the shoulder or extend into the hands and fingers. The muscles classified as primary movers of the elbow are summarized in Table 6-2.

Flexion and Extension

Muscles crossing the anterior side of the elbow are the elbow flexors (Figure 6-18). The muscle strongest in producing flexion at the elbow is the brachialis. Since the distal attachment of the brachialis is the coronoid process of the ulna, the muscle is equally effective when the forearm is in supination or pronation. Because it is the major forearm flexor, the brachialis has been called the workhorse of the elbow (6).

Table 6-2
MAJOR MUSCLES OF THE ELBOW

MUSCLE	PROXIMAL ATTACHMENT	DISTAL ATTACHMENT	PRIMARY ACTIONS
Biceps brachii (long head)	Superior rim of glenoid fossa	Tuberosity of radius	Flexion
Biceps brachii (short head)	Coracoid process of scapula	Tuberosity of radius	Flexion
Brachioradialis	Upper two-thirds lateral supracondylar ridge of humerus	Styloid process of radius	Flexion
Brachialis	Anterior lower half of humerus	Anterior coronoid process of ulna	Flexion
Pronator teres (humeral head)	Medial epicondyle of humerus	Lateral midpoint of radius	Assists with flexion, pronation
Pronator teres (ulnar head)	Coronoid process of ulna	Lateral midpoint of radius	Assists with flexion, pronation
Pronator quadratus	Lower fourth of anterior ulna	Lower fourth of anterior radius	Pronation
Triceps brachii (long head)	Just inferior to glenoid fossa	Olecranon process of ulna	Extension
Triceps brachii (lateral head)	Upper half of posterior humerus	Olecranon process of ulna	Extension
Triceps brachii (medial head)	Lower two thirds of posterior humerus	Olecranon process of ulna	Extension
Anconeus	Posterior, lateral epicondyle of humerus	Lateral olecranon, posterior ulna	Assists with extension
Supinator	Lateral epicondyle of humerus and adjacent ulna	Lateral upper third of radius	Supination

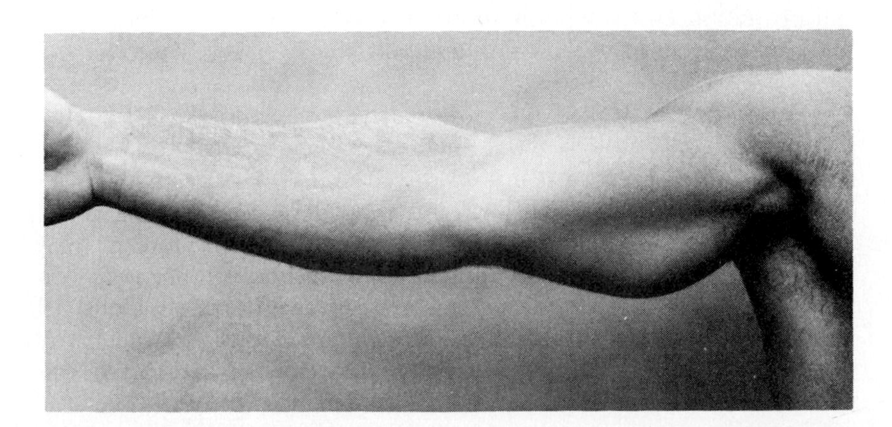

Muscles on the anterior side of the arm produce flexion at the shoulder, elbow, and wrist.

Figure 6-18
The major flexor muscles of the elbow.

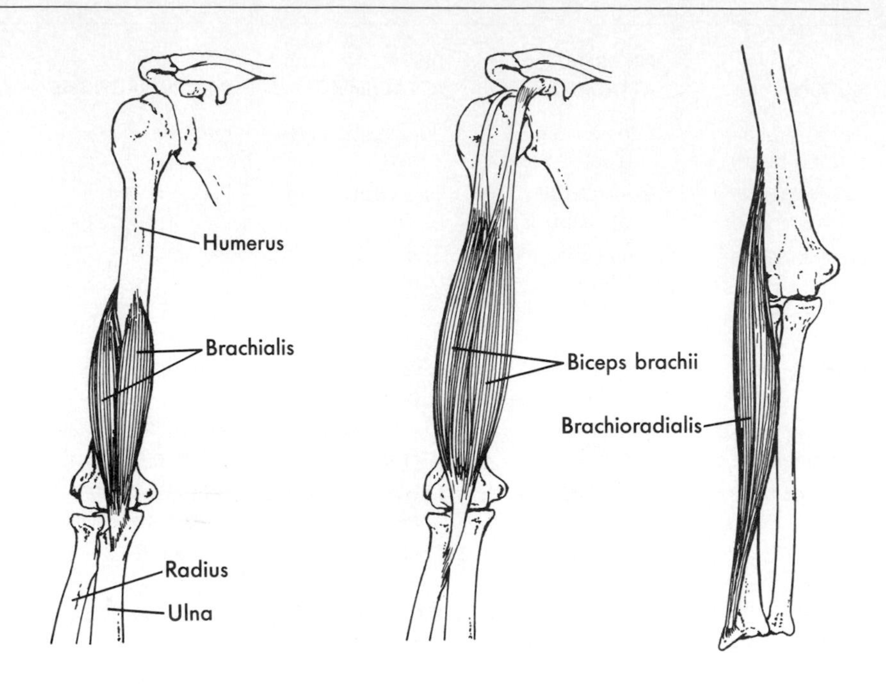

The long and short heads of the biceps brachii join distally to a single tendon attaching to the radial tuberosity. The muscle contributes to flexion at the elbow when the forearm is supinated. When the forearm is pronated, the radius' changed position places the muscle on a slight slack with less resting tension present, so its effectiveness as a flexor at the elbow lessens.

A third contributor to flexion at the elbow is the brachioradialis. Because this muscle attaches distally to the base of the styloid process on the lateral aspect of the radius, it is most effective as an elbow flexor when the forearm is in a neutral position, midway between full pronation and full supination. In this position the muscle is in slight stretch, and the radial attachment is centered in front of the elbow joint.

The major extensor of the elbow is the triceps, which crosses the posterior aspect of the joint. Although the three heads have separate proximal attachments, a common distal tendon attaches to the olecranon process of the ulna. Even though the distal attachment is relatively close to the axis of rotation at the elbow, the size and strength of the muscle make it effective as an elbow extensor. The relatively small anconeus muscle, which courses from the posterior surface of the lateral epicondyle of the humerus to the lateral olecranon and posterior proximal ulna, also assists with extension. The elbow extensors are illustrated in Figure 6-19.

Long head, triceps brachii

Lateral head (medial head not shown)

Ulna

Radius

Humerus

Anconeus

Ulna

Radius

Figure 6-19
The major extensor muscles of the elbow.

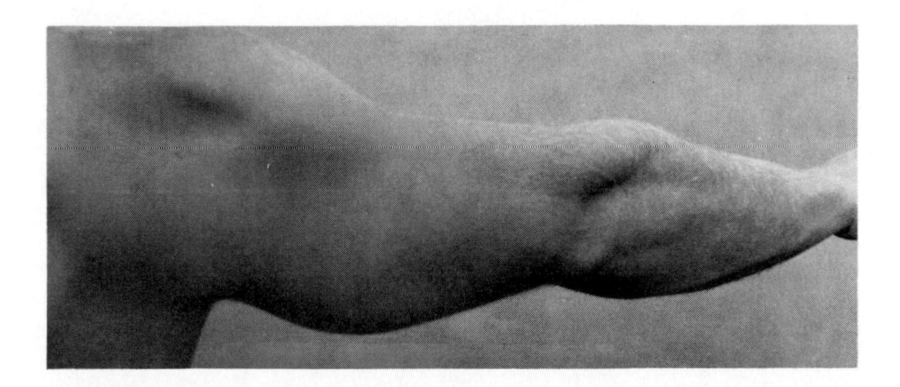

Resisted extension at the shoulder, elbow, and wrist results from action of the posterior arm muscles.

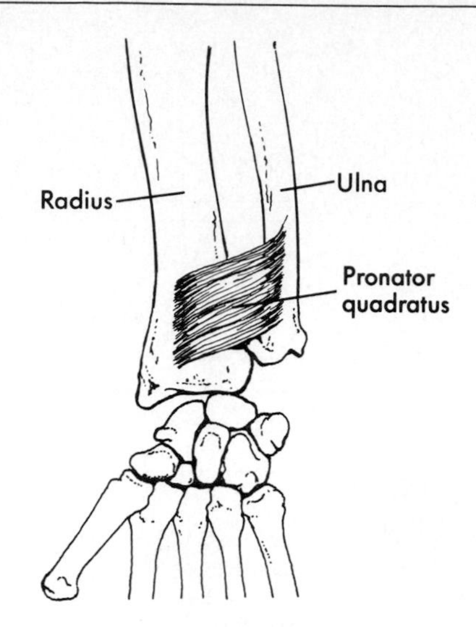

Figure 6-20
The major pronator-muscle is the pronator quadratus.

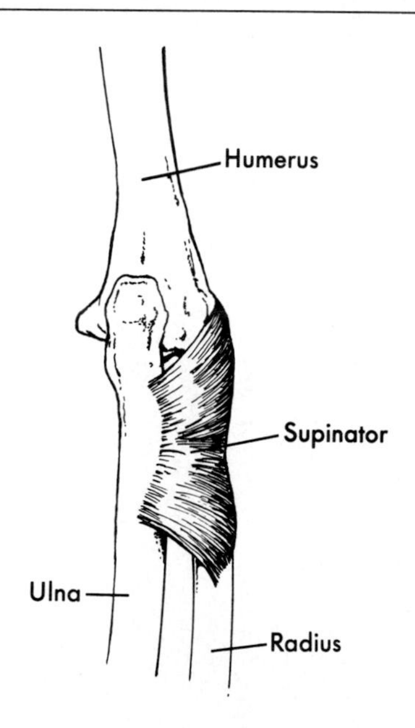

Figure 6-21
The major supinator muscle is the supinator.

Pronation and Supination

Pronation and supination of the forearm involve rotation of the radius around the ulna. There are three radioulnar articulations: the proximal, middle, and distal radioulnar joints. Both the proximal and distal joints are pivot joints, and the middle radioulnar joint is a syndesmosis at which an elastic, interconnecting membrane permits supination and pronation and prevents longitudinal displacement of the bones.

The major pronator is the pronator quadratus, which attaches to the distal ulna and radius (Figure 6-20). When pronation is resisted or rapid, the pronator teres crossing the proximal radioulnar joint assists.

The supinator is the muscle primarily responsible for supination (Figure 6-21). It is attached to the lateral epicondyle of the humerus and to the lateral proximal third of the radius. When the elbow is in flexion, however, tension in the supinator lessens, and the biceps also participates in supination.

COMMON INJURIES OF THE ELBOW

The ranges of motion at the elbow needed for most activities of daily living are 30 to 130 degrees of flexion, approximately 50 degrees of pronation, and 50 degrees of supination (13). When elbow injuries are rehabilitated, strengthening through these ranges of motion is therefore particularly important, although rehabilitation exercises are normally performed throughout the entire range of motion.

Dislocations

Although dislocations of the elbow do not occur as frequently as dislocations of the glenohumeral joint, elbow dislocations occurring in individuals under the age of 30 are most likely to arise during participation in sports (9). Most elbow dislocations are hyperextension injuries. The subsequent stability of a once dislocated elbow is impaired, particularly if the dislocation was accompanied by humeral fracture or rupture of the medial collateral ligament (9).

Overuse Injuries of the Elbow

With the exception of the knee, the elbow is the joint most commonly affected by overuse injuries (7). Stress injuries to the collagenous tissues at the elbow are progressive. The first symptoms are inflammation and swelling, followed by scarring of the soft tissues. If the condition progresses further, calcium deposits accumulate and ossification of the ligaments ensues.

Lateral **epicondylitis** or tennis elbow involves inflammation or microdamage to the tissues on the lateral side of the distal humerus, including the tendinous attachment of the extensor carpi radialis brevis and possibly that of the extensor digitorum. It has been reported that 30% to 40% of tennis players develop the condition, most often when they are between the ages of 35 and 50 (7). The amount of force to which the lateral aspect of the elbow is subjected during tennis play increases with poor technique and improper

epicondylitis
inflammation and sometimes microrupturing of the collagenous tissues on either the lateral or medial side of the distal humerus; believed to be an overuse injury

equipment. For example, hitting off-center shots and using an overstrung racquet increase the amount of force transmitted to the elbow. Lateral epicondylitis also occurs among golfers. Right-handed golfers who grip the club extremely tightly may develop irritation of the left elbow (Figure 6-22) (2). Other activities, such as swimming, fencing, and the act of hammering, can contribute to lateral epicondylitis as well (7).

Medial epicondylitis, which has been called little leaguer's elbow, is the same type of injury to the tissues on the medial aspect of the distal humerus. During pitching the strain imparted to the medial aspect of the elbow during the initial stage when the trunk and shoulder are brought forward ahead of the forearm and hand contributes to development of the condition. Medial epicondyle avulsion fractures have also been attributed to forceful terminal wrist flexion during the follow-through phase of the pitch (7). Throwing curve balls requires added forceful rotation of the arm and hand during

Figure 6-22
Golfer's elbow involves epicondylitis of the nondominant elbow.

the delivery and increases the stresses placed on the elbow and shoulder. For this reason this pitch is commonly contraindicated for Little League pitchers.

STRUCTURE OF THE WRIST

The wrist is composed of radiocarpal and intercarpal articulations (Figure 6-23). The articulation of the radius with three of the proximal carpal bones (the navicular, lunate, and triquetrum) forms the condyloid joint considered to be the major wrist joint. Sagittal and frontal plane motions, as well as circumduction, are permitted at the **radiocarpal joint.** A cartilaginous disc separates the distal head of the ulna from the lunate and triquetral bones and the radius. Although this articular disk is common to both the radiocarpal joint and the distal radioulnar joint, the two articulations have separate joint capsules. The radiocarpal joint capsule is reinforced by the volar radiocarpal, dorsal radiocarpal, radial collateral, and ulnar collateral ligaments. The intercarpal joints are of the gliding type and add only slightly to wrist motion.

■ The radiocarpal joints make up the wrist.

radiocarpal joints
condyloid articulations between the radius and three carpal bones

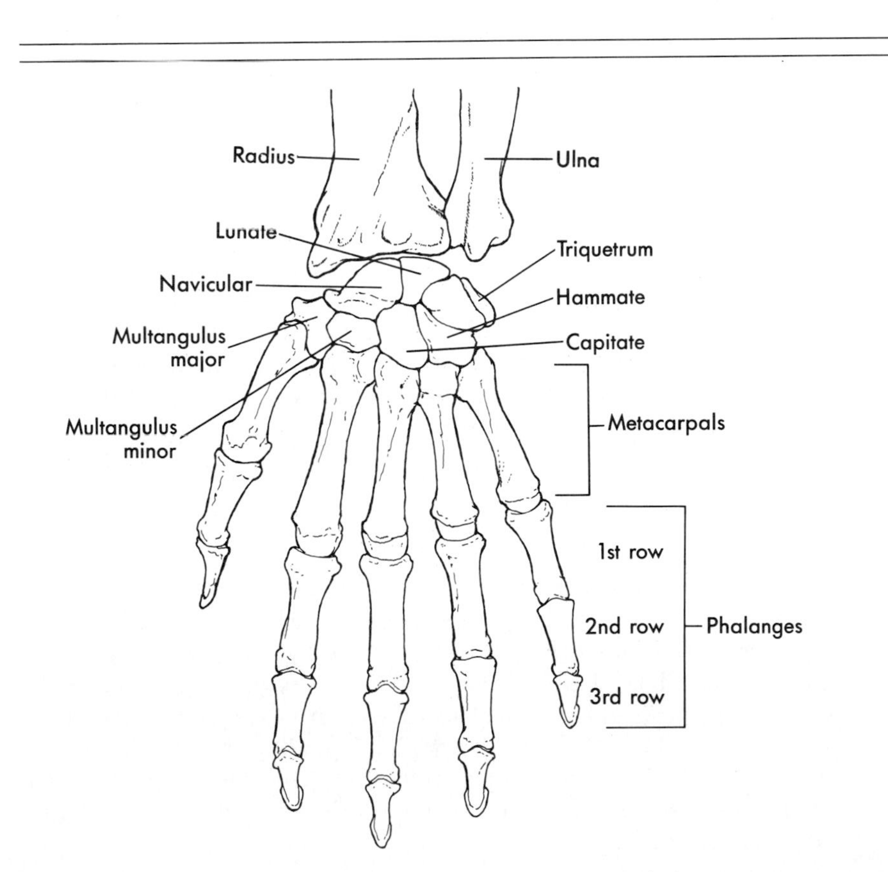

Figure 6-23
The bones of the wrist.

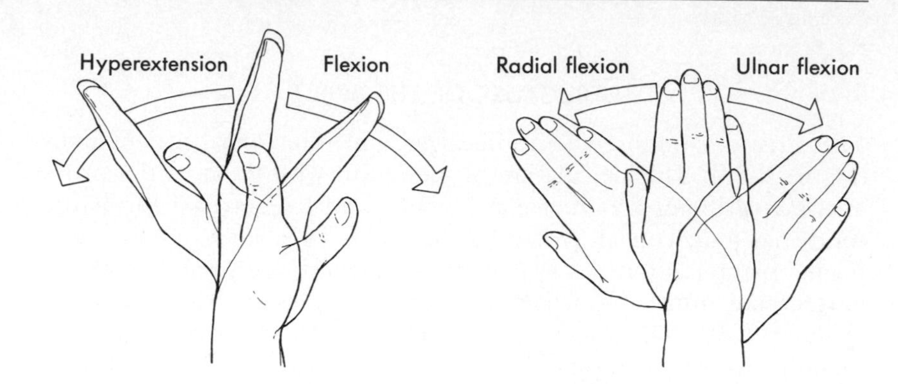

Figure 6-24
Movements occurring at the wrist.

MOVEMENTS OF THE WRIST

Flexion at the wrist consists of the palmar surface of the hand approaching the anterior of the forearm. Extension is the return to anatomical position, and in hyperextension, the dorsal surface of the hand approaches the posterior forearm. Deviation of the hand at the wrist toward the radial side of the arm is radial flexion or abduction. Movement in the opposite direction is designated as ulnar flexion or adduction. Sequential movement of the hand at the wrist through all four directions produces circumduction. The movements of the wrist are illustrated in Figure 6-24.

Flexion

The muscles responsible for flexion at the wrist are the flexor carpi radialis and flexor carpi ulnaris (Figure 6-25). The palmaris longus, a muscle that is often absent in one or both forearms, contributes to flexion when present. All three muscles have proximal attachments on the medial epicondyle of the humerus. The flexor digitorum superficialis and flexor digitorum profundus assist with flexion at the wrist if the fingers are completely extended so that the muscles have sufficient tension to cause movement of the bones when contracting.

Extension

Extension and hyperextension at the wrist result from contraction of the extensor carpi radialis longus, extensor carpi radialis brevis, and extensor carpi ulnaris (Figure 6-26). These muscles originate on the lateral epicondyle of the humerus. The other muscles crossing the wrist on the posterior side may also assist with extension movements, particularly when the metacarpophalangeal and interphalangeal joints are in flexion. Included in this group are the extensor pollicis longus, extensor indicis, extensor digiti minimi, and extensor digitorum (Figure 6-27).

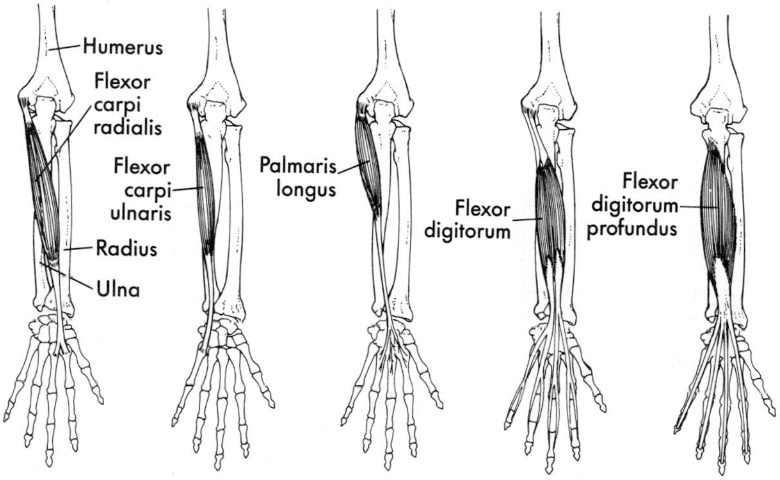

Humerus
Flexor carpi radialis
Flexor carpi ulnaris
Radius
Ulna
Palmaris longus
Flexor digitorum
Flexor digitorum profundus

Figure 6-25
The major flexor muscles of the wrist.

Humerus
Extensor carpi radialis longus
Radius
Ulna
Extensor carpi radialis
Extensor carpi ulnaris

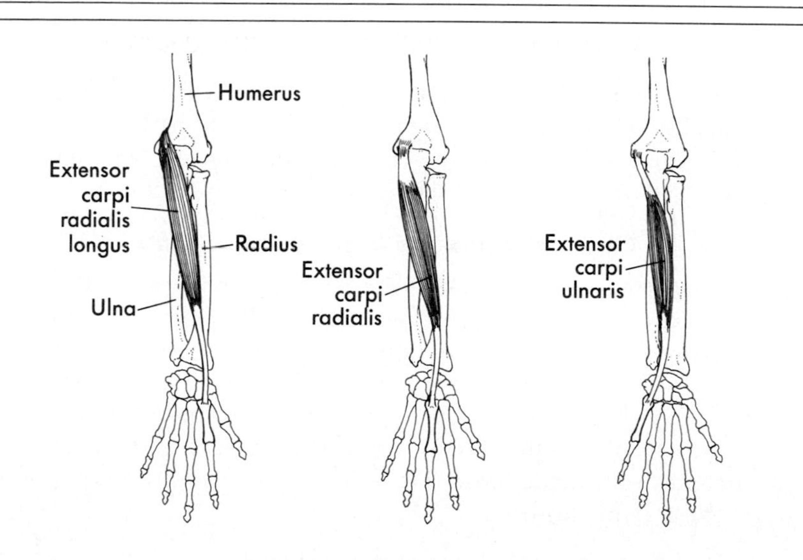

Figure 6-26
The major extensor muscles of the wrist.

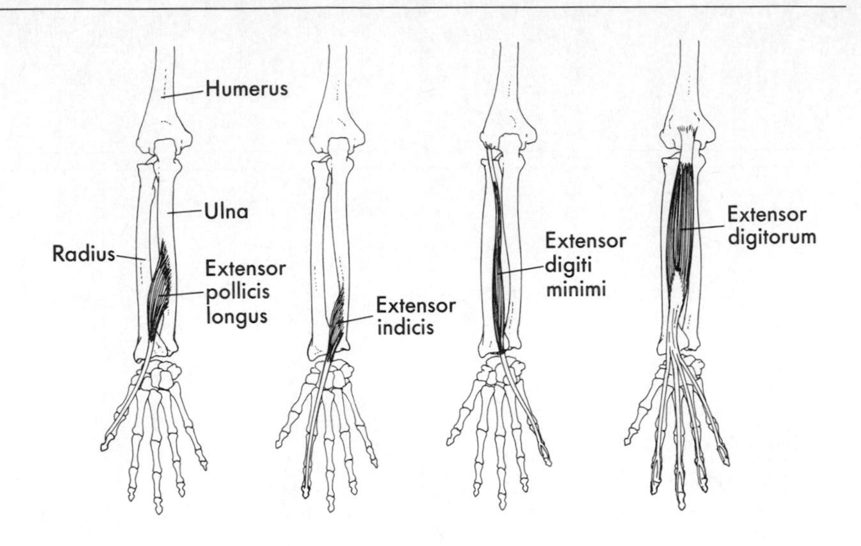

Figure 6-27
Muscles that assist with extension of the wrist.

Radial and Ulnar Flexion

Cooperative action of both flexor and extensor muscles produces lateral deviation of the hand at the wrist. The flexor carpi radialis and extensor carpi radialis contract to produce radial flexion, and the flexor carpi ulnaris and extensor carpi ulnaris cause ulnar flexion.

STRUCTURE OF THE JOINTS OF THE HAND

The joints of the hand include the carpometacarpal (CM), intermetacarpal, metacarpophalangeal (MP), and interphalangeal (IP) joints (Figure 6-28). The fingers are referred to as digits one through five, with the thumb being the first digit.

Carpometacarpal and Intermetacarpal Joints

The first carpometacarpal joint, the articulation between the trapezium and the first metacarpal, is a classic saddle joint. The other carpometacarpal joints are generally regarded as gliding joints, although some researchers have described them as modified saddle joints (11). All carpometacarpal joints are surrounded by joint capsules, which are reinforced by the dorsal, volar, and interosseous carpometacarpal ligaments. The irregular intermetacarpal joints share these joint capsules.

Metacarpophalangeal Joints

The metacarpophalangeal joints are the condyloid joints between the rounded distal heads of the metacarpals and the concave proximal ends of the phalanges. These joints form the knuckles of the

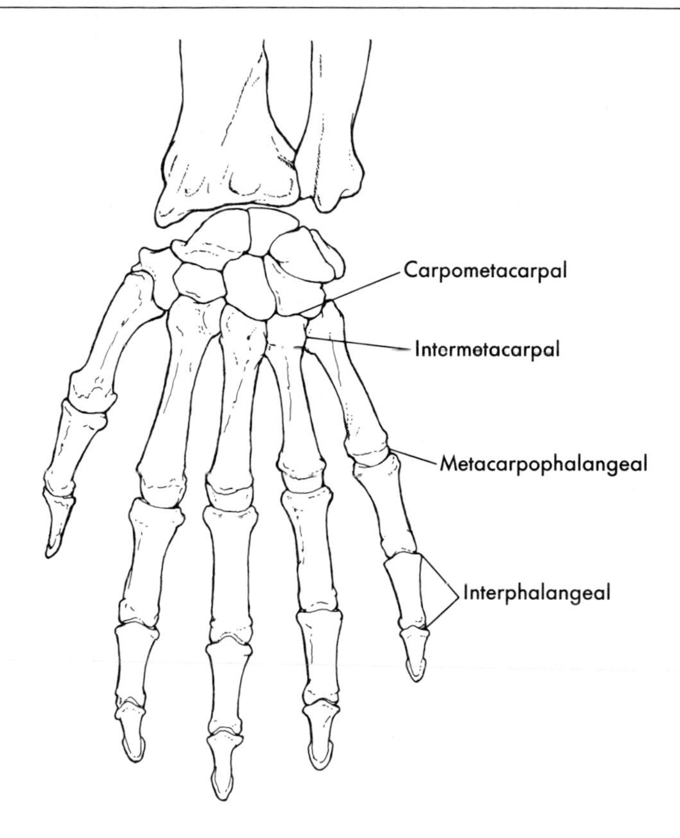

Figure 6-28
The bones of the hand.

hand. Each joint is enclosed in a capsule, with strong collateral ligaments present. A dorsal ligament strengthens the metacarpophalangeal joint of the thumb.

Interphalangeal Joints

The proximal and distal interphalangeal joints are hinge joints. Joint capsules are present, as are collateral and volar ligaments.

MOVEMENTS OF THE HAND

The carpometacarpal joint of the thumb permits a range of movement similar to that of a ball and socket joint (Figure 6-29). Motion at carpometacarpal joints two through four is slight due to constraining ligaments, with somewhat more motion permitted at the fifth carpometacarpal joint.

■ The large range of movement of the thumb compared to that of the fingers is derived from the structure of the thumb's carpometacarpal joint.

The metacarpophalangeal joints allow flexion, extension, abduction, adduction, and circumduction for digits two through five, with abduction defined as movement away from the middle finger and adduction being movement toward the middle finger. The motions

Figure 6-29
Movements of the thumb.

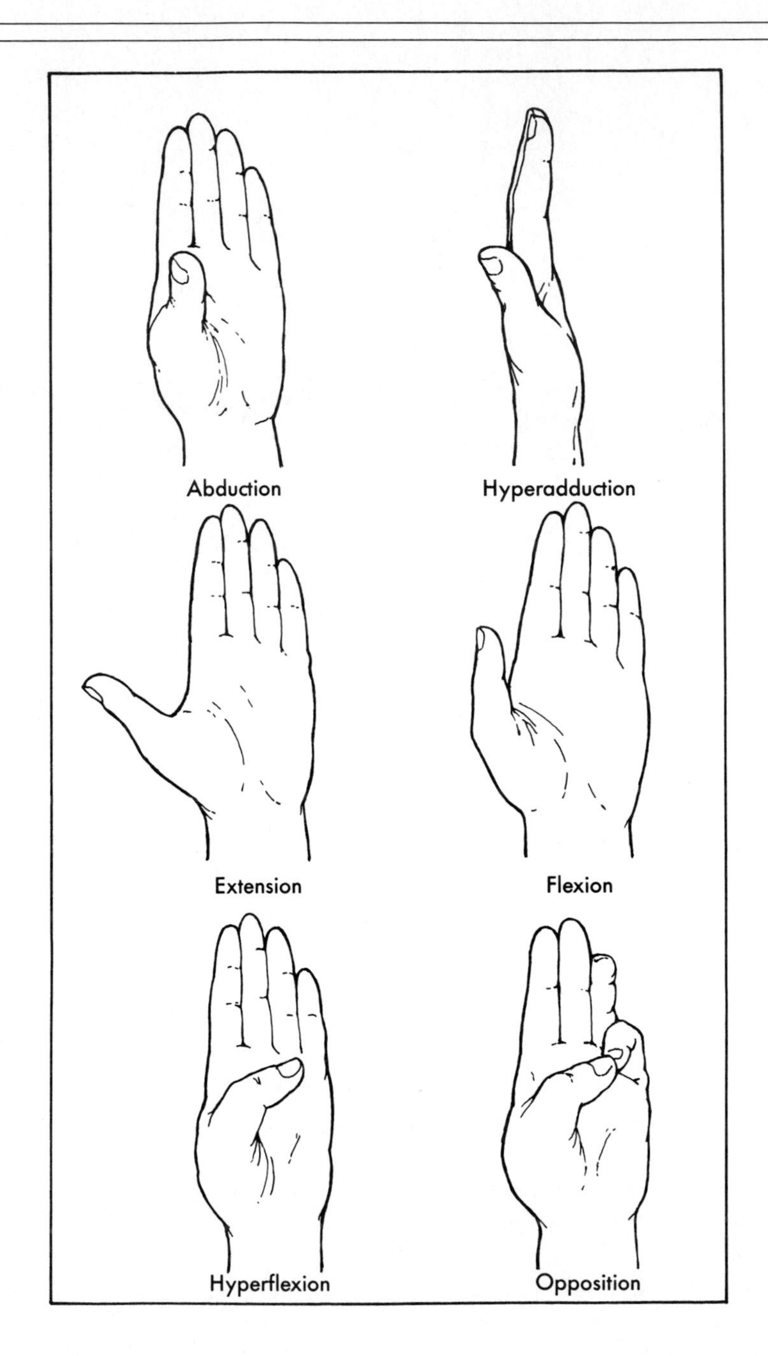

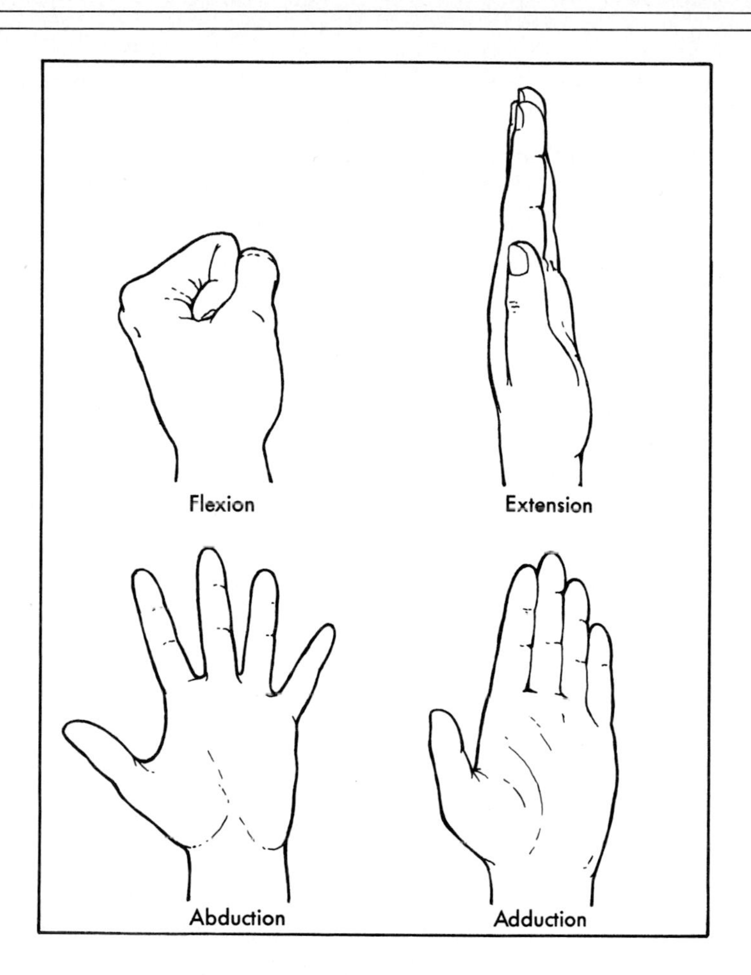

Figure 6-30
Movements of the fingers.

of the fingers are illustrated in Figure 6-30. The articulating bone surfaces at the metacarpophalangeal joint of the thumb are relatively flat, so the joint functions more as a hinge joint with primary movements of flexion and extension.

The interphalangeal joints are classic hinge joints, with only flexion and extension and, in some individuals, slight hyperextension, are permitted.

A relatively large number of muscles are responsible for the many precise movements performed by the hand and fingers. A total of 9 muscles crossing the wrist contribute to the movement of the hand and fingers. These are known as the **extrinsic muscles.** The **intrinsic muscles** are 10 muscles having both attachments distal to the wrist. The attachment sites and actions of the major muscles of the hand and fingers are summarized in Table 6-3.

extrinsic muscles
muscles with proximal attachments located proximal to the wrist and distal attachments located distal to the wrist

intrinsic muscles
muscles with both attachments distal to the wrist

Table 6-3

MAJOR MUSCLES OF THE HAND AND FINGERS

MUSCLE	PROXIMAL ATTACHMENT	DISTAL ATTACHMENT	PRIMARY ACTIONS
Extrinsic muscles			
Extensor pollicis longus	Middle dorsal ulna	Dorsal distal phalanx of thumb	Extension at MP and IP joints of thumb
Extensor pollicis brevis	Middle dorsal radius	Dorsal proximal phalanx of thumb	Extension at MP and CM joints of thumb
Flexor pollicis longus	Middle palmar radius	Palmar distal phalanx of thumb	Flexion at IP and MP joints of thumb
Abductor pollicis longus	Middle dorsal ulna and radius	Radial base of first metacarpal	Abduction of CM joint of thumb
Extensor indicis	Distal dorsal ulna	Ulnar side of extensor digitorum tendon	Extension at MP joint of second digit
Extensor digitorum	Lateral epicondyle of humerus	Base of second and third phalanges, digits two through five	Extension at MP, proximal and distal IP joints, digits two through five
Extensor digiti minimi	Proximal tendon of extensor digitorum	Tendon of extensor digitorum distal to fifth MP joint	Extension at fifth MP joint
Flexor digitorum profundus	Proximal three fourths of ulna	Base of distal phalanx, digits two through five	Flexion at distal and proximal IP joints and MP joints, digits two through five
Flexor digitorum superficialis	Medial epicondyle of humerus	Base of middle phalanx, digits two through five	Flexion at proximal IP and MP joints, digits two through five

MUSCLE	PROXIMAL ATTACHMENT	DISTAL ATTACHMENT	PRIMARY ACTIONS
Intrinsic muscles			
Flexor pollicis brevis	Ulnar side, first metacarpal	Ulnar, palmar base of proximal phalanx of thumb	Flexion at MP joint of thumb
Abductor pollicis brevis	Multangulus major and navicular bones	Radial base of first phalanx of thumb	Abduction at first CM joint
Opponens pollicis	Multangulus major	Radial side of first metacarpal	Opposition at CM joint of thumb
Adductor pollicis	Capitate, distal second and third metacarpals	Ulnar proximal phalanx of thumb	Adduction and flexion at CM joint of thumb
Abductor digiti minimi	Pisiform bone	Ulnar base of proximal phalanx, fifth digit	Abduction and flexion at fifth MP joint
Flexor digiti minimi brevis	Hamate bone	Ulnar base of proximal phalanx, fifth digit	Flexion at fifth MP joint
Opponens digiti minimi	Hamate bone	Ulnar metacarpal of fifth metacarpal	Opposition at fifth CM joint
Dorsal interossei (four muscles)	Sides of metacarpals, all digits	Base of proximal phalanx, all digits	Abduction at second and fourth MP joints, radial and ulnar deviation of third MP joint, flexion of second, third, and fourth MP joints
Palmar interossei (three muscles)	Second, fourth, and fifth metacarpals	Base of proximal phalanx, digits two, four, and five	Adduction and flexion at MP joints, digits two, four, and five
Lumbricales (four muscles)	Tendons of flexor digitorum profundus, digits two through five	Tendons of extensor digitorum, digits two through five	Flexion at MP joints of digits two through five

COMMON INJURIES OF THE WRIST AND HAND

The hand is used almost continuously in daily activities and in many sports. Wrist sprains or strains are fairly common and are occasionally accompanied by dislocation of a carpal bone or the distal radius. These types of injuries often result from the natural tendency to sustain the force of a fall on the hyperextended wrist. Fractures of the scaphoid and navicular bones are relatively common for the same reason.

Certain hand/wrist injuries are characteristic of participation in a given sport. Examples are metacarpal (boxer's) fractures and mallet or drop finger deformity resulting from injury at the distal interphalangeal joints among football receivers and baseball catchers. Forced abduction of the thumb leading to ulnar collateral ligament injury often results from wrestling, football, hockey, and skiing (8).

Both recreational and professional golfers are prone to hand and wrist problems including carpal tunnel syndrome (impingement of the nerves supplying the hands and fingers) and De Quervains disease (tendinitis of the extensor pollicis brevis and the abductor pollicis longus) (2). The wrist extensors can also be strained when a golf club hits the ground forcefully.

SUMMARY

The shoulder is the most complex joint in the human body with four different articulations contributing to movement. The glenohumeral joint is a loosely structured ball and socket joint in which range of movement is substantial and stability is minimal. The sternoclavicular joint enables some movement of the bones of the shoulder girdle, clavicle, and scapula. Movements of the shoulder girdle contribute to optimal positioning of the glenohumeral joint for different humeral movements. Small movements are also provided by the acromioclavicular and coracoclavicular joints.

The humeroulnar articulation controls flexion and extension at the elbow. Pronation and supination of the forearm occur at the proximal and distal radioulnar joints.

The structure of the condyloid joint between the radius and three carpal bones controls motion at the wrist. Flexion, extension, radial flexion, and ulnar flexion are permitted. The joints of the hand at which most movements occur are the carpometacarpal joint of the thumb, the metacarpophalangeal joints, and the hinges at the interphalangeal articulations.

INTRODUCTORY PROBLEMS

1. Construct a chart listing all muscles crossing the glenohumeral joint according to whether they are superior, inferior, anterior, or posterior to the joint center. Note that some muscles may fall into more than one category.

2. Identify the action or actions performed by muscles in the four categories in Problem 1.
3. Construct a chart listing all muscles crossing the elbow joint according to whether they are medial, lateral, anterior, or posterior to the joint center with the arm in anatomical position. Note that some muscles may fall into more than one category.
4. Identify the action or actions performed by muscles in the four categories listed in Problem 3.
5. Construct a chart listing all muscles crossing the wrist joint according to whether they are medial, lateral, anterior, or posterior to the joint center with the arm in anatomical position. Note that some muscles may fall into more than one category.
6. Identify the action or actions performed by muscles in the four categories listed in Problem 5.
7. Identify the activities that stabilize the scapula against forces directed upward, downward, laterally, and medially. Which muscles are used for scapular stabilization during each of these activities?
8. Which three types of bones in the upper extremity are most susceptible to fracture? Why?
9. Pushups are often used to test upper body strength. Which muscles are used as agonists during the performance of a pushup? Which major muscles of the upper extremity are not used?
10. Select a familiar sport activity and identify the muscles of the upper extremity that are used as agonists during the activity.

ADDITIONAL PROBLEMS

1. Identify the sequence of movements that occur at the shoulder, elbow, and wrist joints during the performance of an overhand throw.
2. Which muscles are most likely to serve as agonists to produce the movements identified in your answer to Problem 1?
3. Identify the sequence of movements that occur at the shoulder, elbow, and wrist joints during the execution of a serve in tennis. How does your list compare to the list of movements identified for Problem 1?
4. Which muscles are most likely to serve as agonists to produce the movements identified in your answer to Problem 3?
5. Select a familiar racket sport and identify the sequence of movements that occur at the shoulder, elbow, and wrist joints during the execution of forehand and backhand strokes.
6. Which muscles are most likely to serve as agonists to produce the movements identified in your answer to Problem 5?

7. List the muscles that are active during the performances of pull-ups done with an overhand grip and with an underhand grip.
8. Select five resistance-based exercises for the upper extremity and identify which muscles are the primary movers and which muscles assist during the performance of each exercise.
9. How could you modify the exercises identified in your answer to Problem 8 so that different muscles become the prime movers? Identify the prime movers in each of your modified exercises.
10. Compare and contrast the upper extremity joint movements and the major agonists involved during the swing of a baseball bat and the swing of a golf club (driver).

REFERENCES

1. Doody SG, Freedman L, and Waterland JC: Shoulder movement during abduction in the scapular plane, Arch Phys Med Rehabil 51:595, 1970.
2. Duda M: Golf injuries: they really do happen, Physician Sportsmed 15:191, 1987.
3. Ferretti A, Cerullo G, and Russo G: Subscapular neuropathy in volleyball players, J Bone Joint Surg 69-A:260, 1987.
4. Goldberg B and Boiardo R: Profiling children for sports participation, Clin Sports Med 3:153, 1984.
5. Hay JG and Reid JG: The anatomical and mechanical bases of human motion, Englewood Cliffs, NJ, 1982, Prentice-Hall, Inc.
6. Hinson MM: Kinesiology, ed 2, Dubuque, Ia, 1981, Wm C Brown Group.
7. Jobe FW and Nuber G: Throwing injuries of the elbow, Clin Sports Med 5:621, 1986.
8. Johnson RE and Rust RJ: Sports related injury: an anatomic approach, part 2, Minn Med 68:829, 1985.
9. Josefsson PO and Nilsson BE: Incidence of elbow dislocation, Acta Orthop Scand 57:537, 1986.
10. Kent BE: Functional anatomy of the shoulder complex: a review, J Am Phys Ther Assoc 51:867, 1971.
11. Luttgens K and Wells KF: Kinesiology: scientific basis of human motion, ed 7, Philadelphia, 1982, WB Saunders Co.
12. McLeod WD and Andrews JR: Mechanisms of shoulder injuries, Phys Ther 66:1901, 1986.

13. Morrey BF et al: A biomechanical study of normal functional elbow motion, J Bone Joint Sur [Am] 63:872, 1981.
14. Nash HL: Rotator cuff damage: reexamining the causes and treatments, Physician Sportsmed 16:129, 1988.
15. Nuber GW et al: Fine wire electromyography analysis of muscles of the shoulder during swimming, Am J Sports Med 14:7, 1986.
16. Poppen KN and Walker PS: Normal and abnormal motion of the shoulder, J Bone Joint Surg [Am] 58:195, 1976.
17. Rafii M et al: Athlete shoulder injuries: CT arthrographic findings, Radiology 162:559, 1987.
18. Renstrom P and Johnson RJ: Overuse injuries in sports, Sports Med 2:316, 1985.
19. Richardson AB, Jobe FW, and Collins HR: The shoulder in competitive swimming, Am J Sports Med 8:159, 1980.
20. Soderberg GL: Kinesiology: application to pathological motion, Baltimore, 1986, Williams & Wilkins.
21. Zuckerman JD and Matsen FA: Biomechanics of the elbow. In Nordin M and Frankel VH, eds: Basic biomechanics of the skeletal system, Philadelphia, 1989, Lea & Febiger.
22. Zuckerman JD and Matsen FA: Biomechanics of the shoulder. In Nordin M and Frankel VH, eds: Basic biomechanics of the skeletal system, Philadelphia, 1989, Lea & Febiger.

ANNOTATED READINGS

Brunet ME, Hoddad RJ, and Porche EB: Rotator cuff impingement syndrome in sports, Physician Sportsmed 11:79, 1983.

Discusses common causes, symptoms, and treatment modalities for the rotator cuff impingement syndrome.

Soderberg GL: Shoulder. In Soderberg GL: Kinesiology: application to pathological motion, Baltimore, 1986, Williams & Wilkins.

Includes an interesting discussion of evolutionary changes in the anatomical structures of the shoulder and a section on pathokinesiology of the shoulder.

Zuckerman JD and Matsen FA: Biomechanics of the elbow. In Nordin M and Frankel VH, eds: Basic biomechanics of the skeletal system, Philadelphia, 1989, Lea & Febiger.

Describes kinematics and kinetics of the elbow, with sample calculations of joint reaction forces.

Zuckerman JD and Matsen FA: Biomechanics of the shoulder. In Nordin M and Frankel VH, eds: Basic biomechanics of the skeletal system, Philadelphia, 1989, Lea & Febiger.

Describes kinematics and kinetics of the joints of the shoulder complex.

7 MOVEMENT

The Biomechanics of the Lower Extremity

After reading this chapter, the student will be able to:

Explain the ways in which anatomical structure affects the movement capabilities of lower extremity articulations.

Identify the factors that determine the relative mobility and stability present at lower extremity articulations.

Explain the ways in which the lower extremity is adapted to its weight-bearing function.

Identify the muscles that are active during specific lower extremity movements.

Describe the biomechanical contributors to common injuries of the lower extremity.

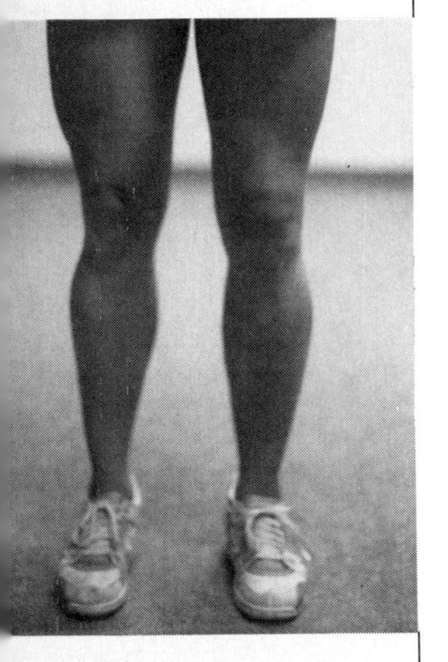

The lower extremity is well structured for its functions of weight bearing and locomotion.

Although there are some similarities between the joints of the upper and lower extremities, the upper extremity is more specialized for activities requiring large ranges of motion. In contrast, the anatomical structure of the lower extremity is well equipped for its functions of weight bearing and locomotion. Beyond these basic functions the executions of a successful field goal in football, the performance of the jumping events in track, and the maintenance of balance on pointe by ballet dancers reveal the more specialized capabilities of the lower extremity. In this chapter the joint and muscle functions contributing to lower extremity movements are examined.

STRUCTURE OF THE HIP

The hip is a ball and socket joint (Figure 7-1). The ball is the head of the femur, which forms approximately two thirds of a sphere. The socket is the concave acetabulum, which is angled obliquely in an anterior, lateral, and inferior direction. Joint cartilage covers both articulating surfaces. The cartilage on the acetabulum is thicker around its periphery where it merges with a rim or labrum of fibrocartilage that contributes to the stability of the joint. The acetabulum also provides a much deeper socket than the glenoid fossa of the shoulder joint, and the bony structure of the hip is therefore much more stable or less likely to dislocate than that of the shoulder.

Several large, strong ligaments also contribute to the stability of the hip (Figure 7-2). The extremely strong iliofemoral or Y ligament and the pubofemoral ligament strengthen the joint capsule anteriorly, with posterior reinforcement from the ischiofemoral ligament. Inside the joint capsule the ligamentum teres supplies a direct attachment from the rim of the acetabulum to the head of the femur. As with the shoulder joint, several bursae are present in the surrounding tissues to assist with lubrication.

MOVEMENTS AT THE HIP

Motion of the Femur

Although movements of the femur are primarily due to rotation occurring at the hip joint, the **pelvic girdle** functions similar to the shoulder girdle in positioning the hip joint for effective limb move-

pelvic girdle
the fused ilium, ischium, and pubis, which can be rotated forward, backward, and laterally to optimize positioning of the hip joint

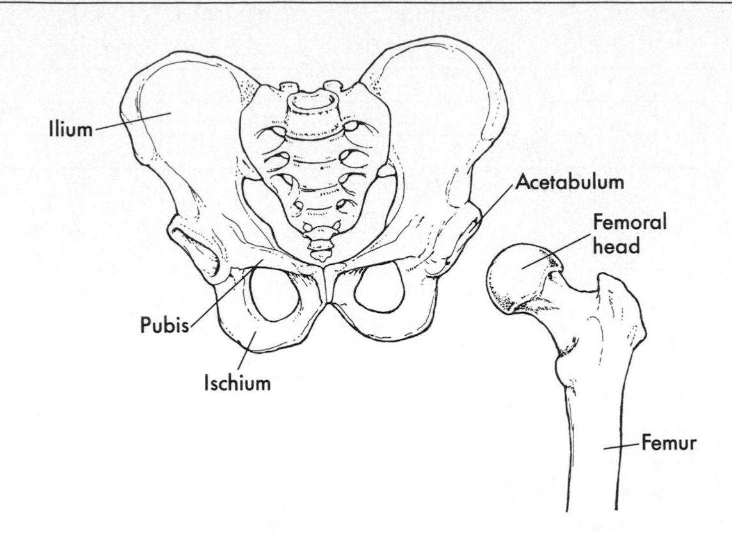

Figure 7-1
The bony structure of the hip.

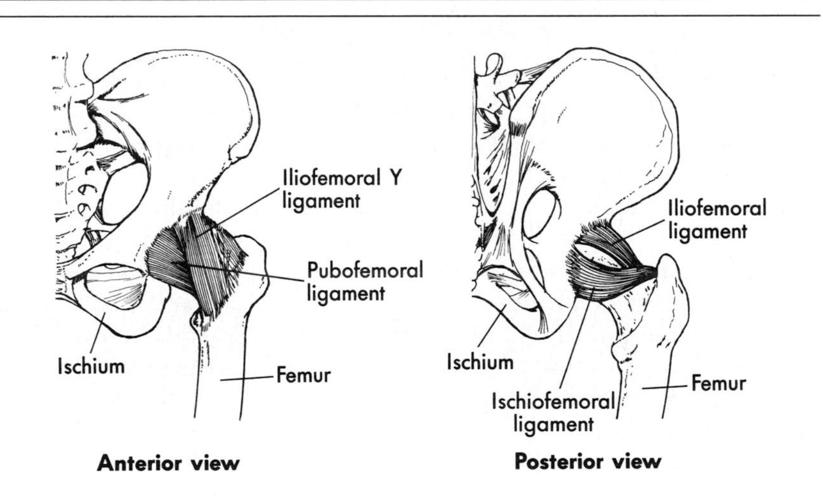

Figure 7-2
The ligaments of the hip.

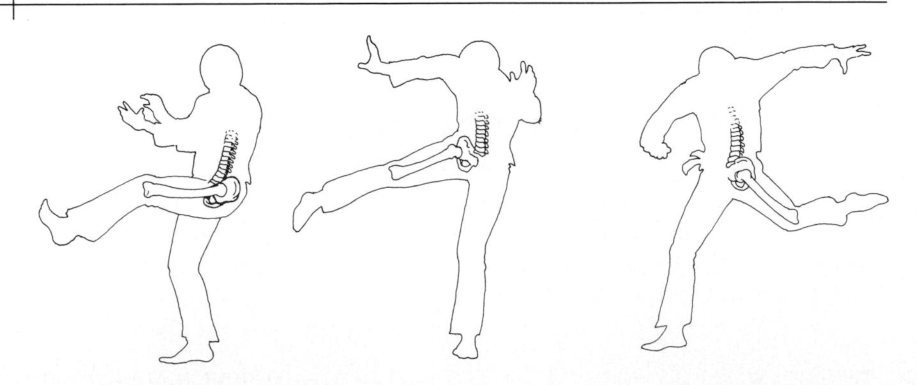

Figure 7-3
Movement of the pelvis positions the hip joint for optimal movement of the femur.

ment. Unlike the shoulder girdle, the pelvis is a single nonjointed structure, but it can rotate in all three planes of movement. The pelvis facilitates movement of the femur by rotating so that the acetabulum is positioned toward the direction of impending femoral movement (Figure 7-3). The backward tilt of the pelvis enhances flexion at the hip, its forward tilt assists in femoral extension, and its lateral tilt toward the opposite side facilitates lateral movements of the femur. Movement of the pelvic girdle also coordinates with certain movements of the spine (see Chapter 8).

Muscles of the Hip

A number of large muscles cross the hip, further contributing to its stability. The muscles of the hip are summarized in Table 7-1 on p. 184.

Flexion

The six muscles primarily responsible for flexion at the hip are those crossing the joint anteriorly—the iliacus, psoas major, pectineus, rectus femoris, sartorius, and tensor fascia latae. Of these, the large iliacus and psoas major (often referred to jointly as the **iliopsoas** because of their common attachment to the femur) are the major hip flexors (Figure 7-4). Other major hip flexors are shown in Figure 7-5. The rectus femoris, known as the *kicking muscle,* is a two-joint muscle that is active during both hip flexion and knee extension. Consequently, it functions more effectively as a hip flexor when the knee is

iliopsoas
the psoas major and iliacus muscles with a common insertion on the lesser trochanter of the femur

■ Two-joint muscles function more effectively at one joint when the position of the other joint stretches the muscle slightly.

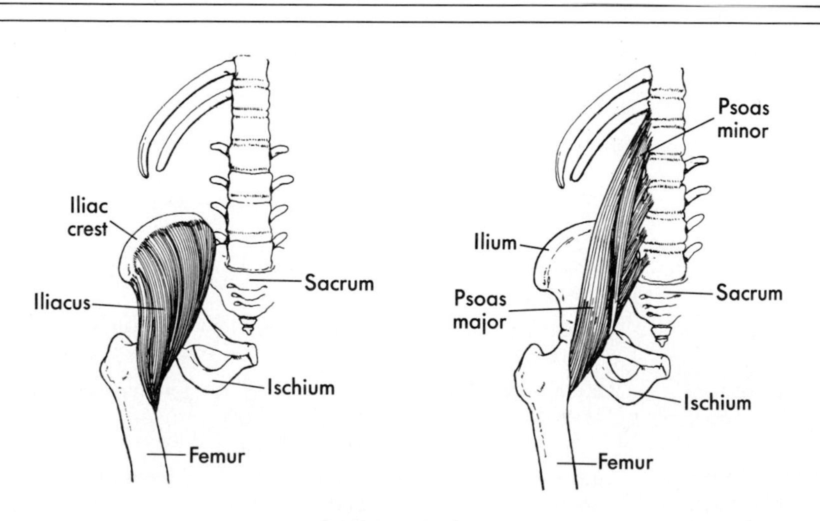

Figure 7-4
The iliopsoas complex is the major flexor of the hip.

Table 7-1

MUSCLES OF THE HIP

MUSCLE	PROXIMAL ATTACHMENT	DISTAL ATTACHMENT	PRIMARY ACTIONS
Rectus femoris	Anterior inferior iliac spine	Patella	Flexion
Iliopsoas (iliacus)	Iliac fossa and adjacent sacrum	Lesser trochanter of femur	Flexion
Iliopsoas (psoas)	Twelfth thoracic and lumbar vertebrae and lumbar discs	Lesser trochanter of femur	Flexion
Sartorius	Anterior superior iliac spine	Upper medial tibia	Assists with flexion, abduction, outward rotation
Pectineus	Pectineal crest of pubic ramus	Medial femur	Flexion, adduction
Tensor fascia lata	Crest of ilium	Iliotibial band	Assists with flexion, abduction, inward rotation
Gluteus maximus	Posterior ilium, iliac crest, sacrum, coccyx	Gluteal tuberosity of femur and iliotibial band	Extension, outward rotation
Gluteus medius	Between posterior and anterior gluteal lines on posterior ilium	Superior, lateral greater trochanter	Abduction
Gluteus minimus	Between anterior and inferior gluteal lines on posterior ilium	Anterior surface of greater trochanter	Inward rotation
Gracilis	Anterior, inferior symphysis pubis	Medial, proximal tibia	Adduction
Adductor magnus	Inferior ramus of pubis and ischium	Entire linea aspera	Adduction
Adductor longus	Anterior pubis	Middle linea aspera	Adduction
Adductor brevis	Inferior ramus of pubis	Upper linea aspera	Adduction
Semitendinosus	Medial ischial tuberosity	Proximal, medial tibia	Extension
Semimembranosus	Lateral ischial tuberosity	Proximal, medial tibia	Extension
Biceps femoris (long head)	Lateral ischial tuberosity	Posterior lateral condyle of tibia, head of fibula	Extension
Biceps femoris (short head)	Lateral linea aspera	Posterior lateral condyle of tibia, head of fibula	Extension
Six outward rotators	Sacrum, ilium, ischium	Posterior greater trochanter	Outward rotation

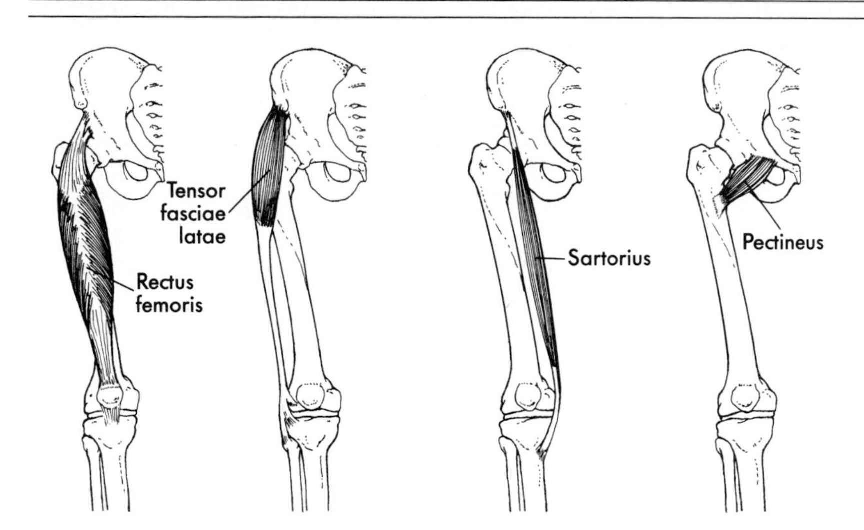

Figure 7-5
Other flexor muscles of the hip.

in flexion. The thin, straplike sartorius, or *tailor's muscle,* is also a two-joint muscle. Crossing from the superior anterior iliac spine to the medial tuberosity of the tibia, the sartorius is the longest muscle in the body.

Extension

The hip extensors include the gluteus maximus and the three **hamstrings**—the biceps femoris, semitendinosus and semimembranosus (Figure 7-6). The gluteus maximus is a massive, powerful muscle and is typically active only when the hip is in a substantial amount of flexion, such as during stair climbing or cycling, or when extension at the hip is resisted (Figure 7-7). The hamstrings derive their name from their prominent tendons, which can readily be palpated on the posterior aspect of the knee. These two-joint muscles contribute to both extension at the hip and flexion at the knee and are active during standing, walking, and running.

hamstrings
the biceps femoris, semimembranosus, and semitendinosus

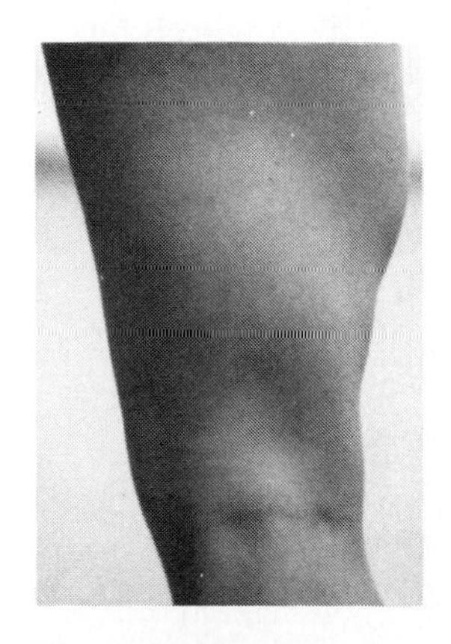

The hamstring muscles can be traced upward from their prominent tendons on the posterior lateral and medial aspects of the knee.

Abduction

The gluteus medius is the major abductor acting at the hip, with the gluteus minimus assisting. These muscles stabilize the pelvis when an individual stands on one leg, during the support phase of walking and running, and during the abduction of the femur in dance movements such as the *grande ronde jambe.*

Adduction

The main adductors at the hip cross the joint medially and include the adductor longus, adductor brevis, adductor magnus, and gracilis

Figure 7-6
The hamstrings are major hip extensors and knee flexors.

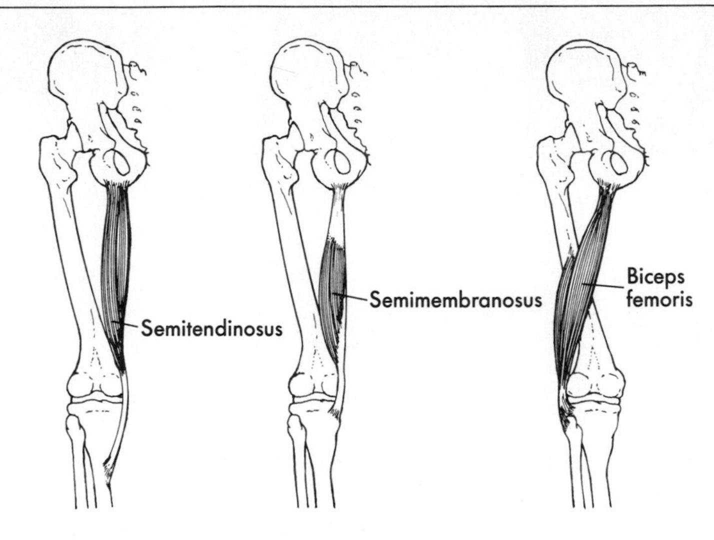

Figure 7-7
The three gluteal muscles.

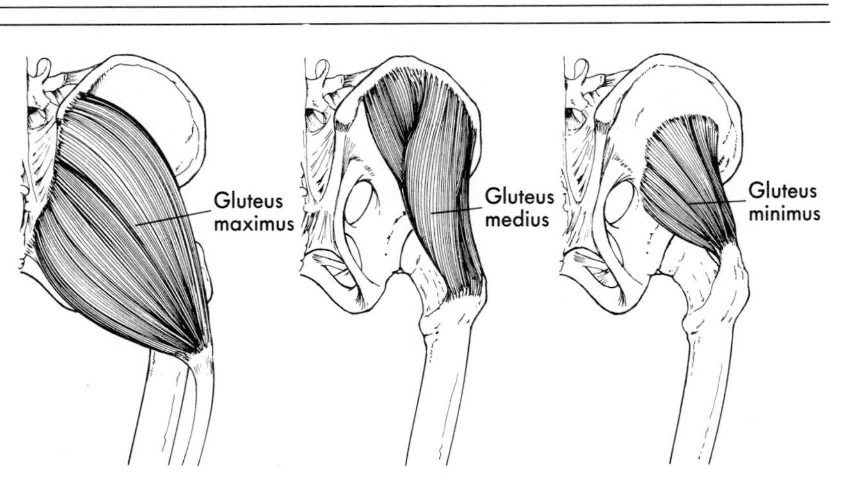

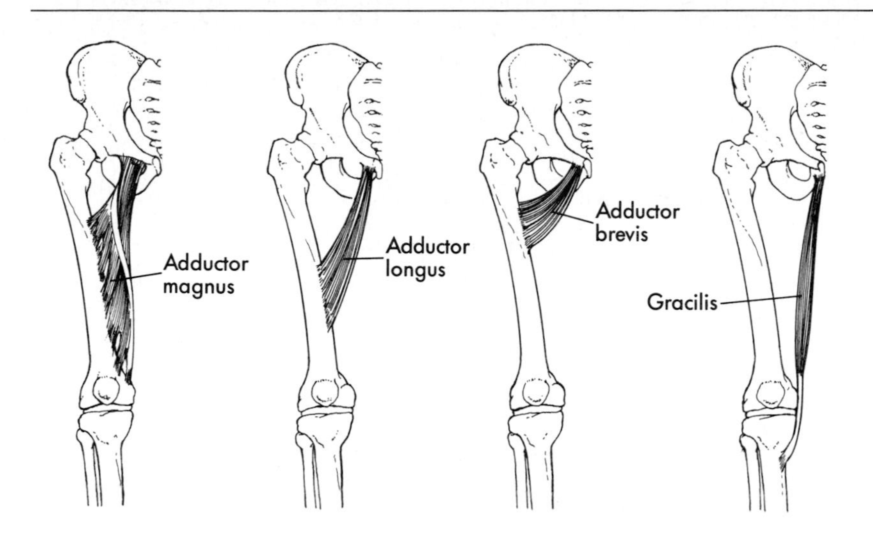

Figure 7-8
Abductor muscles of the hip.

(Figure 7-8). The hip adductors are regularly active during the swing phase of the gait cycle because foot placement during the support phase is typically beneath the body's center of gravity. The gracilis, a long, relatively weak strap muscle, and the three adductor muscles also contribute to flexion and internal rotation at the hip, particularly when the femur is externally rotated.

Medial and Lateral Rotation of the Femur

Although a number of muscles contribute to lateral rotation of the femur, six muscles function solely as lateral rotators. These are the piriformis, gemellus superior, gemellus inferior, obturator internus, obturator externus, and quadratus femoris (Figure 7-9). Although we tend to think of walking and running as involving strictly sagittal plane movement at the joints of the lower extremity, outward rotation of the femur also occurs with every step to accommodate the rotation of the pelvis.

Medial rotation of the femur is usually not a resisted motion requiring a substantial amount of muscular force. The medial rotators' strength is approximately one third that of the lateral rotators (7). The muscle primarily active during inward rotation of the femur is the gluteus minimus, with assistance provided by other muscles, including the tensor fascia latae, semitendinosus, semimembranosus, gluteus medius, and the four adductor muscles.

During the gait cycle, lateral and medial rotation of the femur occur in coordination with pelvic rotation.

Horizontal Abduction and Adduction

Horizontal abduction and adduction of the femur occur when the hip is in 90 degrees of flexion while the femur is either abducted or adducted. These actions require the simultaneous, coordinated ac-

Figure 7-9
The lateral rotator muscles of the femur.

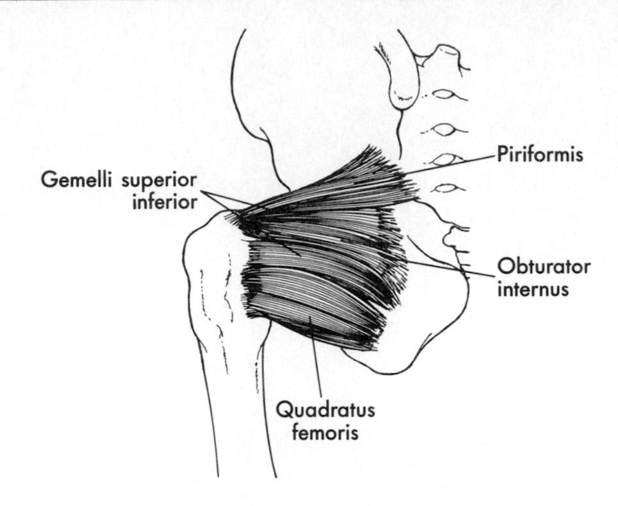

This dancer's pose includes horizontal abduction of the thigh.

tions of several muscles. Tension is required in the hip flexors for elevation of the femur. The hip abductors can then produce horizontal abduction, and from a horizontally abducted position the hip adductors can produce horizontal adduction. The muscles located on the posterior aspect of the hip are more effective as horizontal abductors and adductors than the muscles on the anterior aspect because the former are stretched when the femur is in 90 degrees of flexion, whereas tension in the anterior muscles is usually reduced with the femur in this position.

COMMON INJURIES OF THE HIP

Fractures

Although the pelvis and femur are large, strong bones, the hip is subjected to high, repetitive loads ranging from 4 to 7 times body weight during locomotion (7). Fractures of the femoral neck frequently occur during the support phase of walking among elderly individuals with osteoporosis, a condition of reduced bone mineralization and strength (see Chapter 2). These femoral neck fractures often result in loss of balance and a fall. A common misconception is that the fall always causes the fracture rather than the reverse, which may also be true. When the hip bones have good health and mineralization, they can sustain tremendous loads as illustrated during many weight-lifting events.

■ Fracture of the femoral neck (broken hip) is a seriously debilitating injury that occurs frequently among elderly individuals with osteoporosis.

Contusions

The muscles on the anterior aspect of the thigh are in a prime location for sustaining blows during participation in contact sports. The resulting internal hemorrhaging and appearance of bruises vary from mild to severe.

Strains

Since most daily activities do not require simultaneous hip flexion and knee extension, the hamstrings are rarely stretched unless exercises are performed for that specific purpose. The resulting loss of extensibility makes the hamstrings particularly susceptible to strain. Strains to these muscles most commonly occur during sprinting, particularly if the individual is fatigued and neuromuscular coordination is impaired. Researchers have proposed different theories to explain the mechanism or mechanisms of hamstring strains (18). According to one hypothesis, hamstring strains result from the overstretching of the muscle group, such as during overstriding. An alternative theory is that they result when the fully elongated muscle group must develop maximum tension. Strains to the groin area are also relatively common among athletes in sports in which forceful thigh abduction movements may overstretch the adductor muscles.

■ Excessive tightness of the hamstring muscles is a common problem in many modern societies.

STRUCTURE OF THE KNEE

tibiofemoral joint
dual condyloid joints between the medial and lateral condyles of the tibia and the femur composing the main hinge joint of the knee

patellofemoral joints
the articulation between the patella and the femur

The structure of the knee permits the bearing of tremendous loads as well as the mobility required for locomotor activities. The knee is a large and complex joint that includes three articulations—the two condylar articulations of the **tibiofemoral joint** and the single articulation of the **patellofemoral joint.** Although not a part of the knee, the proximal tibiofibular joint has soft tissue connections that also slightly influence knee motion (16).

Tibiofemoral Joint

▪ The bony anatomy of the knee requires a small amount of medial rotation of the femur to accompany full extension.

The medial and lateral condyles of the tibia and of femur articulate to form two side-by-side condyloid joints (Figure 7-10). Because of their close proximity, these two condyloid joints function as a single hinge joint. The condyles of the tibia, known as the *tibial plateaus,* form slight depressions separated by a region known as the *intercondylar eminence.* The medial and lateral condyles of the femur differ somewhat in size, shape, and orientation, which causes the femur to rotate slightly medially on the tibia as the knee is moved into full extension. This phenomenon has been described as a locking or screwing-home mechanism. Full extension is the close-packed position of greatest stability for the knee.

Menisci

menisci
cartilaginous discs located between the tibial and femoral condyles that are also known as the semilunar cartilage

Discs of fibrocartilage are located on the peripheries of each tibial plateau (Figure 7-11). These **menisci,** which are also known as *semilunar cartilages,* are thickest at their peripheral borders where fibers from the joint capsule solidly anchor them to the tibia (1). The

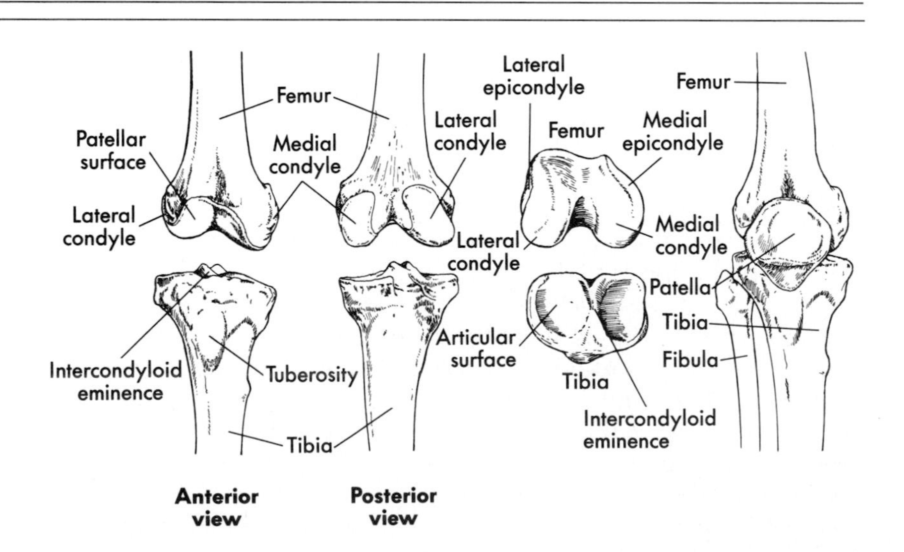

Figure 7-10
The bony structure of the tibiofemoral joint.

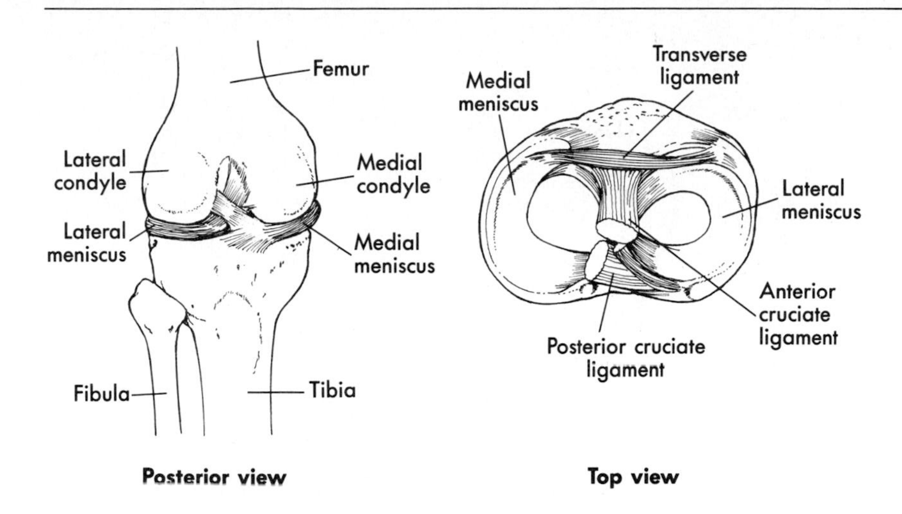

Figure 7-11
The menisci of the knee.

medial semilunar disc is also directly attached to the medial collateral ligament. Medially, both menisci taper down to paper thinness, with the inner edges unattached to the bone.

The menisci deepen the articulating depressions of the tibial plateaus and assist with the absorption of force at the knee. The internal structure of the medial two thirds of each meniscus is particularly well suited to resisting compression (1). The stress on the tibiofemoral joint can be an estimated three times higher during load bearing if the menisci have been removed (15). Injured knees in which part or all of the menisci have been removed may still function adequately but undergo increased wear on the articulating surfaces, significantly increasing the likelihood of the development of degenerative conditions at the joint.

Ligaments

Many ligaments cross the knee, significantly enhancing its stability (Figure 7-12). The location of each ligament determines the direction in which it is capable of resisting the dislocation of the knee.

The medial and lateral **collateral ligaments** prevent lateral motion at the knee as do the collateral ligaments at the elbow. They are also referred to as the *tibial* and *fibular collateral ligaments* respectively after their distal attachments.

collateral ligaments
major ligaments that cross the medial and lateral aspects of the knee

The anterior and posterior **cruciate ligaments** limit the forward and backward sliding of the femur on the tibial plateaus during knee flexion and extension and also limit knee hyperextension. They are named cruciate ligaments because they cross each other. The anterior cruciate ligament courses from the anterior intercondyloid fossa

cruciate ligaments
major ligaments that cross each other in connecting the anterior and posterior aspects of the knee

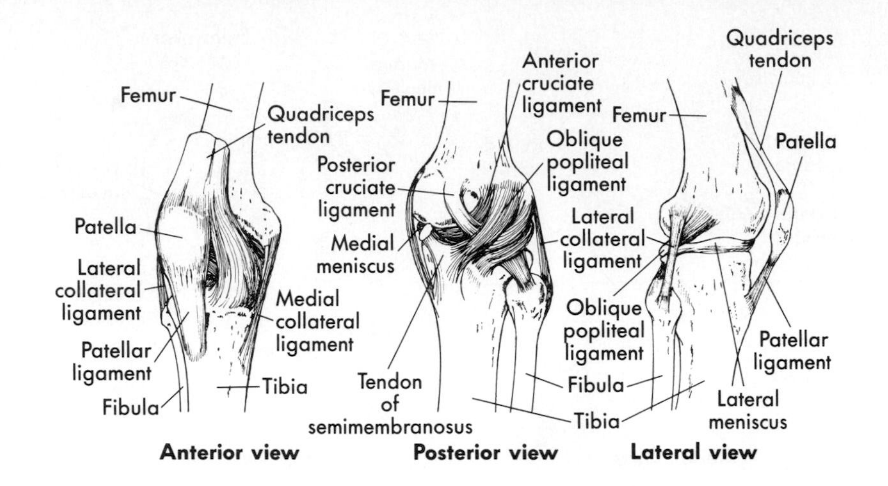

Figure 7-12
The ligaments of the knee.

of the tibia upward and backward to the posterior medial surface of the lateral condyle of the femur. The shorter posterior cruciate ligament attaches from the posterior intercondyloid fossa of the tibia upward and forward to the lateral anterior part of the medial condyle of the femur.

Several other ligaments contribute to the integrity of the knee. The oblique popliteal ligament crosses the knee posteriorly, and the transverse ligament connects the two semilunar discs internally. Another restricting tissue is the **iliotibial band** or tract, a broad, thickened band of the fascia lata with attachments to the lateral condyle of the femur and the lateral tubercle of the tibia, which has been hypothesized to function as an anterolateral ligament of the knee (20).

iliotibial band
a thick, strong band of tissue connecting the tensor fascia lata to the lateral condyle of the femur and the lateral tuberosity of the tibia

Patellofemoral Joint

The patellofemoral joint consists of the articulation of the patella, encased in the patellar tendon, with the femoral condyles (Figure 7-13). The patella's primary function is to increase the angle of insertion of the patellar tendon on the tibia, thereby improving the effectiveness of the four quadriceps muscles in producing knee extension. The patella protects the anterior aspect of the knee slightly.

MOVEMENTS AT THE KNEE

Muscles Crossing the Knee

Similar to the elbow, the knee is crossed by a number of two-joint muscles. The primary actions of the muscles crossing the knee are summarized in Table 7-2.

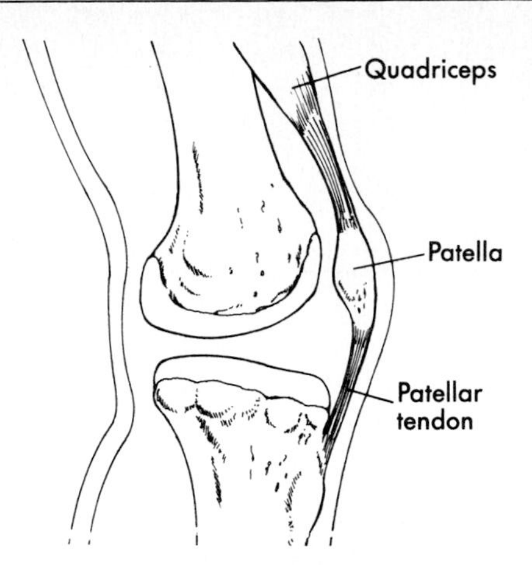

Figure 7-13

The patella increases the angle between the patellar tendon and the tibia, thereby increasing the effectiveness of the muscle group.

Table 7-2

MUSCLES OF THE KNEE

MUSCLE	PROXIMAL ATTACHMENT	DISTAL ATTACHMENT	PRIMARY ACTIONS
Rectus femoris	Anterior inferior iliac spine	Patella	Extension
Vastus lateralis	Greater trochanter and lateral linea aspera	Patella	Extension
Vastus intermedius	Anterior femur	Patella	Extension
Vastus medialis	Medial linea aspera	Patella	Extension
Semitendinosus	Medial ischial tuberosity	Proximal, medial tibia	Flexion, inward rotation
Semimembranosus	Lateral ischial tuberosity	Proximal, medial tibia	Flexion, inward rotation
Biceps femoris (long head)	Lateral ischial tuberosity	Posterior lateral condyle of tibia, head of fibula	Flexion, outward rotation
Biceps femoris (short head)	Lateral linea aspera	Posterior lateral condyle of tibia, head of fibia	Flexion, outward rotation
Sartorius	Anterior superior iliac crest	Upper medial tibia	Assists with flexion, inward rotation
Gracilis	Anterior, inferior symphysis pubis	Medial, proximal tibia	Assists with flexion, inward rotation
Popliteus	Lateral condyle of femur	Medial, posterior tibia	Inward rotation
Gastrocnemius	Posterior medial and lateral condyles of femur	Tuberosity of calcaneus by Achilles tendon	Assists with flexion
Plantaris	Distal, posterior femur	Tuberosity of calcaneus by Achilles tendon	Assists with flexion

Figure 7-14
The popliteus muscle is the unlocker of the knee.

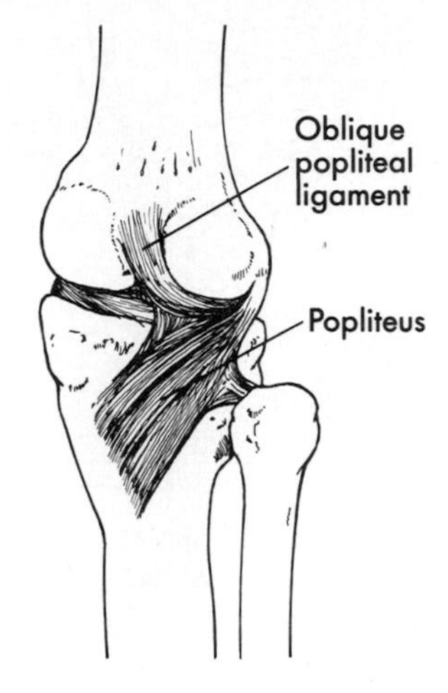

Figure 7-15
The quadriceps muscles extend the knee.

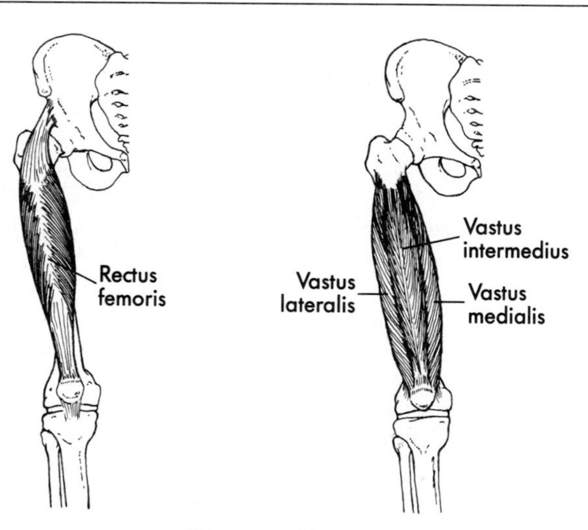

Flexion and Extension

When the knee undergoes a significant amount of flexion, the femur must slide forward on the tibia to prevent rolling off the tibial plateaus. Likewise, the femur must slide backwards on the tibia during extension.

The three hamstring muscles are the primary flexors acting at the knee. Muscles that assist with knee flexion are the gracilis, sartorius, popliteus, and gastrocnemius. When the knee is fully extended, it must first be unlocked from the close-packed position before flexion can be initiated. The **popliteus** unlocks the knee by either lateral rotation of the femur or medial rotation of the tibia (Figure 7-14).

popliteus
the muscle known as the unlocker of the knee because its action is lateral rotation of the femur with respect to the tibia

The **quadriceps** muscles, consisting of the rectus femoris, vastus lateralis, vastus medialis, and vastus intermedius, are the extensors of the knee (Figure 7-15). The rectus femoris is the only one of these muscles that also crosses the hip joint. All four muscles attach distally to the patellar tendon, which inserts on the tibia.

quadriceps
the rectus femoris, vastus lateralis, vastus medialis, and vastus intermedius

Rotation

Rotation of the tibia relative to the femur is possible when the knee is in flexion and not bearing weight. Tension development in the semimembranosus, semitendinosus, and popliteus produces medial rotation of the tibia, with the gracilis and sartorius assisting. The biceps femoris is solely responsible for lateral rotation of the tibia.

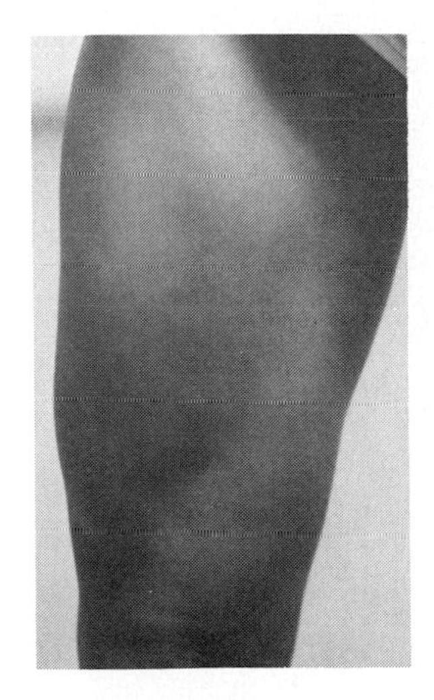

The powerful vastus lateralis, vastus medius, and vastus medialis are prominent anterior thigh muscles.

COMMON INJURIES OF THE KNEE AND LOWER LEG

The location of the knee between the long bones of the lower extremity, combined with its weight bearing and locomotion functions, make it susceptible to injury, particularly during participation in contact sports. A common injury mechanism involves the stretching or tearing of soft tissues on one side of the joint when a blow is sustained from the opposite side during weight bearing.

Ligament Damage

Forces sustained by the knee from the anterior direction may damage the posterior cruciate ligament, and forces directed from the posterior of the knee may damage the anterior cruciate ligament. Among wrestlers, forced flexion of the knee combined with internal rotation of the tibia on the femur can cause injury to the posterior cruciate ligament (17).

Blows to the lateral side of the knee are much more common than blows to the medial side because the opposite leg commonly protects the medial side of the joint. When the foot is planted and a lateral blow of sufficient force is sustained, the result is sprain or rupture of the medial collateral ligament (Figure 7-16). Although in contact

■ In contact sports, blows to the knee are most commonly sustained on the lateral side, with injury occurring to the stretched tissues on the medial side.

Figure 7-16
A medially directed force sustained on the lateral aspect of the knee often creates a tensile strain in the medial ligaments of the knee.

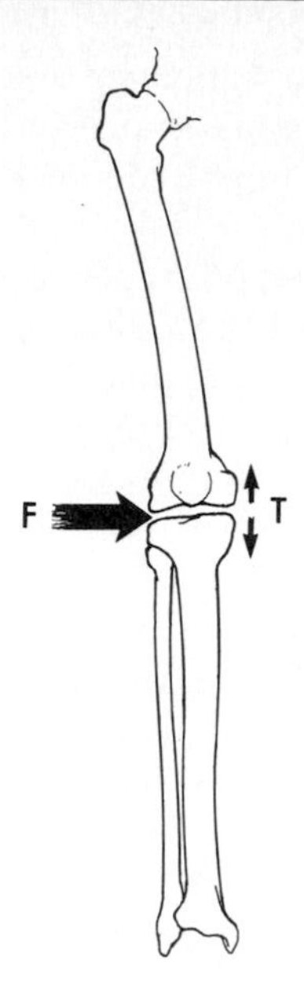

sports such as football the medial collateral ligament is more frequently injured, both medial and lateral collateral ligament sprains occur among wrestlers (22).

Torn Menisci

Because the medial collateral ligament attaches to the medial meniscus, stretching or tearing of the ligament can also result in damage to the meniscus. A torn meniscus is the most common knee injury, with damage to the medial meniscus approximately 10 times as frequent as damage to the lateral meniscus. The mechanism of injury frequently involves the foot being planted during weight bearing while the body undergoes rotation. The condition is problematic in that the unattached cartilage often slips from its normal position, interfering with normal joint mechanics. Symptoms include pain, which is sometimes accompanied by intermittent bouts of locking or buckling of the joint (10).

Iliotibial Band Friction Syndrome

The tensor fascia lata develops tension to assist with stabilization of the pelvis when the knee is in flexion during weight bearing. This increases the friction of the attached iliotibial band against the lateral condyle of the femur during flexion/extension of the knee and may result in symptoms of pain and tenderness over the lateral aspect of

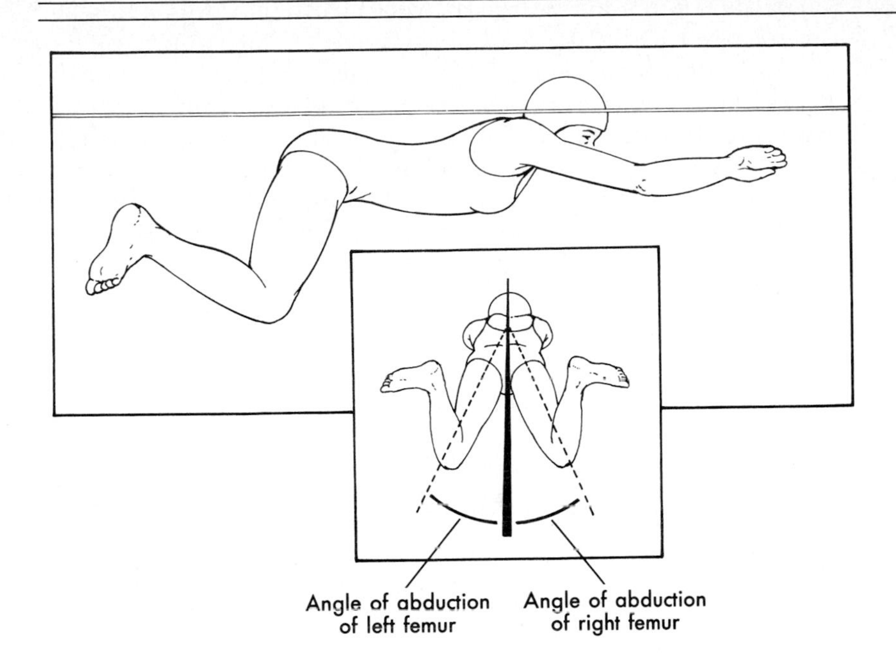

Figure 7-17
The likelihood of acquiring breaststroker's knee is much lower if the angle of hip abduction is between 37 and 42 degrees at the beginning of the propulsive phase of the kick.

the knee. This condition is an overuse syndrome relatively common among runners and is sometimes referred to as *runner's knee* (13). Excessive pronation of the foot results in increased internal rotation of the tibia during locomotion and contributes in some cases of iliotibial band friction syndrome (3). Other factors that may contribute to the syndrome involve tibial alignment and the size of the lateral femoral condyle (12).

Breaststroker's Knee

A condition of pain and tenderness localized on the medial aspect of the knee is often associated with performance of the whip kick—the kick used with the breaststroke. The forceful whipping together of the lower legs that provides the propulsive thrust of the kick often forces the lower leg into slight abduction at the knee, with subsequent irritation to the medial collateral ligament and the medial border of the patellar tract. A survey of 391 competitive swimmers revealed incidences of knee pain among 73% of the breaststroke specialists and 48% of nonbreaststrokers (21). In a study of breaststroke kinematics, it was found that angles of hip abduction of less than 37 degrees or greater than 42 degrees at the initiation of the kick resulted in a dramatically increased incidence of knee pain (21). The angle of abduction at the hip during execution of the whip kick is illustrated in Figure 7-17.

Chondromalacia

An imbalance in tension on the medial and lateral sides of the patella can cause laterally deviated tracking of the patella during knee flexion/extension movements. Tissue irritation and pain result. This condition, known as *chondromalacia,* often results from an imbalance between the strength of the vastus medialis and the vastus lateralis. The vastus medialis is commonly significantly weaker than the lateralis. Femoral torsion, a condition involving inward rotation of the femur with respect to the tibia, can also be a contributing factor. If the patellar misalignment is not extremely severe, simple strengthening of the quadriceps muscles (particularly the vastus medialis) can alleviate or even eliminate the symptoms. Patellofemoral alignment problems are more frequent in females than in males (23).

■ Shin splints is a term often ascribed to any pain emanating from the anterior aspect of the lower leg. It refers specifically to irritation of the attachment of the tibialis posterior to the deep mid third of the medial border of the tibia (9).

Shin Splints

Generalized pain along the medial border of the tibia is commonly known as *shin splints.* This is an overuse injury associated with running that may involve microdamage to the muscle attachments on the deep mid third of the medial border of the tibia and/or inflammation of the periosteum (6). Common causes of the condition include running on a hard surface or running uphill. A change in running conditions or rest usually alleviates shin splints (10).

STRUCTURE OF THE ANKLE

The ankle region includes the tibiotalar, fibulotalar, and distal tibiofibular joints (Figure 7-18). Most motion at the ankle occurs at the tibiotalar hinge joint, where the convex surface of the superior talus articulates with the concave surface of the distal tibia. All three articulations are enclosed in a joint capsule that is thickened on the medial side and extremely thin on the posterior side. Three ligaments—the anterior and posterior talofibular and the calcaneofibular—reinforce the joint capsule laterally. The four bands of the deltoid ligament contribute to joint stability on the medial side. The ligamentous structure of the ankle is displayed in Figure 7-19.

MOVEMENTS OF THE ANKLE

The axis of rotation at the ankle is primarily frontal, although it is slightly oblique and its orientation changes somewhat as rotation occurs at the joint. As shown in Figure 7-20, the movements at the ankle joint are largely flexion/extension movements that are termed *dorsiflexion* and *plantar flexion* (see Chapter 2).

■ Muscles with tendons passing anterior to the malleoli are dorsiflexors; those posterior to the malleoli serve as plantar flexors.

The medial and lateral malleoli serve as pulleys that channel the tendons of many muscles crossing the ankle so that they are either posterior or anterior to the axis of rotation. The forces produced can therefore contribute to dorsiflexion or plantar flexion.

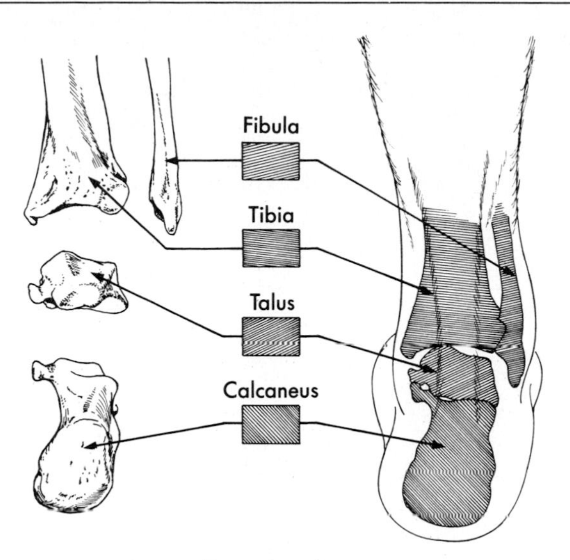

Figure 7-18
The bony structure of the ankle.

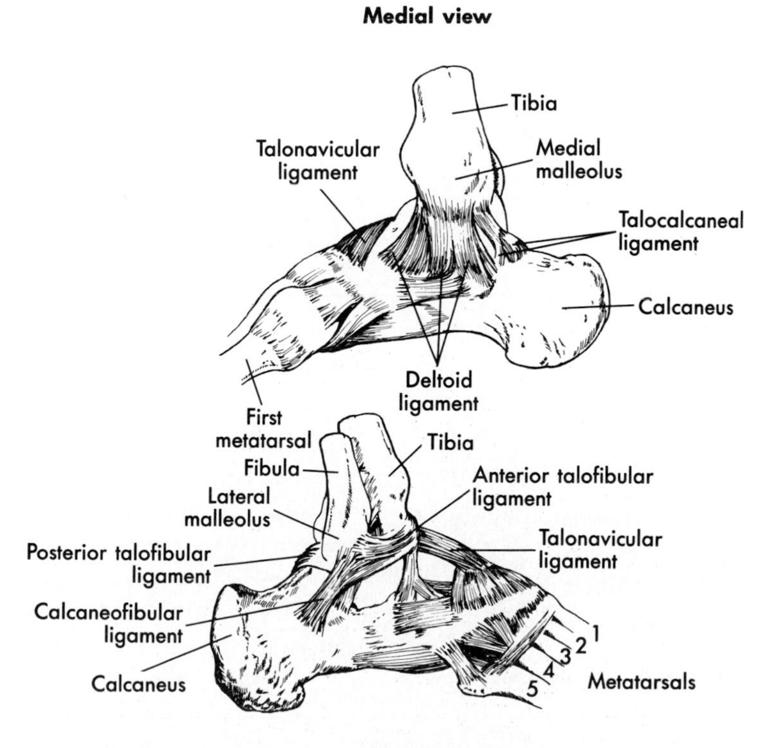

Figure 7-19
The ligaments of the ankle.

Figure 7-20

Sagittal plane movements of the ankle.

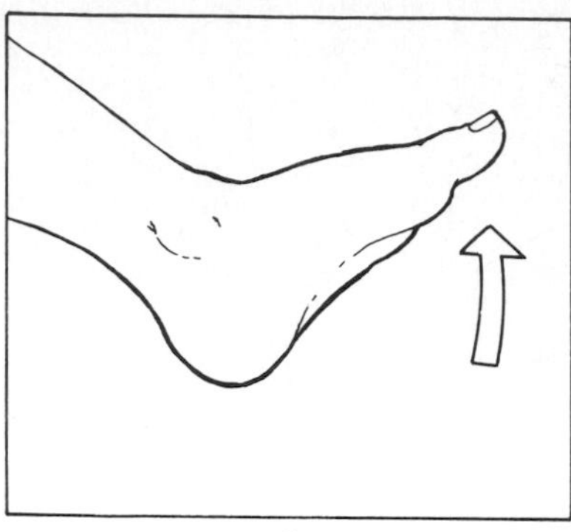
Dorsiflexion

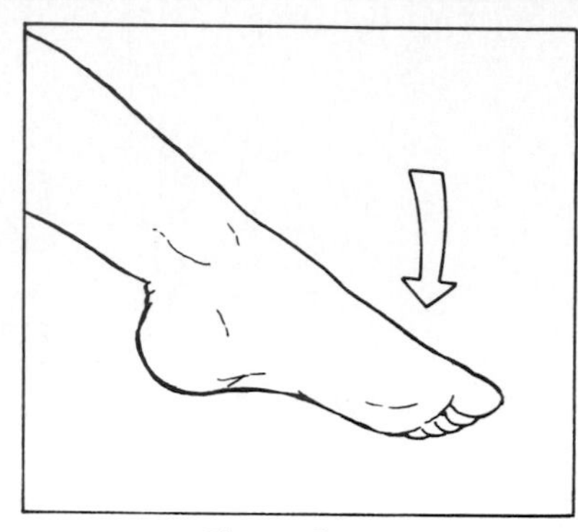
Plantar flexion

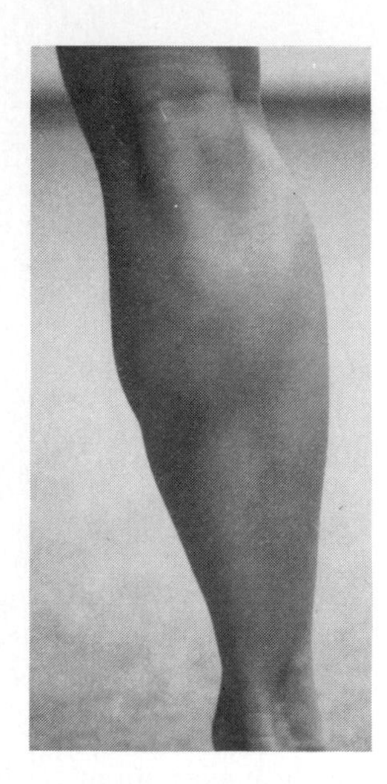

The two heads of the gastrocnemius dominate the posterior side of the lower leg.

Dorsiflexion

The tibialis anterior, extensor digitorum longus, and peroneus tertius are the prime dorsiflexors of the foot. The extensor hallucis longus assists in dorsiflexion (Figure 7-21).

Plantar Flexion

The major plantar flexors are the two heads of the powerful two-joint gastrocnemius and the soleus, which lies beneath the gastrocnemius (Figure 7-22). Assistant plantar flexors include the tibialis posterior, peroneus longus, peroneus brevis, plantaris, flexor hallucis longus, and flexor digitorum longus (Figure 7-23).

Figure 7-21

The dorsiflexors of the ankle.

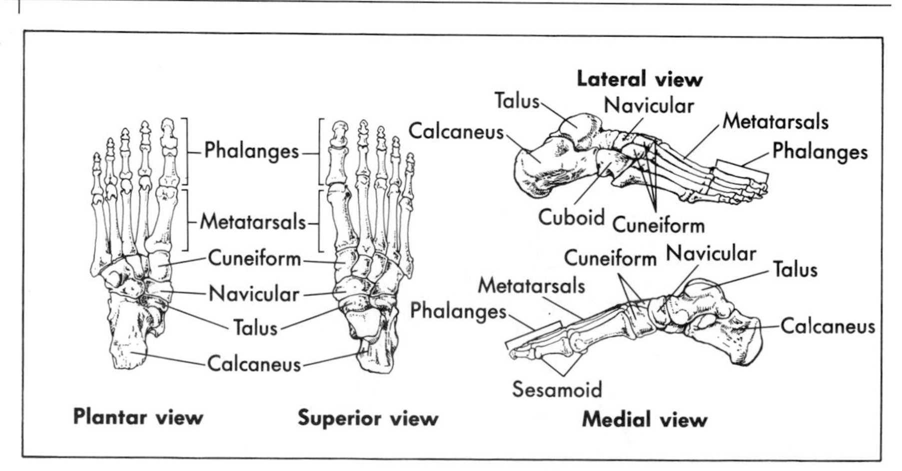

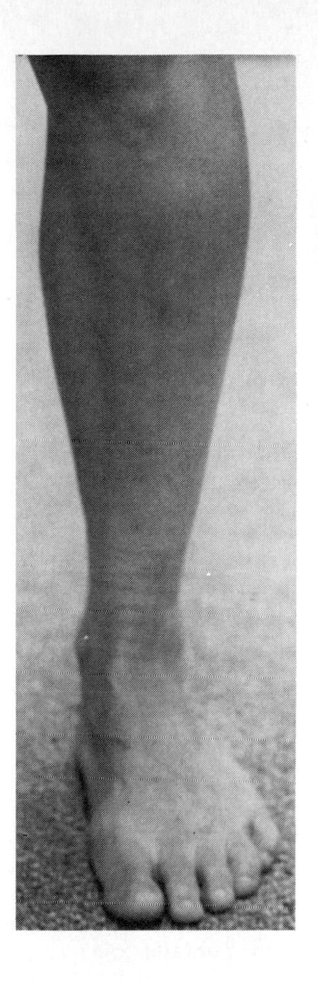

The tibialis anterior is easy to locate on the anterior lower leg.

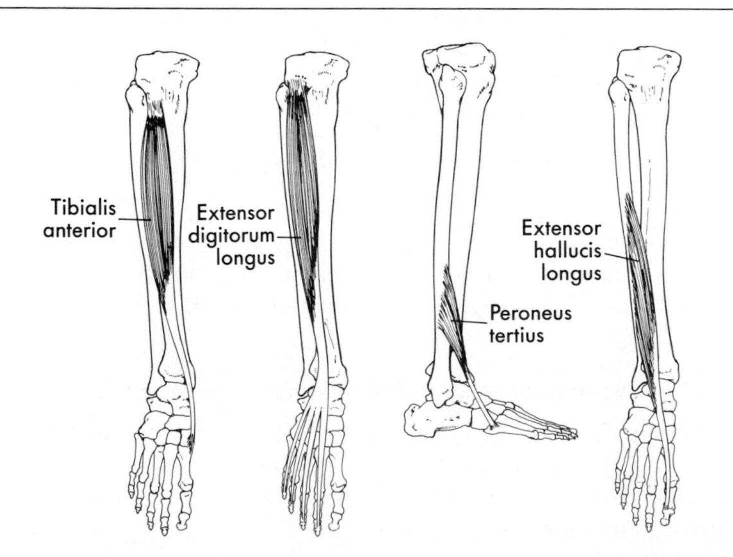

Figure 7-22
The major plantar flexors of the ankle.

Figure 7-23
Muscles with tendons passing posterior to the malleoli assist with plantar flexion of the ankle.

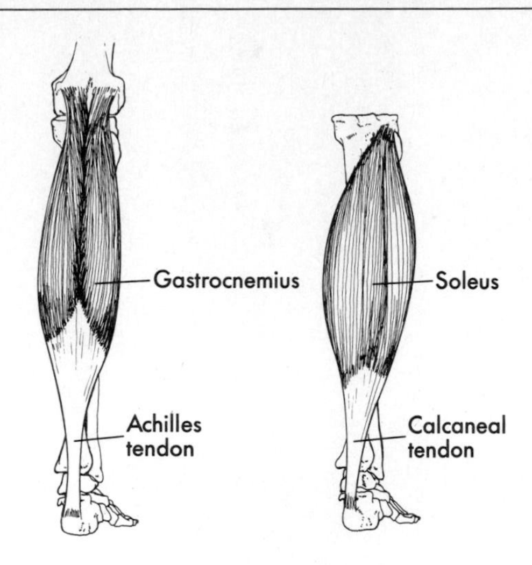

STRUCTURE OF THE FOOT

Like the hand, the foot is a multibone structure containing 26 bones with numerous articulations (Figure 7-24). Included are the subtalar and midtarsal joints and several tarsometatarsal, intermetatarsal, metatarsophalangeal, and interphalangeal joints. These joints function together to produce the rather complex combination of movements of the foot.

Subtalar Joint

Anterior and posterior facets on the inferior side of the talus articulate with the superior calcaneus to form the subtalar joint. Four talocalcaneal ligaments join the talus and the calcaneus. The joint is essentially uniaxial, with an alignment slightly oblique to the conventional descriptive planes of motion.

Tarsometatarsal and Intermetatarsal Joints

Both the tarsometatarsal and intermetatarsal joints are nonaxial, with the bone shapes and the restricting ligaments permitting only gliding movements. These joints enable the foot to function as a semirigid unit or to adapt flexibly to uneven surfaces during weight bearing.

Metatarsophalangeal and Interphalangeal Joints

The metatarsophalangeal and interphalangeal joints are similar to their counterparts in the hand, with the former being condyloid joints and the latter being hinge joints. Numerous ligaments provide reinforcement for these joints.

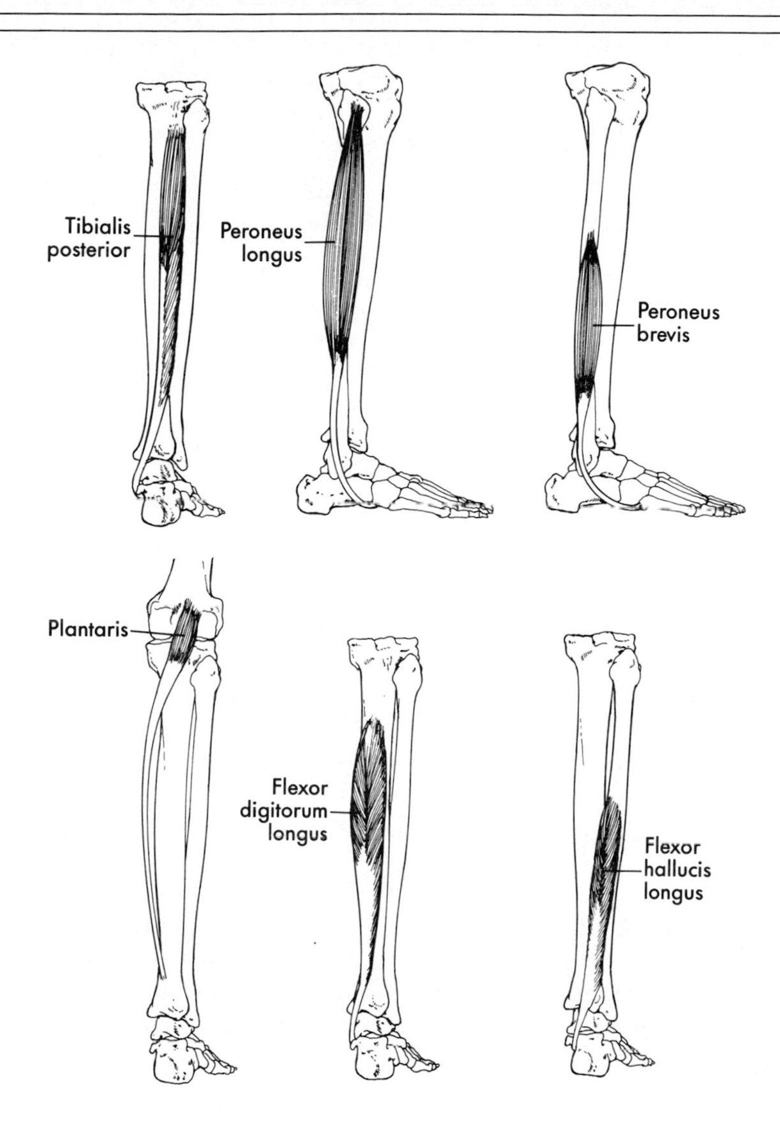

Figure 7-24
The foot is composed of numerous articulating bones.

Plantar Fascia

Thick, fibrous, interconnected bands of connective tissue known as the **plantar fascia** extend over the plantar surface of the foot (Figure 7-25). Tension in the plantar fascia supports the longitudinal arch of the foot. When the plantar fascia is stretched during weight bearing, it functions as a spring to store mechanical energy that is used to push the foot off the support surface.

plantar fascia
thick bands of fascia that cover the plantar aspect of the foot

Figure 7-25
The plantar fascia.

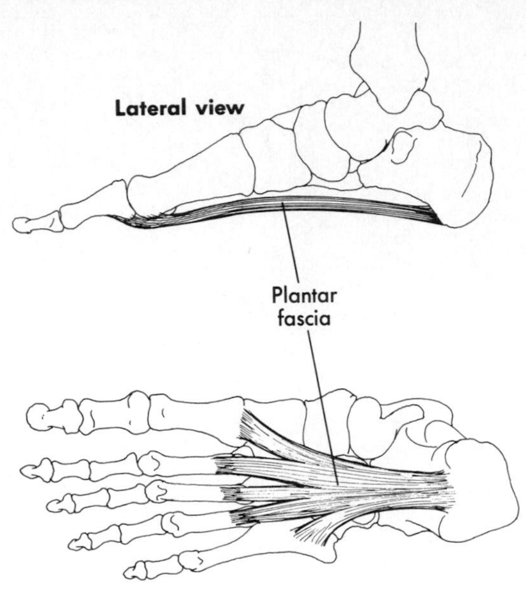

MOVEMENTS OF THE FOOT

Muscles of the Foot

The locations and primary actions of the major muscles of the ankle and foot are summarized in Table 7-3.

Flexion and Extension

Flexion involves the curling under of the toes. The muscles responsible for this action are the flexor hallucis longus and flexor digitorum longus. Extension and hyperextension of the toes result from the actions of the extensor hallucis longus and extensor digitorum longus.

Inversion and Eversion

Rotational movements of the foot in the frontal plane are termed inversion and eversion (see Chapter 2). These movements occur largely at the subtalar joint, although gliding actions among the intertarsal and tarsometatarsal joints also contribute (2). Inversion results in the sole of the foot turning inward toward the midline of the body. The tibialis posterior and tibialis anterior are the main muscles involved. Turning the sole of the foot outward is termed *eversion*. The muscles primarily responsible for eversion are the peroneus longus and peroneus brevis, both with long tendons coursing around the lateral malleolus. The peroneus tertius assists. Inversion and eversion are shown in Figure 7-26.

Table 7-3
MUSCLES OF THE ANKLE AND FOOT

MUSCLE	PROXIMAL ATTACHMENT	DISTAL ATTACHMENT	PRIMARY ACTIONS
Tibialis anterior	Upper two-thirds lateral tibia	Medial surface of first cuneiform and first metatarsal	Dorsiflexion, inversion
Extensor digitorum longus	Upper three-fourths anterior fibula	Second and third phalanges of four lesser toes	Dorsiflexion, eversion
Peroneus tertius	Lower third anterior fibula	Dorsal surface of fifth metatarsal	Dorsiflexion, eversion
Extensor hallucis longus	Middle anterior fibula	Dorsal surface of distal phalanx of great toe	Extension
Gastrocnemius	Posterior medial and lateral condyles of femur	Tuberosity of calcaneus by Achilles tendon	Plantar flexion
Plantaris	Distal, posterior femur	Tuberosity of calcaneus by Achilles tendon	Assists with plantar flexion
Soleus	Posterior upper fibula, middle tibia	Tuberosity of calcaneus by Achilles tendon	Plantar flexion
Peroneus longus	Lateral upper two-thirds fibula	Lateral surface of first cuneiform and first metatarsal	Eversion
Peroneus brevis	Distal two-thirds fibula	Lateral fifth metatarsal	Eversion
Flexor digitorum longus	Posterior tibia	Distal phalanx of four lesser toes	Flexion
Flexor hallucis longus	Lower two-thirds posterior fibula	Distal phalanx of the great toe	Flexion
Tibialis posterior	Posterior upper two-thirds tibia, fibula	Cuboid, navicular, calcaneus, cuneiforms	Inversion

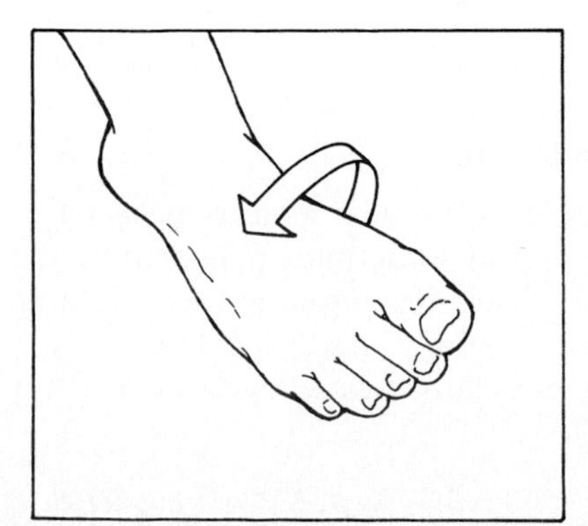

Inversion

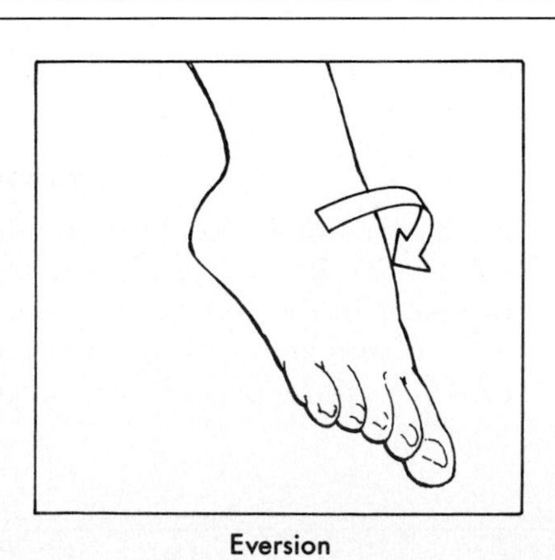
Eversion

Figure 7-26
Inversion and eversion are largely frontal plane movements.

Pronation and Supination

supination
combined conditions of plantar flexion, inversion, and adduction

pronation
combined conditions of dorsiflexion, eversion, and abduction

During walking and running the foot and ankle undergo a cyclical sequence of movements. As the heel contacts the ground, the rear portion of the foot typically inverts to some extent. When the foot rolls forward and the forefoot contacts the ground, plantar flexion of the ankle occurs. The combination of inversion, plantar flexion, and adduction of the foot is known as **supination** (see Chapter 2). While the foot supports the weight of the body during midstance, eversion and abduction often occur as the ankle moves into dorsiflexion. These movements are known collectively as **pronation.**

Pronation and supination are widely used terms in the popular running literature. Excessive pronation or supination are potential contributors to running-related injuries, and many styles of running shoes incorporate features designed to control these motions.

COMMON INJURIES OF THE ANKLE AND FOOT

Because of the crucial roles played by the ankle and foot during locomotion, injuries to this region can greatly limit mobility. Injuries of the lower extremity, especially those of the foot and ankle, may result in weeks or even months of lost training time for athletes, particularly runners. Among dancers, the foot and ankle are the most common sites of both chronic and acute injuries (4).

Ankle Injuries

■ Ankle sprains typically occur on the lateral side because of weaker ligamentous support than is present on the medial side.

Because the joint capsule and ligamentous support is stronger on the medial side of the ankle, inversion sprains involving stretching or rupturing of the lateral ligaments are more common than eversion sprains of the medial ligaments. Because of the protection by the opposite limb on the medial side, fractures in the ankle region also occur more often on the lateral than on the medial side. Ankle injuries make up 20% to 25% of all injuries associated with running and jumping sports resulting in a loss of training time (6). Among classic ballet and modern dancers, ankle sprains are the most common traumatic injury, with the usual cause being forced inversion of the ankle during incorrect landing from a jump while the foot is plantar flexed (4).

Overuse Injuries

Achilles tendinitis has been identified as the most common injury in sports (13). It involves inflammation and sometimes microrupturing of tissues in the Achilles tendon, typically accompanied by swelling. The inflammation of any tendon is usually a stress-related injury caused by the repeated sustenance of relatively low levels of force by the tendon. Two possible mechanisms for tendinitis have been proposed (13). The first is that repeated tension development results in fatigue and decreased flexibility in the muscle, increasing tensile load on the tendon even during relaxation of the muscle. The second theory is that repeated loading actually leads to failure or rupturing of the collagen threads in the tendon. Achilles tendinitis is usually associated with running and jumping activities and is extremely common among theatrical dancers (4). It has also been reported in skiers. Complete rupturing of the Achilles tendon occurs almost exclusively in male skiers, although incidence of the injury has decreased with the advent of high, rigid ski boots and effective release bindings (11).

Mechanical fatigue fractures, commonly known as *stress fractures,* occur relatively frequently in the bones of the lower extremity. Among a group of 320 athletes with bone-scan-positive stress fractures, the bone most frequently injured was the tibia (49.1%), followed by the tarsals (25.3%), metatarsals (8.8%), femur (7.2%), fibula (6.6%), and pelvis (1.6%) (9). The sites most frequently injured in the older athletes in the group were the femur and the tarsals, with the fibula and tibia most often injured among the younger athletes. Among runners, factors associated with stress fractures include forefoot striking (toe-heel gait), running on hard surfaces such as concrete, and alignment anomalies of the trunk and/or lower extremity (13). Stress fractures among dancers occur most frequently to the second and third metatarsals and appear to be related to dancing on overly hard surfaces (4).

Repetitive stretching of the plantar fascia can result in plantar fascitis, a condition characterized by microtears and inflammation of the plantar fascia near its attachment to the calcaneus. The symptoms are pain in the heel and/or arch of the foot. The condition is the fourth most common cause of pain among runners and also occurs with some frequency among basketball players, tennis players, gymnasts, and dancers (8). Anatomical contributors to the likelihood of plantar fascitis include pes planus (flat foot), a rigid cavus (high arch) foot, and a tight Achilles tendon, all of which generally reduce the foot's shock-absorbing capability (19).

Although advertising often addresses the concept of proper footwear for the prevention of lower extremity injuries, Robbins and Hanna (14) have stated that the intrinsic muscles of the foot remain inactive during running with shoes but become active and may serve as effective shock absorbers during barefoot running. Because of the rough surfaces and temperature extremes present on man-made surfaces, however, they advocate the development of footwear promoting activation of the foot muscles during running rather than the performance of barefoot running.

Alignment Anomalies of the Foot

varus
a condition of inward deviation in alignment from the proximal to the distal end of a body segment

valgus
a condition of outward deviation in alignment from the proximal to the distal end of a body segment

Different misalignments of the foot may lead to abnormal gait patterns contributing to the development of overuse injuries. Although walking normally involves approximately 6 to 8 degrees of pronation at the subtalar joint, individuals with pes planus undergo 10 to 12 degrees of pronation (13). Excessive pronation has been documented among 60% of one group of injured runners (5). Overpronation of the foot produces a compensatory inward rotation of the tibia, which results in increased stress within the Achilles tendon. **Varus** and **valgus** conditions (inward and outward lateral deviation of a body segment respectively) may be present among all of the major links of the lower extremity. Forefoot varus and forefoot valgus refer to inversion and eversion misalignments of the metatarsals, and rearfoot varus and valgus involve inversion and eversion misalignments present at the subtalar joint (Figure 7-27). Varus and valgus conditions are sometimes present in both the tibia and femur and can alter the normal kinematics and kinetics of joint motion because they result in added tensile stress on the stretched side of the affected joint or joints. A combination of femoral varus and tibial valgus (a *knock-knee* condition) places added tension on the medial aspect of the knee (Figure 7-28). In contrast, *bow-legged* condition of femoral valgum and tibial varus stresses the lateral aspect of the knee and is therefore a predisposing factor for iliotibial band friction syndrome. Varus and valgus conditions may be congenital or may arise from an imbalance in muscular strength. Unfortunately, lateral misalignments at one joint of the lower extremity are typically

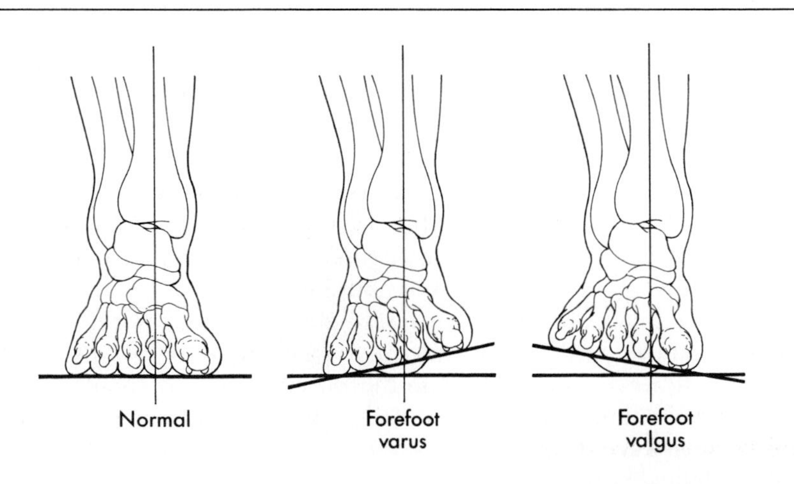

Figure 7-27
Varus and valgus conditions of the forefoot.

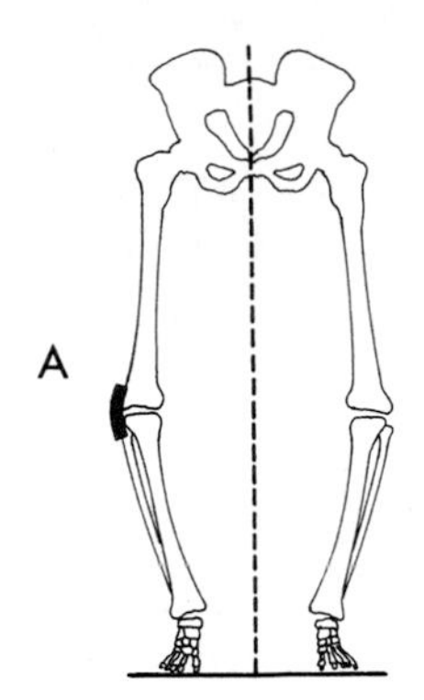

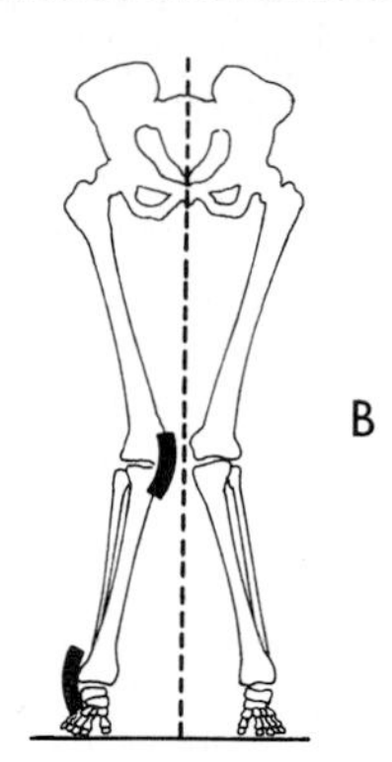

Figure 7-28
A, Femoral valgus and tibial varus. **B,** Femoral varus and tibial valgus.

accompanied by compensatory misalignments at other lower extremity joints because of the nature of joint loading during weight bearing.

Depending on the cause of the misalignment problem, correctional procedures may involve exercises to strengthen specific muscles and/or ligaments of the lower extremity and the use of orthotics, specially designed inserts worn inside the shoe to provide added support for a portion of the foot.

▪ Misalignment at a lower extremity joint typically results in compensatory misalignments at one or more other joints because of the lower extremity's weight-bearing function.

SUMMARY

Whereas the upper extremity is specialized for activities requiring large ranges of motion, the lower extremity is well adapted to its functions of weight bearing and locomotion. This is particularly evident at the hip, where the bony structure and several large, strong ligaments provide considerable joint stability. The hip is a typical ball and socket joint, with flexion, extension, abduction, adduction, horizontal abduction, horizontal adduction, medial and lateral rotation, and circumduction of the femur permitted.

The knee is a large, complex joint composed of two side-by-side condyloid articulations. Medial and lateral semilunar cartilages (menisci) improve the fit between the articulating bone surfaces and assist in absorbing forces transmitted across the joint. Because of differences in the sizes, shapes, and orientations of the medial and lateral articulations, medial rotation of the tibia accompanies full knee extension. A number of ligaments cross the knee and restrain its mobility. The primary movements allowed at the knee are flexion and extension, although some rotation of the tibia is also possible when the knee is in flexion and not bearing weight.

The ankle includes the articulations of the tibia and fibula with the talus. This is a hinge joint, which is reinforced both laterally and medially by ligaments. Movements at the ankle joint are dorsiflexion and plantar flexion.

Like the hand, the foot is composed of numerous small bones and their articulations. Movements of the foot include inversion and eversion, abduction and adduction, and flexion and extension of the toes.

INTRODUCTORY PROBLEMS

1. Construct a chart listing all muscles crossing the hip joint according to whether they are anterior, posterior, medial, or lateral to the joint center. Note that some muscles may fall into more than one category.
2. Identify the action or actions performed by the muscles listed in the four categories in Problem 1.
3. Construct a chart listing all muscles crossing the knee joint according to whether they are anterior, posterior, medial, or lateral to the joint center. Note that some muscles may fall into more than one category.
4. Identify the action or actions performed by the muscles listed in the four categories in Problem 3.
5. Construct a chart listing all muscles crossing the ankle joint according to whether they are anterior, posterior, medial, or lateral to the joint center. Note that some muscles may fall into more than one category.
6. Identify the action or actions performed by the muscles listed in the four categories in Problem 5.
7. Compare the structure of the hip (including bones, ligaments, and muscles) to the structure of the shoulder. What are the relative advantages and disadvantages of the two joint structures?
8. Compare the structure of the knee (including bones, ligaments, and muscles) to the structure of the elbow. What are the relative advantages and disadvantages of the two joint structures?
9. Identify five activities requiring either medial or lateral rotation of the femur and discuss the importance of this rotation to successful performance of each activity.
10. Describe sequentially the movements of the lower extremity that occur during kicking a ball. Include sufficient detail that the reader of your answer can visualize the movement.

ADDITIONAL PROBLEMS

1. Identify the sequence of movements occurring at the hip, knee, and ankle joints during the performance of a vertical jump.
2. Which muscles are most likely to serve as agonists to produce each of the movements identified in your answer to Problem 1?
3. What are the differences in motion at the joints of the lower extremity during a soccer style kick and a straight-on kick?
4. What differences are most likely to occur in the actions of the muscles serving as agonists to produce the differences in joint motion identified in your answer to Problem 3?
5. Explain the roles of two-joint muscles in the lower extremity, providing specific examples. How does the orientation of the limbs articulating at one joint influence the action of a two-joint muscle at the other joint?
6. Which muscles of the lower extremity are called on more for running uphill than for running on a level surface? For running downhill as compared to running on a level surface? Explain why.
7. What differences would you expect to find between the feet of people of modernized societies in which rigid shoes are habitually worn and the feet of primitive peoples?
8. Controversy exists as to whether the performance of deep knee bend exercises is beneficial or harmful. Explain the rationale for both sides of this issue based on your knowledge of the anatomical structure of the knee and the properties of collagenous tissues.
9. Describe the sequencing of joint actions in the ankle and foot during the support phase of walking or running.

REFERENCES

1. Beaupre A et al: Knee menisci: correlation between microstructure and biomechanics, Clin Orthop 208:72, 1986.
2. Gowitzke BA and Milner MM: Scientific bases of human movement, ed 3, Baltimore, 1988, Williams & Wilkins.
3. Grady JF, O'Connor KJ, and Bender J: Iliotibial band syndrome, J Am Podiatr Med Assoc 76:558, 1986.
4. Hardaker WT, Margello S, and Goldner JL: Foot and ankle injuries in theatrical dancers, Foot Ankle 6:59, 1985.
5. James SL, Bates BT, and Osternig LR: Injuries to runners, Am J Sports Med 6:40, 1978.
6. Johnson RE and Rust RJ: Sports related injury: an anatomic approach, part 2, Minn Med 68:829, 1985.

7. Johnston RC: Mechanical considerations of the hip joint, Arch Surg 107:411, 1973.
8. Leach RE, Seavey MS, and Salter DK: Results of surgery in athletes with plantar fascitis, Foot Ankle 7:156, 1986.
9. Matheson GO et al: Stress fractures in athletes, Am J Sports Med 15:46, 1987.
10. Mercier LR: Practical orthopedics, ed 2, Chicago, 1987, Year Book Medical Publishers, Inc.
11. Oden RR: Tendon injuries about the ankle resulting from skiing, Clin Orthop 216:63, 1987.
12. Olson DT: Iliotibial band friction syndrome, Athletic Training 21:32, 1986.
13. Renstrom P and Johnson RJ: Overuse injuries in sports: a review, Sports Med 2:316, 1985.
14. Robbins SE and Hanna AM: Running-related injury prevention through barefoot adaptations, Med Sci Sports Exerc 19:148, 1987.
15. Seedhom BB, Dowson D, and Wright V: The load-bearing function of the menisci: a preliminary study. In Ingwerson OS et al, eds: The knee joint: recent advances in basic research and clinical aspects, Amsterdam, 1974, Excerpta Medica.
16. Soderberg GL: Kinesiology: application to pathological motion, Baltimore, 1986, Williams & Wilkins.
17. Stanish WD et al: Posterior cruciate ligament tears in wrestlers Can J A Sport Sci 11:173, 1986.
18. Sutton G: Hamstrung by hamstring strains: a review of the literature, J Orthop Sports Phys Ther 5:184, 1984.
19. Tanner SM and Harvey JS: How we manage plantar fascitis, Physician Sportsmed 16:39, 1988.
20. Terry GC, Hughston JC, and Norwood LA: The anatomy of the iliopatellar band and iliotibial tract, Am J Sports Med 14:39, 1986.

21. Vizsolyi P et al: Breaststroker's knee, Am J Sports Med 15:63, 1987.
22. Wroble RR et al: Patterns of knee injuries in wrestling: a six year study, Am J Sports Med 14:55, 1986.
23. Yates C and Grana WA: Patellofemoral pain: a prospective study, Orthopedics, 9:663, 1986.

ANNOTATED READINGS

Bates BT et al: Lower extremity function during the support phase of running. In Asmussen E and Jorgensen K, eds: Biomechanics VI-B: proceedings of the sixth international symposium on biomechanics, Baltimore, 1978, University Park Press.

Reports results of research on dynamic functional relationships during the support phase of running, particularly the nature, duration, and extent of pronation occurring.

Nigg BM et al: Gait analysis and sport-shoe construction. In Asmussen E and Jorgensen K, eds: Biomechanics VI-B: proceedings of the sixth international symposium on biomechanics, Baltimore, 1978, University Park Press.

Reports results of research based on film and force measurements on the influence of shoe arch support on gait characteristics.

Nordin M and Frankel VH: Biomechanics of the hip. In Nordin M and Frankel VH, eds: Basic biomechanics of the musculoskeletal system, ed 2, Philadelphia, 1989, Lea & Febiger.

Discusses kinematics and kinetics of hip function, including a section on solution of unknown coplanar forces in joint reactions.

Sammarco GJ: Biomechanics of the foot. In Nordin M and Frankel VH, eds: Basic biomechanics of the musculoskeletal system, ed 2, Philadelphia, 1989, Lea & Febiger.

Describes the functioning of the various joints of the foot, the motion of the foot during gait, the function of the plantar fascia, and muscle control in the foot.

8 MOVEMENT

The Biomechanics of the Spine and Pelvis

After reading this chapter, the student will be able to:

Explain the ways in which anatomical structure affects the movement capabilities of the spine.

Identify the factors that determine the relative mobility and stability of different regions of the spine.

Identify the functions of the spine and pelvic girdle and explain the ways in which the spine and pelvic girdle are adapted to carry out these functions.

Explain the relationship between muscle location and the nature and effectiveness of muscle action in the trunk.

Describe the biomechanical contributors to common injuries of the spine.

The spine is the most complex and functionally significant segment of the human body. Providing the linkage between the upper and lower extremities, the spine enables motion in all three planes yet still functions as a bony protector of the delicate spinal cord.

To many researchers and clinicians, the lumbar region of the spine is of particular interest because low back pain is a major medical and socioeconomic problem of modern times. Low back problems are especially common among certain populations of athletes, including female gymnasts and football players, particularly linemen (9, 17, 25).

STRUCTURE OF THE SPINE

Vertebral Column

The spine consists of a curved stack of 33 vertebrae divided structurally into five regions (Figure 8-1). Proceeding from superior to inferior, there are 7 cervical vertebrae, 12 thoracic vertebrae, 5 lumbar vertebrae, 5 fused sacral vertebrae, and 4 small, fused coccygeal vertebrae. There may be one extra vertebra or one less, particularly in the lumbar region.

Because of structural differences, varying amounts of movement are permitted between adjacent vertebrae in the cervical, thoracic, and lumbar portions of the spine. Within these regions, two adjacent vertebrae and the soft tissues between them are known as a **motion segment.** The motion segment is considered to be the functional unit of the spine (Figure 8-2).

motion segment
two adjacent vertebrae and the intervening soft tissues of the joint; the functional unit of the spine

Figure 8-1
Vertebral structure and spinal curvature determine the five regions of the spine.

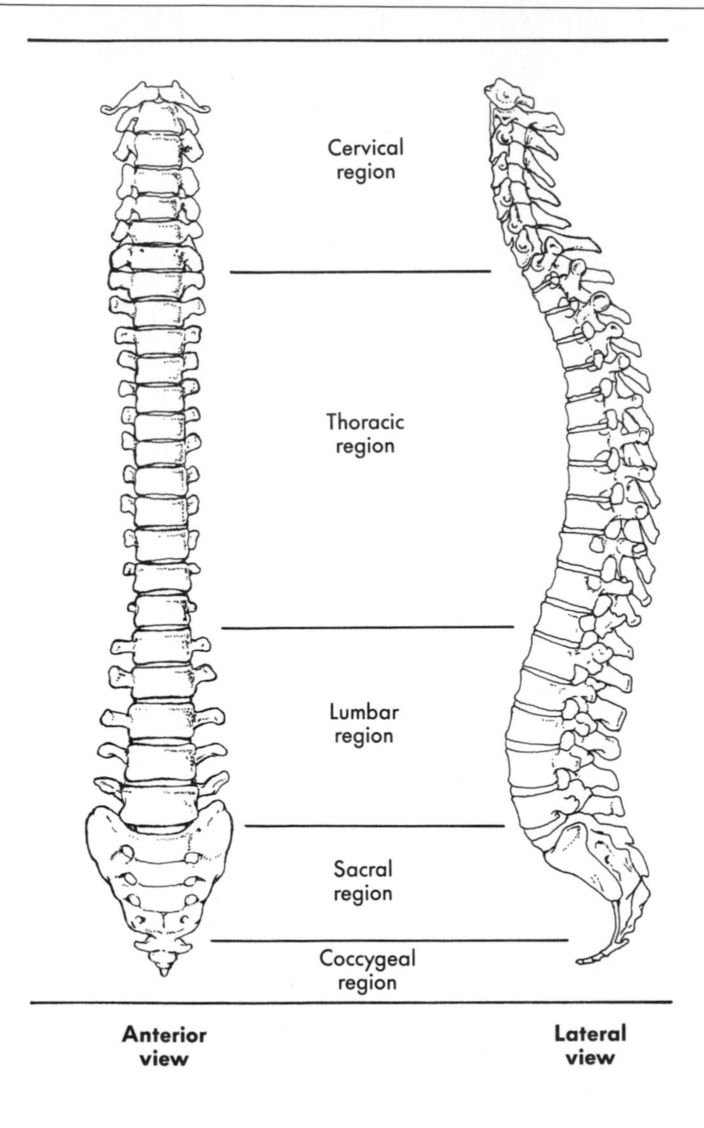

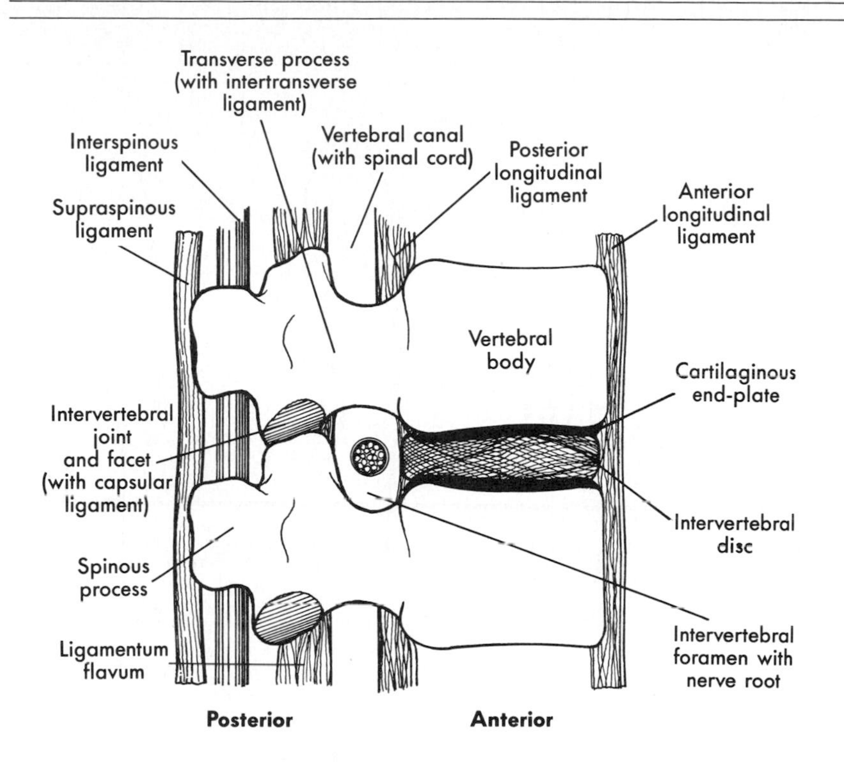

Figure 8-2
The motion segment, composed of two adjacent vertebrae and the soft tissues of the joint, is the functional unit of the spine.

Spinal Curves

As viewed in the sagittal plane, the spine contains four normal curves. The thoracic and sacral curves, which are concave anteriorly, are present at birth and are referred to as **primary curves.** The lumbar and cervical curves, which are concave posteriorly, develop from supporting the body in an upright position after young children begin to sit up and stand. Since these curves are not present at birth, they are known as the **secondary spinal curves.** Although the cervical and thoracic curves change little during the growth years, the curvature of the lumbar spine increases approximately 10% between the ages of 7 and 17 (37). Spinal curvature (posture) is influenced by heredity, pathological conditions, an individual's mental state, and the forces to which the spine is habitually subjected.

primary spinal curves
curves that are present at birth

secondary spinal curves
the cervical and lumbar curves, which do not develop until the weight of the body begins to be supported in sitting and standing positions

As discussed in Chapter 3, bones are constantly modeled or shaped in response to the magnitudes and directions of the forces acting on them. Similarly, the four spinal curves can become distorted when the spine is habitually subjected to asymmetrical forces. Distorted spinal curvatures are shown in Figure 8-3.

Figure 8-3
Abnormal spinal curvatures.

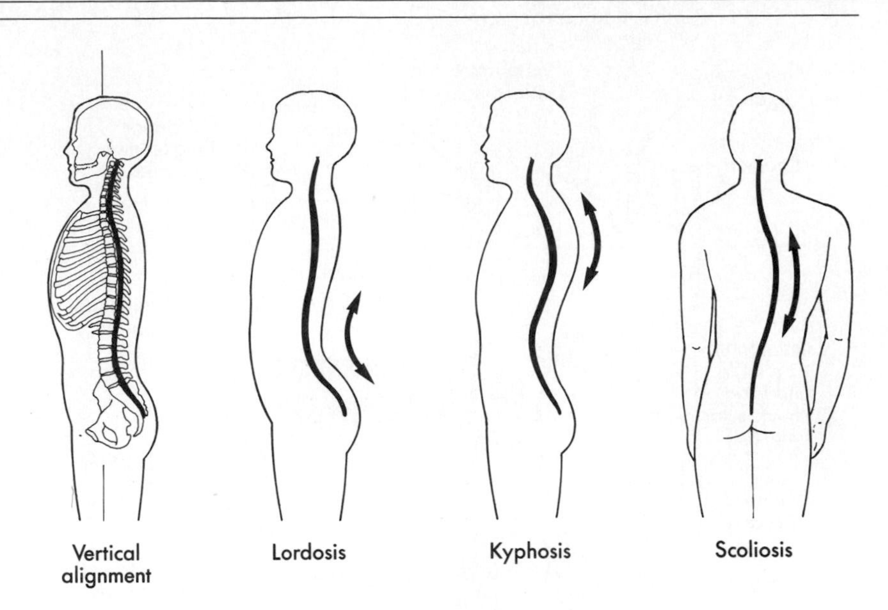

The lumbar curve does not reach full development until approximately age 17.

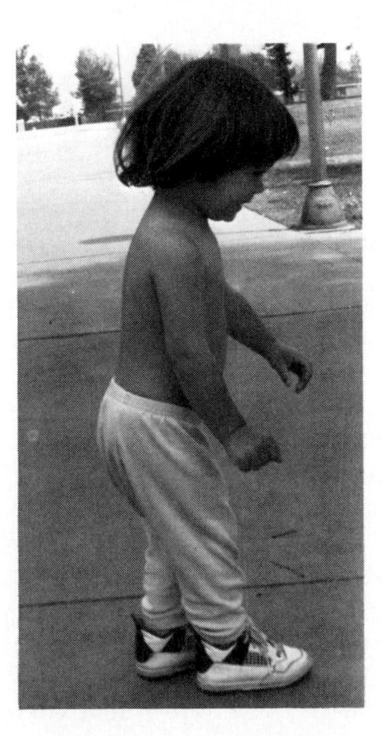

The condition of accentuated lumbar curvature is known as **lordosis.** It typically results from a strength imbalance between the lower back muscles and weakened abdominal muscles. Anterior tilt of the pelvis frequently accompanies lordosis and further contributes to the stretching of the abdominal muscles. This condition is the most common cause of postural low back pain (7). Lordosis accompanied by anterior pelvic tilt predisposes many adolescent athletes for the development of low back pain (14).

lordosis
an extreme curvature in the lumbar region of the spine

Another abnormality in spinal curvature is **kyphosis** (exaggerated thoracic curvature). Kyphosis often results from Scheuermann's disease in which one or more wedge-shaped vertebrae develop because of abnormal epiphyseal plate behavior. The condition has been called *swimmer's back* because it is frequently seen in adolescents who have trained heavily with the butterfly stroke (19). Occasionally the growth plate irregularities of Scheuermann's disease are found in the lumbar vertebrae as well. The condition is typically treated through bracing. Kyphosis often develops in elderly women with osteoporosis (36).

kyphosis
an extreme curvature in the thoracic region of the spine

Lateral deviation or deviations in spinal curvature are referred to as **scoliosis.** Small lateral deviations in spinal curvature are relatively common and may result from a habit such as carrying books or a heavy purse on one side of the body every day. Severe scoliosis is characterized by extreme lateral deviation and localized rotation of the spine, can be painful and deforming, and is treated clinically.

scoliosis
a lateral spinal curvature

Vertebrae

A typical vertebra consists of a body and, posterior to the body, a bony ring known as the neural arch (Figure 8-4). The interior surfaces of the neural arches and the attached vertebral bodies form a protective passageway for the spinal cord. From the exterior surface of each neural arch, several bony processes protrude. The spinous and transverse processes are attachment sites for muscles. The superior and inferior articular processes mate with the corresponding reciprocal articular processes of the vertebrae above and below to form the facet joints.

■ Although all vertebrae have the same basic shape, there is a progressive superior-inferior increase in the size of the vertebral bodies and a progression in the size and orientation of the articular processes.

From the cervical through the lumbar regions, there is a progression in vertebral size and in the orientation of the articular facets (Figure 8-5). For example, the bodies of the lumbar vertebrae are larger and thicker than those located in the more superior regions of the spine. Because each vertebra must support the weight of the portion of the body above it, the increased surface area of the lumbar vertebrae reduces the amount of stress to which these vertebrae would otherwise be subjected. The size and orientation of the spinous and transverse processes and of the articular facets also varies with vertebral location. The shapes of the articulating surfaces of the facet joints physically limit the ranges of movement possible at

■ The orientation of the facet joints determines the movement capabilities of the motion segment.

■ Although most of the load sustained by the spine is borne by the symphysis joints, the facet joints may play an assistive role, particularly when the spine is in hyperextension and when disc degeneration has occurred.

Figure 8-4
Vertebral structure, as seen in a midthoracic vertebra.

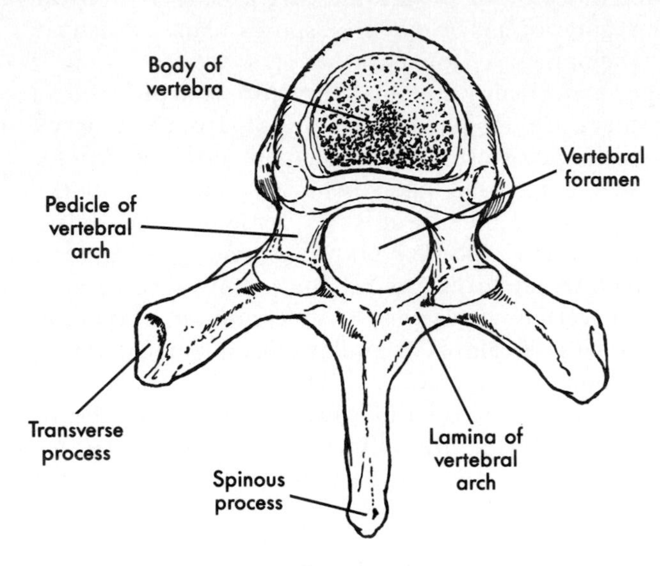

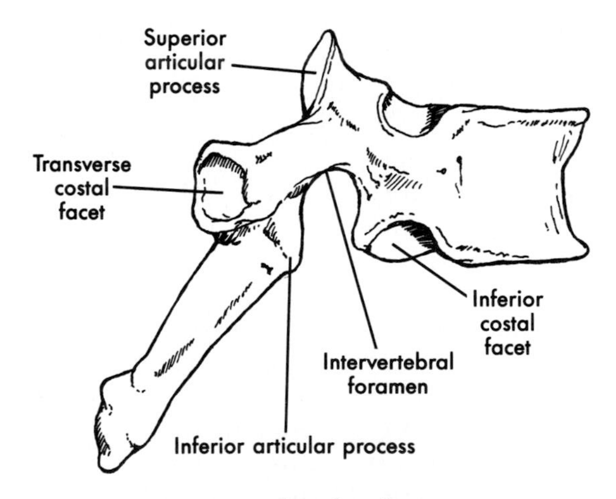

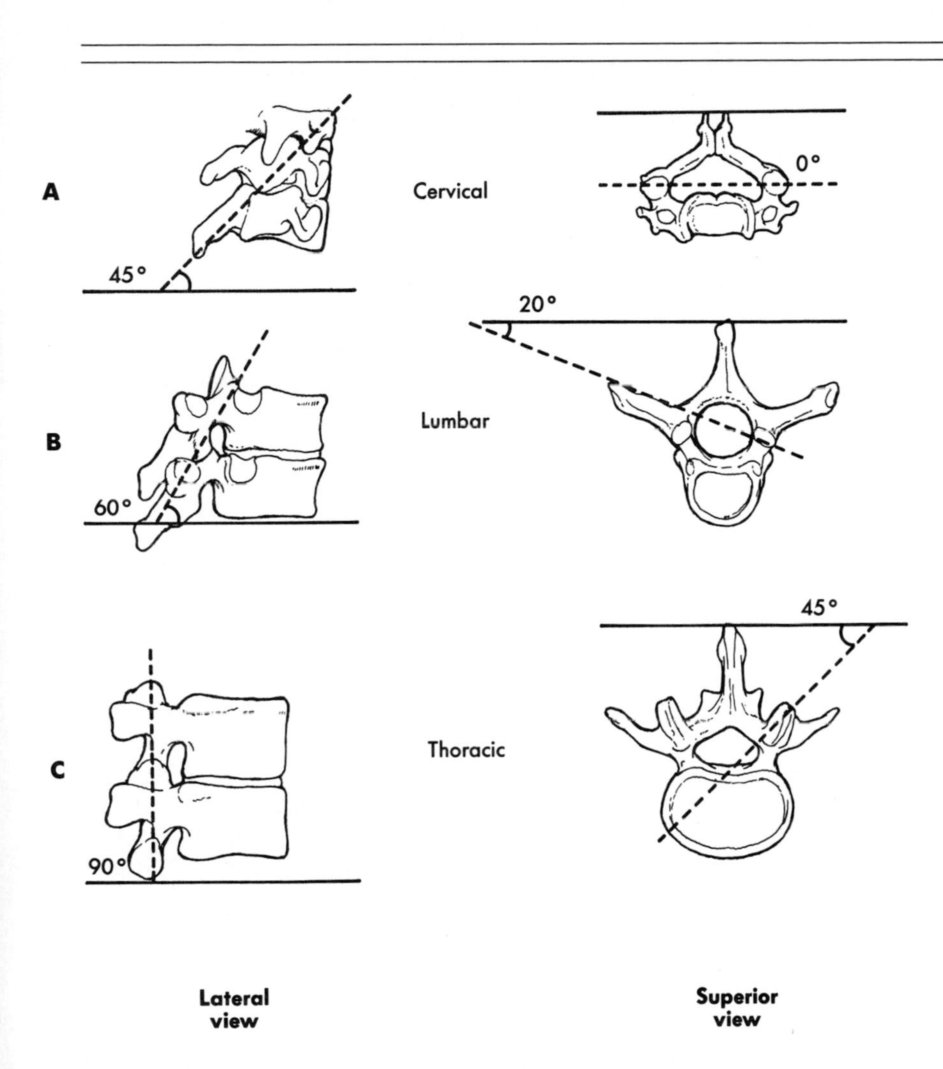

Figure 8-5

Approximate orientations of the facet joints. **A,** Lower cervical spine, with facets oriented 45 degrees to the transverse plane and parallel to the frontal plane. **B,** Thoracic spine, with facets oriented 60 degrees to the transverse plane and 20 degrees to the frontal plane. **C,** Lumbar spine, with facets oriented 90 degrees to the transverse plane and 45 degrees to the frontal plane.

Figure 8-6
Hyperextension of the lumbar spine creates compression at the facet joints.

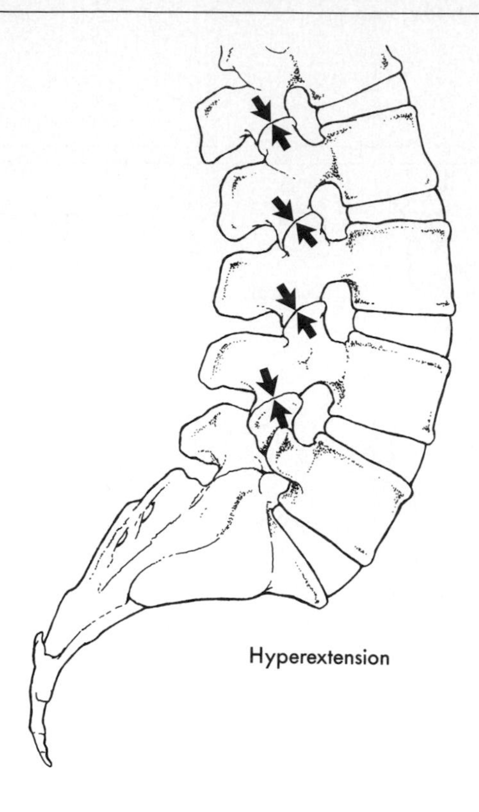

■ The spine may be viewed as a triangular stack of articulations, with symphysis joints between vertebral bodies on the anterior side and two gliding diarthrodial facet joints on the posterior side.

different levels of the spine. In addition to channeling the movement of the motion segment, the facet joints assist in sustaining approximately 30% of the loads on the spine, particularly when the spine is in hyperextension (Figure 8-6) (24).

Intervertebral Discs

The articulations between adjacent vertebral bodies are symphysis joints with intervening fibrocartilaginous discs. These discs act as cushions between the vertebrae. Healthy intervertebral discs in an adult account for approximately one fourth of the height of the spine. When the trunk is erect, the differences in the anterior and posterior thicknesses of the discs produce the locations and sizes of the lumbar, thoracic, and cervical curves of the spine.

The intervertebral disc is composed of two functional structures: A thick outer ring composed of fibrous cartilage called the **annulus fibrosus** or annulus surrounds a central gelatinous material known as the **nucleus pulposus** or nucleus (Figure 8-7). The collagen fibers of the annulus crisscross vertically, making the structure especially

annulus fibrosus
a thick, fibrocartilaginous ring that forms the exterior portion of the intervertebral disc

nucleus pulposus
a colloidal gel with a high fluid content, located inside the annulus fibrosus of the intervertebral disc

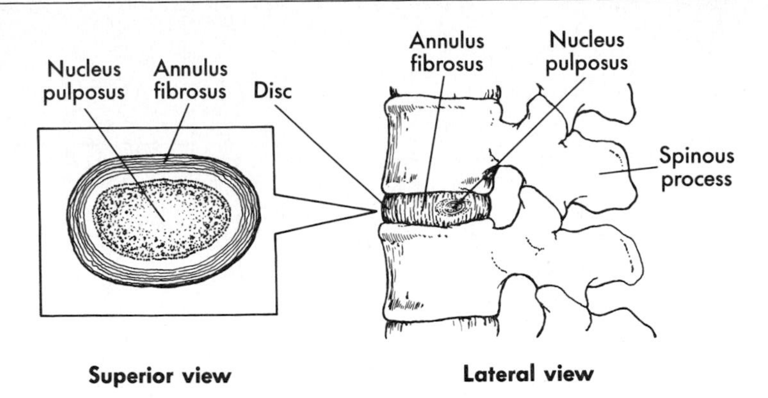

Figure 8-7
The fibrocartilaginous intervertebral disc cushions adjacent vertebrae.

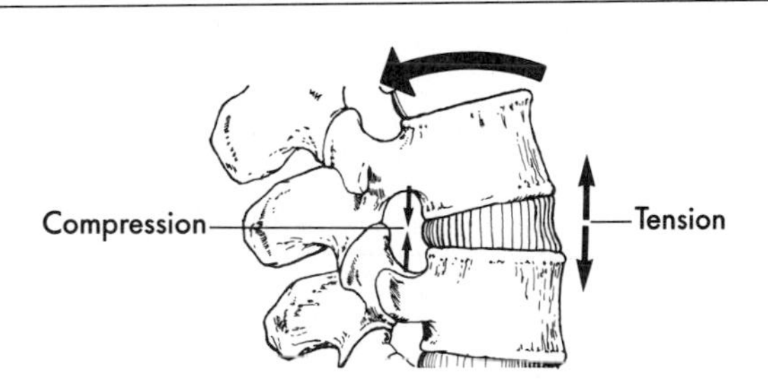

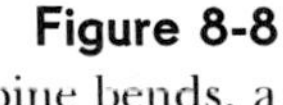

Figure 8-8
When the spine bends, a tensile load is created on one side of the discs and a compressive load is created on the other.

resistant to forces resulting from spinal bending and torsion. The nucleus has an extremely high fluid content that makes it resistant to compression. As shown in Figure 8-8, spinal flexion, extension, and lateral flexion produce compressive stress on one side of the discs and tensile stress on the other, whereas spinal rotation creates shear stress in the discs (Figure 8-9) (24). However, compression is the form of loading to which the spine is most commonly subjected during upright posture.

When a disc is loaded in compression, it tends to simultaneously lose water and absorb sodium and potassium until its internal electrolyte concentration is sufficient to prevent further water loss (23). When this chemical equilibrium is achieved, internal disc pressure is equal to the external pressure (4). Continued loading over a period of several hours results in a further slight decrease in disc hydration

Figure 8-9
Spinal rotation creates a shear force within the discs.

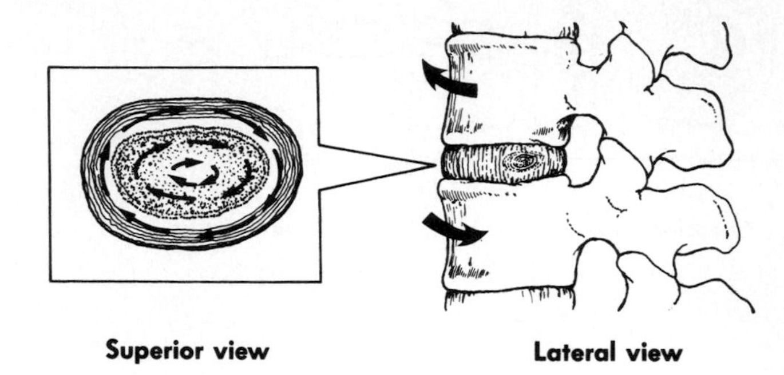

Figure 8-10
Flexion of the lumbar spine relieves compression of the facet joints.

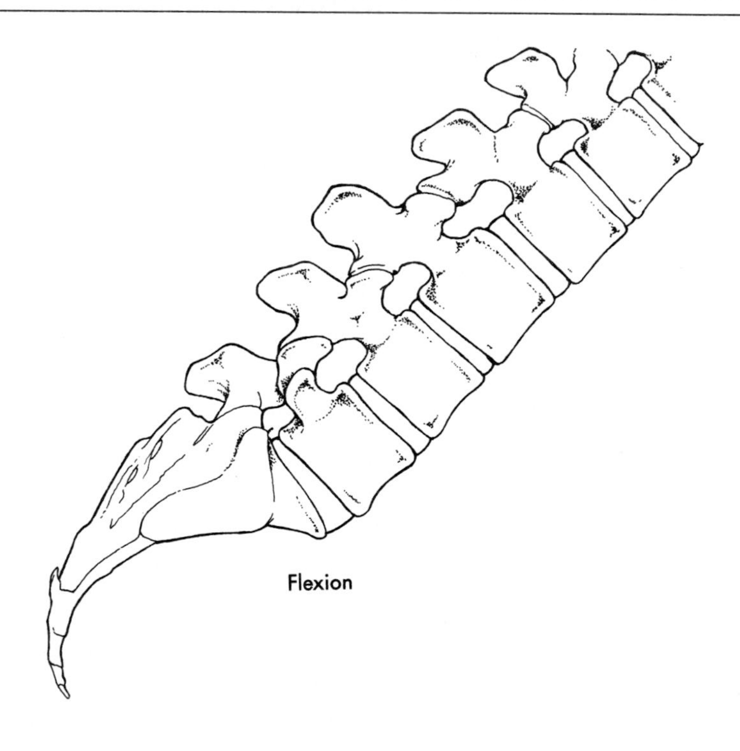

(1). For this reason, the average person undergoes a decrease in standing height of approximately 1 cm over the course of a day (10). Once pressure on the discs is relieved, the discs quickly reabsorb water and disc volumes and heights are increased (23). Astronauts experience a temporary increase in spine height of approximately 5 centimeters while free from the influence of gravity (27).

Because the intervertebral disc is an avascular structure, it must rely on a mechanically based means for maintaining a healthy nutritional status. Intermittent changes in posture and body position alter internal disc pressure, causing a *pumping action* in the disc (26). The influx and outflux of water transports nutrients in and flushes out metabolic waste products, basically fulfilling the same function that the circulatory system provides for vascularized structures within the body. Maintaining an unchanging body position that curtails this pumping action over a period of time may negatively affect disc health.

Injury and the chronological aging process irreversibly reduce the water-absorption capacity of the discs, with a concomitant decrease in shock-absorbing capability. The fluid content of the discs begins to diminish around the second decade of life (4). A typical geriatric disc has a fluid content that is reduced by approximately 35% (38). As this normal degenerative change occurs, more of the compressive, tensile, and shear loads on the spine must be assumed by other structures—particularly the facets and joint capsules. Results include a reduced height of the spinal column, often accompanied by degenerative changes in the spinal structures that are forced to assume the discs' loads. Postural alterations may also occur. The normal lordotic curve of the lumbar region may be reduced as an individual attempts to relieve compression on the facet joints by maintaining a posture of spinal flexion (Figure 8-10) (38).

Ligaments of the Spine

Ligaments contribute to the stability of the motion segments. The anterior and posterior longitudinal ligaments, which extend from the base of the skull to the sacrum, link the vertebral bodies. The **supraspinous ligament** runs the length of the entire spinal column, with each of the spinous processes attached to it. This ligament is enlarged in the cervical region, where it is referred to as the *ligamentum nuchae* or *ligament of the neck* because of its prominence (Figure 8-11). Adjacent vertebrae have additional connections between spinous processes, transverse processes, and laminae, supplied respectively by the interspinous ligaments, the intertransverse ligaments, and the ligamenta flava.

Another major ligament, the **ligamentum flavum,** connects the pedicles of adjacent vertebrae. Although most spinal ligaments are composed primarily of collagen fibers that stretch only minimally,

supraspinous ligament
the ligament attaching to the spinous processes throughout the spine

■ The enlarged cervical portion of the supraspinous ligament is the ligamentum nuchae or ligament of the neck.

ligamentum flavum
the ligament that connects the laminae of adjacent vertebrae; distinguished by its elasticity

Figure 8-11

The supraspinous ligament is well developed in the cervical region, where it is referred to as the ligamentum nuchae.

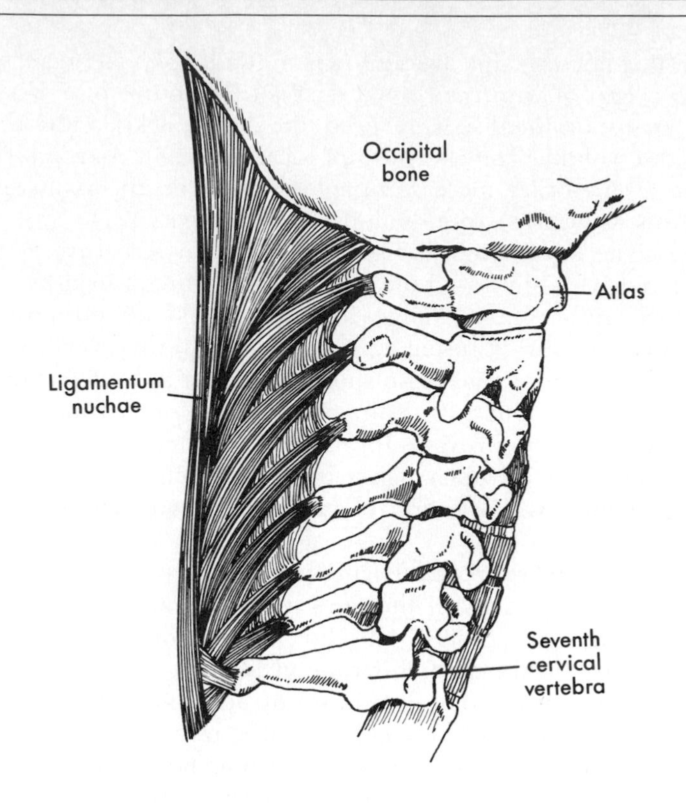

Figure 8-12

The major ligaments of the spine.

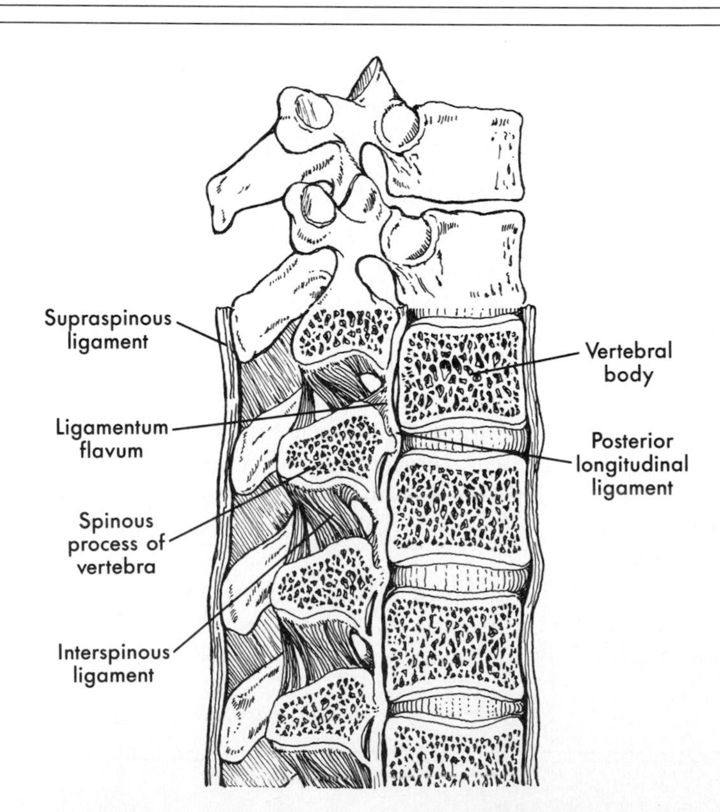

the ligamentum flavum contains a high proportion of elastic fibers, which lengthen when stretched during spinal flexion and shorten during spinal extension. The elasticity of the ligament is sufficient to keep it constantly in tension, even when the spine is in anatomical position, thus enhancing the spine's stability. This tension creates a slight, constant compression in the intervertebral discs, referred to as *prestress.* The ligaments of the spine are shown in Figure 8-12.

MOVEMENTS OF THE SPINE

As a unit the spine allows motion in all three planes of movement. Movement between adjacent vertebrae, however, is small, and spinal movements always involve a number of motion segments. The directions and ranges of motion of the individual motion segments differ according to the anatomical constraints in the respective regions of the spine. Spinal range of motion is related to age, with a decrease of approximately 50% occurring from adolescence to old age (24). However, the combined actions of many motion segments enable a relatively large range of motion for the trunk, with circumduction and planar movements allowed.

■ The movement capabilities of the spine as a unit are those of a ball and socket joint, with movement in all three planes, as well as circumduction, allowed.

Flexion, Extension, and Hyperextension

Spinal flexion should not be confused with the anterior tilt of the pelvis or with hip flexion. During an activity such as touching the toes, all three occur. The range of motion for flexion/extension of

Female gymnasts undergo extreme lumbar hyperextension during many commonly performed skills.

Figure 8-13
The range of motion in flexion/extension varies throughout the length of the spinal column.

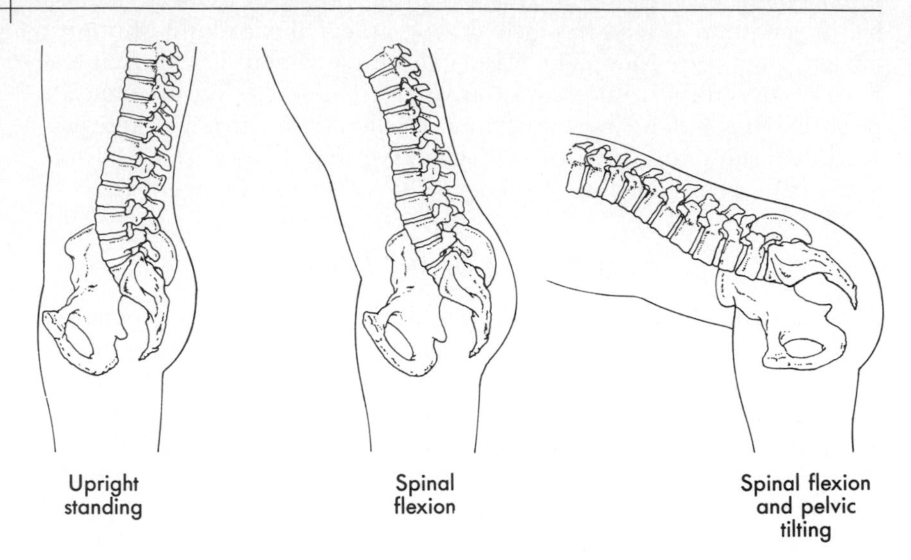

motion segments is considerable in the cervical and lumbar regions, with representative values as high as 17 degrees at the C5,6 vertebral joint and 20 degrees at L5,S1. Because of the orientation of the facets in the thoracic spine, the range of motion spans approximately 4 degrees at T1,2 to approximately 12 degrees at T11,12 (24).

Sagittal plane movement of the spine in a backward direction past anatomical position is hyperextension. Lumbar hyperextension is particularly evident during the performance of gymnastic routines. For example, during the execution of a back handspring the curvature normally present in the lower lumbar region may increase twentyfold (15). Spinal flexion and hyperextension are shown in Figure 8-13.

Lateral Flexion and Rotation

Frontal plane movement of the spine away from the anatomical position is termed *lateral flexion* (Figure 8-14). The largest range of motion for lateral flexion occurs in the cervical region, with approximately 11 to 12 degrees of motion allowed at C4,5. Somewhat less lateral flexion is allowed in the thoracic region, where the range of motion between adjacent vertebrae is approximately 6 degrees, except in the lower segments in which lateral flexion capability may be as high as 8 to 9 degrees. Lateral flexion in the lumbar spine is also approximately 6 degrees, except at L5,S1 where it is reduced to only 3 degrees (24). Because of the normal anteroposterior spinal curvatures and the slants of the articular facets, some vertebral rotation must accompany lateral flexion in the cervical region. However, this rotation is only observable with x rays.

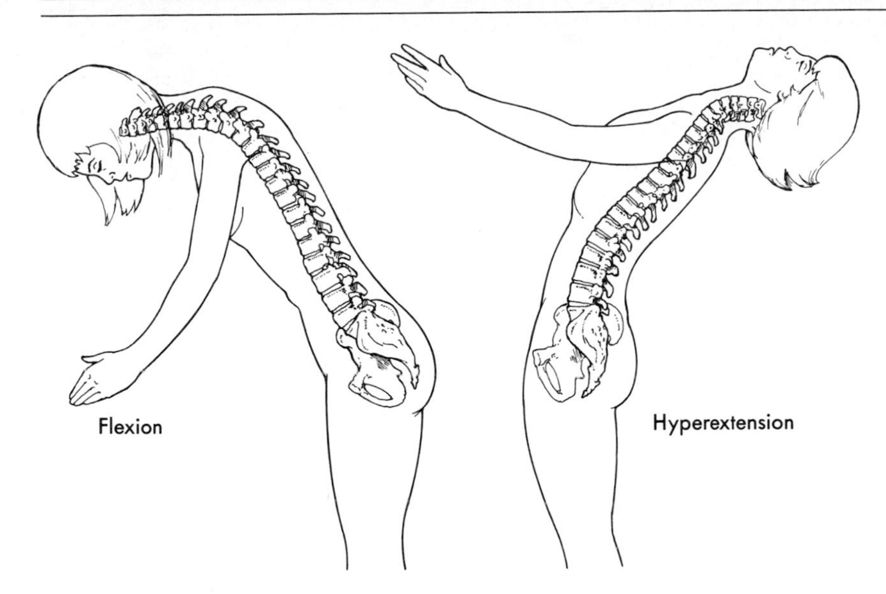

Figure 8-14
The range of motion for lateral flexion varies throughout the length of the spine.

Spinal rotation in the transverse plane is most free in the cervical region of the spine, with over 30 degrees of motion allowed at C1,2. In the thoracic region, approximately 9 degrees of rotation is permitted in the upper motion segments. From T7,8, however, the range of rotational capability progressively decreases, with only about 2 degrees of motion allowed in the lumbar spine because of the interlocking of the articular processes. At the lumbosacral joint, approximately 5 degrees of rotation are allowed (24). Since the structure of the cervical spine causes lateral flexion and rotation to be coupled, a slight lateral flexion to the same side accompanies rotation, although this motion is not observable with the naked eye.

■ Many muscles of the neck and trunk cause lateral flexion when contracting unilaterally but either flexion or extension when contracting bilaterally.

STRUCTURE OF THE PELVIC GIRDLE

The sacral region of the spine is fused with three other bones—the ilium, the ischium, and the pubis—to form the pelvic girdle or pelvis (Figure 8-15). The joints between these bones are synarthroses, with no movement allowed. They form a protective basin around the internal organs of the lower trunk and transfer loads from the upper extremity and trunk to the lower extremity through the hip joint.

■ During pregnancy and parturition, the normally motionless symphysis pubis and sacroiliac joints may separate slightly because of hormonal influence.

MOVEMENTS OF THE PELVIS

The pelvis can be moved in all three planes. Sagittal plane movements are termed *anterior* and *posterior tilt.* Movement in the frontal plane is known as *lateral tilt* to the right or to the left, according to

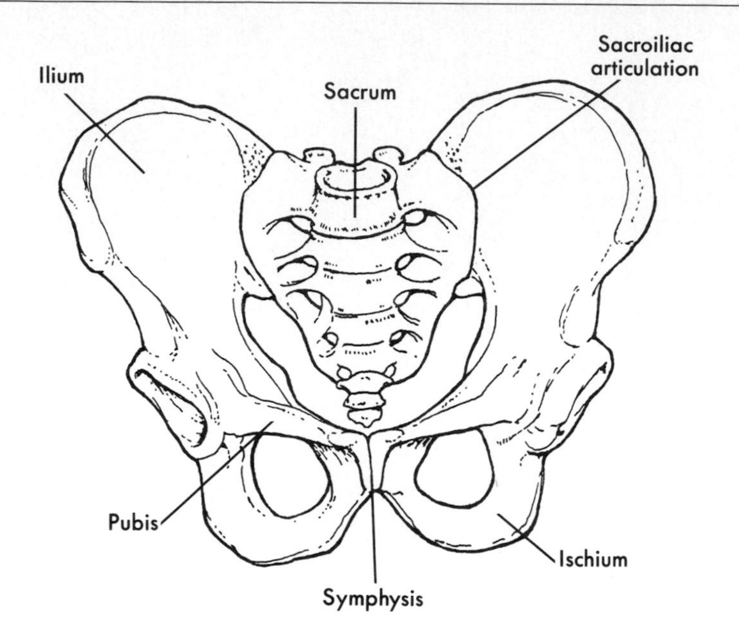

Figure 8-15
The pelvis.

the side of the pelvis that lowers. Pelvic rotation can also occur in the transverse plane, as it does during walking and running.

Primary Pelvic Movements

Pelvic movement results from changes in the orientation of the lumbosacral joint, the femurs, or both. Purposeful pelvic movements, termed *primary movements of the pelvis,* are employed in relatively few movement activities. Hawaiian dancing and dancing the twist are examples. More commonly, pelvic motion accompanies and facilitates the movement of the spine or hip. This is known as *secondary movement of the pelvis.*

Secondary Pelvic Movements

The cooperative movement of the pelvis is needed during the performance of many activities that are usually considered spinal or hip movements. For example, touching the toes is typically initiated with 50 to 60 degrees of flexion of the lumbar spine, with the remainder of the movement due primarily to anterior tilt of the pelvis (11). Figure 8-16 shows the increased range of motion permitted when anterior pelvic tilt accompanies spinal flexion. Similarly, pelvic movements increase the lateral flexion and rotational movements of which the trunk is capable. Secondary movements of the pelvis that optimize the position of the hip joint for movement of the femur are summarized along with other pelvic movements in Table 8-1.

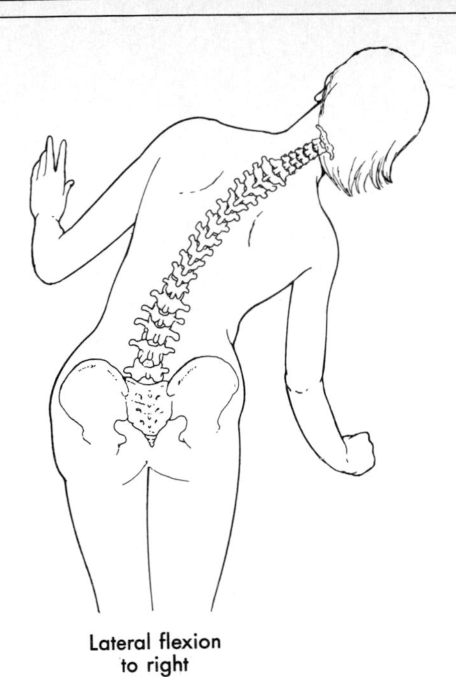

Figure 8-16

When the trunk is flexed, the first 50 to 60 degrees of motion occurs in the lumbar spine, with additional motion resulting from anterior pelvic tilt.

Table 8-1

EFFECTS OF PELVIC MOVEMENTS

PRIMARY PELVIC MOVEMENTS

PELVIS	LUMBAR SPINAL JOINTS	HIP JOINTS
Anterior tilt	Hyperextension	Flexion
Posterior tilt	Slight flexion	Extension
Right lateral tilt	Slight left lateral flexion	Abduction to right, Adduction to left
Right rotation	Left rotation	Right outward rotation, Left inward rotation

PELVIC MOVEMENTS SECONDARY TO SPINAL MOVEMENTS

SPINE	PELVIS
Flexion	Posterior tilt
Hyperextension	Anterior tilt
Right lateral flexion	Right lateral tilt
Right rotation	Right rotation

PELVIC MOVEMENTS SECONDARY TO FEMORAL MOVEMENTS

FEMUR	PELVIS
Flexion	Posterior tilt
Extension	Anterior tilt
Right abduction	Left lateral tilt
Right rotation	Right rotation
Left rotation	Left rotation

Table 8-2

MUSCLES OF THE SPINE

MUSCLE	PROXIMAL ATTACHMENT	DISTAL ATTACHMENT	PRIMARY ACTIONS
Prevertebral muscles (rectus capitis anterior, rectus capitis lateralis, longus capitis, longus coli)	Anterior aspect of occipital bone and cervical vertebrae	Anterior surfaces cervical and first 3 thoracic vertebrae	Flexion, lateral flexion, rotation to opposite side
Rectus abdominis	Costal cartilage of ribs 5 to 7	Pubic crest	Flexion, lateral flexion
External oblique	External surface of lower 8 ribs	Linea alba and anterior iliac crest	Flexion, lateral flexion, rotation to opposite side
Internal oblique	Linea alba and lower 4 ribs	Inguinal ligament, iliac crest, lumbodorsal fascia	Flexion, lateral flexion, rotation to same side
Splenii (splenius capitis, cervicis)	Mastoid process of temporal bone, transverse processes of first 3 cervical vertebrae	Lower half of ligamentum nuchae, spinous processes of seventh cervical, upper 6 thoracic vertebrae	Extension, lateral flexion, rotation to same side
Suboccipitals (obliquus capitus superior and inferior, rectus capitis posterior major and minor)	Occipital bone, transverse process of first cervical vertebra	Posterior surfaces of first 2 cervical vertebrae	Extension, lateral flexion, rotation to same side
Sacrospinalis (spinalis, longissimus, iliocostalis)	Lower part of the ligamentum nuchae; posterior cervical, thoracic, lumbar spine; lower 9 ribs; iliac crest; posterior sacrum	Mastoid process of temporal bone; posterior cervical, thoracic, lumbar spine; 12 ribs	Extension, lateral flexion, rotation to opposite side

MUSCLE	PROXIMAL ATTACHMENT	DISTAL ATTACHMENT	PRIMARY ACTIONS
Semispinalis (capitis, cervicis, thoracis)	Occipital bone, spinous processes of thoracic vertebrae 2 to 4	Transverse processes of thoracic and seventh cervical vertebrae	Extension, lateral flexion, rotation to opposite side
Deep spinal muscles (multifidi, rotatores, interspinales, intertransversarii, levatores costarum)	Posterior processes of all vertebrae, posterior sacrum	Spinous and transverse processes and laminae of vertebrae below those of proximal attachment	Extension, lateral flexion, rotation to opposite side
Sternocleidomastoid	Mastoid process of temporal bone	Superior sternum, inner third of clavicle	Flexion of neck, extension of head, lateral flexion, rotation to opposite side
Levator scapulae	Transverse process of first 4 cervical vertebrae	Vertebral border of scapula	Lateral flexion
Scaleni (scalenus anterior, medius, posterior)	Transverse processes of cervical vertebrae	Upper 2 ribs	Flexion, lateral flexion
Quadratus lumborum	Last rib, transverse processes of first 4 lumbar vertebrae	Iliolumbar ligament, adjacent iliac crest	Lateral flexion
Psoas major	Sides of twelfth thoracic, all lumbar vertebrae	Lesser trochanter of femur	Flexion

MUSCLES OF THE SPINE AND PELVIC GIRDLE

Many muscles of the neck and trunk are named in pairs, with one located on the left and the other on the right side of the body. These muscles cause lateral flexion or rotation when they contract unilaterally and contribute to spinal flexion or extension when bilateral contractions occur. The primary functions of the major muscles of the spine are summarized in Table 8-2.

The rectus abdominis and the external obliques are prominent abdominal muscles.

Anterior Aspect

The major anterior muscle groups of the cervical region are the prevertebral muscles, including the rectus capitis anterior, rectus capitis lateralis, longus capitis, longus colli, and the eight pairs of hyoid muscles. The locations of these muscles are shown in Figures 8-17 and 8-18. Bilateral tension development by these muscles results in flexion of the head, although the main function of the hyoid muscles appears to be to move the hyoid bone during the act of swallowing. Unilateral tension development in the prevertebrals contributes to lateral flexion of the head toward the contracting muscles or to rotation of the head away from the contracting muscles, depending on which other muscles are functioning as neutralizers.

The main abdominal muscles are the rectus abdominis, the external obliques, and the internal obliques. The locations of these muscles are shown in Figures 8-19 to 8-21. Functioning bilaterally, these muscles are the major spinal flexors and also reduce anterior pelvic tilt. Unilateral tension development by the muscles produces lateral

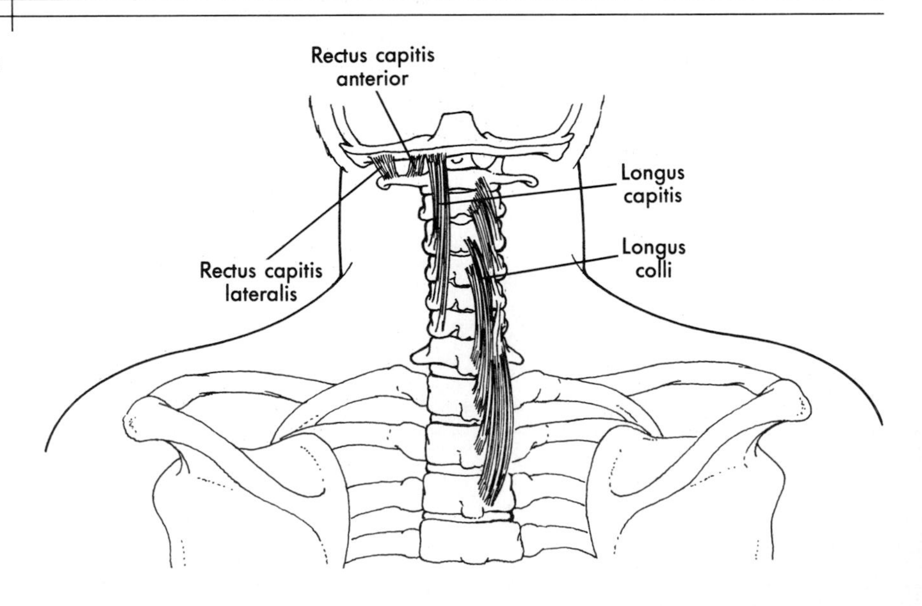

Figure 8-17
Anterior muscles of the cervical region.

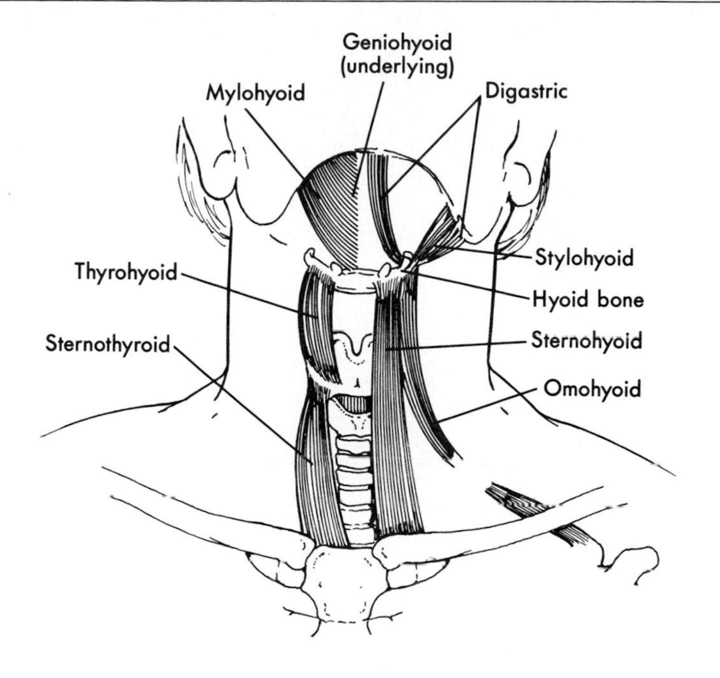

Figure 8-18
The hyoid muscles.

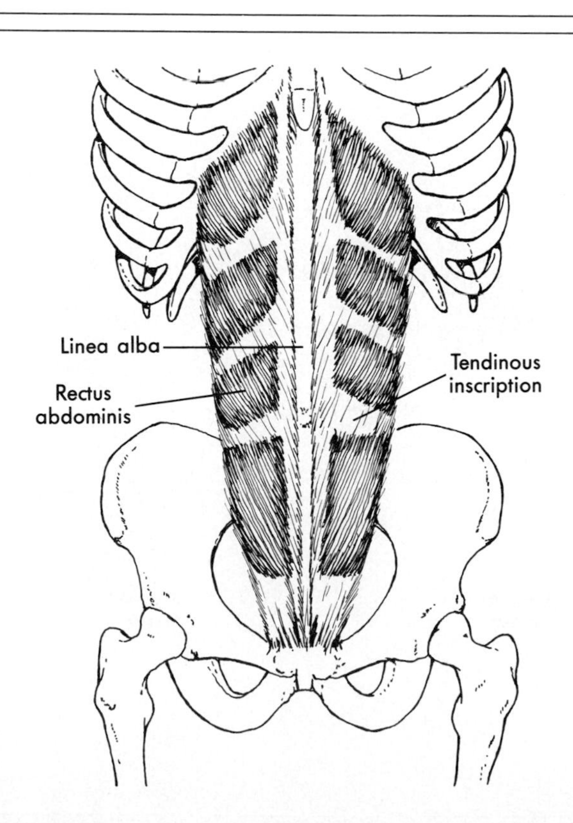

Figure 8-19
The rectus abdominis.

Figure 8-20
The external obliques.

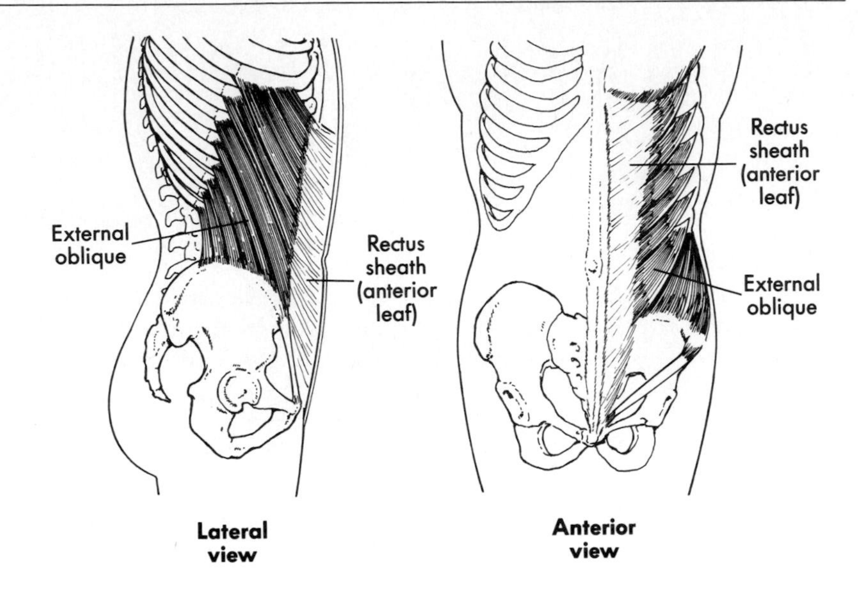

Figure 8-21
The internal obliques.

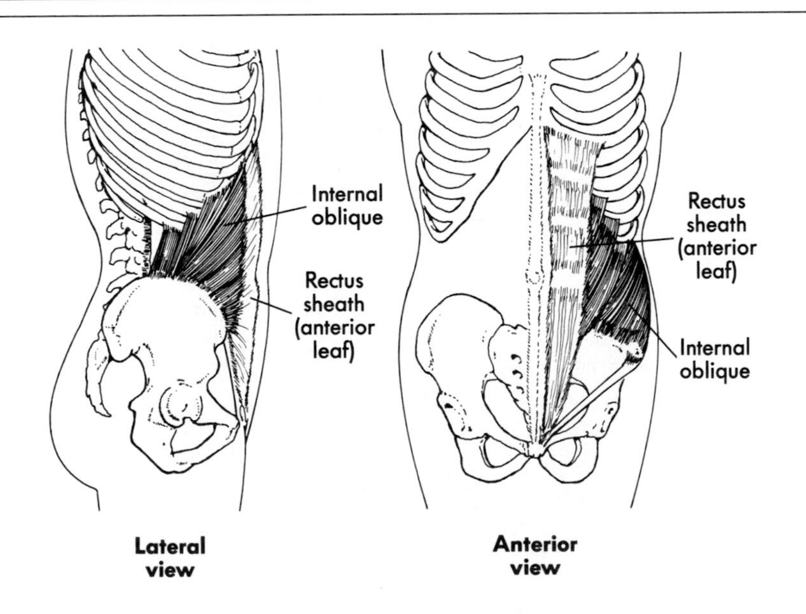

flexion of the spine toward the tensed muscles. Tension development in the internal obliques causes rotation of the spine toward the same side. Tension development by the external obliques results in rotation toward the opposite side. If the spine is fixed, the internal obliques produce pelvic rotation toward the opposite side, with the external obliques producing rotation of the pelvis toward the same side. These muscles also form the major part of the abdominal wall, which protects the internal organs of the abdomen.

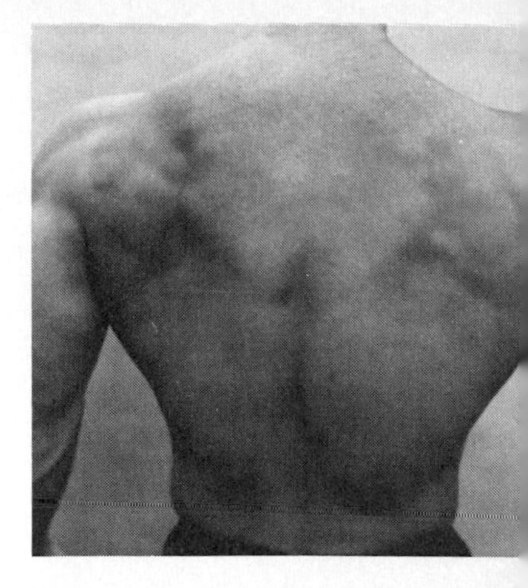

The superficial muscles of the posterior trunk.

Posterior Aspect

The splenius capitis and splenius cervicis are the primary cervical extensors (Figure 8-22) (28). Bilateral tension development in the four suboccipitals—the rectus capitis posterior major and minor and the obliquus capitis superior and inferior—assist (Figure 8-23). When these posterior cervical muscles develop tension on one side only, they laterally flex or rotate the head toward the side of the contracting muscles.

The posterior thoracic and lumbar region muscle groups are the massive erector spinae (sacrospinalis), the semispinalis, and the deep spinal muscles. As shown in Figure 8-24, the erector spinae group includes the spinalis, longissimus, and iliocostalis muscles. The semispinalis, with its capitis, cervicis, and thoracis branches, is shown in Figure 8-25. The deep spinal muscles, including the multifidi, rota-

■ The prominent erector spinae muscle group—the major extensor and hyperextensor of the trunk—is the muscle group of the trunk most often strained.

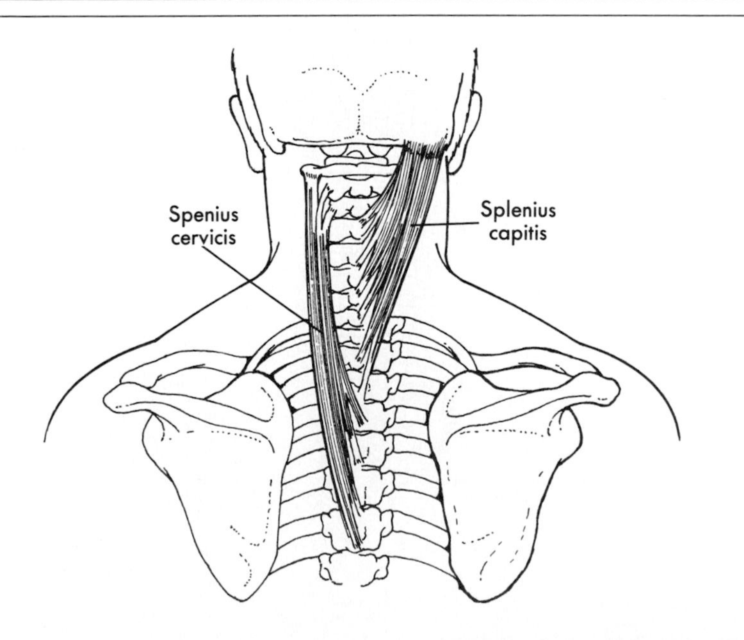

Figure 8-22
The major cervical extensors.

Figure 8-23
The suboccipital muscles.

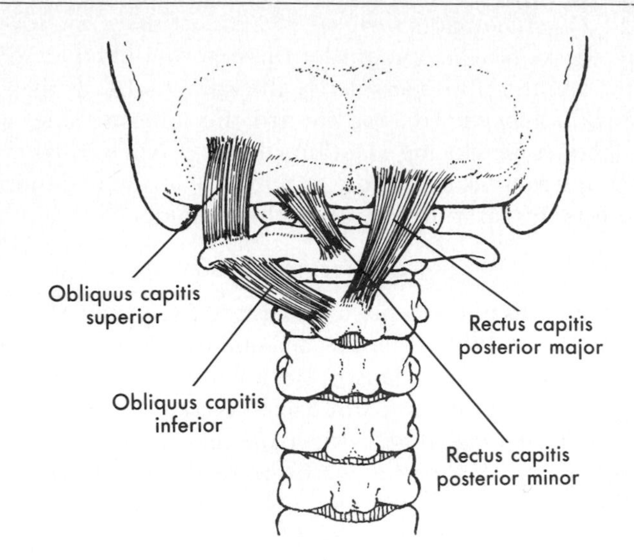

Figure 8-24
The sacrospinalis (erector spinae) muscles.

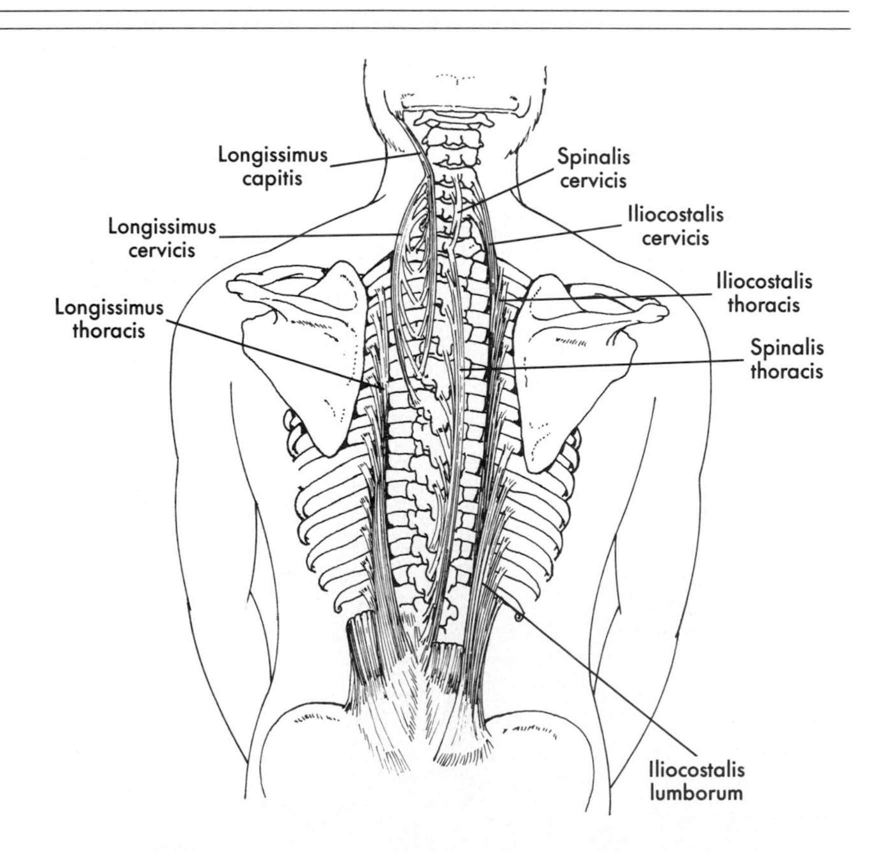

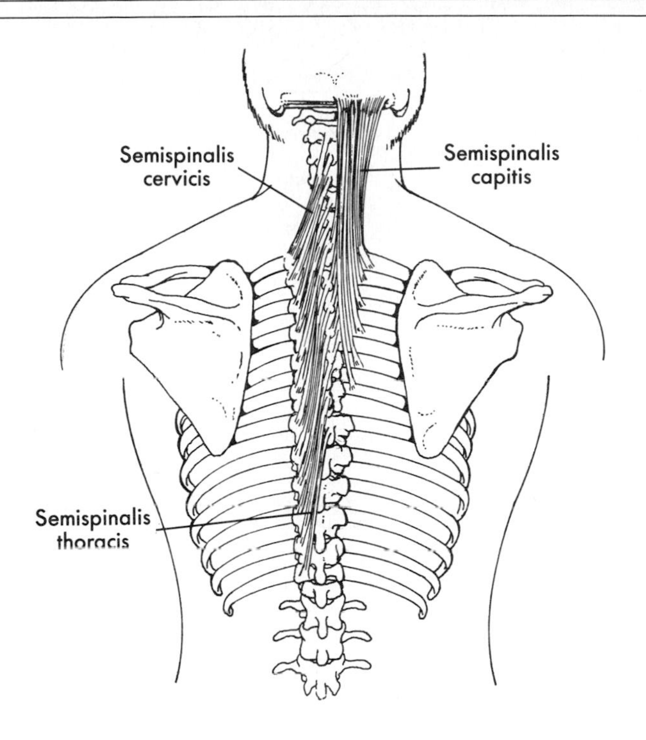

Figure 8-25
The semispinalis group.

tores, interspinales, intertransversarii, and levatores costarum, are represented in Figure 8-26. The sacrospinalis groups are the major extensors and hyperextensors of the trunk. All posterior trunk muscles contribute to extension and hyperextension when contracting bilaterally and to lateral flexion or rotation to the opposite side when contracting unilaterally.

Lateral Aspect

Muscles on the lateral aspect of the neck include the prominent sternocleidomastoid, the levator scapulae, and the scalenus anterior, posterior, and medius. These muscles are shown in Figures 8-27 through 8-29. Bilateral tension development in the sternocleidomastoid may result in either flexion of the neck or extension of the head, with unilateral contraction producing lateral flexion to the same side or rotation to the opposite side. The levator scapulae can also contribute to lateral flexion of the neck when contracting unilaterally with the scapula stabilized (see Chapter 6). The three scalenes assist with flexion and lateral flexion of the neck, depending on whether tension development is bilateral or unilateral.

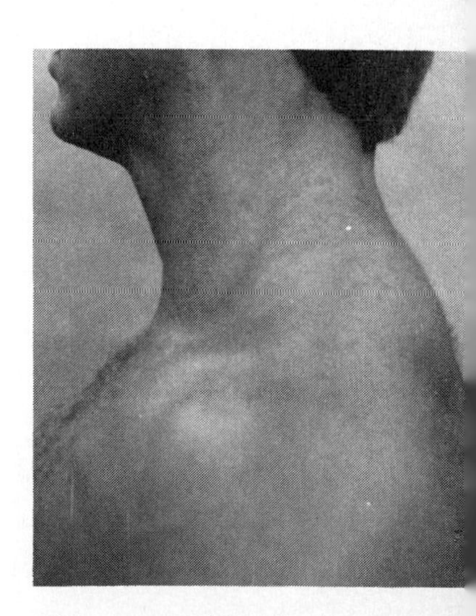

The sternocleidomastoid dominates the lateral aspect of the neck.

Figure 8-26
The deep spinal muscles.

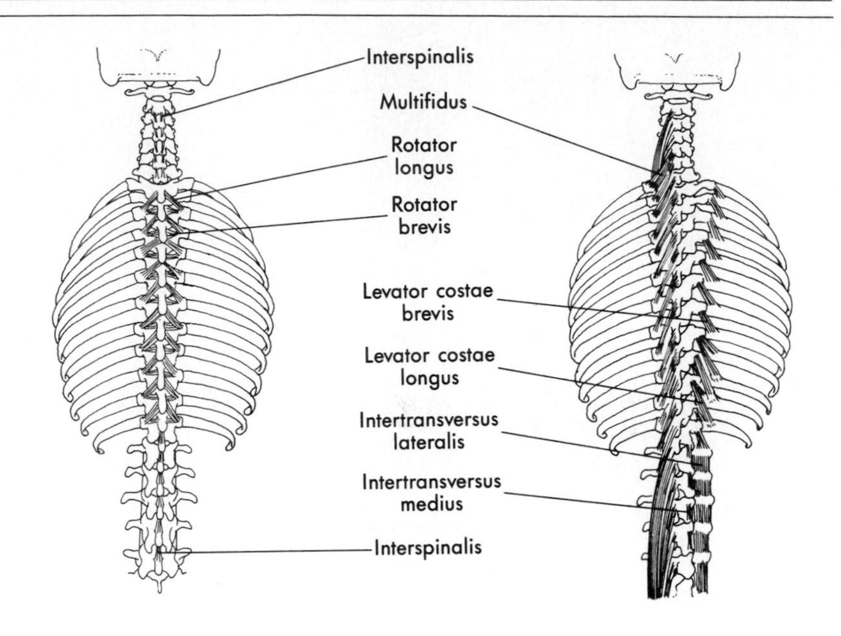

Figure 8-27
The sternocleidomastoid.

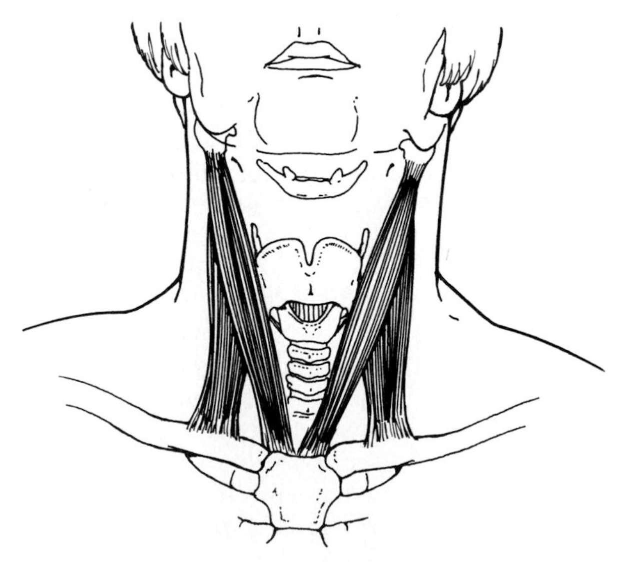

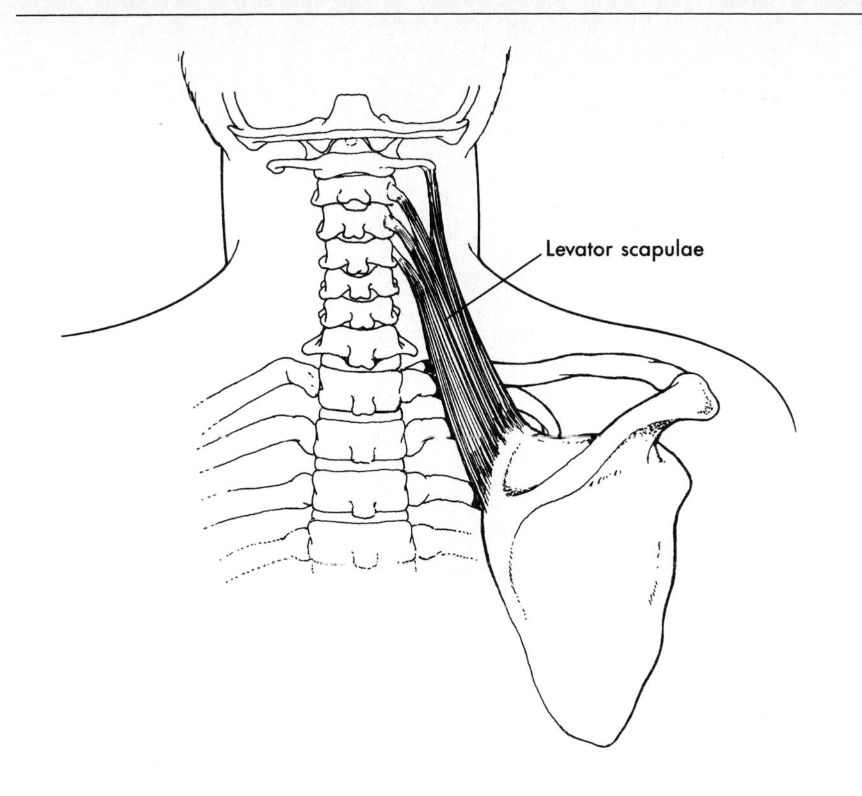

Figure 8-28
The levator scapulae.

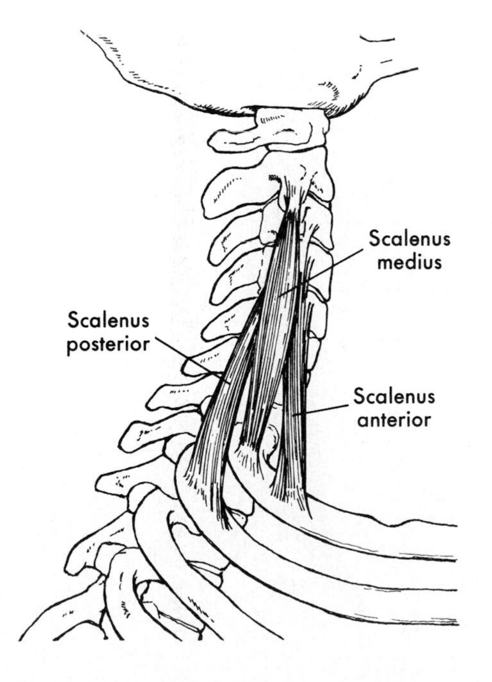

Figure 8-29
The scaleni muscles.

Figure 8-30
The quadratus lumborum.

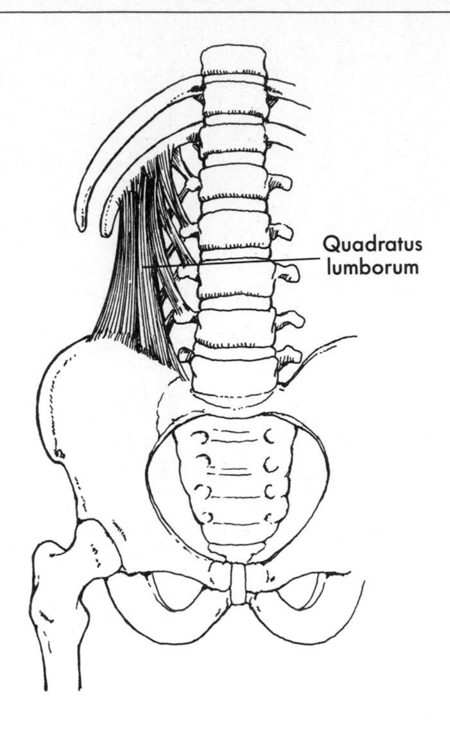

Figure 8-31
The psoas major.

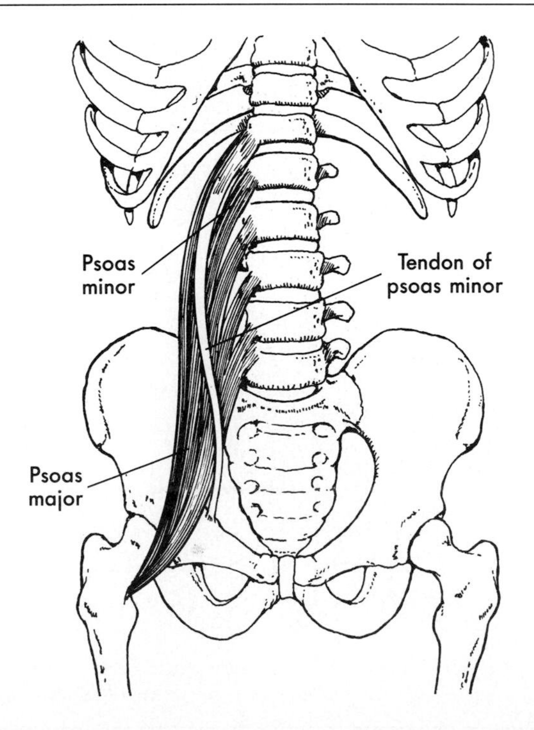

In the lumbar region, the quadratus lumborum and psoas major are large, laterally oriented muscles (Figures 8-30 and 8-31). The quadratus lumborum acts bilaterally to stabilize the lumbar region of the spine and the pelvis and unilaterally to flex the lumbar spine laterally. Two major functions of the psoas major are bilaterally to stabilize and unilaterally to laterally flex the lumbar spine.

COMMON INJURIES OF THE BACK AND NECK

Low Back Pain

An estimated 70% to 80% of people experience low back pain at some time during life. Low back pain is second only to the common cold in causing absences from the workplace (12), and back injuries are the most frequent and the most expensive of all worker's compensation claims in the United States (20). Most back injuries involve the lumbar or low back region (Figure 8-32). Low back pain is also the diagnosis in 10% of all chronic health problems, as well as being the eleventh ranked cause of hospitalization in the United States (8, 34).

Although psychological and social components are a factor in some low back pain cases, mechanical stress typically plays a significant causal role in the development of low back pain (13). Some personal and occupational factors are associated with higher incidences of low back pain. Although back pain is the leading cause of disability for all persons aged 20 to 45 in the United States, the most common victim is a man approximately 35 years of age. Perhaps because

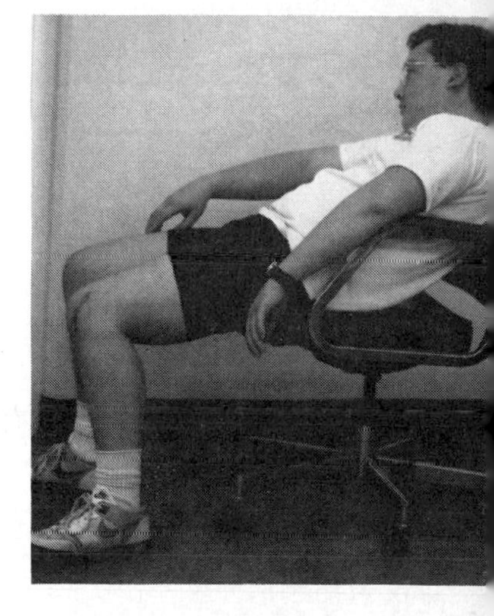

Sitting in a slouched position can increase the pressure exerted on the lumbar intervertebral discs.

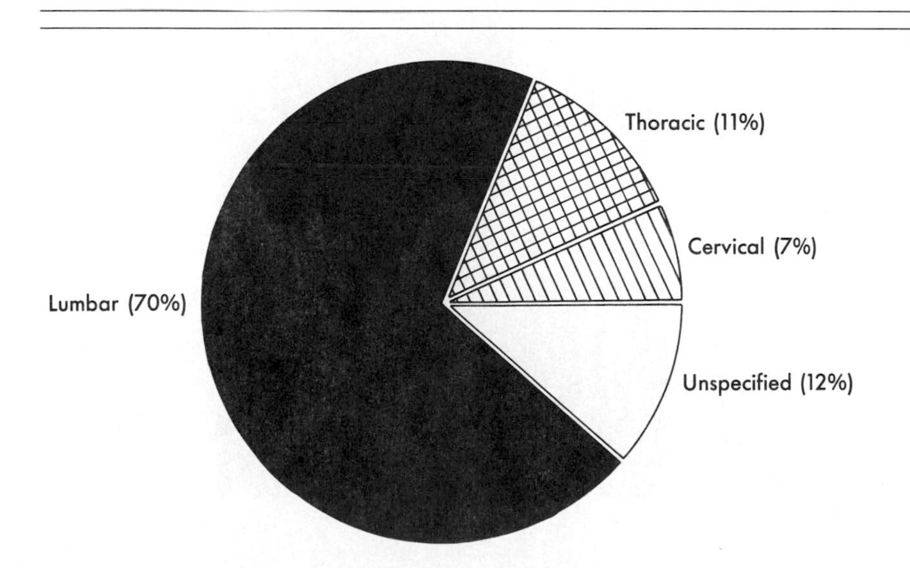

Figure 8-32
The majority of back injuries that result in lost work time involve the lumbar region.

Weight training with improper technique can result in potentially injurious stresses on the joints of the human body, including those of the lumbar spine.

Many activities of daily living can place high levels of mechanical stress on the low back. The execution of lifts such as the one shown should be slow and controlled to minimize the risk of low back injury.

Lifting while twisting requires assymetrical involvement of the low back muscles, which may promote low back muscle strain.

of their predominance in occupations involving heavy materials handling, men file approximately three fourths of all low back claims (34). However, some female-dominated groups such as nurses' aides register higher rates of low back injury than male workers in general (35). High-risk occupations for the development of low back pain (in order of frequency) include miscellaneous laborers, truck drivers, garbage collectors, warehouse workers, miscellaneous mechanics, nursing aides, material handlers, lumber workers, practical nurses, and construction laborers (34).

Although some known pathologies may cause low back pain, the majority of cases are nondiagnosable (26). The inability to specifically identify the anatomical structure or structures that are the source of the pain makes it more difficult to determine the biomechanical factors causing the development of pain. Unlike other major health problems that have declined in incidence in response to medical efforts, the number of low back pain-disabled individuals in the United States increased at a rate 14 times the rate of population growth between 1971 and 1981 (26).

Soft Tissue Injuries

Contusions, muscle strains, and ligament sprains collectively compose the most common athletic injury of the back (3). These types of injuries typically result from either sustaining a blow or overloading the muscles, particularly those of the lumbar region. Painful spasms and knotlike contractions of the back muscles may also develop as a sympathetic response to spinal injuries and may actually be only symptoms of the underlying problem. Researchers believe that a biochemical mechanism is responsible for these sympathetic muscle spasms (5), which act as a protective mechanism to immobilize the injured area (30).

Fractures

Differences in the causative force or forces determine the type of vertebral fracture incurred. Transverse or spinous process fractures may result from extremely forceful contraction of the attached muscles or from the sustenance of a hard blow to the back of the spine, which may occur during participation in contact sports such as football, rugby, soccer, basketball, hockey, and lacrosse (30). The most common cause of cervical fractures is indirect trauma, that is, the sustenance of force by the head or trunk rather than by the cervical region itself (6). Fractures of the cervical vertebrae frequently occur from impacts to the head when people dive into shallow water or engage in gymnastics or trampolining activities without appropriate supervision (32, 33).

Large compressive loads (such as those encountered in the sport of weight lifting or in heavy materials handling) typically cause fractures of the vertebral end plates. High levels of impact force may result in anterior compression fractures of the vertebral bodies. This type of injury is generally associated with vehicular accidents, although it can also result from hitting the boards during ice hockey, head-on blocks or tackles in football, or from impacts during tobogganing, snowmobiling, and hot air ballooning (3). When a snowmobile drops 4 feet, the forces generated far exceed the level known to cause a compression fracture (18).

Because one function of the spine is to protect the spinal cord, acute spinal fractures are extremely serious, with possible outcomes including paralysis and death. Unfortunately, the recent growth in leisure time activities is associated with an increased prevalence of spinal injuries (31). Whenever a spinal fracture is a possibility, only trained personnel should move the victim.

▪ Stress-related fractures of the pars interarticularis, the weakest section of the neural arch, are unusually common among participants in sports involving repeated hyperextension of the lumbar spine.

The most common type of vertebral fracture results from smaller forces that are sustained repeatedly. These stress fractures occur most often in the weakest portion of the neural arch, the pars interarticularis, which is the region between the superior and inferior articular facets (Figure 8-33). Unlike most stress fractures, these pars defects usually persist, sometimes necessitating abstention from mechanically strenuous activities. The most common site of this injury is the lumbosacral joint, and repeated hyperextension of the lumbar spine may be one causal factor. Unusually high incidences of pars fractures have been documented among female gymnasts, interior

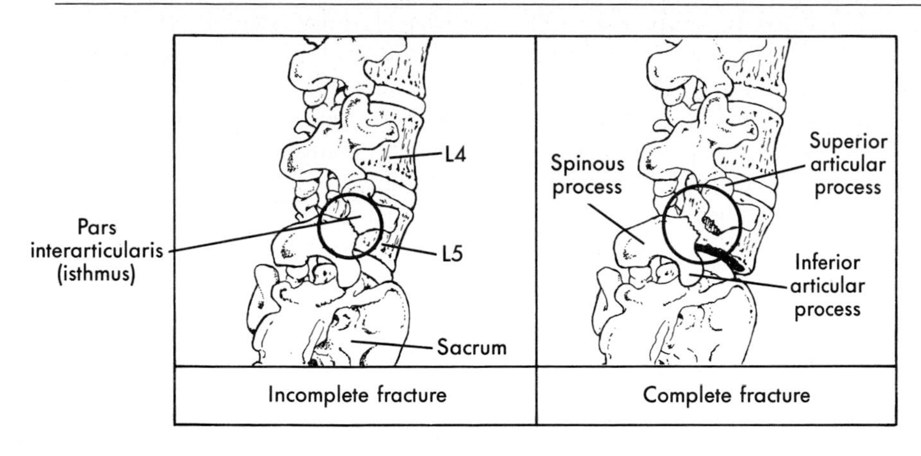

Figure 8-33
Stress fractures of the pars interarticularis may occur unilaterally or bilaterally and may or may not result in a complete separation.

football linemen, and weight lifters, with increased incidences also observed among volleyball players, pole vaulters, wrestlers, and rowers (3).

Fractures of the ribs are generally caused from blows received during accidents or participation in contact sports. Rib fractures are extremely painful because pressure is exerted on the ribs with each inhalation. Damage to the underlying soft tissues is a potentially serious complication with this type of injury.

Disc Herniations

The source of a common cause of back pain is a herniated disc, which consists of the protrusion of part of the nucleus pulposa from the annulus. Disc herniations may be either traumatic or stress related and typically involve a disc that shows signs of previous degeneration (2). The most common sites of protrusions are between the fifth and sixth and the sixth and seventh cervical vertebrae and between the fourth and fifth lumbar vertebrae and the fifth lumbar vertebra and the first sacral vertebra (21, 22). Most occur on the posterior or posterior-lateral aspect of the disc (4). Greater height in men and women and greater body weight in men have been associated with an increased incidence of herniated lumbar discs in one investigation (16), although others have found anthropometric characteristics to be unrelated (5). Shorter and faster runners are more likely to develop lumbar spine problems, perhaps because they tend to overstride, forcing their lumbar spines into hyperextension (29).

■ The term *slipped disc* is often used to refer to a herniated or prolapsed disc. It is a misnomer because the discs as intact units do not slip around.

Although the disc itself is not innervated and therefore incapable of generating a sensation of pain, sensory nerves do supply the anterior and posterior longitudinal ligaments, the vertebral bodies, and the articular cartilage of the facet joints (7). If the herniation presses on one or more of these structures, on the spinal cord, or on a spinal nerve, pain or numbness may result.

SUMMARY

The spine is composed of 33 vertebrae that are divided structurally into five regions—cervical, thoracic, lumbar, sacral, and coccygeal. Although most vertebrae adhere to a characteristic shape, there is a progression in vertebral size and in the orientation of the articular facets throughout the length of the spinal column.

Within the cervical, thoracic, and lumbar regions, each pair of adjacent vertebrae with the intervening soft tissues is called a motion segment. The motion segment is the functional unit of the spine. Three joints interconnect the vertebrae of each motion segment. The anterior joint is a symphysis joint, with the bodies of the vertebrae separated by a prominent intervertebral disc. The discs function as shock absorbers. The other spinal joints are the facet joints in which the two superior articular processes of one vertebra articulate with the two inferior articular processes of the vertebra above it. While the anterior joints bear most of the weight of the trunk and upper extremity, the varying orientations of facet joints significantly influence the movement capabilities of the motion segments at different levels of the spine.

The pelvic girdle is composed of three bones—the ilium, the ischium, and the pubis. The joints between these bones are synarthrodial types, with no movement allowed. Movement of the pelvis extends the range of movement of the trunk beyond the motion provided by the spine and also optimally positions the hip joint for more effective movements of the femur.

Because the spine serves as the protector of the spinal cord, spinal injuries are serious. Low back pain is also a major health problem and a leading cause of lost working days.

INTRODUCTORY PROBLEMS

1. Identify the relative movement capabilities of the spine in all three planes in the cervical, thoracic, and lumbar regions. Of what practical value are the movement capability differences among these three spinal regions?
2. Construct a chart listing the muscles of the cervical region of the trunk according to whether they are anterior, posterior, medial, or lateral to the joint center. Note that some muscles may fall into more than one category.
3. Identify the action or actions performed by the muscles in Problem 2.
4. Construct a chart listing the muscles of the thoracic region of the trunk according to whether they are anterior, posterior, medial, or lateral to the joint center. Note that some muscles may fall into more than one category.
5. Identify the action or actions performed by the muscles in Problem 4.

6. Construct a chart listing the muscles of the lumbar region of the trunk according to whether they are anterior, posterior, medial, or lateral to the joint center. Note that some muscles may fall into more than one category.
7. Identify the action or actions performed by the muscles in Problem 6.
8. What are the postural consequences of having extremely weak abdominal muscles?
9. How would trunk movement capability be affected if the pelvis were immovable?
10. What exercises strengthen the muscles on the anterior, lateral, and posterior aspects of the trunk?

ADDITIONAL PROBLEMS

1. Formulate a theory explaining why osteoporosis is often associated with increased thoracic kyphosis.
2. What exercises should be prescribed for individuals with scoliosis? Lordosis? Kyphosis?
3. Should sit-ups be prescribed as an exercise for a low back pain patient? Explain why or why not.
4. Why do individuals who work at a desk all day develop low back pain?
5. Identify the major muscle groups used when lifting an object from the floor with the trunk erect and with the trunk oriented at a 45 degree angle to the floor. What can be concluded from the two lists?
6. Compare and contrast the major muscles that serve as agonists during performances of straight-leg and bent-knee sit-ups.
7. Formulate a theory explaining why a loss in spinal flexibility of approximately 50% is a result of the aging process.
8. What are the consequences of the loss in intervertebral disc hydration that accompanies the aging process?
9. What spinal exercises are appropriate for senior citizens? Provide a rationale for your choices.
10. Weight training is used in conjunction with conditioning for numerous sports. What would you advise regarding spinal posture during weight training?

REFERENCES

1. Adams MA and Hutton WC: The effect of posture on the fluid content of lumbar intervertebral discs, Spine 8:665, 1983.
2. Adams MA and Hutton WC: Gradual disc prolapse, Spine 10:524, 1985.
3. Alexander MJL: Biomechanical aspects of lumbar spine injuries in athletes: a review, Can J Applied Sport Sci 10:1, 1985.

4. Ashton-Miller JA and Schultz AB: Biomechanics of the human spine and trunk, Exerc Sport Sci Rev 16:169, 1988.
5. Brennan GP et al: Physical characteristics of patients with herniated intervertebral lumbar discs, Spine 12:699, 1987.
6. Byun HS, Cantos EL, and Patel PP: Severe cervical injury due to break dancing: a case report, Orthopedics 9:550, 1986.
7. Cailliet R: Low back pain syndrome, ed 3, Philadelphia, 1981, FA Davis Co.
8. Cypress B: Characteristics of physician visits for back symptoms: a national perspective, Am J Public Health 73:389, 1983.
9. Day AL, Friedman WA, and Indelicato PA: Observations on the Treatment of Lumbar Disc Disease in College Football Players, Am J Sports Med 15:72, 1987.
10. Ecklund JAE and Corlett EN: Shrinkage as a measure of the effect of load on the spine, Spine 9:189, 1984.
11. Farfan HF: Muscular mechanism of the lumbar spine and the position of power and efficiency, Orthop Clin North Am 6:135, 1975.
12. Frymoyer JW and Mooney V: Current concepts review: occupational orthopedics, J Bone Joint Surg 68A:469, 1986.
13. Frymoyer JW and Pope M: The role of trauma in low back pain: a review, J Trauma 18:628, 1978.
14. Goldberg B and Boiardo R: Profiling children for sports participation, Clin Sports Med 3:153, 1984.
15. Hall SJ: Mechanical contribution to lumbar stress injuries in female gymnasts, Med Sci Sports Exerc 18:599, 1986.
16. Heliovaara M: Body height, obesity, and risk of herniated lumbar intervertebral disc, Spine 12:469, 1987.
17. Jackson DW, Wiltse LL, and Cirincione RJ: Spondylolysis in the female gymnast, Clin Orthop 117:68, 1976.
18. Keene JS: Thoracolumbar fractures in winter sports, Clin Orthop 216:39, 1987.
19. Keene JS and Drummond DS: Mechanical back pain in the athlete, Compr Ther 11:7, 1985.
20. Kelsey J and White A: Epidemiology and impact of low back pain, Spine 5:133, 1980.
21. Kelsey JL et al: Acute prolapsed lumbar intervertebral disc: an epidemiological study with special reference to driving automobiles and cigarette smoking, Spine 9:608, 1984.
22. Kelsey JL et al: An epidemiological study of acute prolapsed cervical intervertebral disc, J Bone Joint Surg 66A:907, 1984.
23. Kraemer J, Kolditz D, and Gowin R: Water and electrolyte content of human intervertebral discs under variable load, Spine 10:69, 1985.
24. Lindh M: Biomechanics of the lumbar spine. In Nordin M and Frankel VH: Basic biomechanics of the musculoskeletal system, Philadelphia, 1989, Lea & Febiger.
25. McCarroll JR, Miller JM, and Ritter MA: Lumbar spondylolysis and spondylolisthesis in college football players, Am J Sports Med 14:404, 1986.
26. Mooney V: Where is the pain coming from? Spine 12:754, 1987.
27. Nixon J: Intervertebral disc mechanics: a review, J World Soc Med 79:100, 1986.

28. Nolan JP and Sherk HH: Biomechanical evaluation of the extensor musculature of the cervical spine, Spine 13:9, 1988.
29. Roncarati A, Bucciarelli J, and English D: The incidence of low back pain in runners: a descriptive study, Corporate Fitness Recreation p 33, Feb/Mar, 1986.
30. Rovere GD: Low back pain in athletes, Physician Sportsmed 15:105, 1987.
31. Silver JR: Spinal injuries as a result of sporting accidents, Paraplegia 25:16, 1987.
32. Silver JR, Silver DD, and Godfrey JJ: Injuries of the spine sustained during gymnastic activities, Br Med J 293:861, 1986.
33. Silver JR, Silver DD, and Godfrey JJ: Trampolining injuries of the spine, Injury 17:117, 1986.
34. Snook S, Fine L, and Silverstein B: Musculoskeletal disorders. In Levy B and Wegman D, eds: Occupational health: recognizing and preventing work-related disease, Boston, 1988, Little, Brown & Co, Inc.
35. Spengler D et al: Back injury in industry: a retrospective study, Spine 11:241, 1986.
36. Thevenon A et al: Relationship between kyphosis, scoliosis, and osteoporosis in the elderly population, Spine 12:744, 1987.
37. Voutsinas SA and MacEwan GD: Sagittal profiles of the spine, Clin Orthop 210:235, 1986.
38. Wiesel SW, Bernini P, and Rothman RH: The aging lumbar spine, Philadelphia, 1982, WB Saunders Co.

ANNOTATED READINGS

Alexander MJL: Biomechanical aspects of lumbar spine injuries in athletes: a review, Can J Sport Sci 10:1, 1985.

Reviews lumbar spine anatomy, describes techniques for calculating forces acting on the lumbar spine, and describes lumbar injuries common to participants in gymnastics, football, weight lifting, running, rowing, ballet, polevaulting, swimming and diving, and the military.

Anderson GBL et al: Biomechanical analysis of loads on the lumbar spine in sitting and standing postures. In Matsui H and Kobayashi K: Biomechanics VIII-A: proceedings of the eighth international congress on biomechanics, Champaign, Ill, 1983, Human Kinetics Publishers, Inc.

Summarizes experiments in which mechanical models were used to analyze the loads on the low spine during static poses.

Kazarian L: Injuries to the human spinal column: biomechanics and injury classification, Exerc Sport Sci Rev 9:297, 1981.

Thoroughly reviews spinal anatomy, mechanics, and types of injuries.

Lindh M: Biomechanics of the lumbar spine. In Nordin M and Frankel VH: Basic biomechanics of the musculoskeletal system, Philadelphia, 1989, Lea & Febiger.

Discusses the kinematics and kinetics of the lumbar spine, with explanations of factors affecting lumbar spine loads during various postures and movements.

White AA and Panjabi MM: Clinical biomechanics of the spine, Philadelphia, 1978, JB Lippincott Co.

Serves as an extensive and well-illustrated reference work on the biomechanics of spinal function.

9 MOVEMENT

Linear Kinematics

After reading this chapter, the student will be able to:

Discuss the interrelationships among the primary kinematic variables.

Correctly associate linear kinematic quantities with their units of measure.

Identify and describe the effects of the factors governing the trajectory of a projectile.

Explain the reason projectile motion is typically analyzed according to its horizontal and vertical components.

Distinguish between average and instantaneous quantities and identify the circumstances under which each is a quantity of interest.

Select and use appropriate equations to solve problems related to linear kinematics.

As discussed in Chapter 1, biomechanics involves the application of mechanical principles in the study of living organisms. This chapter introduces the study of movement mechanics addressing the kinematic characteristics of a pure form of movement—linear motion.

LINEAR KINEMATIC QUANTITIES

kinematics
the form, pattern, or sequencing of movement with respect to time

Kinematics is the study of the geometry, pattern, or form of motion with respect to time. Kinematics, which describes the appearance of motion, is distinguished from kinetics, the study of the forces associated with motion. Linear kinematics involves the study of the shape, form, pattern, and sequencing of linear movement through time, without particular reference to the force or forces that cause or result from the motion.

The kinematics of a golf swing.

Kinematics spans qualitative and quantitative forms of analysis. For example, a qualitative kinematic description of a soccer kick may involve identification of the major joint actions occurring, including hip flexion, knee extension, and plantar flexion at the ankle. A more detailed qualitative kinematic analysis may also describe the precise sequencing and timing of the body segments' movements, which translates to the degree of skill evident on the part of the kicker. The progressive modification of kinematics reflects the learning process when students and athletes acquire a new skill, and kinematic changes accompany the process of rehabilitation of an injured joint.

A kinematic description of a soccer kick may include identification of joint actions, body segment displacements, velocities, accelerations, movement time, and initial ball velocity.

Although careful qualitative kinematic analyses of performance are invaluable for clinicians and teachers and coaches of physical skills, biomechanic researchers typically employ quantification in kinematic analyses of human motion. For example, an actual quantitative kinematic comparison of an expert professional football place kicker and a club soccer player involved quantification of variables such as knee extension time, with averages of 0.67 and 0.49 seconds, and resultant ball velocity, with averages of 29.5 and 30.7 m/s, for the expert and club player respectively (12). Sport biomechanists often study different sports to discover the kinematic features that characterize an elite performance or to determine the biomechanical factors that may limit the performance of a particular athlete (3, 4). This type of analysis sometimes results in the construction of a model that details the kinematic characteristics of superior performance for practical use by coaches and athletes (5).

However, most kinematic studies are performed on nonelite subjects. Biomechanists working in conjunction with motor development specialists have studied the kinematic characteristics of the various stages of gait development and throwing ability in young children. In collaboration with adapted physical education specialists, biomechanists have also documented the characteristic kinematic patterns associated with several relatively common disabling conditions such as cerebral palsy, Down's syndrome, and stroke.

Biomechanists commonly use high-speed cinematography or videography to perform quantitative kinematic analyses. The process involves taking a carefully planned film or video of a performance,

Figure 9-1

Stick figure representation of a single stride taken by a runner from which kinematic quantities of interest can be measured by a researcher.

with subsequent computerized or computer-assisted analysis of the performance on a picture-by-picture basis. A typical approach involves the reduction of the images of the human body to stick figure representations of the body, with lines connecting the joint centers. Precise measurements can then be made of changes in body segment positions. Figure 9-1 shows a computer-generated stick figure representation of a runner's stride that a researcher could use to assess kinematic quantities of interest. Kinematic descriptions of human movement may incorporate vector and/or scalar quantities depending on the goals and focus of the analysis.

Distance and Displacement

Units of distance and displacement are units of length. In the metric system, the most commonly used unit of distance and displacement is the **meter** (m). A kilometer (km) is 1000 m, a centimeter (cm) is 1⁄100 m, and a millimeter (mm) is 1⁄1000 m. In the English system, common units of length are the inch, the foot, the yard, and the mile.

meter
most common international unit of length on which the metric system is based

When a runner completes 1½ laps around a 400 m track, the distance that the runner has covered is equal to 600 (400 + 200) m. **Displacement** is measured in a straight line from the starting position to the final position. At the end of 1½ laps around the track, the runner's displacement is the length of the straight imaginary line that transverses the field, connecting the runner's starting position to the runner's final position halfway around the track (see Problem 1). At the completion of 2 laps around the track, the distance run is 800 m. Because starting and finishing positions are the same, the runner's displacement is 0. When a skater moves around an ice rink, the distance the skater travels may be measured along the tracks left by the skates. The skater's displacement is measured along a straight line from beginning to finishing positions on the ice (Figure 9-2).

displacement
change in position

■ The metric system is the predominant standard of measurement in every major country in the world except for the United States.

Another difference is that distance is a scalar quantity and displacement is a vector quantity. Consequently, displacement includes

Figure 9-2
The distance that a skater travels may be measured from the track on the ice. The skater's displacement is measured in a straight line from starting position to final position.

Figure 9-3

The course for the Hawaiian Ironman Triathlon includes a 2.4 mile swim, a 112 mile bike course, and a 26.2 mile run. A competitor's displacement at the end of the race, however, is close to zero.

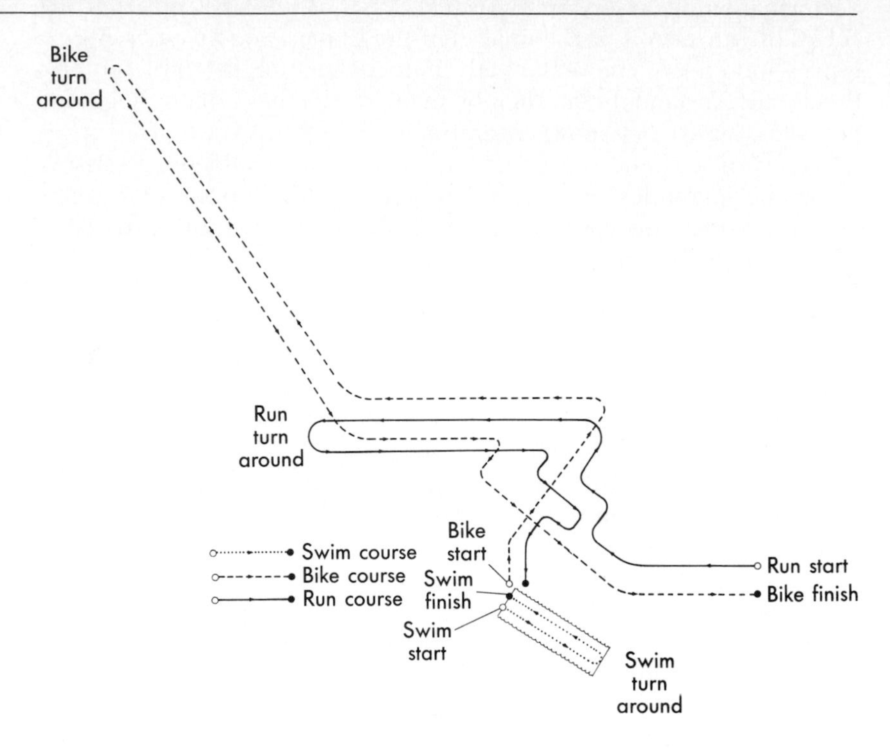

more than just the length of the line between two positions. Of equal importance is the *direction* in which the displacement occurs. The direction of a displacement relates the final position to the starting position. For example, the displacement of a yacht that has sailed 900 m on a tack due south would be identified as 900 m to the south.

The direction of a displacement may be indicated in several different, equally acceptable ways. Compass directions, such as south or northwest, the terms *left* and *right* or *up* and *down,* or the use of *positive* to refer to one direction and *negative* to refer to the opposite are all appropriate labels. Upwards is typically defined as the positive direction and downwards is regarded as the negative direction in studying the motion of projectiles. This enables indication of direction using plus and minus signs. Whatever the system or convention adopted for indication of direction, it should be used consistently in describing a given situation. It would be confusing to describe a displacement as 500 m north followed by 300 m to the right.

Either distance or displacement may be the more important quantity of interest depending on the situation. Many 5 km and 10 km race courses are set up so that the finish line is only a block or two from the starting line. This is also the case for the Ironman Triathlon course (Figure 9-3). Participants in these races are usually interested in the number of kilometers of distance covered or the num-

SAMPLE PROBLEM 1

Figure 9-4

A swimmer crosses a lake that is 0.9 km wide in 30 minutes. What was his average velocity? Can his average speed be calculated?

Known

After reading the problem carefully, the next step is to sketch the problem situation, showing all quantities that are known or may be deduced from the problem statement:

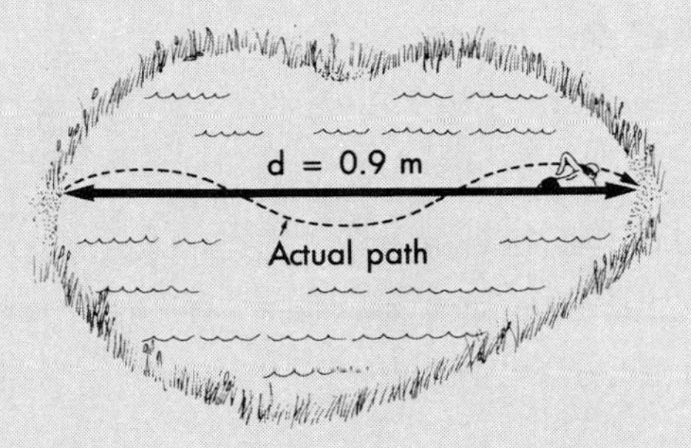

t = 30 min (0.5 hr)

Solution

In this situation, we know that the swimmer's displacement is 0.9 km. However, we know nothing about the exact path that the swimmer may have followed. The next step is to identify the appropriate formula to use to find the unknown quantity, which is velocity:

$$v = \frac{d}{t}$$

The known quantities can now be filled in to solve for velocity:

$$v = \frac{0.9 \text{ km}}{0.5 \text{ hr}}$$

$$\boxed{v = 1.8 \text{ km/hr}}$$

Speed is calculated as distance divided by time. Although we know the time taken to cross the lake, we do not know nor can we surmise from the information given the exact distance covered by the swimmer. Therefore, the swimmer's speed cannot be calculated.

ber of kilometers left to cover as they progress along the race course. Knowledge of displacement is not particularly valuable during this type of event. In other situations, displacement is more important. For example, triathlon competitions may involve a swim across a lake. Because swimming in a perfectly straight line when crossing a lake is virtually impossible, the actual distance a swimmer covers is always somewhat greater than the width of the lake (Figure 9-4).

However, the course is set up so that the identified length of the swim course is the length of the deplacement between the entry and exit points on the lake.

■ Distance covered and displacement may be equal for a given movement or distance may be greater than displacement, but the reverse is never true.

The magnitude of displacement and the distance covered during a particular movement can be identical. If a cross-country skier travels down a straight path through the woods, both distance covered and the skier's displacement are equal. If the path of motion is not rectilinear, the magnitudes of the distance traveled and the resultant displacement will differ.

Speed and Velocity

velocity
change in position with respect to time

Two quantities that parallel distance and displacement are speed and **velocity.** These terms are often used synonymously in general conversation, but in mechanics they have precise and different meanings. Speed, a scalar quantity, is defined as the distance covered divided by the time taken to cover it:

■ Displacement and velocity are vector equivalents of the scalar quantities distance and speed.

$$\text{speed} = \frac{\text{length (or distance)}}{\text{change in time}}$$

Velocity (v) is the change in position or the displacement that occurs during a given period of time:

$$v = \frac{\text{change in position}}{\text{change in time}}$$

$$v = \frac{\text{displacement}}{\text{change in time}}$$

Because the capital Greek letter delta (Δ) is commonly used in mathematical expressions to mean *change in,* a shorthand version of the relationship expressed above follows, with t representing the amount of time elapsed during the velocity assessment:

$$v = \frac{\Delta \text{ position}}{\Delta \text{ time}}$$

$$v = \frac{d}{\Delta t}$$

Another way to express change in position is $\text{position}_2 - \text{position}_1$ in which position_1 represents the body's position at one point in time and position_2 represents the body's position at a later point:

$$\text{velocity} = \frac{\text{position}_2 - \text{position}_1}{\text{time}_2 - \text{time}_1}$$

■ Units of speed and velocity are always units of length divided by units of time.

Because velocity is based on displacement, it is also a vector quantity. Consequently, description of velocity must include an indication of both the direction and magnitude of the motion. If the direction of the motion is positive, velocity is positive; if the direction is negative, velocity is a negative quantity. A change in a body's velocity may represent a change in its speed, movement direction, or both.

Whenever two or more velocities act, the laws of vector algebra govern the ultimate speed and direction of the resultant motion. For example, the path actually taken by a swimmer crossing a river is determined by the vector sum of the swimmer's speed in the intended direction and the velocity of the river's current (Figure 9-5). The sample problem shown in Figure 9-6 provides an illustration of this situation.

Units of speed and velocity are units of length divided by units of time. In the metric system, common units for speed and velocity are meters per second (m/s) and kilometers per hour (km/hr).

Any unit of length divided by any unit of time yields an acceptable unit of speed or velocity. For example, a speed of 5 m/s can also be expressed as 5000 mm/s or 18,000 m/hr. It is usually most practical to select units that will result in expression of the quantity in the smallest, most manageable form.

Since maximizing speed is the objective of all racing events, sport biomechanists have focused on the kinematic features that appear to accompany fast performances in running, skiing, skating, cycling, swimming, and rowing events. Two kinematic variables that have been studied extensively are stride length and stride frequency. In running, speed is the product of these two parameters. As running speed increases, characteristic changes in stride length and stride

Running speed is the product of stride length and stride frequency.

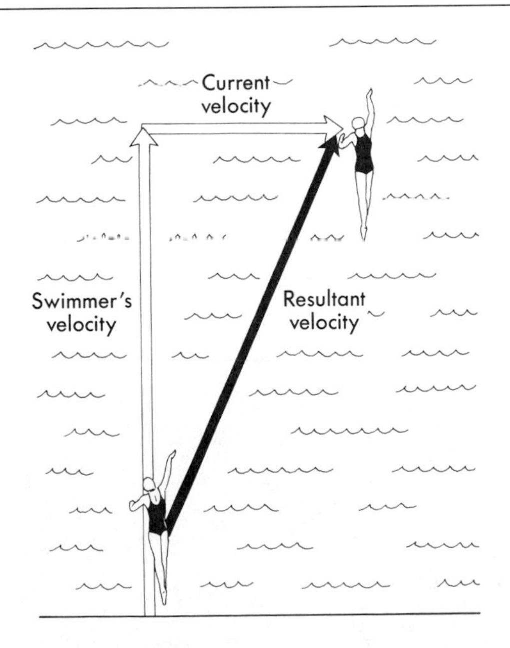

Figure 9-5
The velocity of a swimmer in a river is the vector sum of the swimmer's velocity and the velocity of the current.

Figure 9-6

SAMPLE PROBLEM 2

A swimmer orients herself perpendicular to the parallel banks of a river. If the swimmer's velocity is 2 m/s and the velocity of the current is 0.5 m/s, what will be the swimmer's resultant velocity? How far will the swimmer actually have to swim to get to the other side if the banks of the river are 50 m apart?

Solution

A diagram showing vector representations of the velocities of the swimmer and the current is drawn:

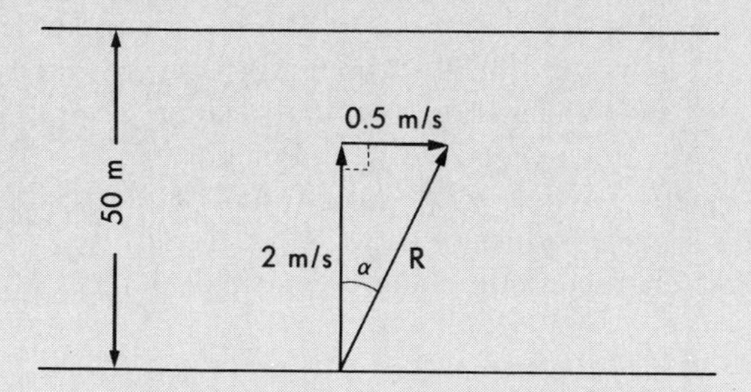

The resultant velocity can be found graphically by measuring the length and the orientation of the vector resultant of the two given velocities:

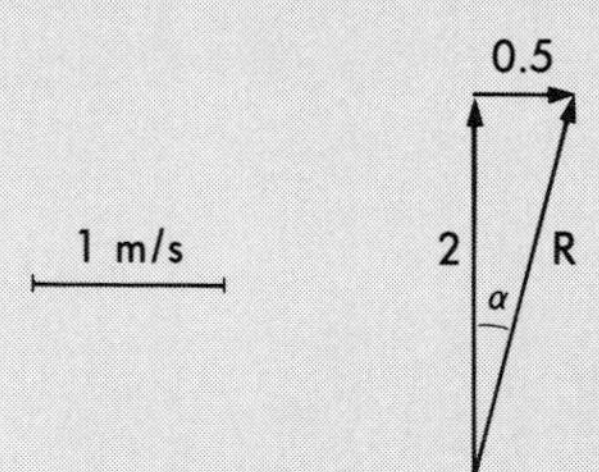

R is approximately 2.1 m/s
α is approximately 15 degrees

The resultant velocity can also be found using trigonometric relationships. The magnitude of the resultant velocity may be calculated using the Pythagorean theorem:

$$R^2 = 2^2 + 0.5^2$$

$$R = \sqrt{2^2 + 0.5^2}$$

$$\boxed{R = 2.06 \text{ m/s}}$$

The direction of the resultant velocity may be calculated using the cosine relationship:

$$R \cos \alpha = 2$$

$$2.06 \cos \alpha = 2$$

$$\alpha = \arccos\left(\frac{2}{2.06}\right)$$

$$\boxed{\alpha = 14 \text{ degrees}}$$

If the swimmer travels in a straight line in the direction of her resultant velocity, the cosine relationship may be used to calculate her resultant displacement:

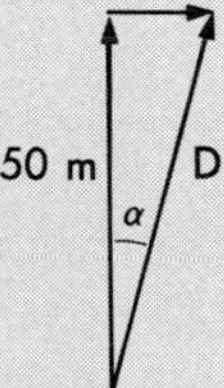

$$D \cos \alpha = 50 \text{ m}$$

$$D \cos 14 = 50 \text{ m}$$

$$\boxed{D = 51.5 \text{ m}}$$

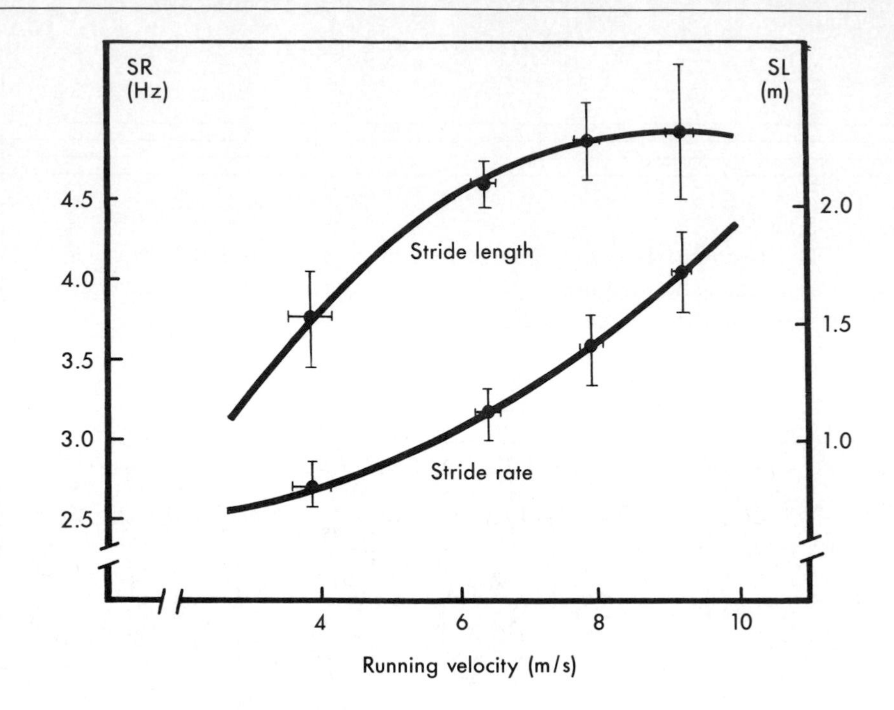

Figure 9-7
Changes in stride length and stride rate with running velocity. (From Luhtanen P and Komi PV: Mechanical factors influencing running speed. In Asmussen E and Jorgensen K, eds: Biomechanics VI-B, Baltimore, 1978, University Park Press.)

frequency often occur (Figure 9-7). Research has shown that the best male and female sprinters are distinguished from their less-skilled counterparts by extremely high stride frequencies and short ground contact times, although their stride lengths are usually only average or slightly greater than average (4). In contrast, the fastest cross-country skiers have longer-than-average cycle lengths, with cycle rates that are only average (13). Research on skating kinematics has shown that better ice skaters appear to excel because of higher stride rates (10), whereas elite roller skaters are distinguished by longer strides (15).

When racing performances are analyzed, comparisons are usually based on pace rather than speed or velocity. Pace is the inverse of speed. Rather than units of distance divided by units of time, pace is presented as units of time divided by units of distance. Pace is the time taken to cover a given distance and is commonly quantified as minutes per km or minutes per mile.

Acceleration

acceleration
the rate of change in velocity

We are well aware that the consequence of pressing down or letting up on the accelerator pedal of an automobile is usually a change in the automobile's speed (and velocity). **Acceleration** (a) is defined as the rate of change in velocity or the change in velocity occurring over a given time interval, with t representing the amount of time elapsed during the velocity assessment:

$$a = \frac{\text{change in velocity}}{\text{change in time}}$$

$$a = \frac{\Delta v}{\Delta t}$$

Another way to express change in velocity is $v_2 - v_1$, in which v_1 represents velocity at one point in time and v_2 represents velocity at a later point:

$$a = \frac{v_2 - v_1}{\Delta t}$$

Units of acceleration are units of velocity divided by units of time. If a car increases its velocity by 1 km/hr each second, its acceleration is 1 km/hr/s. If a skier increases velocity by 1 m/s each second, the acceleration is 1 m/s/s. In mathematical terms, it is simpler to express the skier's acceleration as 1 m/s squared (1 m/s^2). A common unit of acceleration in the metric system is m/s^2.

Acceleration is the rate of change in velocity or the degree with which velocity is changing with respect to time. For example, a body accelerating in a positive direction at a constant rate of 2 m/s^2 is increasing its velocity by 2 m/s each second. If the body's initial velocity was 0, a second later its velocity would be 2 m/s, a second after that its velocity would be 4 m/s, and a second after that its velocity would be 6 m/s.

In general usage, the term *accelerating* means speeding up or increasing in velocity. If v_2 is greater than v_1, acceleration is a positive number, and the body in motion may have speeded up during the time period in question. However, because it is sometimes appropriate to label the direction of motion as positive or negative, a positive value of acceleration may not mean that the body is speeding up.

If the direction of motion is described in terms other than positive or negative, a positive value of acceleration does indicate that the body being analyzed has speeded up. For example, if a sprinter's velocity is 3 m/s on leaving the blocks and is 5 m/s one second later, calculation of the acceleration that has occurred will yield a positive number. Because $v_1 = 3$ m/s, $v_2 = 5$ m/s, and $t = 1$ s:

$$a = \frac{v_2 - v_1}{\Delta t}$$

$$a = \frac{5\ \text{m/s} - 3\ \text{m/s}}{1\ \text{s}}$$

$$a = 2\ \text{m/s}^2$$

Sliding into a base results in negative acceleration of the base runner.

Whenever the direction of motion is described in terms other than positive or negative and v_2 is greater than v_1, the value of acceleration will be a positive number and the object in question is speeding up.

Acceleration can also assume a negative value. As long as the direction of motion is described in terms other than positive or negative, negative acceleration indicates that the body in motion is slowing down or that its velocity is decreasing. For example, when a base runner slides to a stop over home plate, acceleration is negative. If a base runner's velocity is 4 m/s when going into a 0.5 s slide that stops the motion, $v_1 = 4$ m/s, $v_2 = 0$, and $t = 0.5$ s. Acceleration may be calculated as the following:

$$a = \frac{v_2 - v_1}{t}$$

$$a = \frac{0 - 4 \text{ m/s}}{0.5 \text{ s}}$$

$$a = -8 \text{ m/s}^2$$

Whenever v_1 is greater than v_2 in this type of situation, acceleration will be negative. The sample problem in Figure 9-8 provides another example of a situation involving negative acceleration.

Understanding acceleration is more complicated when one direction is designated as positive and the opposite direction is designated as negative. In this situation a positive value of acceleration can indicate either that the object is speeding up in a positive direction or that it is slowing down in a negative direction (Figure 9-9).

Consider the case of a ball being dropped from a hand. As the ball falls faster and faster because of the influence of gravity, it is gaining speed, for example 0.3 m/s to 0.5 m/s to 0.8 m/s. Because the

SAMPLE PROBLEM 3

Figure 9-8

A soccer ball is rolling down a field. At t = 0, the ball has an instantaneous velocity of 4 m/s. If the acceleration of the ball is constant at $-0.3\ m/s^2$, how long will it take the ball to come to a complete stop?

Known

After reading the problem carefully, the next step is to sketch the problem situation, showing all quantities that are known or given in the problem statement.

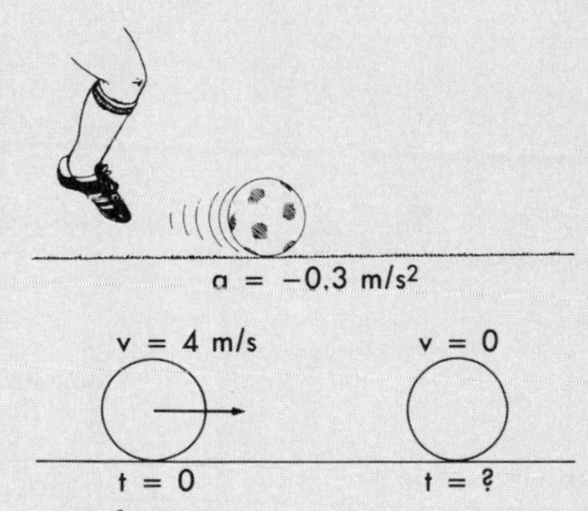

Solution

The next step is to identify the appropriate formula to use to find the unknown quantity:

$$a = \frac{v_2 - v_1}{t}$$

The known quantities can now be filled in to solve for the unknown variable (time):

$$-0.3\ m/s^2 = \frac{0 - 4\ m/s}{t}$$

Rearranging the equation, we have the following:

$$t = \frac{0 - 4\ m/s}{-0.3\ m/s^2}$$

Simplifying the expression on the right side of the equation, we have the solution:

$$\boxed{t = 13.3\ s}$$

Figure 9-9
Right is regarded as the positive direction, and *left* is the negative direction. Acceleration may be positive, negative, or equal to zero, based on the direction of the motion and the direction of the change in velocity.

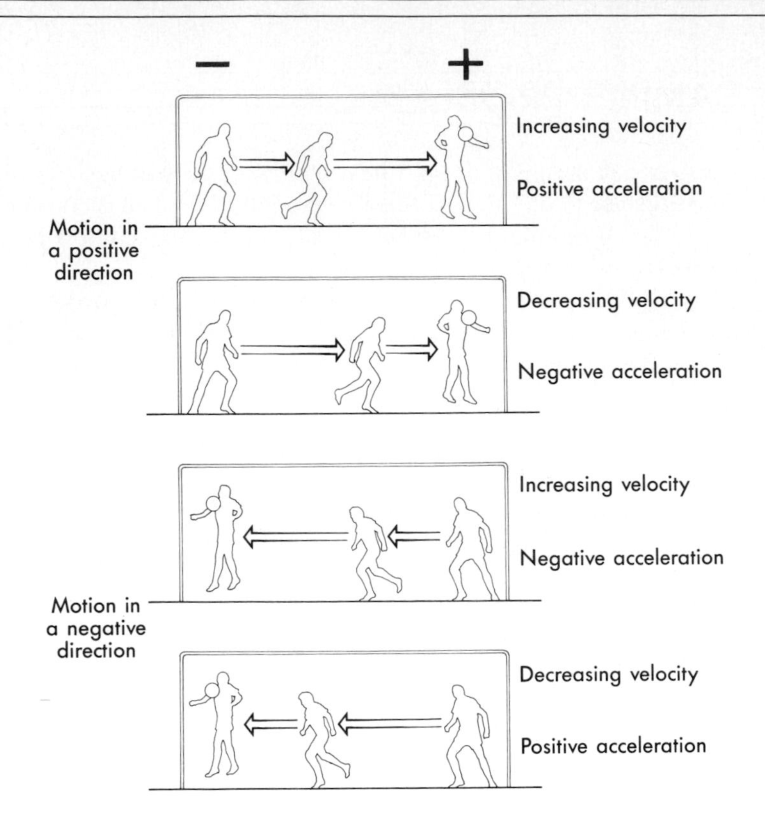

downward direction is considered as the negative direction, the ball's velocity is actually −0.3 m/s to −0.5 m/s to −0.8 m/s. If $v_1 = -0.3$ m/s, $v_2 = -0.5$ m/s, and $t = 0.02$ s, acceleration is calculated as follows:

$$a = \frac{v_2 - v_1}{t}$$

$$a = \frac{-0.5 \text{ m/s} - -0.3 \text{ m/s}}{0.02 \text{ s}}$$

$$a = -10 \text{ m/s}^2$$

In this situation the ball is speeding up, yet its acceleration is negative because it is speeding up in a negative direction. If acceleration is negative, velocity may be either increasing in a negative direction or decreasing in a positive direction. Alternatively, if acceleration is positive, velocity may be either increasing in a positive direction or decreasing in a negative direction.

■ When acceleration is 0, velocity is constant.

The third alternative is for acceleration to be equal to 0. Acceleration is 0 whenever velocity is constant, that is, when v_1 and v_2 are the same. In the middle of a 100 m sprint, a sprinter's acceleration should be close to 0 because at that point the runner should be running at a constant, near maximum velocity.

KINEMATICS OF PROJECTILE MOTION

projectile
a body in free fall that is subject only to the forces of gravity and air resistance

Bodies projected into the air are **projectiles.** A basketball, a discus, a high jumper, and a sky diver are all projectiles as long as they are moving through the air unassisted. Depending on the projectile, different kinematic quantities are of interest. The resultant horizontal displacement of the projectile determines the winner of the contest in field events such as the shot put, discus throw, or javelin throw. High jumpers and pole vaulters maximize ultimate vertical displacement to win events. Sky divers manipulate both horizontal and vertical components of velocity to land as close as possible to a target on the ground.

■ The force of gravity produces a constant acceleration on bodies near the surface of the earth equal to approximately -9.81 m/s^2.

However, not all objects that fly through the air are projectiles. A projectile is a body in free fall that is subject only to the forces of gravity and air resistance. Therefore, objects such as airplanes and rockets do not qualify as projectiles because they are also influenced by the forces generated by their engines.

■ The three mechanical factors that determine a projectile's motion are projection angle, projection speed, and relative height of projection.

Three factors influence the **trajectory** (flight path) of a projectile: the angle of projection, projection speed, and the relative height of projection (Figure 9-10).

trajectory
the flight path of a projectile

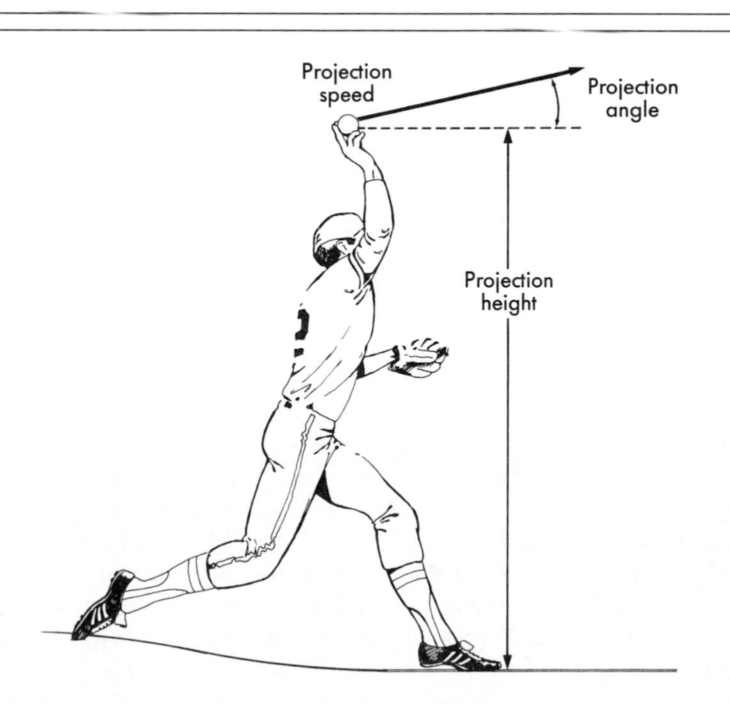

Figure 9-10
Factors affecting the trajectory of a projectile include projection angle, projection speed, and relative height of projection.

Figure 9-11
The effect of projection angle on projectile trajectory with projection speed held constant.

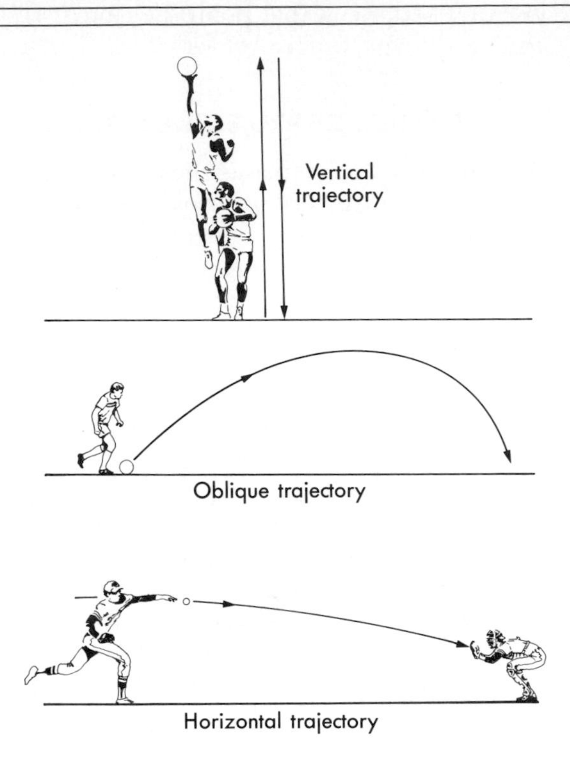

Projection angle is particularly important in the sport of basketball. A common error among novice players is shooting the ball with too flat a trajectory.

angle of projection
the angle at which a body is projected with respect to the horizontal

Influence of Projection Angle

The **angle of projection** and the effects of air resistance govern the shape of a projectile's trajectory. Changes in projection speed influence the size of the trajectory, but trajectory shape is solely dependent on projection angle. In the absence of air resistance, the trajectory of a projectile assumes one of three general shapes, depending on the angle of projection. If the projection angle is perfectly vertical (90 degrees to the horizontal), the trajectory is also perfectly vertical, with the projectile following the same path straight up and then straight down again. If the projection angle is oblique (at some angle between 0 and 90 degrees), the trajectory is *parabolic,* which means shaped like a parabola. A parabola is symmetrical, so its right and left halves are mirror images of each other. A body projected perfectly horizontally (at an angle of 0 degrees) will follow a trajectory resembling one half of a parabola (Figure 9-11). Figure 9-12 displays scaled, theoretical trajectories for an object projected at different angles at a given speed. A ball thrown upward at a projection angle of 80 degrees to the horizontal follows a relatively high and narrow trajectory, achieving more height than horizontal distance. A ball projected upward at a 10 degree angle to the horizontal follows a trajectory that is flat and long in shape.

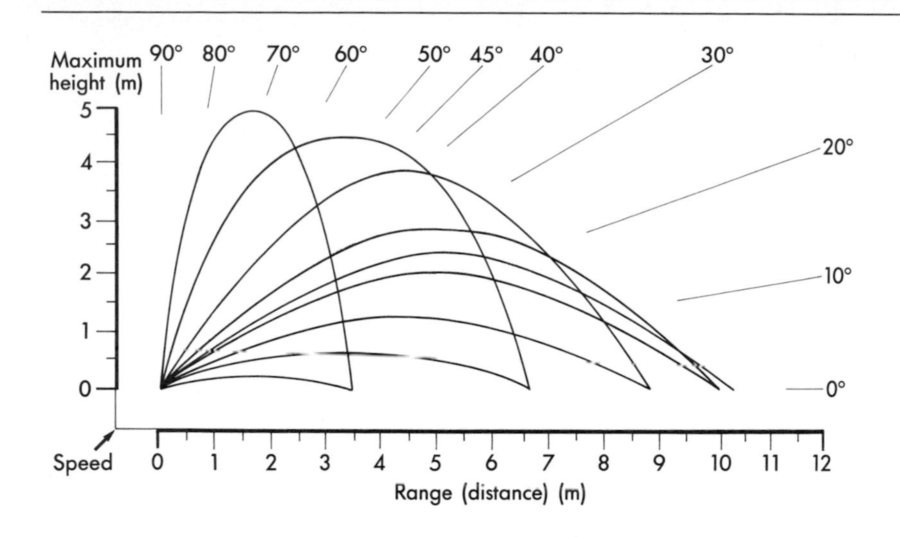

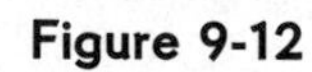

Figure 9-12

The size and shape of projectile trajectories vary with the angle and speed of projection.

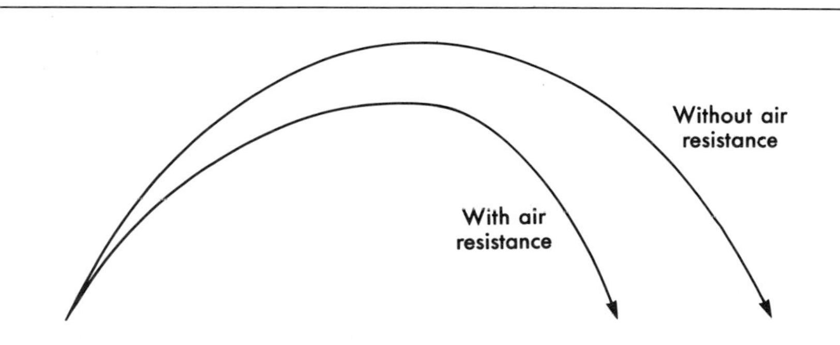

Figure 9-13

In real-life situations, air resistance causes a projectile to deviate from its theoretical parabolic trajectory.

In reality, air resistance may create irregularities in the shape of a projectile's trajectory. A typical modification in trajectory caused by air resistance is displayed in Figure 9-13. For purposes of simplification, the effects of aerodynamic forces will be disregarded in the discussion of projectile motion.

Influence of Projection Speed

When projection angle and other factors are constant, the **projection speed** determines the length or size of a projectile's trajectory. For example, when a body is projected vertically upward, the projectile's initial speed determines the height of the **apex** (highest point in the trajectory). For a body that is projected at an oblique angle, the speed of projection determines both the height and horizontal length of the trajectory (Figure 9-14). The combined effects of pro-

projection speed
the magnitude of projection velocity

apex
the highest point in the trajectory of a projectile

Figure 9-14

The effect of projection speed on projectile trajectory with projection angle held constant.

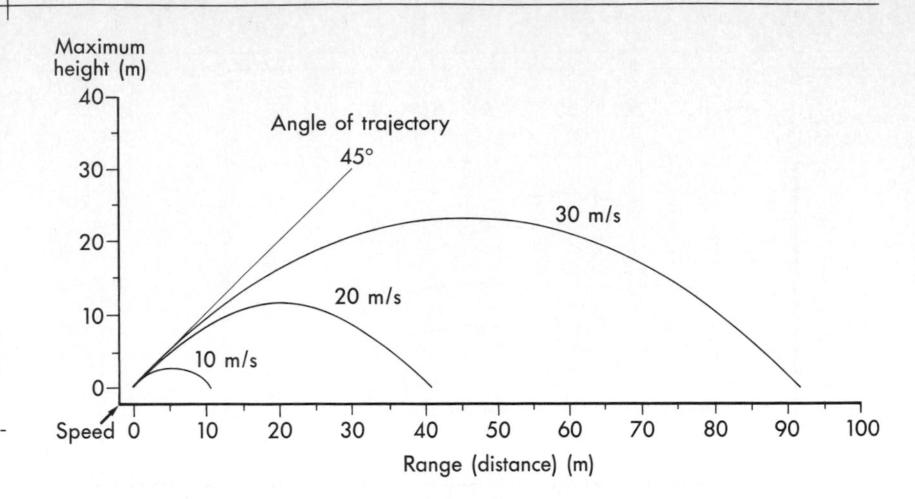

■ A projectile's flight time is increased by increasing the vertical component of projection velocity or by increasing the relative projection height.

Table 9-1

EFFECT OF PROJECTION ANGLE ON RANGE

PROJECTION SPEED (m/s)	PROJECTION ANGLE (degrees)	RANGE (m)
10	10	3.49
10	20	6.55
10	30	8.83
10	40	10.04
10	45	10.19
10	50	10.04
10	60	8.83
10	70	6.55
10	80	3.49
20	10	13.94
20	20	26.21
20	30	35.31
20	40	40.15
20	45	40.77
20	50	40.15
20	60	35.31
20	70	26.21
20	80	13.94
30	10	31.38
30	20	58.97
30	30	79.45
30	40	90.35
30	45	91.74
30	50	90.35
30	60	79.45
30	70	58.98
30	80	31.38

Relative projection height equals 0.

jection speed and projection angle on the horizontal displacement or **range** of a projectile are shown in Table 9-1.

range
the horizontal displacement of a projectile at landing

Performance in the execution of a vertical jump on a flat surface is entirely dependent on takeoff speed; that is, the greater the vertical velocity at takeoff, the higher the jump, and the higher the jump, the greater the amount of time the jumper is airborne (see margin).

HEIGHT (cm)	TIME (s)
5	0.2
11	0.3
20	0.4
31	0.5
44	0.6
60	0.7
78	0.8
99	0.9

The time required for the performance of a vertical jump can be an important issue for dance choreographers. The incorporation of vertical jumps into a performance must be planned carefully (8). If the tempo of the music necessitates that vertical jumps be executed within one third of a second, the height of the jumps is restricted to approximately 12 cm. The choreographer must be aware that under these circumstances the larger footed dancers may not have sufficient floor clearance to point their toes during jump execution.

Influence of Relative Projection Height

The third major factor influencing the kinematics of projectile motion is the **relative projection height.** This is the difference in the height from which the body is initially projected and the height at which it either lands on a surface or is stopped. When a discus is released by a thrower from a height of 1½ m above the ground, the relative projection height is 1½ m because the projection height is 1½ m greater than the height of the field on which the discus lands. If a driven golf ball becomes lodged in a tree, the relative projection height is negative because the landing height is greater than the projection height. Figure 9-15 illustrates the concept of relative projection height. When projection velocity is constant, the greater the relative projection height, the longer the flight time, and the greater the horizontal displacement of the projectile.

relative projection height
the difference between projection height and landing height

Optimal Projection Conditions

For events in which achieving maximum horizontal displacement or maximum vertical displacement of the projectile is the objective, one goal is to maximize the speed of projection. In the throwing events with all other factors being equal, another goal is to maximize release height because greater relative projection height produces longer flight time, which results in greater horizontal displacement of the projectile. However, it is not prudent for a thrower to sacrifice release speed for added release height in most cases.

The factor that varies the most, both with the event and with the anthropometric characteristics of the performer, is the optimum angle of projection. When projection and landing heights are equal and projection speed is constant, the angle of projection that results in maximum horizontal displacement of a projectile is 45 degrees with respect to the horizontal. As relative projection height increases, this optimum angle of projection decreases, and as relative

The human body becomes a projectile during the airborne phase of a jump.

Figure 9-15
The relative projection height.

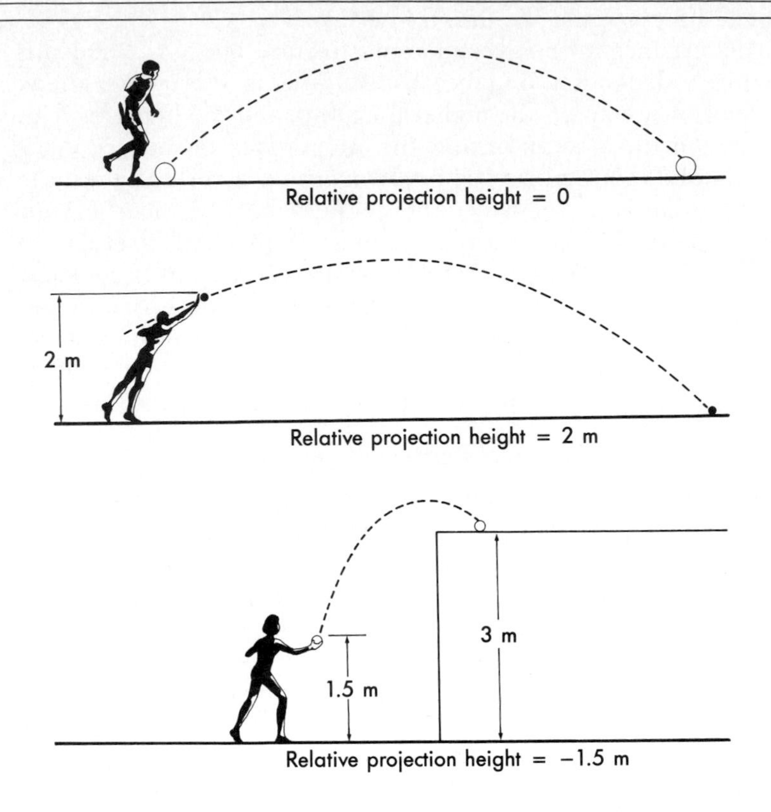

Figure 9-16
When projection speed is constant and aerodynamics are not considered, the optimum projection angle is based on the relative height of projection. When the relative projection height is 0, an angle of 45 degrees is optimum. As the relative projection height increases, optimum projection angle decreases. As the relative projection height becomes increasingly negative, the optimum projection angle increases.

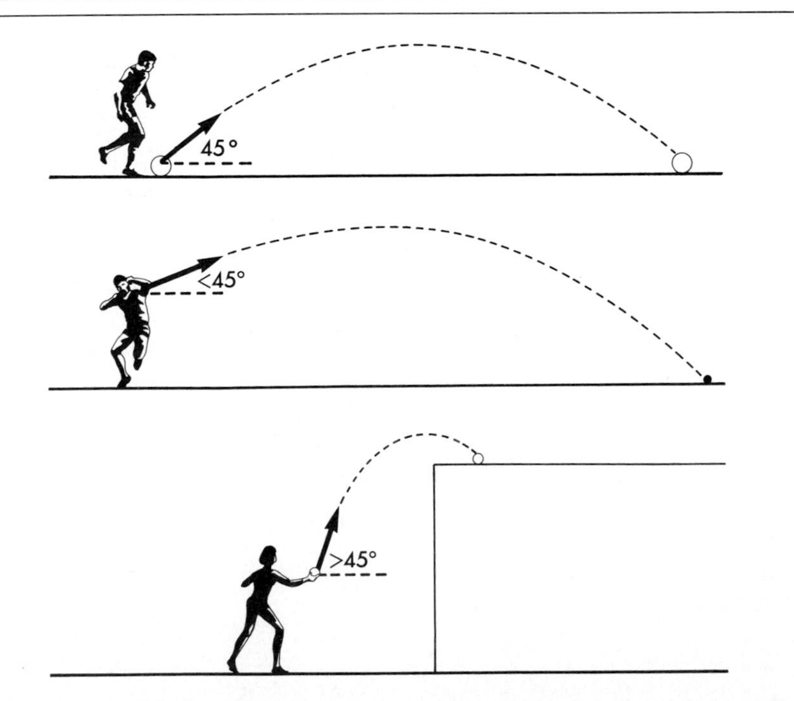

projection height decreases, the optimum angle increases (Figure 9-16). Optimum projection angles for human performances are more difficult to analyze because the angle at which the body is projected at takeoff can affect projection speed.

In the performance of the long jump, for example, because the takeoff and landing heights are the same, the theoretically optimum angle of takeoff is 45 degrees with respect to the horizontal. However, it has been estimated by Hay (6) that to obtain the theoretically optimum takeoff angle, long jumpers would decrease the horizontal velocity they could otherwise obtain by approximately 50%. The actual takeoff angles employed by elite long jumpers range from approximately 18 to 27 degrees (6). Takeoff angles during all three phases of the triple jump are even smaller among elite performers than those used in the long jump (11). In the ski jump, athletes have the advantage of a large relative height between takeoff and landing, and takeoff angles of 4.6 to 6.2 degrees with respect to the horizontal have been reported (14). In an event such as the high jump in which the goal is to maximize the jumper's vertical displacement, takeoff angles among skilled Fosbury Flop style jumpers range from 40 to 48 degrees (2).

The instantaneous velocity of the shot at the moment of release primarily determines the ultimate horizontal displacement of the shot.

In the throwing events, the aerodynamic characteristics of the projected implements also influence the trajectory. In these events (shot, discus, javelin, and hammer), only the trajectory of the shot is not appreciably affected by aerodynamic forces. The concept that the optimum angle of release must not restrict release speed is still a paramount consideration for performance in the shot put. The release angles reported among elite throwers in the shot are approximately 36 to 37 degrees (7).

VECTOR COMPONENTS OF PROJECTILE MOTION

initial velocity
vector quantity incorporating both angle and speed of projection

Because velocity is a vector quantity, the **initial velocity** of a projectile incorporates both the initial speed (magnitude) and the angle of projection (direction) into a single quantity. When the initial velocity of a projectile is resolved into horizontal and vertical components, the horizontal component has a certain speed or magnitude in a horizontal direction, and the vertical component has a speed or magnitude in a vertical direction (Figure 9-17). The magnitudes of the horizontal and vertical components are always quantified so that if they were added together through the process of vector composition, the resultant velocity vector would be equal in magnitude and direction to the original initial velocity vector. The horizontal and vertical components of initial velocity may be quantified either graphically or trigonometrically (Figure 9-18).

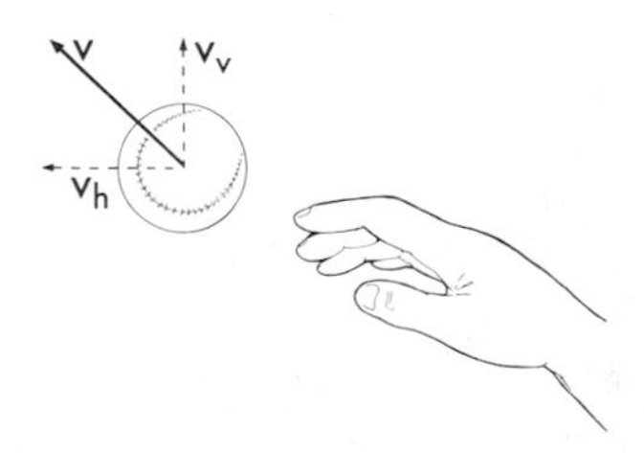

Figure 9-17
The vertical and horizontal components of projection velocity.

Just as it is more convenient to analyze general motion in terms of its linear and angular components, it is usually more meaningful to

Figure 9-18

SAMPLE PROBLEM 4

A basketball is released with an initial speed of 8 m/s at an angle of 60 degrees. Find the horizontal and vertical components of the ball's initial velocity, both graphically and trigonometrically.

Known

A diagram showing a vector representation of the initial velocity is drawn using a scale of 1 cm = 2 m/s:

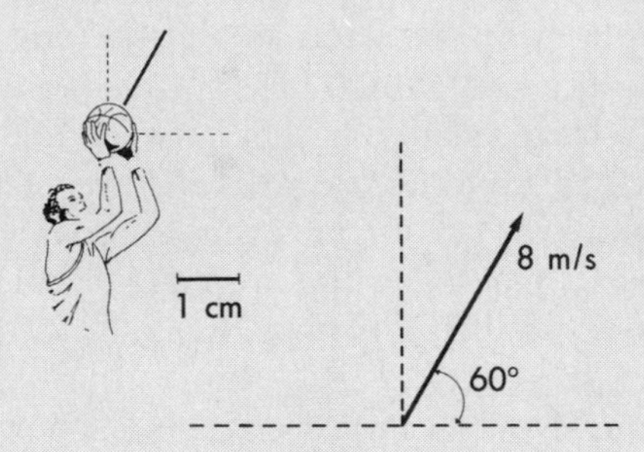

Solution

The horizontal component is drawn in along the horizontal line to a length that is equal to the length that the original velocity vector extends in the horizontal direction. The vertical component is then drawn in the same fashion in a direction perpendicular to the horizontal line:

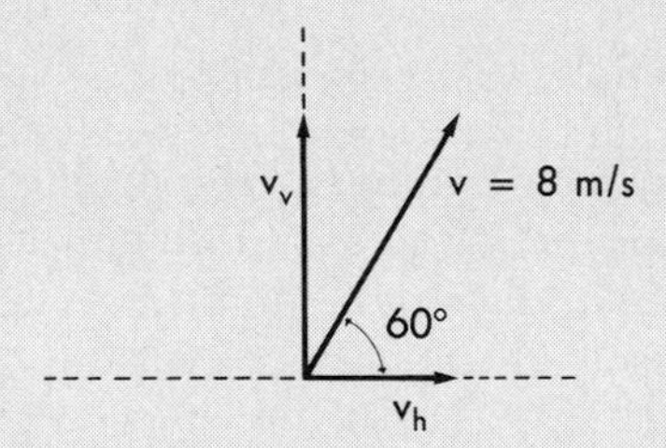

The lengths of the horizontal and vertical components are then measured:

$$\text{length of horizontal component} = 2 \text{ cm}$$

$$\text{length of vertical component} = 3.5 \text{ cm}$$

To calculate the magnitudes of the horizontal and vertical components, use the scale factor of 2 m/s/cm:
Magnitude of horizontal component:

$$v_h = 2 \text{ cm} \times 2 \text{ m/s/cm}$$

$$\boxed{v_h = 4 \text{ m/s}}$$

Magnitude of vertical component:

$$v_v = 3.5 \text{ cm} \times 2 \text{ m/s/cm}$$

$$\boxed{v_v = 7 \text{ m/s}}$$

To solve for v_h and v_v trigonometrically, construct a right triangle with the sides being the horizontal and vertical components of initial velocity and the initial velocity represented as the hypotenuse:

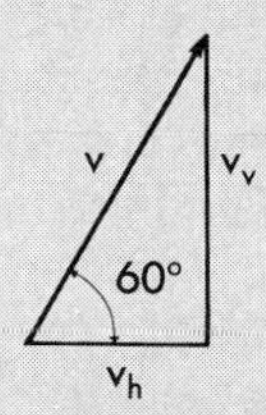

The sine and cosine relationships may be used to quantify the horizontal and vertical components:

$$v_h = 8 \text{ m/s} \times \cos 60$$

$$\boxed{v_h = 4 \text{ m/s}}$$

$$v_v = 8 \text{ m/s} \times \sin 60$$

$$\boxed{v_v = 6.9 \text{ m/s}}$$

Note that the magnitude of the horizontal component is *always* equal to the magnitude of the initial velocity multiplied by the cosine of the projection angle. Similarly, the magnitude of the initial vertical component is *always* equal to the magnitude of the initial velocity multiplied by the sine of the projection angle.

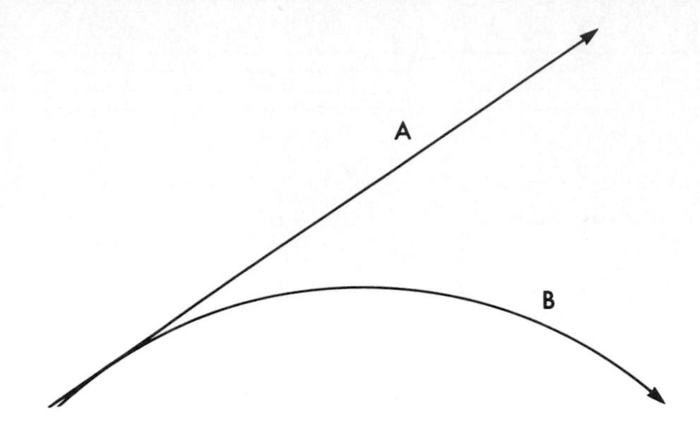

Figure 9-19
The influence of gravitational force on projectile trajectories *A*, without and *B*, with gravitational influence.

■ A projectile's range is the product of its horizontal speed and flight time.

analyze the horizontal and vertical components of projectile motion separately. This is true for two reasons: First, the vertical component is influenced by gravity, whereas no force (neglecting air resistance) affects the horizontal component. Second, the horizontal component of motion relates to the distance the projectile travels, and the vertical component relates to the maximum height achieved by the projectile. Once a body has been projected into the air, its overall (resultant) velocity is constantly changing because of the forces acting on it. When examined separately, the horizontal and vertical components of projectile velocity change predictably.

The product of a projectile's horizontal velocity and flight time determine its range (neglecting air resistance). In most real-life situations, air resistance affects the horizontal component of projectile velocity. A ball thrown with a certain initial velocity in an outdoor area will travel much farther if it is thrown with a tail wind rather than into a head wind. If an object were projected in a vacuum in which air resistance is not a factor, the horizontal component of its velocity would remain exactly the same throughout the flight. For purposes of simplification, the horizontal component of a given projectile's velocity will be regarded as an unchanging (constant) quantity.

When a projectile drops vertically through the air in a typical real-life situation, its velocity at any point is also related to air resistance. A skydiver's velocity, for example, is much smaller after the opening of the parachute than before its opening.

A major factor that influences the vertical but not the horizontal component of projectile motion is the force of gravity, which accelerates bodies in a vertical direction toward the surface of the earth (Figure 9-19). Unlike aerodynamic factors that may vary with the ve-

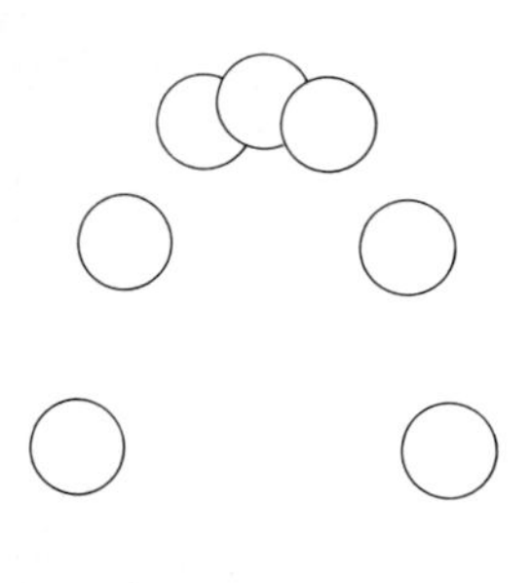

Figure 9-20
The pattern of change in the vertical velocity of a projectile is symmetrical about the apex of the trajectory.

locity of the wind, gravitational force is a constant, unchanging force that produces a constant vertical acceleration. Because the vertically upward direction is conventionally designated as the positive direction and downward as the negative direction, the acceleration of gravity is treated as a negative quantity (-9.81 m/s^2). This acceleration remains constant regardless of the size, shape, or weight of the projectile. The vertical component of the initial projection velocity determines the maximum vertical displacement achieved by a body projected from a given relative projection height.

■ Neglecting air resistance, the horizontal speed of a projectile remains constant throughout the trajectory.

Figure 9-20 illustrates the influence of gravity on projectile flight in the case of a ball tossed vertically upwards into the air by a juggler. The ball leaves the juggler's hand with a certain vertical velocity. As the ball travels higher and higher, the magnitude of its velocity decreases because it is undergoing a negative acceleration (the acceleration of gravity in a downward direction). At the apex of the trajectory, which is that instant between going up and coming down, vertical velocity is 0. As the ball falls downward, its speed progressively increases, again because of gravitational acceleration. Since the direction of motion is downward, the ball's velocity is becoming progressively more negative. If the ball is caught at the same height from which it was tossed, the ball's speed is exactly the same as its initial speed, although its direction is now reversed. Graphs of the displacement, velocity, and acceleration of a vertically tossed ball are shown in Figure 9-21.

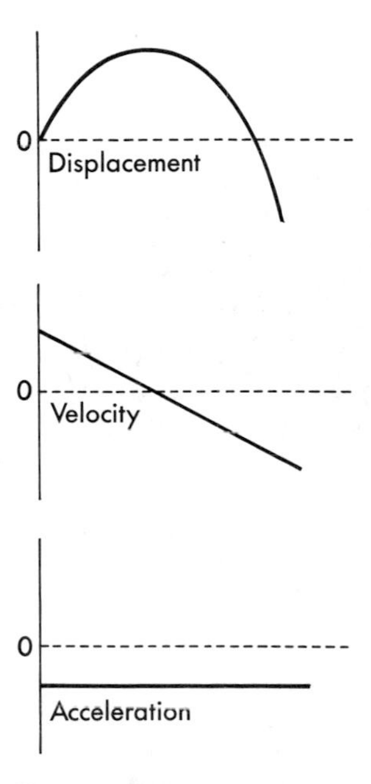

Figure 9-21
Displacement, velocity, and acceleration graphs for a ball tossed into the air that falls to the ground.

Horizontal and vertical components of projectile motion are independent of each other. In the example shown in Figure 9-22 a baseball is dropped from a height of 1 m at the same instant that a second ball is horizontally struck by a bat at a height of 1 m resulting in a line drive. Both balls land on the level field simultaneously because the vertical components of their motions are identical. However, because the line drive also has a horizontal component of motion, it also undergoes some horizontal displacement.

For purposes of analyzing the motion of projectiles, it will be assumed that the horizontal component of projectile velocity is con-

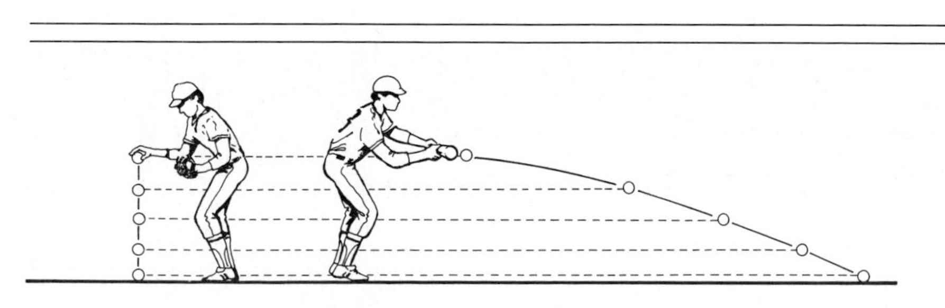

Figure 9-22
The vertical and horizontal components of projectile motion are independent.

Figure 9-23
The horizontal and vertical components of projectile velocity. Notice that the horizontal component is constant, and the vertical component is constantly changing.

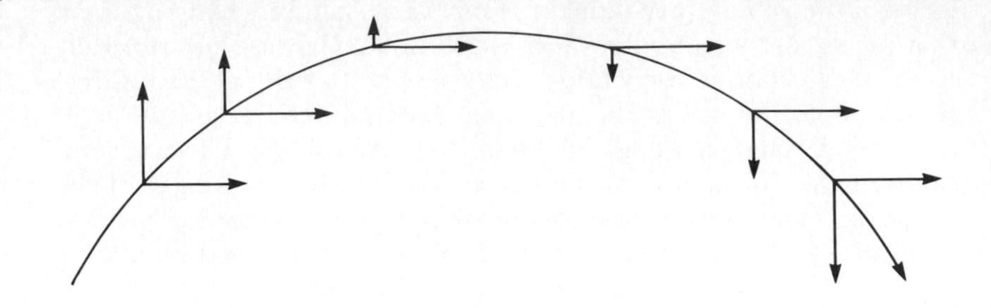

■ The vertical speed of a projectile is constantly changing because of gravitational acceleration.

■ The horizontal acceleration of a projectile is always 0.

stant throughout the trajectory and that the vertical component of projectile velocity is constantly changing because of the influence of gravity (Figure 9-23). Since horizontal projectile velocity is constant, horizontal acceleration is equal to the constant of 0 throughout the trajectory. The vertical acceleration of a projectile is equal to the constant -9.81 m/s^2.

EQUATIONS OF CONSTANT ACCELERATION

laws of constant acceleration
three formulas relating displacement, velocity, acceleration, and time when acceleration is unchanging

When a body is moving with a constant acceleration (positive, negative, or equal to 0), certain interrelationships are present among the kinematic quantities associated with the motion of the body. These interrelationships may be expressed using three mathematical equations originally derived by Galileo, which are known as the **laws of constant acceleration** or the laws of uniformly accelerated motion. Using the variable symbols d, v, a, and t (representing displacement, velocity, acceleration, and time, respectively) and with the subscripts 1 and 2 (representing first or initial and second or final points in time), the equations are the following:

$$v_2 = v_1 + at \qquad (1)$$

$$d = v_1 t + (1/2)\, at^2 \qquad (2)$$

$$v_2^2 = v_1^2 + 2ad \qquad (3)$$

It is instructive to examine these relationships as applied to the horizontal component of projectile motion in which $a = 0$. In this case, each term containing acceleration may be removed from the equation. The equations then appear as the following:

$$v_2 = v_1 \qquad (1H)$$

$$d = v_1 t \qquad (2H)$$

$$v_2^2 = v_1^2 \qquad (3H)$$

SAMPLE PROBLEM 5

Figure 9-24

The score was tied at 20 to 20 in the final 1987 AFC playoff game between the Denver Broncos and the Cleveland Browns. During the first overtime period, Denver had the opportunity to kick a field goal, with the ball placed on the tee at a distance of 29 m from the goal posts. If the ball was kicked with the horizontal component of initial velocity being 18 m/s and a flight time of 2 seconds, was the kick long enough to make the field goal?

Known

$$v_h = 18 \text{ m/s}$$

$$t = 2 \text{ s}$$

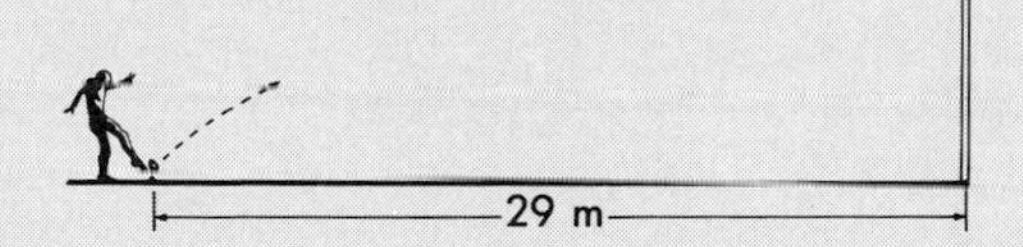

Solution

The formula 2H is selected to solve the problem since two of the variables contained in the formula (v_h and t) are known quantities and since the unknown variable (d) is the quantity we wish to find:

$$d_h = v_h t$$

$$d = 18 \text{ m/s} \times 2 \text{ s}$$

$$\boxed{d = 36 \text{ m}}$$

The ball did travel a sufficient distance for the field goal to be good, and the field goal was good, advancing Denver to Super Bowl XXI.

Equations 1H and 3H reaffirm that the horizontal component of projectile velocity is a constant. Equation 2H indicates that horizontal displacement is equal to the product of horizontal velocity and time (Figure 9-24).

When the constant acceleration relationships are applied to the vertical component of projectile motion, acceleration is equal to -9.81 m/s^2, and the equations cannot be simplified by the deletion of the acceleration term. However, in analysis of the vertical component of projectile motion, the initial velocity (v_1) is equal to 0 in certain cases. For example, when an object is dropped from a stationary position, the initial vertical velocity of the object is 0. When this is the

Figure 9-25

SAMPLE PROBLEM 6

A volleyball is deflected vertically by a player in a game housed in a high school gymnasium where the ceiling clearance is 10 m. If the initial velocity of the ball is 15 m/s, will the ball contact the ceiling?

Known

$$v_1 = 15 \text{ m/s}$$

$$a = -9.81 \text{ m/s}$$

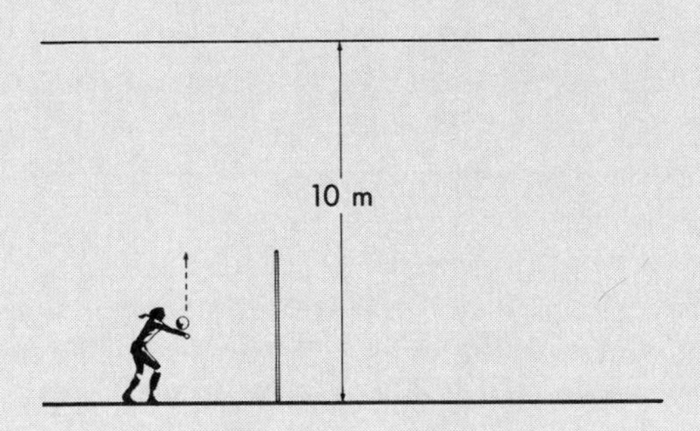

Solution

The equation selected for use in solving this problem must contain the variable d for vertical displacement. Equation 2 contains d but also contains the variable t, which is an unknown quantity in this problem. Equation 3 contains the variable d and, recalling that vertical velocity is zero at the apex of the trajectory, equation 3A can be used to find d:

$$v_2^2 = v_1^2 + 2ad \qquad (3)$$

$$0 = v_1^2 + 2ad \qquad (3A)$$

$$0 = (15 \text{ m/s})^2 + 2 \times (-9.81 \text{ m/s}^2) \times d$$

$$19.62 \text{ m/s}^2 \times d = 225 \text{ m}^2/\text{s}^2$$

$$\boxed{d = 11.47 \text{ m}}$$

Therefore, the ball has sufficient velocity to contact the 10 m ceiling.

case, the equations of constant acceleration may be expressed as the following:

$$v_2 = at \qquad (1V)$$

$$d = (½)at^2 \qquad (2V)$$

$$v_2^2 = 2ad \qquad (3V)$$

When an object is dropped, equation 1V relates that the object's velocity at any instant is the product of gravitational acceleration and the amount of time the object has been in free fall. Equation 2V indicates that the vertical distance through which the object has fallen can be calculated from gravitational acceleration and the amount of time the object has been falling. Equation 3V expresses the relationship between the object's velocity and vertical displacement at a certain time and gravitational acceleration. Notice that each of the equations contains a unique combination of three of the four kinematic quantities—displacement, velocity, acceleration, and time. This provides considerable flexibility for the task of solving problems in which two of the quantities are known and the objective is to solve for a third.

It is often useful in analyzing projectile motion to remember that at the apex of a projectile's trajectory the vertical component of velocity is 0. If the goal is to determine the maximum height achieved by a projectile, v_2 in equation 3 may be set equal to 0:

$$0 = v_1^2 + 2ad \qquad (3A)$$

An example of this use of equation 3A is shown in Figure 9-25. If the problem is to determine the total flight time, one approach is to calculate the time it takes to reach the apex, which is one half of the total flight time if the projection and landing heights are equal. In this case, v_2 in equation 1 for the vertical component of the motion may be set equal to 0 because vertical velocity is 0 at the apex:

$$0 = v_1 + at \qquad (1A)$$

The sample problem in Figure 9-26 illustrates this use of equation 1A.

When using the equations of constant acceleration, it is important to remember that they may be applied to the horizontal component of projectile motion or to the vertical component of projectile motion but *not* to the resultant motion of the projectile. If the horizontal component of motion is being analyzed, $a = 0$, but if the vertical component is being analyzed, $a = -9.81$ m/s^2. The equations of constant acceleration and their special variations are summarized in the box on p. 283.

Figure 9-26

SAMPLE PROBLEM 7

A ball is kicked at a 35 degree angle, with an initial speed of 12 m/s. How high and how far does the ball go?

Solution

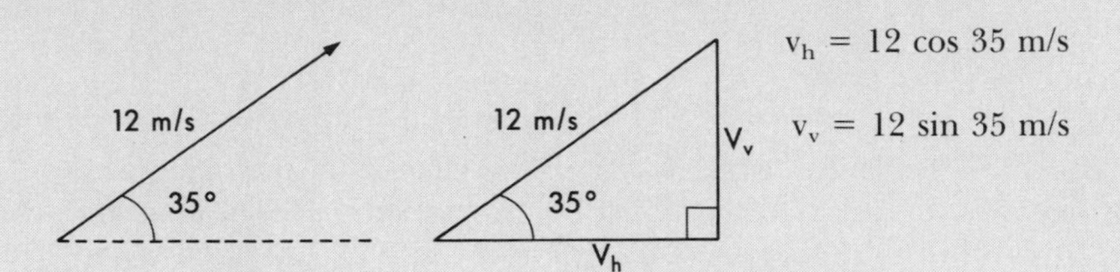

$v_h = 12 \cos 35$ m/s

$v_v = 12 \sin 35$ m/s

How high does the ball go?

Equation 1 cannot be used because it does not contain d. Equation 2 cannot be used unless t is known. Since vertical velocity is 0 at the apex of the ball's trajectory, equation 3A is selected:

$$0 = v_1^2 + 2ad \qquad (3A)$$

$$0 = (12 \sin 35 \text{ m/s})^2 + 2 \times (-9.81 \text{ m/s}^2) \times d$$

$$(19.62 \text{ m/s}^2) = 47.37 \text{ m}^2/\text{s}^2$$

$$\boxed{d = 2.41 \text{ m}}$$

How far does the ball go?

Equation 2H for horizontal motion cannot be used because t for which the ball was in the air is not known. Equation 1A can be used to solve for the time it took the ball to reach its apex:

$$0 = v_1 + at \qquad (1A)$$

$$0 = 12 \sin 35 \text{ m/s} + (-9.81 \text{ m/s}^2)\, t$$

$$t = \frac{6.88 \text{ m/s}}{9.81 \text{ m/s}^2}$$

$$t = 0.70 \text{ s}$$

Recalling that the time to reach the apex is one half of the total flight time, total time is the following:

$$t = 0.70 \text{ s} \times 2$$

$$t = 1.40 \text{ s}$$

Equation 2H can then be used to solve for the horizontal distance the ball traveled:

$$d_h = v_h t \qquad (2H)$$

$$d_h = 12 \cos 35 \text{ m/s} \times 1.40 \text{ s}$$

$$\boxed{d_h = 13.76 \text{ m}}$$

FORMULAS RELATING TO PROJECTILE MOTION

Equations of constant acceleration

These equations may be used to relate linear kinematic quantities whenever a is a constant, unchanging value:

$$v_2 = v_1 + at \qquad (1)$$

$$d = v_1t + (\tfrac{1}{2})at^2 \qquad (2)$$

$$v_2^{\,2} = v_1^{\,2} + 2ad \qquad (3)$$

Special applications

For the horizontal component of projectile motion, with $a = 0$:

$$d = v_1t \qquad (2H)$$

For the vertical component of projectile motion, with $v_1 = 0$:

$$v_2 = at \qquad (1V)$$

$$d = (\tfrac{1}{2})at^2 \qquad (2V)$$

$$v_2^{\,2} = 2ad \qquad (3V)$$

For the vertical component of projectile motion, with $v_2 = 0$:

$$0 = v_1 + at \qquad (1A)$$

$$0 = v_1^{\,2} + 2ad \qquad (3A)$$

Average and Instantaneous Quantities

It is often of interest to determine the velocity or acceleration of an object or body segment at a particular time. For example, the **instantaneous** velocity of a shot or a discus at the moment the athlete releases it greatly affects the distance that the implement will travel. It is sometimes sufficient to quantify the **average** speed or velocity of the entire performance.

instantaneous
occurring during a small interval of time

average
occurring over a designated time interval

When speed or velocity is calculated, the procedures depend on whether the average or the instantaneous value is the quantity of interest. Average velocity is calculated as the final displacement divided by the total time period. Average acceleration is calculated as the difference in the final and initial velocities divided by the entire time interval. Calculation of instantaneous values can be approximated by dividing differences in velocities over an extremely small time interval. With calculus, velocity can be calculated as the derivative of displacement and acceleration as the derivative of velocity.

Selection of the time interval over which speed or velocity is quantified is important when analyzing the performance of athletes in racing events. Many athletes can maintain world-record paces for the first one half or three fourths of the event but slow during the last leg because of fatigue. In a study involving female high school sprinters performing the 100 m run, it was found that maximum running speeds of 8.0 to 8.4 m/s were reached 23 to 37 m from the start and that an average of 7.3% of maximum speed was lost when the runners entered the final 10 m (1). Alternatively, some athletes may intentionally perform at a controlled pace during some segments of a race and then achieve maximum speed at the end. The longer the event is, the more information that is potentially lost or concealed when only the final time or average speed is reported.

SUMMARY

Linear kinematics is the study of the form or sequencing of linear motion with respect to time. Linear kinematic quantities include the scalar quantities of distance and speed and the vector quantities of displacement, velocity, and acceleration. Depending on the motion being analyzed, either a vector quantity or its scalar equivalent and either an instantaneous or an average quantity may be of interest.

A projectile is a body in free fall that is affected only by gravity and air resistance. The factors that determine the height and distance the projectile achieves are the projection angle, the projection speed, and the relative projection height. When the effects of air resistance are disregarded, the trajectory's shape depends on the angle of projection and the trajectory's size depends on projection speed.

Projectile motion is typically analyzed in terms of its horizontal and vertical components. The two components are completely independent of each other; the vertical component is influenced by gravitational force, and horizontal acceleration is assumed to be 0 (disregarding air resistance).

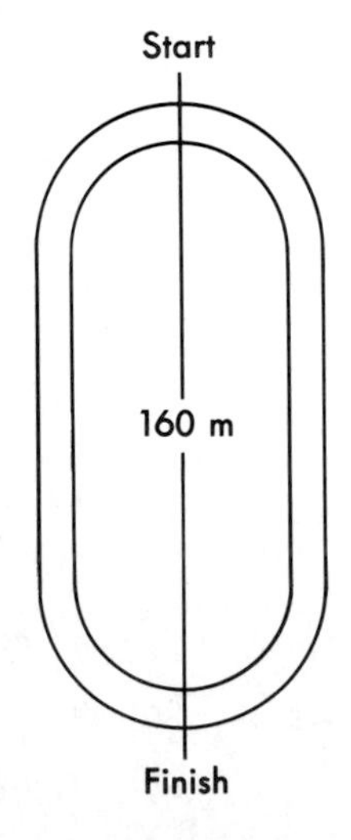

INTRODUCTORY PROBLEMS

Note: Some problems require vector algebra (see Chapter 2).

1. A runner completes 6½ laps around a 400 m track during a 12 minute (720 s) run test.
   Calculate the following quantities:
   a. The distance the runner covered
   b. The runner's displacement at the end of 12 minutes
   c. The runner's average speed
   d. The runner's average velocity
   e. The runner's average pace
   (Answer: a. 2.6 km; b. 160 m; c. 3.6 m/s; d. 0.22 m/s; e. 4.6 min/km)

2. A ball rolls with an acceleration of -0.5 m/s^2. If it stops after 7 seconds, what was its initial velocity? (Answer: 3.5 m/s)
3. A wheelchair marathoner has a velocity of 5 m/s after rolling down a small hill in 1.5 s. If the wheelchair underwent a constant acceleration of 3 m/s^2 during the descent, what was the marathoner's velocity at the top of the hill? (Answer: 0.5 m/s)
4. An orienteer runs 400 m directly east and then 500 m to the northeast (at a 45 degree angle from due east and from due north). Provide a graphic solution to show final displacement with respect to the starting position.
5. An orienteer runs north at 5 m/s for 120 seconds and then west at 4 m/s for 180 seconds. Provide a graphic solution to show the orienteer's resultant displacement.
6. Why are the horizontal and vertical components of projectile motion analyzed separately?
7. A soccer ball is kicked with an initial horizontal speed of 5 m/s and an initial vertical speed of 3 m/s. Assuming that projection and landing heights are the same and neglecting air resistance, identify the following quantities:
   a. The ball's horizontal velocity 0.5 seconds into its flight
   b. The ball's horizontal velocity midway through its flight
   c. The ball's horizontal velocity immediately before contact with the ground
   d. The ball's vertical velocity at the apex of the flight
   e. The ball's vertical velocity midway through its flight
   f. The ball's vertical velocity immediately before contact with the ground
8. If a baseball, basketball, and 71.2 N shot were dropped simultaneously from the top of the Empire State Building (air resistance was not a factor), which would hit the ground first? Why?
9. A tennis ball leaves a racket during the execution of a perfectly horizontal ground stroke with a velocity of 22 m/s. If the ball is in the air for 0.7 seconds, what distance does it travel? (Answer: 15.4 m)
10. A trampolinist springs vertically upward with an initial velocity of 9.2 m/s. How high above the trampoline will the trampolinist go? (Answer: 4.31 m)

ADDITIONAL PROBLEMS

1. A timing light on a high-speed movie camera is set to generate a blip of light on the margin of the film once every 0.1 seconds. When the film is processed, a light blip appears every 9½ frames. At what speed was the film transported past the light generator? (Answer: 95 frames/s)
2. Provide a trigonometric solution for Problem 4. (Answer: D = 832 m; ∠ = 25° north of due east)

3. Provide a trigonometric solution for Problem 5. (Answer: D = 937 m; ∠ = 50° west of due north)
4. A buoy marking the turn in the ocean swim leg of a triathlon becomes unanchored. If the current carries the buoy southward at 0.5 m/s and the wind blows the buoy westward at 0.7 m/s, what is the resultant displacement of the buoy after 5 minutes? (Answer: 258 m; ∠ = 54.5° west of due south)
5. A sailboat is being propelled westerly by the wind at a speed of 4 m/s. If the current is flowing at 2 m/s to the northeast, where will the boat be in 10 minutes with respect to its starting position? (Answer: D = 1.8 km; ∠ = 29° north of due west)
6. A Dallas Cowboy carrying the ball straight down the near sideline with a velocity of 8 m/s crosses the 50 yard line at the same time that the last Pittsburgh Steeler who can possibly hope to catch him starts running from the 50 yard line at a point that is 13.7 m from the near sideline. What must the Steeler's velocity be if he is to catch the Cowboy just short of the goal line? (Answer: 8.35 m/s)
7. A soccer ball is kicked from the playing field at a 45° angle. If the ball goes 46 m and is in the air for 3 seconds, what is the maximum height achieved? (Answer: 12.0 m)
8. A ball is kicked a horizontal distance of 45.8 m. If it reaches a maximum height of 24.2 m with a flight time of 2.1 seconds, was the ball kicked at a projection angle less than, greater than, or equal to 45 degrees? Provide a rationale for your answer based on the appropriate calculations. (Answer: < 45°)
9. A badminton shuttlecock is struck by a racquet at a 35° angle, giving it an initial speed of 10 m/s. How high will it go? How far will it travel horizontally before being contacted by the opponent's racquet at the same height from which it was projected? (Answer: d_v = 1.68 m; d_h = 9.58 m)
10. An archery arrow is shot with a speed of 45 m/s at an angle of 10 degrees. How far horizontally can the arrow travel before hitting a target at the same height from which it was released? (Answer: 70.6 m)

REFERENCES

1. Chow JW: Maximum speed of female high school runners, Int J Sport Biomech 3:110, 1987.
2. Dapena J: Mechanics of translation in the fosbury flop, Med Sci Sports Exerc 12:37, 1980.
3. Dapena J: The connection of basic and applied research in sports biomechanics. In Dapena J: Biomechanics—kinanthropometry and sports medicine, exercise science scientific program abstracts, 1984 Olympic Scientific Congress, Eugene, Ore, 1984, Microform Academic Publishers.

4. Dillman CJ: Overview of the United States Olympic Committee sports medicine biomechanics program. In Butts NK, Gushiken TT, and Zarins BT, eds: The elite athlete, New York, 1985, Spectrum Publications, Inc.
5. Groppel JL: The application of quantitative-qualitative biomechanical research to the athlete and coach. In Groppel JL: Biomechanics—kinanthropometry and sports medicine, exercise science scientific program abstracts, 1984 Olympic Scientific Congress, Eugene, Ore, 1984, Microform Academic Publishers.
6. Hay JG: The biomechanics of the long jump, Exerc Sport Sci Rev 14:401, 1986.
7. Hubbard M: The throwing events in track and field. In Vaughan CL, ed: Biomechanics of Sport, Boca Raton, Fla, 1989, CRC Press, Inc.
8. Laws K: The physics of dance, New York, 1984, Schirmer Books.
9. Mann R: Biomechanical analysis of the elite sprinter and hurdler. In Butts NK, Gushiken TT, and Zarins BT, eds: The elite athlete, New York, 1985, Human Kinetic Publishers, Inc.
10. McCaw ST and Hoshizaki TB: A kinematic comparison of novice, intermediate, and elite ice skaters. In Jonsson B, ed: Biomechanics X-B, Champaign, Ill, 1977, Human Kinetics Publishers, Inc.
11. Miller JA and Hay JG: Kinematics of a world record and other world-class performances in the triple jump, Int J Sport Biomech 2:272, 1986.
12. Phillips SJ: Invariance of elite kicking performance. In Winter DA et al, eds: Biomechanics IX-B, Champaign, Ill, 1985, Human Kinetics Publishers, Inc.
13. Smith GA, McNitt-Gray J, and Nelson RC: Kinematic analysis of alternate stride skating in cross-country skiing, Int J Sport Biomech, 4:49, 1988.
14. Watanabe K: Ski-jumping, alpine-, cross-country-, and nordic combination skiing. In Vaughan CL, ed: Biomechanics of sport, Boca Raton, Fla, 1989, CRC Press, Inc.
15. Wilson BD, McDonald M, and Neal RJ: Roller skating sprint technique. In Jonsson B, ed: Biomechanics X-B, Champaign, Ill, 1987, Human Kinetics Publishers, Inc.

ANNOTATED READINGS

Maugh TH: Physics of basketball: those golden arches, Science 1:106, 1981.
Describes practical considerations for shooting a basketball to maximize the probability of its going through the hoop.

Townend MS: Throwing. In Townend MS: Mathematics in sport, New York, 1984, John Wiley & Sons, Inc.
Covers an in-depth mathematical analysis of the mechanical principles of relevance in projecting the shot, hammer, discus, javelin, and basketball.

Williams KR: Biomechanics of running, Exerc Sport Sci Rev 13:389, 1985.
Reviews the literature on running mechanics, including an extensive section on running kinematics.

Zatsiorsky VM, Lanka GE, and Shalmanov AA: Biomechanical analysis of shot putting technique, Exerc Sport Sci Rev 9:353, 1982.
Summarizes the results of experimental investigations on the biomechanics of shot putting.

10 MOVEMENT

Angular Kinematics

After reading this chapter, the student will be able to:

Distinguish angular motion from rectilinear and curvilinear motion.

Discuss the relationships among the primary angular kinematic variables.

Correctly associate angular kinematic quantities with their respective units of measure.

Explain the relationships between angular and linear displacement, angular and linear velocity, and angular and linear acceleration.

Solve quantitative problems involving angular kinematic quantities and the relationships between angular and linear kinematic quantities.

Angular motion is rotational motion around an axis. The axis of rotation is an imaginary line that is aligned perpendicular to the plane in which the rotation occurs, similar to the axle for the wheels of a cart. Like linear motion, angular motion is a basic component of general motion.

An understanding of angular motion is particularly important for the student of human movement because most volitional human movement involves rotation of one or more body segments around the joints at which they articulate. When flexion occurs at the elbow during a forearm curl exercise, the radius and the ulna are collectively rotating around an imaginary frontal axis of rotation passing through the elbow joint. During the performance of jumping jacks, both the arms and legs rotate around imaginary sagittal axes passing through the shoulder and hip joints. The angular motion of sport implements such as golf clubs, baseball bats, and hockey sticks, as well as household and garden tools, is also often of interest.

MEASURING ANGLES

As reviewed in Appendix A, an angle is composed of two sides that intersect at a vertex. Kinematic analysis sometimes involves the projection of contour images of the human body onto a piece of paper, with joint centers marked with dots and the dots connected with segmental lines representing the longitudinal axes of the body segments (Figure 10-1). A protractor can be used to make hand measurements of angles of interest from this representation, with the joint centers forming the vertices of the angles between adjacent body segments. (The procedure for measuring angles with a protractor is reviewed in Appendix A.) Biomechanic researchers use this same basic procedure to evaluate the angles present at the joints of the human body and the angular orientations of the body segments. The angle assessments are usually done with computer software from stick figure representations of the human body constructed in computer memory. A computer-generated representation of the motion of a cyclist's legs derived from projections of high-speed film images is shown in Figure 10-2.

For direct measurement of joint angles on a live human subject, different instruments are available. One device is the goniometer, which is a protractor with two long arms attached. One arm is fixed so that it extends from the protractor at an angle of 0 degrees. The other arm extends from the center of the protractor and is free to rotate. The center of the protractor is aligned over the joint center, and the two arms are aligned over the longitudinal axes of the two body segments that connect at the joint. The angle at the joint is then read at the intersection of the freely rotating arm and the protractor scale. The accuracy of the reading depends on the accuracy of the positioning of the goniometer. Knowledge of the underlying joint anatomy is essential for proper location of the joint center of rotation. Placing marks on the skin to identify the location of the center of rotation at the joint and the longitudinal axes of the body segments before aligning the goniometer is sometimes helpful.

Figure 10-1
For the human body, joint centers form the vertices of body segment angles.

Figure 10-2
Angular changes at the hip, knee, and ankle joints of a cyclist.

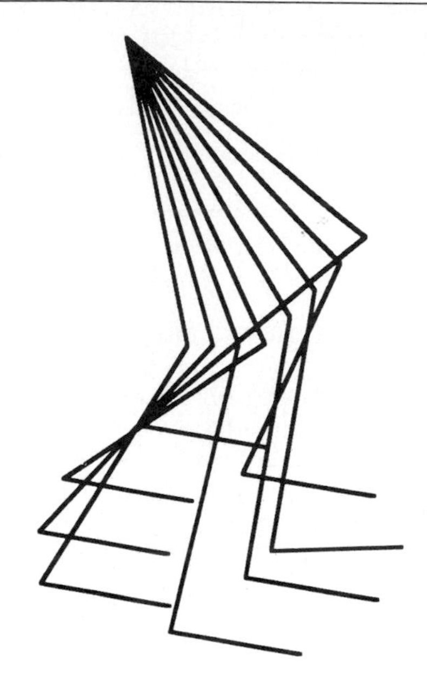

A goniometer is used to measure joint angles.

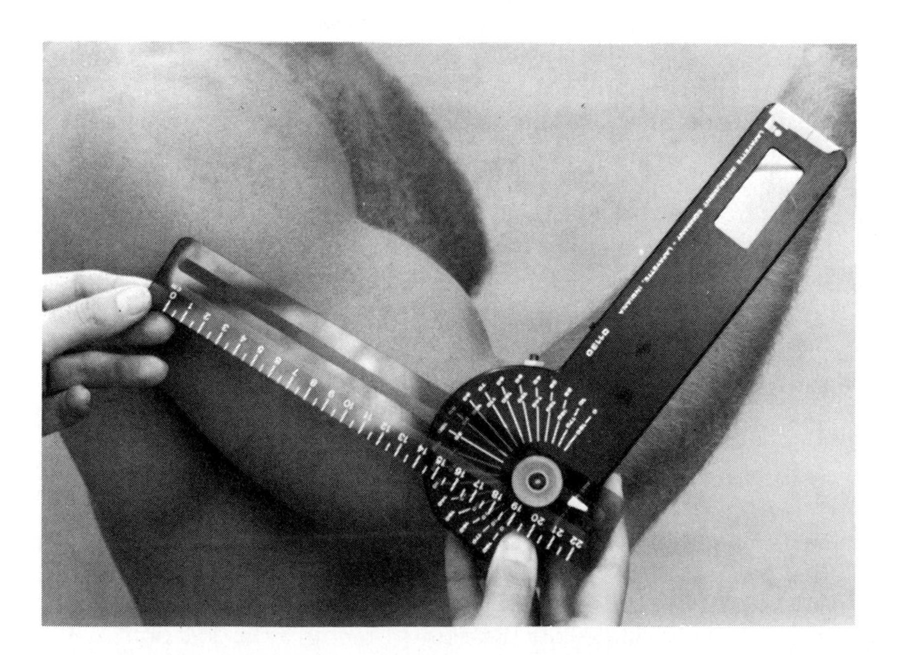

Other instruments available for quantifying angles relative to the human body are the electrogoniometer and the Leighton flexometer. The electrogoniometer (referred to as an *elgon*) was developed by Karpovich in the late 1950s (1). An elgon is simply a goniometer with an electrical potentiometer at its vertex. When the arms of the elgon are attached with tape or velcro straps over a joint center, changes in the relative angle at the joint cause proportional changes in the electrical current emitted by the elgon. These changes are usually recorded on a moving strip chart, which marks the angle present at the joint with respect to time during a movement of interest. Another device used for direct assessment of human body angles is the Leighton flexometer, a gravitationally based instrument that identifies the angle of orientation of the body segment to which it is strapped.

Instant Center of Rotation

Quantification of joint angles is complicated by the fact that one bone often displaces slightly with respect to the other bone or bones articulating at the joint. This phenomenon is caused by normal asymmetries in the shapes of the articulating bone surfaces. One example is the tibiofemoral joint, at which medial rotation and anterior displacement of the femur on the tibial plateau accompany flexion (Figure 10-3). As a result, the location of the exact center of rotation

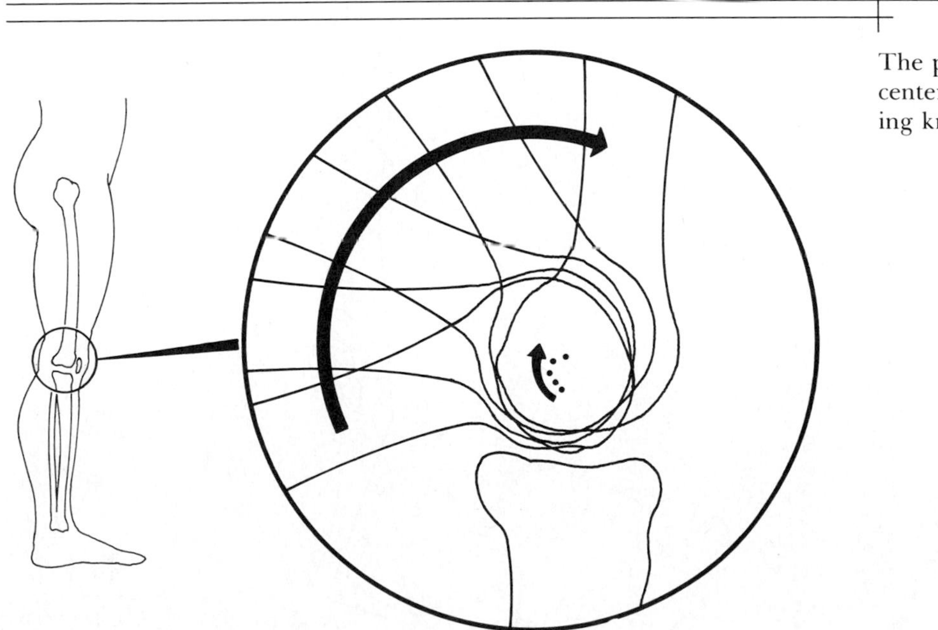

Figure 10-3
The path of the instant center at the knee during knee flexion.

at the joint changes slightly when the angle present at the joint changes. The center of rotation at a given joint angle, or at a given instant in time during a dynamic movement, is called the **instant center.** The exact location of the instant center for a given joint may be determined through measurements taken from roentgenograms (x rays), which are usually taken at 10 degree intervals throughout the range of motion at the joint (11). Although it is not practical to x ray a joint to determine the exact location of its center of rotation when assessing a relative joint angle, joint center of rotation locations *do* vary with joint angles.

instant center
the precisely located center of rotation at a joint at a given instant in time

Relative Versus Absolute Angles

Often an angular quantity of interest is the amount of flexion, hyperextension, abduction, or other movement present at a joint. For example, when exercises such as squats or knee bends are described,

■ Relative angles should consistently be measured on the same side of a given joint.

Figure 10-4
Angles measured at joints are the angles of one adjacent body segment relative to the other.

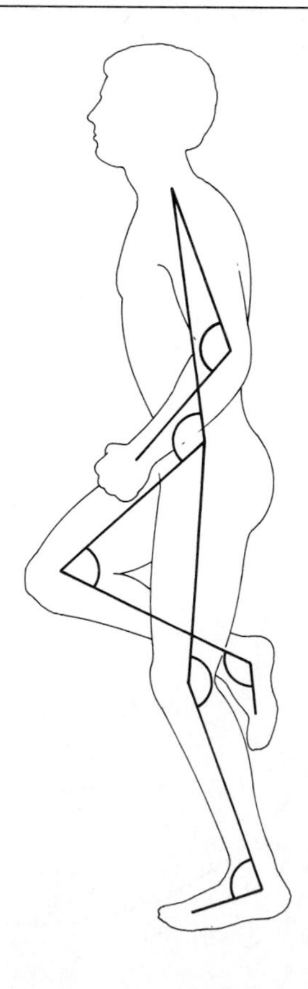

the minimum angle that should be sustained at the knee is often noted because the ligaments of the knee are most vulnerable to injury at the point of maximum knee flexion. Assessing the angle at a joint involves measuring the angle of one body segment relative to the other body segment articulating at the joint. The **relative angle** at the knee is the angle formed between the longitudinal axis of the thigh and the longitudinal axis of the lower leg (Figure 10-4).

relative angle
the angle formed between the longitudinal axes of two adjacent body segments articulating at a joint

absolute angle
the angular orientation of a single body segment with respect to a fixed line of reference

When researchers study the coordinated movements of the human body, they often examine an angle-angle diagram in which one angle is plotted against another angle of interest (4). Figure 10-5 shows angle-angle diagrams of the knee-thigh and knee-ankle for one running stride.

Relative angles are those present at joints, and absolute angles describe body segment positions.

Other times the angular kinematic quantity of interest relates not to a joint but to one or more of the body segments themselves. When the trunk is inclined forward in flexion, the angle of inclination of the trunk directly affects the amount of force that must be generated by the muscles of the lumbar region to support the trunk in the position assumed (8). The angle of inclination of a body segment, referred to as its **absolute angle,** may be measured with respect to either the horizontal or the vertical. Figure 10-6 shows quantification of segment angles with respect to the right horizontal.

Absolute angles should consistently be measured in the same direction from a single reference—either horizontal or vertical.

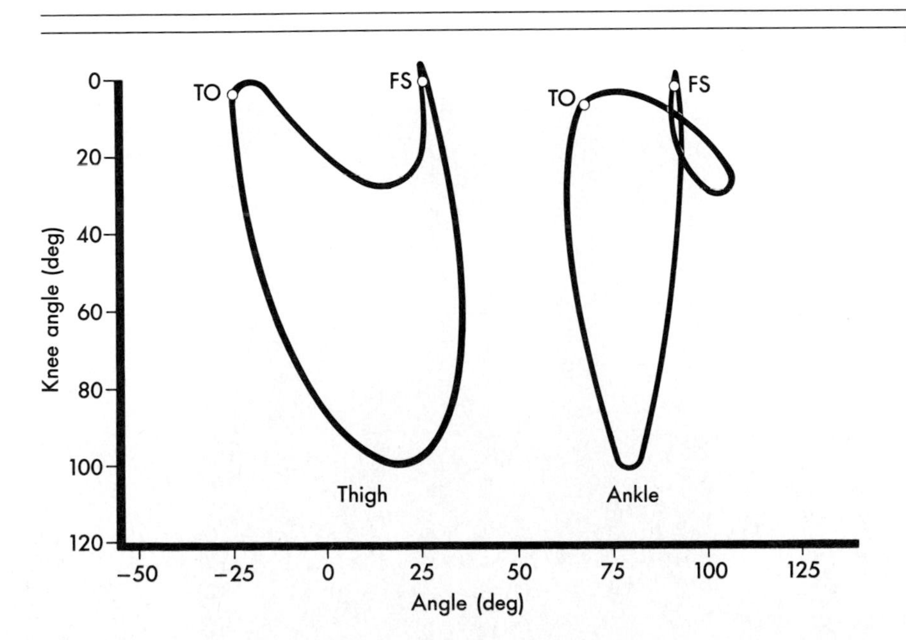

Figure 10-5
Angle-angle diagrams of the knee-thigh and knee-ankle for one stride during running. (*TO*, toe-off; *FS*, foot strike.)

Figure 10-6

Angles of orientation of individual body segments are measured with respect to an absolute (fixed) line of reference.

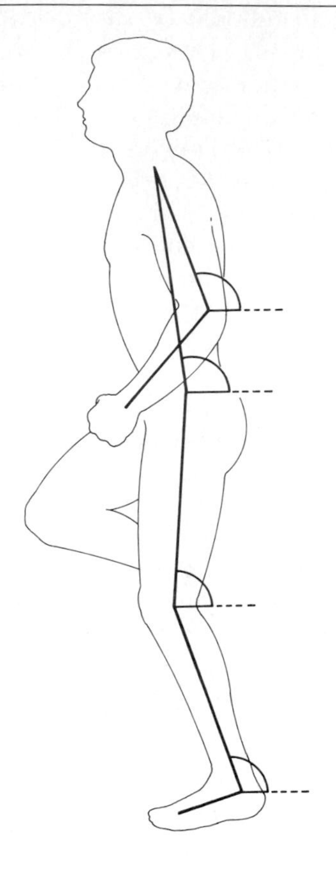

The relative angle at the knee and the absolute angle of the trunk.

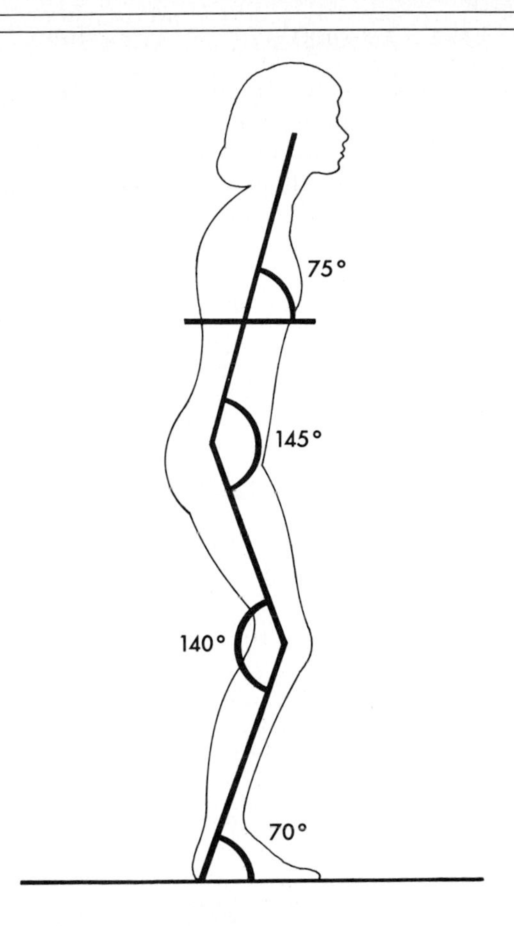

Figure 10-7
Angle measurement convention for identifying the relative angles at the hip, knee, and ankle and the absolute angle of the trunk (2).

Measurement Conventions for Angles

A consistent frame of reference must be used to identify relative and absolute angles with respect to the joints and segments of the human body. Unfortunately, different joint angle quantification systems have been used in the investigations reported in the biomechanics literature, which makes comparing results difficult. Angle conventions advocated by Bates, Morrison, and Hamill (2) for angles commonly quantified in investigations of running kinematics (trunk, thigh, knee, and ankle) are shown in Figure 10-7. As emphasized by Zatsiorsky, Lanka, and Shalmanov (13), identifying both the angle measurement system employed and the exact time at which angles are assessed during dynamic movements is important if the information is to be of value.

ANGULAR KINEMATIC RELATIONSHIPS

The interrelationships among the angular kinematic quantities are similar to those present among the linear kinematic quantities. Although different units of measure are associated with the angular kinematic quantities than with their linear counterparts, the relationships among the angular units also parallel those present among the linear units.

Angular Distance and Displacement

Consider a pendulum swinging back and forth from a point of support. The pendulum is rotating around an axis passing through its point of support perpendicular to the plane of motion. If the pendulum swings through an arc of 60 degrees, it has swung through an angular distance of 60 degrees. If the pendulum then swings back through 60 degrees to its original position, it has traveled an angular distance totaling 120 (60 + 60) degrees. Angular distance is measured as the sum of all angular changes undergone by a rotating body.

■ The straight, fully extended position at a joint is regarded as 180 degrees.

The same procedure may be used for quantifying the angular distances through which the segments of the human body move. If the angle at the elbow joint changes from 180 degrees to 40 degrees during the flexion phase of a forearm curl exercise, the angular dis-

When the forearm returns to its original position at the completion of a curl exercise, the angular displacement at the elbow is 0.

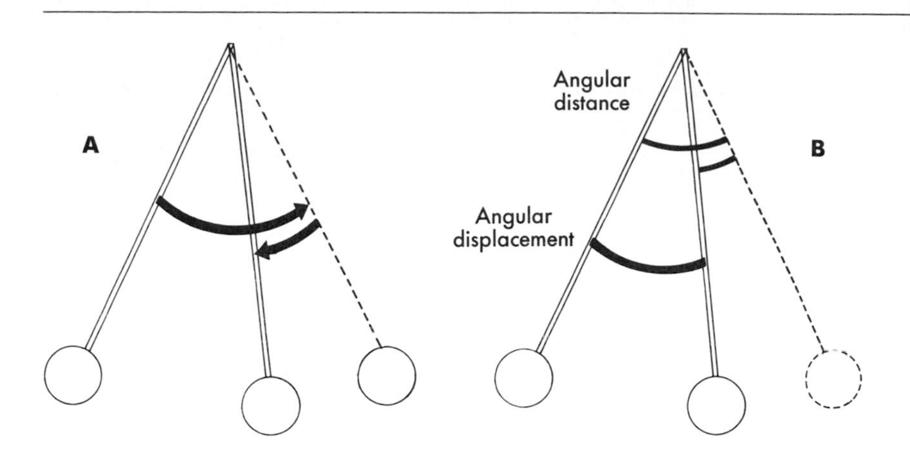

Figure 10-8
A, The path of motion of a swinging pendulum. **B,** The angular distance is the sum of all angular changes that have occurred, the angular displacement is the angle between the initial and final positions.

tance covered is 140 (180-40) degrees. If the extension phase of the curl returns the elbow to its original position of 180 degrees, an additional 140 degrees have been covered, resulting in a total angular distance of 280 degrees for the complete curl. If 10 curls are performed, the angular distance transcribed at the elbow is 2800 (10 × 280) degrees.

Just as with its linear counterpart, **angular displacement** is assessed as the difference in the initial and final positions of the moving body. If the angle at the knee of the support leg changes from 180 degrees to 150 degrees during the support phase of a running stride, the angular distance and the angular displacement at the knee are 30 degrees. If extension occurs at the knee, returning the joint to its original 180 degree position, angular distance totals 60 (30 + 30) degrees but angular displacement is 0 because the final position of the joint is the same as its original position. The relationship between angular distance and angular displacement is represented in Figure 10-8.

angular displacement
the change in angular position

Angular displacement is a vector quantity possessing both direction and magnitude. Since rotation as viewed from a perpendicular perspective occurs in either a clockwise or a counterclockwise direction, the direction of angular displacement may be indicated using these terms. The counterclockwise direction is conventionally designated as positive (+), and the clockwise direction as negative (−) (Figure 10-9).

■ In quantifying angular movement, the counterclockwise direction is regarded as positive, and the clockwise direction is regarded as negative.

Three units of measure are commonly used to represent angular distance and angular displacement. The most familiar of these units is the degree. A complete circle of rotation transcribes an arc of 360 degrees, an arc of 180 degrees subtends a straight line, and 90 de-

Figure 10-9
Since angular displacement is a vector quantity, its direction must be indicated. Direction is commonly identified as counterclockwise (or positive) versus clockwise (or negative).

Figure 10-10
Angles measured in degrees.

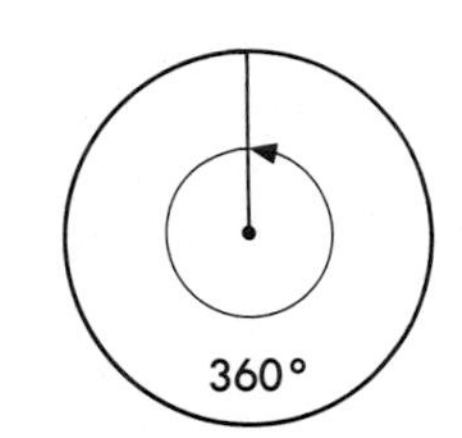

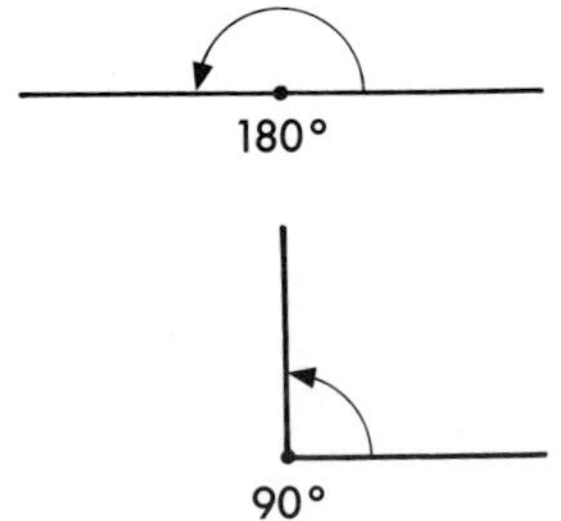

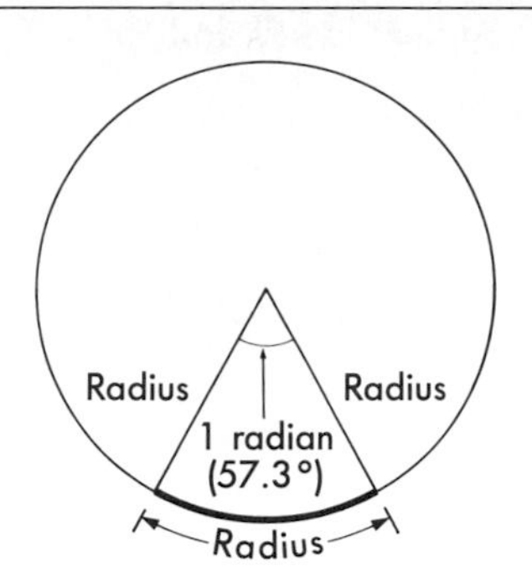

Figure 10-11
A radian is defined as the size of the angle subtended at the center of a circle by an arc equal in length to the radius of the circle.

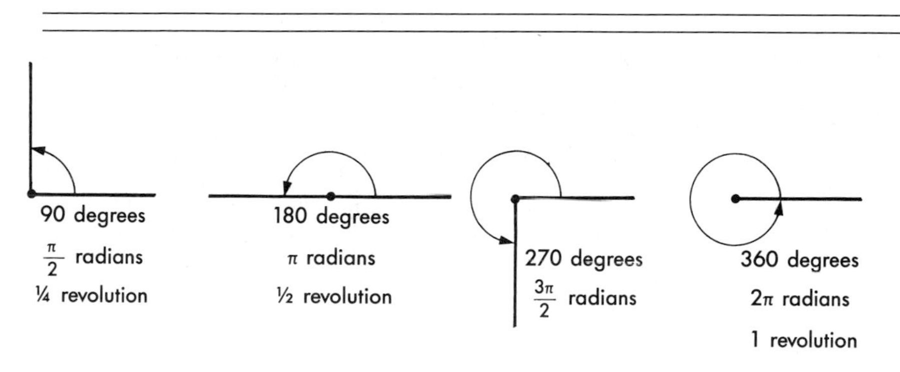

Figure 10-12
Comparison of degrees, radians, and rotations.

grees forms a right angle between perpendicular lines (Figure 10-10).

Another unit of angular measure commonly used in biomechanical analyses is the **radian.** A line connecting the center of a circle to any point on the circumference of the circle is a radius. A radian is defined as the size of the angle subtended at the center of a circle by an arc equal in length to the radius of the circle (Figure 10-11). One radian is equivalent to 57.3 degrees. Angular quantities must be expressed in radian-based units when angular-linear conversions are made. Because a radian is much larger than a degree, it is a more convenient unit for the representation of extremely large angular distances or displacements. Radians are often quantified in multiples of pi (π). One complete circle is an arc of 2π radians.

radian
the unit of angular measure used in angular-linear kinematic quantity conversions that is equal to 57.3 degrees

The third unit sometimes used to quantify angular distance or displacement is the revolution. One revolution transcribes an arc equal to a circle. Dives and some gymnastic skills are often described by the number of revolutions the human body undergoes during their execution. The one and a half forward somersault dive is a descriptive example. Figure 10-12 illustrates the way in which degrees, radians, and revolutions compare as units of angular measure.

■ Pi is a mathematical constant equal to approximately 3.14, which is the ratio of the circumference to the diameter of a circle.

Angular Speed and Velocity

Angular speed is a scalar quantity and is defined as the angular distance covered divided by the time interval over which the motion occurred:

$$\text{angular speed} = \frac{\text{angular distance}}{\text{change in time}}$$

$$\sigma = \frac{\phi}{t}$$

The small Greek letter sigma (σ) represents angular speed, the small Greek letter phi (ϕ) represents angular distance, and t represents time.

angular velocity
the rate of change in angular position

Angular velocity is calculated as the change in angular position or the angular displacement that occurs during a given period of time:

$$\text{angular velocity} = \frac{\text{change in angular position}}{\text{change in time}}$$

$$\omega = \frac{\Delta \text{ angular position}}{\Delta t}$$

■ Angular displacement and angular velocity are the vector equivalents of the scalar quantities of angular distance and angular speed.

$$\text{angular velocity} = \frac{\text{angular displacement}}{\text{change in time}}$$

$$\omega = \frac{\Theta}{\Delta t}$$

The small Greek letter omega (ω) represents angular velocity, the capital Greek letter theta (Θ) represents angular displacement, and t represents the time elapsed during the velocity assessment. Another way to express change in angular position is angular position_2 − angular position_1 in which angular position_1 represents the body's position at one point in time and angular position_2 represents the body's position at a later point:

$$\omega = \frac{\text{angular position}_2 - \text{angular position}_1}{\text{time}_2 - \text{time}_1}$$

Because angular velocity is based on angular displacement, it is a vector quantity and must include an identification of the direction (clockwise or counterclockwise, negative or positive) in which the angular displacement on which it is based occurred.

Units of angular speed and angular displacement are units of angular distance or angular displacement divided by units of time. The unit of time most commonly used is the second. Units of angular speed and angular velocity are degrees per second (deg/s), radians per second (rad/s), and revolutions per minute (rpm).

Because moving the body segments at a high rate of angular velocity is a characteristic of skilled performance in many sports, some biomechanists have studied the magnitudes and the sequencing of the segmental angular velocities of skilled performers. Angular velocities at the joints of the throwing arm in major league baseball pitchers have been reported to reach 6180 deg/s (107.9 rad/s) of internal rotation at the shoulder and approximately 4595 deg/s (80.2 rad/s) of elbow extension (12). A comparison of the fast ball and curve ball pitches among members of the Australian National pitching squad showed average angular velocities at the wrist of 3.1 rad/s and 5.8 rad/s for the fast ball and curve ball pitches respectively (7). Angular velocity of the racket during serves executed by professional male tennis players has been found to range from 1900 to 2200 deg/s (33.2 to 38.4 rad/s) just before ball impact (5).

■ Skilled performances of high-velocity movements are characterized by precisely coordinated timing of angular body segment movements.

Angular Acceleration

Angular acceleration is the rate of change in angular velocity or the change in angular velocity occurring over a given time. The conventional symbol for angular acceleration is the small Greek letter alpha (α):

angular acceleration
the rate of change in angular velocity

$$\text{angular acceleration} = \frac{\text{change in angular velocity}}{\text{change in time}}$$

$$\alpha = \frac{\Delta}{\Delta t}$$

The calculation formula for angular acceleration is therefore the following:

$$\alpha = \frac{\omega_2 - \omega_1}{t_2 - t_1}$$

In this formula, ω_1 represents angular velocity at an initial point in time, ω_2 represents angular velocity at a second or final point in time, and t_1 and t_2 are the times at which velocity was assessed. Use of this formula is illustrated in the sample problem in Figure 10-13.

Just as with linear acceleration, angular acceleration may be positive, negative, or 0. When angular acceleration is equal to 0, angular velocity is constant over the time interval investigated. Because plus and minus signs are commonly used to indicate angular direction, positive angular acceleration may indicate either increasing angular velocity in the positive direction or decreasing angular velocity in the negative direction. Similarly, a negative value of angular acceleration may represent either decreasing angular velocity in the positive direction or increasing angular velocity in the negative direction.

■ Human movement rarely involves constant velocity or constant acceleration.

Units of angular acceleration are units of angular velocity divided by units of time. Common examples are degrees per second squared

Figure 10-13

SAMPLE PROBLEM 1

A golf club is swung with an average angular acceleration of 1.5 rad/s^2. What is the angular velocity of the club when it strikes the ball at the end of a 0.8 second swing? (Provide an answer in both radian and degree-based units.)

Known

$$\alpha = 1.5 \text{ rad/s}^2$$

$$t = 0.8 \text{ s}$$

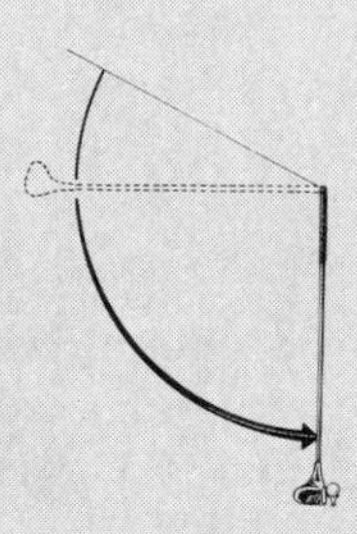

Solution

The formula to be used is the equation relating angular acceleration, angular velocity, and time:

$$\alpha = \frac{\omega_2 - \omega_1}{t}$$

Substituting in the known quantities, yields the following:

$$1.5 \text{ rad/s}^2 = \frac{\omega_2 - \omega_1}{0.8 \text{ s}}$$

It may also be deduced that the angular velocity of the club at the beginning of the swing was 0:

$$1.5 \text{ rad/s}^2 = \frac{\omega_2 - 0}{0.8 \text{ s}}$$

$$1.5 \text{ rad/s}^2 (0.8 \text{ s}) = \omega_2 - 0$$

$$\boxed{\omega_2 = 1.2 \text{ rad/s}}$$

In degree-based units:

$$\omega_2 = (1.2 \text{ rad/s}) (57.3 \text{ deg/rad})$$

$$\boxed{\omega_2 = 68.8 \text{ deg/s}}$$

Table 10-1
COMMON UNITS OF MEASURE

	DISPLACEMENT	VELOCITY	ACCELERATION
Linear	meters	meters/second	meters/second^2
Angular	radians	radians/second	radians/second^2

(deg/s^2), radians per second squared (rad/s^2), and revolutions per second squared (rev/s^2). Units of angular and linear kinematic quantities are compared in Table 10-1.

Angular Motion Vectors

Because representing angular vector quantities using symbols such as curved arrows would be impractical, angular vector quantities are represented with conventional straight vectors using what is called the **right hand rule.** According to this rule, when the fingers of the right hand are curled in the direction of an angular motion, the vector used to represent the motion is oriented perpendicular to the plane of rotation in the direction the extended thumb points (Figure 10-14). The magnitude of the quantity may be indicated through proportionality to the vector's length. With this procedure, angular kinematic quantities can be added together to yield a single resultant vector representing a complex angular motion, such as that of a trampolinist combining a series of somersaults and twists in the air.

right hand rule
the procedure for identifying the direction of an angular motion vector

■ Angular motion vectors may be composed (added) and resolved into perpendicular components.

Figure 10-14
An angular motion vector is oriented perpendicular to the plane of rotation in the direction indicated by the right hand rule.

Average Versus Instantaneous Angular Quantities

Angular speed, velocity, and acceleration may be calculated as instantaneous or average values, depending on the length of the time interval selected. The instantaneous angular velocity of a baseball bat at the instant of contact with a ball is typically of greater interest than the average angular velocity of the swing because the former directly affects the resultant velocity of the ball.

Relationship between Angular and Linear Motions

Why is a driver longer than a putter? Why do batters slide their hands up the handle of the bat to execute a bunt but not to execute a power hit? The relationship between the angular motion of the implement being swung and the resulting linear motion of the ball contacted explains the answers to these questions.

Linear and Angular Displacement

The greater the distance a given point on a rotating body is located from the axis of rotation, the greater the linear displacement under-

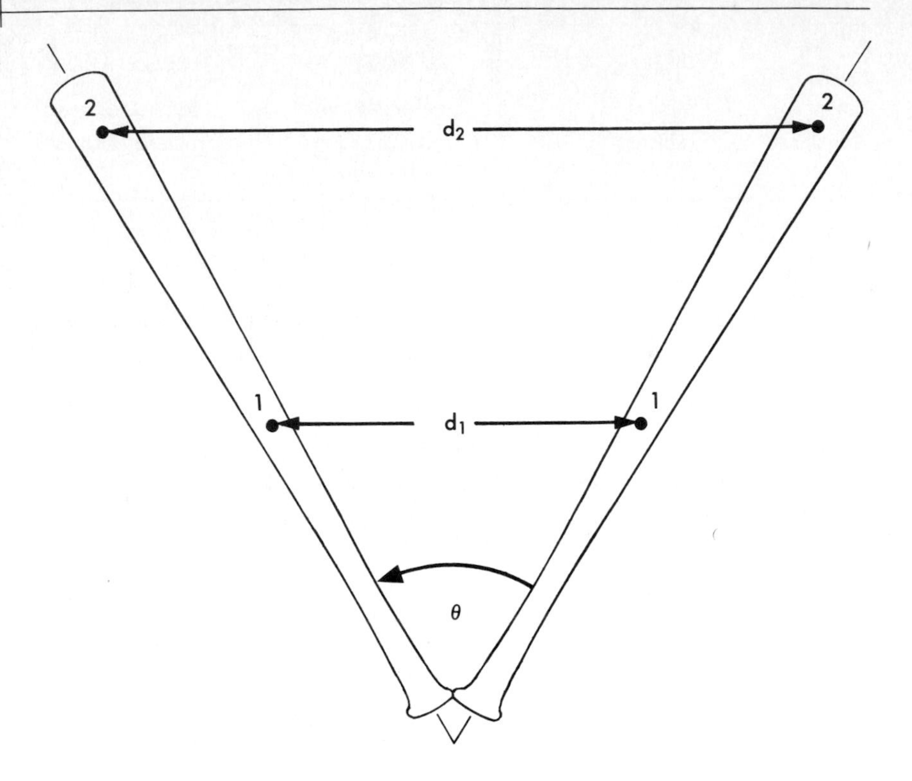

Figure 10-15
The farther a point on a rotating body is from the axis of rotation, the greater the linear displacement that it undergoes during a unidirectional rotation.

gone by that point (Figure 10-15). This observation is expressed in the form of a simple equation:

$$d = r\Theta$$

radius of rotation
the distance from the axis of rotation to a given point of interest on a rotating body

The linear displacement of the point of interest is d, r is the point's **radius of rotation** (distance of the point from the axis of rotation), and Θ is the angular displacement of the rotating body, which is quantified in radians.

For this relationship to be valid, two conditions must be met: The linear displacement and the radius of rotation must be quantified in the same units of length, and angular displacement must be expressed in radians. Although units of measure are normally balanced on opposite sides of an equal sign when a valid relationship is expressed, this is not the case here. When the radius of rotation (expressed in meters) is multiplied by angular displacement in radians, the result is linear displacement in meters. Radians disappear on the right side of the equation in this case because, as may be observed from the definition of the radian, the radian serves as a conversion factor between linear and angular measurements.

Linear and Angular Velocity

The same type of relationship exists between the angular velocity of a rotating body and the linear velocity of a point on that body at a given instant in time. The relationship is expressed as the following:

$$v = r\omega$$

The linear (tangential) velocity of the point of interest is v, r is the radius of rotation for that point, and ω is the angular velocity of the rotating body. For the equation to be valid, angular velocity must be expressed in radian-based units (typically rad/s) and velocity must be expressed in the units of the radius of rotation divided by the appropriate units of time. Radians are again used as a linear-angular conversion factor and are not balanced on opposite sides of the equal sign:

$$m/s = (m)(rad/s)$$

The use of radian-based units for conversions between linear and angular velocities is shown in Figure 10-16.

A comparison of two styles of the soccer throw-in exemplifies the effect of angular velocity on linear velocity. Studies have shown that greater ball velocity and therefore greater throw-in range result from the use of the handspring throw-in as opposed to the standard throw-in (3, 10). This increased ball velocity is associated with greater angular velocity of the fore arm at the instant of ball release (3). Maximizing the range of the throw-in is not always appropriate because it is sometimes more strategic to direct the ball to a near player. However, if strategy calls for throwing the ball as close as possible to the goal (a relatively large distance away), the properly executed handspring throw-in is likely to be advantageous.

During several sport activities an immediate performance goal is to direct an object such as a ball, shuttlecock, or hockey puck accurately while imparting a relatively large amount of velocity to it with a bat, club, racket, or stick. Studies of serves and ground strokes in tennis have consistently shown that skilled performers actually reduce the angular velocities of their rackets just before contact with the ball to enhance the accuracy and control of the shot (4, 6). In baseball batting, the initiation of the bat swing and the angular velocity of the swing must be timed precisely to make contact with the ball and to direct it into fair territory. A 40 m/s pitch reaches the batter 0.41 seconds after leaving the pitcher's hand. It has been estimated that a difference of 0.001 seconds in the time of initiation of the swing can determine whether the ball is directed to center field or down the foul line and that a swing initiated 0.003 seconds too early or too late will result in no contact with the ball (9).

Timing is important in the execution of a groundstroke in tennis. If the ball is contacted too soon or too late, it may be hit out of bounds.

With all other factors held constant, the greater the radius of rotation at which a swinging implement hits a ball, the greater the lin-

Figure 10-16

SAMPLE PROBLEM 2

Two baseballs are consecutively hit by a bat. The first ball is hit 20 cm from the bat's axis of rotation and the second ball is hit 40 cm from the bat's axis of rotation. If the angular velocity of the bat was 30 rad/s at the instant that both balls were contacted, what was the linear velocity of the bat at the two contact points?

Known

$$r_1 = 20 \text{ cm}$$

$$r_2 = 40 \text{ cm}$$

$$\omega_1 = \omega_2 = 30 \text{ rad/s}$$

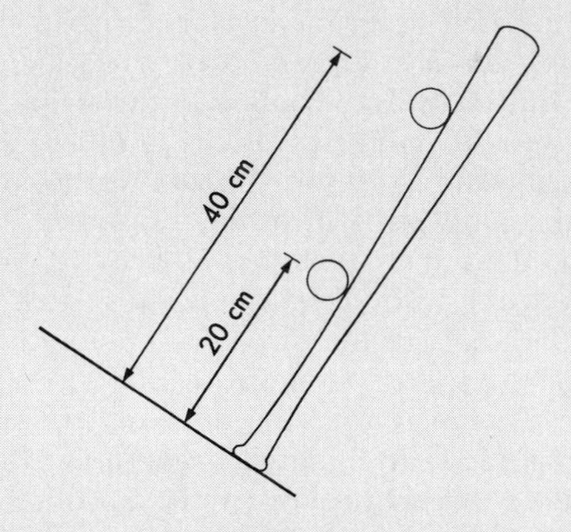

Solution

The formula to be used is the equation relating linear and angular velocities:

$$v = r\Theta$$

For ball 1:

$$v_1 = (0.20 \text{ m})\ (30 \text{ rad/s})$$

$$\boxed{v_1 = 6 \text{ m/s}}$$

For ball 2:

$$v_2 = (0.40 \text{ m})\ (30 \text{ rad/s})$$

$$\boxed{v_2 = 12 \text{ m/s}}$$

ear velocity imparted to the ball. In golf, longer clubs are selected for longer shots, and shorter clubs are selected for shorter shots. However, the magnitude of the angular velocity figures as heavily as the length of the radius of rotation in determining the linear velocity of a point on a swinging implement. Little Leaguers often select long bats, which increase the potential radius of rotation if a ball is contacted but are also too heavy for the young players to swing as quickly as a shorter, lighter bat. The relationship between the radius of rotation of the contact point between a striking implement and a ball and the subsequent velocity of the ball is shown in Figure 10-17.

The linear velocity of a ball struck by a bat, racket, or club is *not* identical to the linear velocity of the contact point on the swinging implement. Other factors, such as the directness of the hit and the elasticity of the impact, also influence ball velocity.

Top view

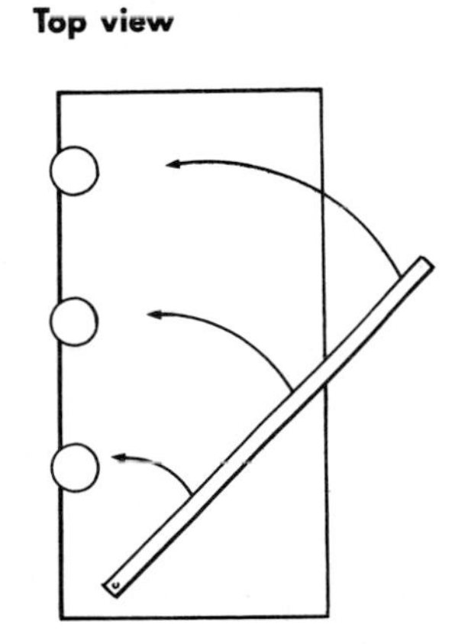

Side view

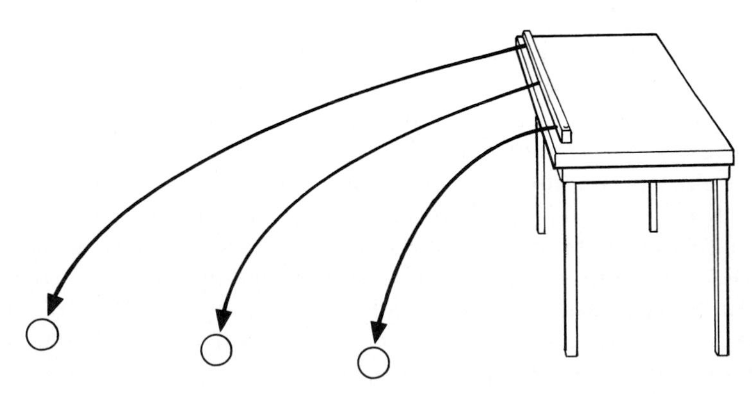

Figure 10-17
A simple experiment demonstrates the significance of the radius of rotation.

The greater the angular velocity of a baseball bat, the farther a struck ball will travel when all other conditions are equal.

Linear and Angular Acceleration

The acceleration of a body in angular motion may be resolved into two perpendicular linear acceleration components. These components are directed along and perpendicular to the path of angular motion at any point in time (Figure 10-18).

The component directed along the path of angular motion takes its name from the term *tangent.* A tangent is a line that touches but does not cross through a curve. The tangential component, known as **tangential acceleration,** represents the change in linear speed for a body traveling on a curved path. The formula for tangential acceleration is the following:

tangential acceleration
the component of angular acceleration directed along a tangent to the path of motion that indicates change in linear speed

$$a_t = \frac{v_2 - v_1}{t}$$

Tangential acceleration is a_t, v_1 is the tangential linear velocity of the moving body at an initial time, v_2 is the tangential linear velocity of the moving body at a second time, and t is the time interval over which the velocities are assessed.

A hammer and a discus are accelerated along a curved path during the throw. The tangential component of hammer and discus acceleration represents the rate of change in the linear speed of each implement. Because the speed of projection greatly affects a projectile's range, tangential velocity should be maximum just before the release of the hammer or discus. Once release occurs, tangential acceleration is 0 because the thrower is no longer applying a force.

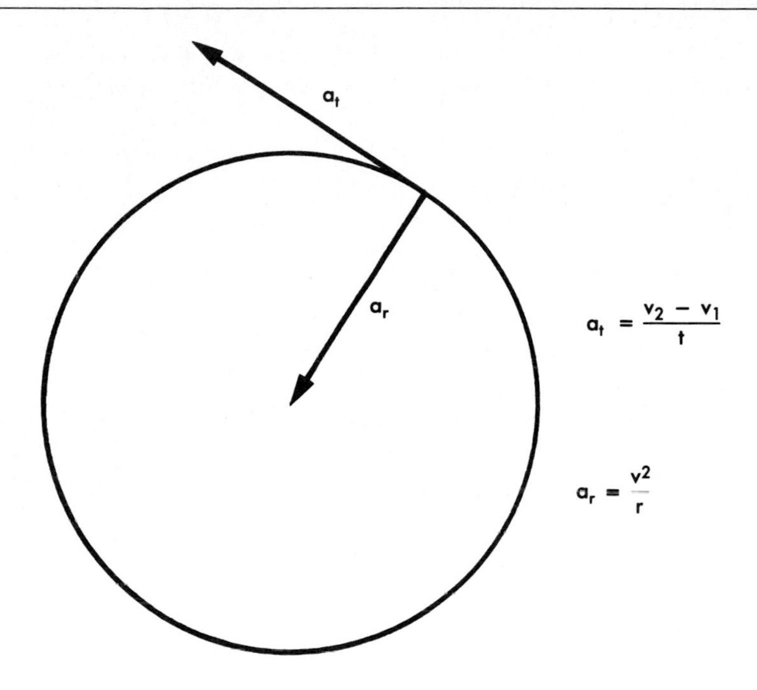

Figure 10-18
Tangential and radial acceleration vectors shown relative to a circular path of motion.

The relationship between tangential acceleration and angular acceleration is expressed as follows:

$$a_t = r\alpha$$

Linear acceleration is a, r is the radius of rotation, and α is angular acceleration. The units of linear acceleration and the radius of rotation must be compatible, and angular acceleration must be expressed in radian-based units for the relationship to be accurate.

Although the linear speed of an object traveling along a curved path may not change, its motion direction is constantly changing. The second component of angular acceleration represents the rate of change in direction of a body in angular motion. This component is called **radial acceleration,** and it is always directed toward the center of curvature. Radial acceleration may be quantified by using the following formula:

radial acceleration
the component of angular acceleration directed toward the center of curvature that indicates change in direction

$$a_r = \frac{v^2}{r}$$

Radial acceleration is a_r, v is the tangential linear velocity of the moving body, and r is the length of the radius of rotation. An increase in linear velocity or a decrease in the radius of curvature increases radial acceleration. Thus, the smaller the radius of curvature (the

Figure 10-19

SAMPLE PROBLEM 3

A windmill style softball pitcher executes a pitch in 0.65 seconds. If her pitching arm is 0.7 m long, what are the magnitudes of the tangential and radial accelerations on the ball just before ball release when tangential ball speed is 20 m/s? What is the magnitude of the total acceleration on the ball at this point?

Known

$$t = 0.65 \text{ s}$$

$$r = 0.7 \text{ m}$$

$$v_2 = 20 \text{ m/s}$$

Solution

To solve for tangential acceleration, use the following formula:

$$a_t = \frac{v_2 - v_1}{t}$$

Substitute in what is known and assume that $v_1 = 0$:

$$a_t = \frac{20 \text{ m/s} - 0}{0.65 \text{ s}}$$

$$\boxed{a_t = 30.8 \text{ m/s}^2}$$

To solve for radial acceleration, use the following formula:

$$a_r = \frac{v^2}{r}$$

Substitute in what is known:

$$a_r = \frac{(20 \text{ m/s})^2}{0.7 \text{ m}}$$

$$\boxed{a_r = 571.4 \text{ m/s}^2}$$

To solve for total acceleration, perform vector composition of tangential and radial acceleration. Since tangential and radial acceleration are oriented perpendicular to each other, the Pythagorean theorem can be used to calculate the magnitude of total acceleration.

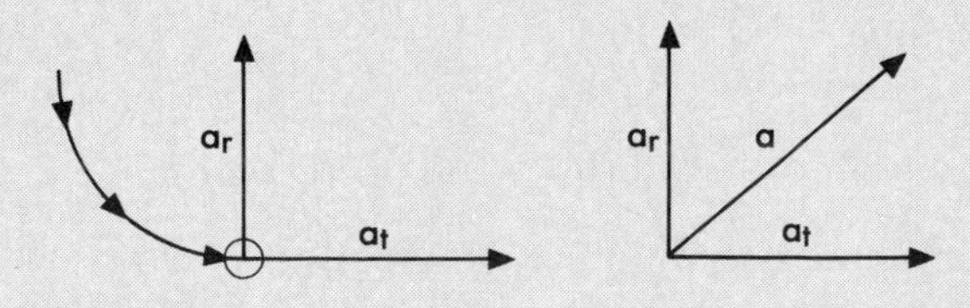

$$a = \sqrt{(30.8 \text{ m/s}^2)^2 + (571.4 \text{ m/s}^2)^2}$$

$$\boxed{a = 572.2 \text{ m/s}^2}$$

tighter the curve), the more difficult it is for a cyclist to negotiate the curve at a high velocity (see Chapter 13).

During execution of the hammer and discus throws, the implements follow a curved path because the thrower's arm and, in the case of the hammer, the cable restrain them. This restraining force causes radial acceleration toward the center of curvature throughout the motion. When the thrower releases the implement, radial acceleration no longer exists and the implement follows the path of the tangent to the curve at that instant. The timing of release is therefore critical; if release occurs too soon or too late, the implement will be directed to the left or the right rather than straight ahead. The sample problem in Figure 10-19 demonstrates the way in which the relationships between angular acceleration and its two components may be quantified.

■ At the instant that a hammer or a discus is released, its tangential and radial accelerations become equal to 0 because a thrower is no longer applying force.

SUMMARY

An understanding of angular motion is an important part of the study of biomechanics because most volitional motion of the human body involves the rotation of bones around imaginary axes of rotation passing through the joint centers at which the bones articulate. The angular kinematic quantities—angular displacement, angular velocity, and angular acceleration—possess the same interrelationships as their linear counterparts, with angular displacement representing change in angular position, angular velocity defined as the rate of change in angular position, and angular acceleration indicating the rate of change in angular velocity during a given time. Depending on the selection of the time interval, either average or instantaneous values of angular velocity and angular acceleration may be quantified.

Angular kinematic variables may be quantified for the relative angle formed by the longitudinal axes of two body segments articulating at a joint or for the absolute angular orientation of a single body segment with respect to a fixed reference line. Different instruments are available for direct measurement of angles on a human subject.

Using the right hand rule, vectors oriented perpendicular to the plane of rotation may be formulated to represent angular kinematic quantities when it is necessary to combine two or more of the quantities using the laws of vector algebra. It is also possible to translate angular motion into the linear motion of a point on the rotating body.

INTRODUCTORY PROBLEMS

1. The relative angle at the knee changes from 180 degrees to 95 degrees during the knee flexion phase of a squat exercise. If 10 complete squats are performed, what is the total angular

distance and the total angular displacement undergone at the knee? (Provide answers in both degrees and radians.) (Answer: σ = 850 deg, 14.8 rad; Θ = 0)

2. Identify the angular displacement, the angular velocity, and the angular acceleration of the second hand on a clock over the time interval in which it moves from the number 12 to the number 6. Provide answers in both degree and radian-based units. (Answer: $\Theta = -180$ deg, $-\pi$ rad; $\omega = -6$ deg/s, $-\pi/30$ rad/s; $\alpha = 0$)
3. How many revolutions are completed by a top spinning with a constant angular velocity of 3 π rad/s during a 20 second time interval? (Answer: 30 rev)
4. A kicker's extended leg is swung for 0.4 seconds in a counterclockwise direction while accelerating at 200 deg/s^2. What is the angular velocity of the leg at the instant of contact with the ball? (Answer: 80 deg/s, 1.4 rad/s)
5. The angular velocity of a runner's thigh changes from 3 rad/s to 2.7 rad/s during a 0.5 second time period. What has been the average angular acceleration of the thigh? (Answer: -0.6 rad/s^2, -34.4 deg/s^2)
6. Identify three movements during which the instantaneous angular velocity at a particular time is the quantity of interest. Explain your choices.
7. Fill in the missing corresponding values of angular measure in the table below.

Degrees	Radians	Revolutions
90		
	1	
180		
		1

8. Measure and record the following angles for the drawing shown below:

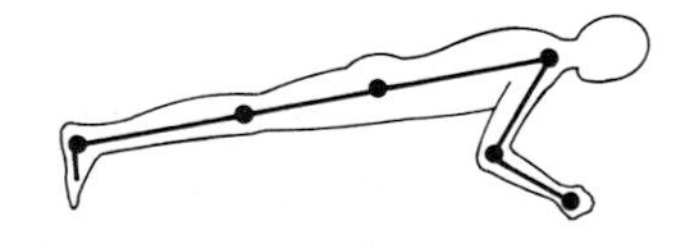

a. The relative angle at the shoulder
b. The relative angle at the elbow
c. The absolute angle of the upper arm
d. The absolute angle of the forearm.

Use the right horizontal as your reference for the absolute angles.

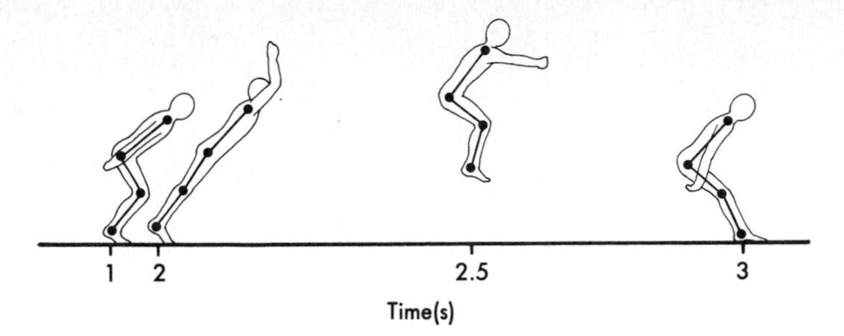

9. Calculate the following quantities for the diagram shown above:
   a. The angular velocity at the hip over each time interval
   b. The angular velocity at the knee over each time interval
   Would it provide meaningful information to calculate the average angular velocities at the hip and knee for the movement shown above? Provide a rationale for your answer.
10. A tennis racket swung with an angular velocity of 12 rad/s strikes a motionless ball at a distance of 0.5 m from the axis of rotation. What is the linear velocity of the racket at the point of contact with the ball? (Answer: 6 m/s)

ADDITIONAL PROBLEMS

1. A 1.2 m golf club is swung in a planar motion by a right-handed golfer with an arm length of 0.76 m. If the initial velocity of the golf ball is 35 m/s, what was the angular velocity of the left shoulder at the point of ball contact? (Assume that the left arm and the club form a straight line and that the ini tial velocity of the ball is the same as the linear velocity of the club head at impact.) (Answer: 17.86 rad/s)
2. David is fighting Goliath. If David's 0.75 m sling is accelerated for 1.5 seconds at 20 rad/s^2, what will be the initial velocity of the projected stone? (Answer: 22.5 m/s)
3. A baseball is struck by a bat 46 cm from the radius of rotation when the angular velocity of the bat is 70 rad/s. If the ball is hit at a height of 1.2 m at a 45 degree angle, will the ball clear a 1.2 m fence 110 m away? (Assume that the initial linear velocity of the ball is the same as the linear velocity of the bat at the point at which it is struck.) (Answer: No, the ball will fall through a height of 1.2 m at a distance of 105.7 m.)
4. A polo player's arm and stick form a 2.5 m rigid segment. If the arm and stick are swung with an angular speed of 1.0 rad/s as the player's horse gallops at 5 m/s, what is the resultant velocity of a motionless ball that is struck head-on? (Assume that

ball velocity is the same as the linear velocity of the end of the stick.) (Answer: 7.5 m/s)

5. Explain how the velocity of the ball in Problem 4 would differ if the stick is swung at a 30 degree angle to the direction of motion of the horse.
6. Midway through one complete pedal revolution, the change in the angle at a cyclist's knee is 1 radian. The cyclist is pedaling at a constant velocity of 1.67 rev/s. At the end of 30.25 seconds, what is the angular displacement, velocity, and acceleration at the knee? (Answer: Θ = 1 rad, ω = 10.5 rad/s, α = 0)
7. A majorette in the Rose Bowl Parade tosses a baton into the air with an initial angular velocity of 2.5 rev/s. If the baton undergoes a constant acceleration while airborne of -0.2 rev/s^2 and its angular velocity is 0.8 rev/s when the majorette catches it, how many revolutions did it make in the air? (Answer: 14 rev)
8. A cyclist enters a curve of 30 m radius at a speed of 12 m/s. As the brakes are applied, speed is decreased at a constant rate of 0.5 m/s^2. What are the magnitudes of the cyclist's radial and tangential accelerations when his speed is 10 m/s? (Answer: a_r = 3.33 m/s^2; a_t = 0.5 m/s^2)
9. A hammer is being accelerated at 15 rad/s^2. Given a radius of rotation of 1.7 m, what are the magnitudes of the radial and tangential components of acceleration when tangential hammer speed is 25 m/s? (Answer: a_r = 367.6 m/s^2; a_t = 25.5 m/s^2)
10. An ice skater increases her speed from 10 m/s to 12.5 m/s over a period of 3 seconds while coming out of a curve of 20 m radius. What are the magnitudes of her radial, tangential, and total accelerations as she leaves the curve? (Remember that a_r and a_t are the vector components of total acceleration.) (Answer: a_r = 7.81 m/s^2; a_t = 0.83 m/s^2; a = 7.85 m/s^2)

REFERENCES

1. Adrian MJ and Cooper JM: Biomechanics of human movement, Indianapolis, 1989, Benchmark Press, Inc.
2. Bates BT, Morrison E, and Hamill J: A comparison between forward and backward running. In Adrian M and Deutsch H, eds: Biomechanics: the 1984 Olympic Scientific Congress proceedings, Eugene, Ore, 1986, Microform Publications.

3. Brown EW, Witten W, and Ahn BH: Biomechanical comparison of the standard and handspring soccer throw-in. In Terauds J, Gowitzke BA, and Holt LE, eds: Biomechanics in sports-III & IV, Del Mar, Calif, 1987, Academic Publishers.
4. Cavanagh PR and Grieve DW: The graphical display of angular movement of the body, Br J Sports Med 7:129, 1973.
5. Elliott BC: Tennis strokes and equipment. In Vaughan CL, ed: Biomechanics of sport, Boca Raton, Fla, 1989, CRC Press, Inc.
6. Elliott B, Marsh T, and Blanksby B: A three-dimensional cinematographical analysis of the tennis serve, Int J Sport Biomech 2:260, 1986.
7. Elliott B et al: A three-dimensional cinematographic analysis of the fastball and curveball pitches in baseball, Int J Sport Biomech 2:20, 1986.
8. Garhammer J: Weight lifting and training. In Vaughan CL, ed: Biomechanics of sport, Boca Raton, Fla, 1989, CRC Press, Inc.
9. Gutman D: The physics of foul play, Discover, p 70, Apr 1988.
10. Messier SP and Brody MA: Mechanics of translation and rotation during conventional and handspring soccer throw-ins, Int J Sport Biomech 2:301, 1986.
11. Nordin M and Frankel VH: Biomechanics of the knee. In Nordin M and Frankel VH, eds: Biomechanics of the musculoskeletal system, ed 2, Philadelphia, 1989, Lea & Febiger.
12. Pappas AM, Zawacki RM and Sullivan TJ: Biomechanics of baseball pitching, Am J Sports Med 13:216, 1985.
13. Zatsiorsky VM, Lanka GE, and Shalmanov AA: Biomechanical analysis of shot putting technique, Exer Sport Sci Rev 9:353, 1982.

ANNOTATED READINGS

Hay JG: Angular kinematics. In Hay JG: The biomechanics of sports techniques, ed 3, Englewood Cliffs, NJ, 1985, Prentice-Hall, Inc.
Provides an in-depth discussion of angular kinematic quantities in the context of sports.

Jacobs HR: Basic ideas and operations. In Mathematics: a human endeavor, ed 2, New York, 1982, WH Freeman & Co Publishers.
Explains simple but important mathematical measurement procedures.

Northrip JW, Logan GA, and McKinney WC: Techniques of biomechanics analysis. In Northrip JW, Logan GA, and McKinney WC: Introduction to biomechanic analysis of sport, ed 2, Dubuque, Ia, 1979, Wm C Brown Group.
Identifies techniques for quantification of angular and linear kinematic quantities from films.

Pons DJ and Vaughan CL: Mechanics of cycling. In Vaughan CL, ed: Biomechanics of sport, Boca Raton, Fla, 1989, CRC Press, Inc.
Includes a detailed description on kinematic and other aspects of cycling.

11 THE USE OF FORCE

Linear Kinetics

After reading this chapter, the student will be able to:

Identify Newton's laws of motion and gravitation and describe practical illustrations of the laws.

Identify the factors that affect friction and discuss the role of friction in sports and daily activities.

Define impulse and momentum and explain the relationship between them.

Identify the factors that govern the outcome of a collision between two bodies.

Discuss the interrelationships among mechanical work, power, and energy.

Solve quantitative problems related to kinetic concepts.

Why do some skis perform better on wet snow whereas others are better on dry snow? Why do some balls bounce higher on one surface than on another? How can football linemen push larger opponents backward? What mechanical factors contribute to an athlete's ability to excel in explosive events such as the long jump or the shot put? In this chapter the topic of kinetics will be introduced with a discussion of some important basic concepts and principles relating to linear kinetics.

NEWTON'S LAWS

Sir Isaac Newton (1642–1727) discovered many of the fundamental relationships that form the foundation for the field of modern mechanics. These principles highlight the interrelationships among the basic kinetic quantities introduced in Chapter 2.

A skater has a tendency to continue gliding with constant speed and direction because of inertia.

The Law of Inertia

Newton's first law of motion is known as the *law of inertia.* This law states the following:

A body will maintain a state of rest or constant velocity unless acted on by an external force that changes the state.

In other words, a motionless object will remain motionless unless there is a net force (a force not counteracted by another force) acting on it. Similarly, a body traveling with a constant speed along a straight path will continue its motion unless acted on by a net force that alters either the speed or the direction of the motion.

It seems intuitively obvious that an object in a static (motionless) situation will remain motionless barring the action of some external force. We assume that a piece of furniture such as a chair will maintain a fixed position unless pushed or pulled by a person exerting a net force to cause its motion. When a body is traveling with a constant velocity, however, the enactment of the law of inertia is not as obvious because external forces do act to reduce velocity in most situations. The law of inertia implies that a skater gliding on ice will

The magnitude and direction of a soccer ball's acceleration is a direct function of the force applied by the kicker.

continue gliding with the same speed and in the same direction barring the action of an external force. In reality, friction and air resistance are two forces normally present that act to slow skaters and other moving bodies.

The Law of Acceleration

Newton's second law of motion is an expression of the interrelationships among force, mass, and acceleration. This law, known as the *law of acceleration,* may be stated as follows:

A force applied to a body causes an acceleration of that body of a magnitude proportional to the force, in the direction of the force, and inversely proportional to the body's mass.

When a ball is thrown, kicked, or struck with an implement, it tends to travel in the direction of the line of action of the applied force. Similarly, the greater the amount of force applied, the more quickly the ball moves. The algebraic expression of the law is a well-known formula that expresses the quantitative relationships among an applied force, a body's mass, and the resulting acceleration of the body:

$$F = ma$$

Thus, if a 1 kg ball is struck with a force of 10 N, the resulting acceleration of the ball is 10 m/s^2. If the ball has a mass of 2 kg, the application of the same 10 N force results in an acceleration of only 5 m/s^2.

The Law of Reaction

The third of Newton's laws of motion states that every applied force is accompanied by a reaction force:

For every action, there is an equal and opposite reaction.

In terms of forces, the law may be stated as follows:

When one body exerts a force on a second, the second body exerts a reaction force that is equal in magnitude and opposite in direction on the first body.

■ The law of reaction holds true even when the bodies in contact are of significantly different masses.

When a person leans with a hand against a rigid wall, the wall pushes back on the hand with a force that is equal and opposite to that exerted by the hand on the wall. The harder the hand pushes against the wall, the greater the amount of pressure felt across the surface of the hand where it contacts the wall. Another illustration of Newton's third law of motion is found in Figure 11-1.

The reaction force supplied by the ground during the takeoff phase of the high jump can contribute to performance (6). At the beginning of the stride before takeoff, good high jumpers are mov-

SAMPLE PROBLEM 1

Figure 11-1

A 90 kg ice hockey player collides head-on with an 80 kg ice hockey player. If the first player exerts a force of 450 N on the second player, how much force is exerted by the second player on the first?

Known

$m_1 = 90$ kg

$m_2 = 80$ kg

$F_1 = 450$ N

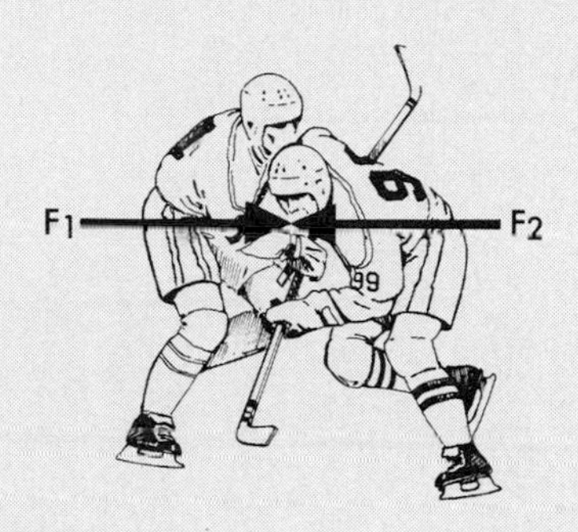

Solution

This problem does not require computation. According to Newton's third law of motion, for every action there is an equal and opposite reaction. If the force exerted by the first player on the second has a magnitude of 450 N and a positive direction, then the force exerted by the second player on the first has a magnitude of 450 N and a negative direction.

−450 N

ing with a large horizontal velocity and a slight downwardly directed vertical velocity. The ground reaction force (GRF) reduces the jumper's horizontal velocity and creates an upwardly directed vertical velocity (Figure 11-2). Skillful high jumpers not only enter the takeoff phase of the jump with high horizontal velocities but also effectively use the GRF to convert horizontal velocity to upward vertical velocity.

Researchers have studied the GRFs sustained with every footfall during running to investigate the causes of running-related injuries (12). The magnitude of the vertical component of the GRF during running is generally two to three times the runner's body weight, with the pattern of force varying with running style. Cavanagh and Lafortune have classified runners as rearfoot, midfoot, or forefoot strikers, according to the portion of the shoe first making contact with the ground. Typical vertical GRF patterns for rearfoot strikers

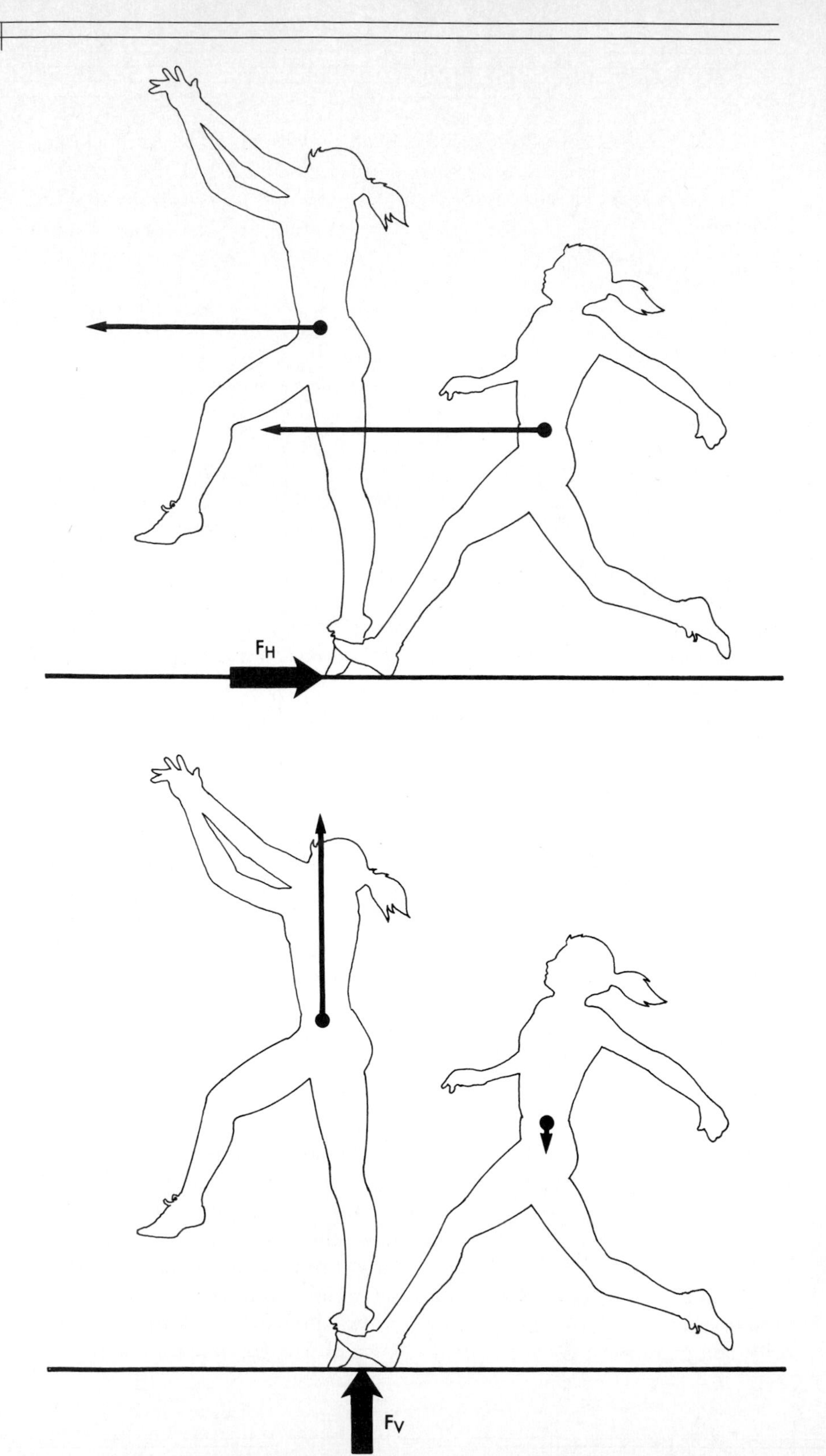

Figure 11-2
During the high jump takeoff, horizontal velocity is reduced and upward vertical velocity is created by the ground reaction force as the takeoff foot contacts the ground.

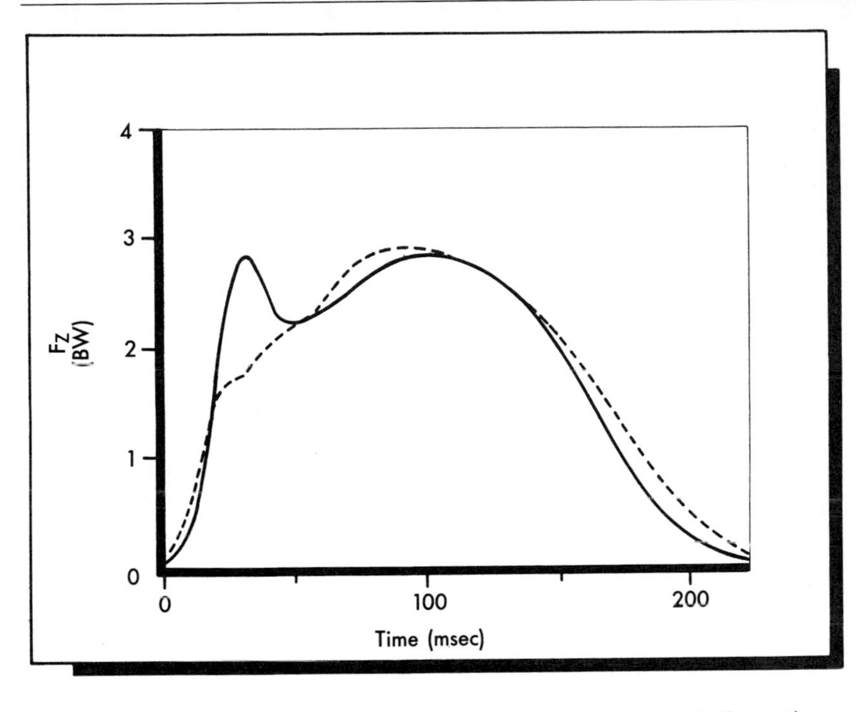

Figure 11-3
Typical ground reaction force patterns for rearfoot strikers and others. Runners may be classified as rearfoot, midfoot, or forefoot strikers according the portion of the shoe that typically contacts the ground first.

and others are shown in Figure 11-3. Factors influencing GRF patterns include running speed, footwear, ground surface, and grade (4). Different shoes and the use of orthotics affect GRF patterns differently for each individual (14).

The Law of Gravitation

Newton's discovery of the law of universal gravitation was one of the most significant contributions to the Scientific Revolution and is considered by many to mark the beginning of modern science (5). According to legend, Newton's thoughts on gravitation were provoked either by his observation of a falling apple or by his actually being struck on the head by a falling apple. In his writings on the subject, Newton used the example of the falling apple to illustrate the principle that every body attracts every other body (3). Newton's law of gravitation states the following:

All bodies are attracted to one another with a force proportional to the product of their masses and inversely proportional to the distance between them.

In accordance with Newton's third law of motion, ground reaction forces are sustained with every footfall during running.

Stated algebraically, the law is the following:

$$F_g = G\frac{m_1 m_2}{d^2}$$

The force of gravitational attraction is F_g, G is a numerical constant, m_1 and m_2 are the masses of the bodies, and d is the distance between the mass centers of the bodies.

For the example of the falling apple, Newton's law of gravitation indicates that just as the earth attracts the apple, the apple attracts the earth, although to a much smaller extent. As the formula for gravitational force shows, the greater the mass of either body, the greater the attractive force between the two. Similarly, the greater the distance between the bodies, the smaller the attractive force between them.

For biomechanical applications, the only gravitational attraction of sufficient magnitude to effect a change is that between the earth and bodies on the surface of the earth. It is the extremely large mass of the earth that makes the force of consequence. The attractive force exerted by the earth on objects on its surface is known as *weight.* The rate of gravitational acceleration at which bodies are attracted toward the surface of the earth (9.81 m/s^2) is based on the earth's mass and the distance to the center of the earth in the continental United States.

MECHANICAL BEHAVIOR OF BODIES IN CONTACT

According to Newton's third law of motion, for every action, there is an equal and opposite reaction. However, consider the case of a horse hitched to a cart. According to Newton's third law, when the horse exerts a force on the cart to cause forward motion, the cart exerts a backward force of equal magnitude on the horse (Figure 11-4). Considering the horse and the cart as a single mechanical system,

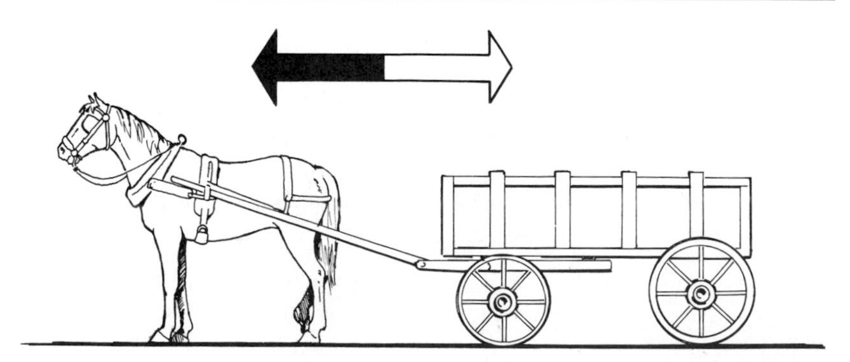

Figure 11-4
When a horse attempts to pull a cart forward, the cart exerts an equal and opposite force on the horse, in accordance with Newton's law of action-reaction.

if the two forces are equal in magnitude and opposite in direction, their vector sum is 0. How does the horse and cart system achieve forward motion? The answer relates to the presence of another force that acts with a different magnitude on the cart than on the horse—the force of friction.

Friction

Friction is a force that acts at the interface of surfaces in contact in the direction opposite the direction of motion or impending motion. Because friction is a force, it is quantified in units of force (N). The magnitude of the generated friction force determines the relative ease or difficulty of motion for two objects in contact.

friction
a force acting at the area of contact between two surfaces in the direction opposite that of motion or motion tendency

Consider the example of a box sitting on a level table top (Figure 11-5). The two forces acting on the undisturbed box are its own weight and a reaction force (R) applied by the table. In this situation the reaction force is equal in magnitude and opposite in direction to the box's weight.

Figure 11-5

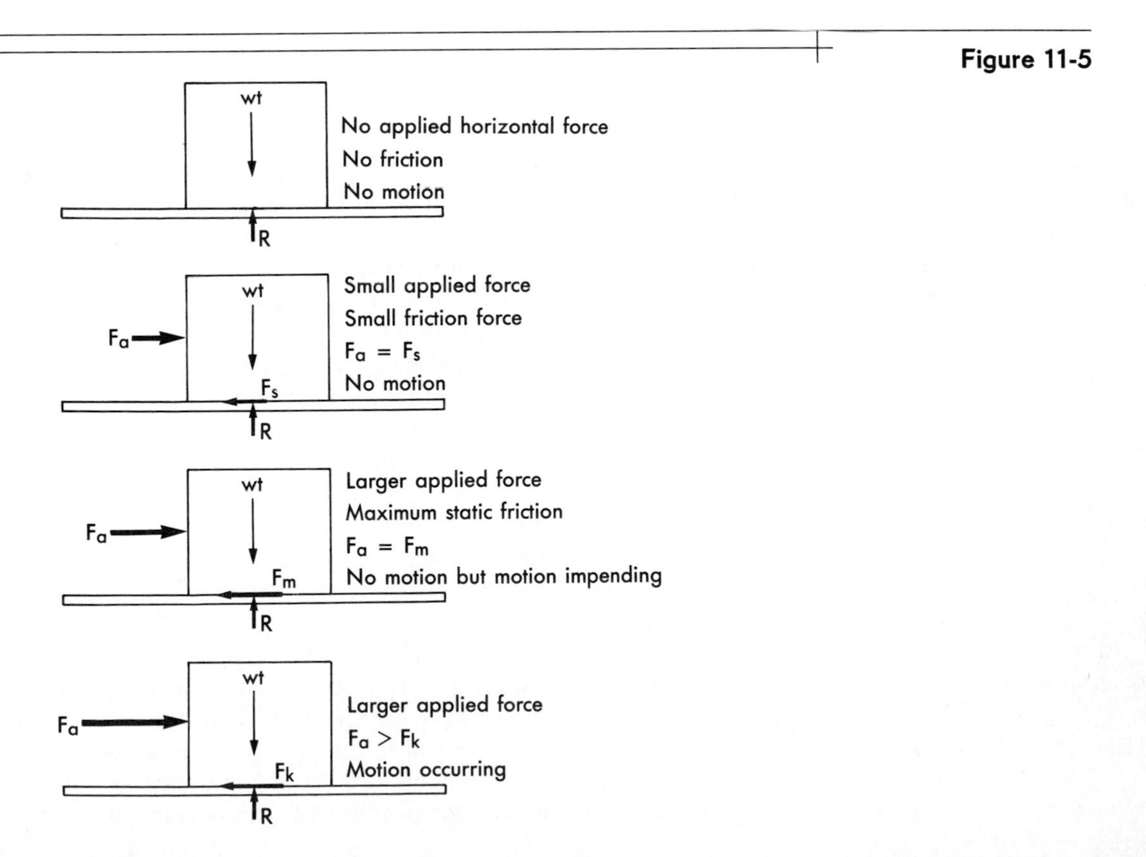

Figure 11-6

As long as a body is static, the magnitude of the friction force developed is equal to that of an applied external force. Once the situation becomes dynamic, the magnitude of the friction force developed remains constant and at a level below that of maximum static friction.

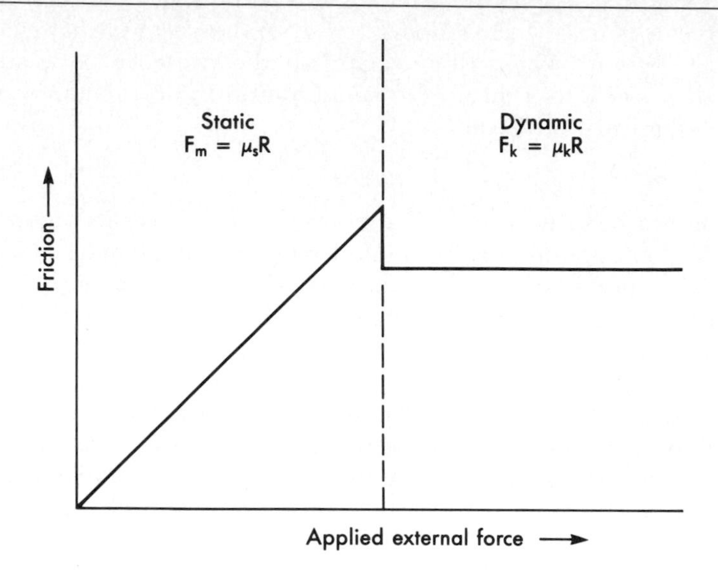

When an extremely small horizontal force is applied to this box, it remains motionless. The box can maintain its static position because the applied force causes the generation of a friction force at the box/table interface that is equal in magnitude and opposite in direction to the small applied force. As the magnitude of the applied force becomes greater and greater, the magnitude of the opposing friction force also increases to a certain critical point. At that point the friction force present is termed **maximum static friction** (F_m). If the magnitude of the applied force is increased beyond this value, motion will occur (the box will slide).

maximum static friction
the maximum amount of friction that can be generated between two static surfaces

kinetic friction
the constant friction generated between two surfaces in contact during motion

Once the box is in motion, an opposing friction force continues to act. The friction force present during motion is referred to as **kinetic friction** (F_k). Unlike static friction, the magnitude of kinetic friction remains at a constant value that is *less than* the magnitude of maximum static friction. Regardless of the amount of the applied force or the speed of the occurring motion, the kinetic friction force remains the same. The changes that occur in the magnitude of the friction force in response to changes in the net applied force are shown in Figure 11-6.

What factors determine the amount of applied force needed to move an object? More force is required to move a refrigerator than to move the empty box in which the refrigerator was delivered. More force is also needed to slide the refrigerator across a carpeted floor than across a smooth linoleum floor. Two factors govern the magnitude of the force of maximum static friction or kinetic friction in any situation: the **coefficient of friction,** represented by the small Greek letter mu (μ), and the **normal** (perpendicular) **reaction force** (R):

coefficient of friction
a number that serves as an index of the interaction between two surfaces in contact

normal reaction force
the force acting perpendicular to two surfaces in contact

$$F = \mu R$$

The coefficient of friction is a unitless number indicating the relative ease of sliding or the amount of mechanical and molecular interaction between two surfaces in contact. Factors influencing the value of μ are the relative roughness and hardness of the surfaces in contact and the type of molecular interaction between the surfaces. The greater the mechanical and molecular interaction, the greater the value of μ. For example, the coefficient of friction between two blocks covered with rough sandpaper is larger than the coefficient of friction between a skate and a smooth surface of ice. The coefficient of friction describes the interaction between two surfaces in contact and is not descriptive of either surface alone. The coefficient of friction for the blade of an ice skate in contact with ice is different from that for the blade of the same skate in contact with a sandy beach.

The coefficient of friction between two surfaces assumes one of two different values, depending on whether the bodies in contact are motionless (static) or in motion (kinetic). The two coefficients are known as the *coefficient of static friction* (μ_s) and the *coefficient of kinetic friction* (μ_k). The magnitude of maximum static friction is based on the coefficient of static friction:

$$F_m = \mu_s R$$

The magnitude of the kinetic friction force is based on the coefficient of kinetic friction:

$$F_k = \mu_k R$$

For any two bodies in contact, μ_k is always smaller than μ_s. Kinetic friction coefficients as low as 0.003 have been reported between the blade of a racing skate and a properly treated ice rink under optimal conditions (17). Use of the coefficients of static and kinetic friction is illustrated in Figure 11-7.

The other factor affecting the magnitude of the friction force generated is the normal reaction force. If weight is the only vertical force acting on a body sitting on a horizontal surface, R is equal in magnitude to the weight. If the object is a tackling sled with a 100 kg coach standing on it, R is equal to the weight of the sled plus the weight of the coach. Other vertically directed forces such as pushes or pulls can also affect the magnitude of R, which is always equal to the vector sum of all forces or force components acting normal to the surfaces in contact (Figure 11-8).

The magnitude of R can be intentionally altered to increase or decrease the amount of friction present in a particular situation. When

Figure 11-7

SAMPLE PROBLEM 2

The coefficient of static friction between a sled and the snow is 0.18, with a coefficient of kinetic friction of 0.15. A 250 N boy sits on the 200 N sled. How much force directed parallel to the horizontal surface is required to start the sled in motion? How much force is required to keep the sled in motion?

Known

$\mu_s = 0.18$

$\mu_k = 0.15$

$\text{wt} = 250\ \text{N} + 200\ \text{N}$

Solution

To start the sled in motion, the applied force must exceed the force of maximum static friction:

$$F_m = \mu_s R$$

$$F_m = (0.18)\ (250\ \text{N} + 200\ \text{N})$$

$$F_m = 81\ \text{N}$$

The applied force must be greater than 81 N.

To maintain motion, the applied force must equal the force of kinetic friction:

$$F_k = \mu_k R$$

$$F_k = (0.15)(250\ \text{N} + 200\ \text{N})$$

$$F_k = 67.5\ \text{N}$$

The applied force must be at least 67.5 N.

a football coach stands on the back of a tackling sled, the normal reaction force exerted by the ground on the sled is increased, with a concurrent increase in the amount of friction generated, making it more difficult for a tackler to move the sled. Alternatively, if the magnitude of R is decreased, friction is decreased and it is easier to initiate motion.

How can the normal reaction force be decreased? Suppose you need to rearrange the furniture in a room. Is it easier to push or pull an object such as a desk to move it? When a desk is pushed, the force exerted is typically directed diagonally downward. In contrast,

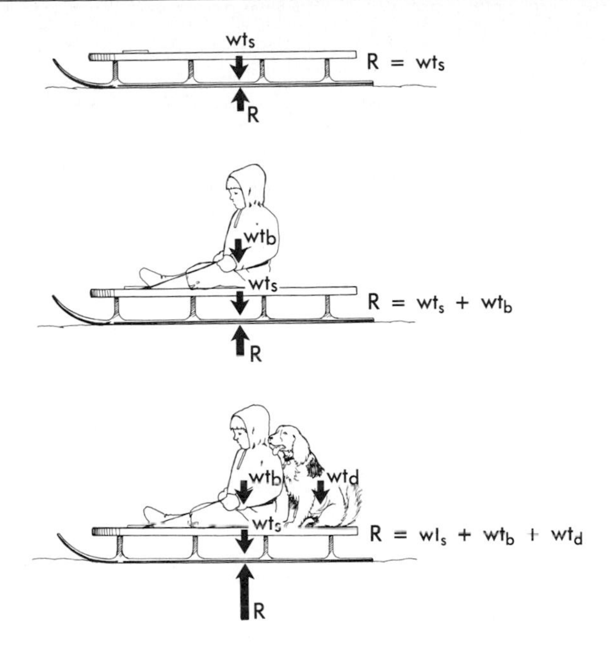

Figure 11-8
Increases in the magnitude of the reaction force translate to increases in friction.

force is usually directed diagonally upward when a desk is pulled. The vertical component of the push or pull either adds to or subtracts from the magnitude of the normal reaction force, thus influencing the magnitude of the friction force generated and the relative ease of moving the desk (Figure 11-9).

■ It is advantageous to pull with a line of force that is directed slightly upward when moving a heavy object.

The amount of friction present between two surfaces can also be changed by altering the coefficient of friction between the surfaces. For example, the use of gloves in sports such as golf and racquetball increases the coefficient of friction between the hand and the grip of the club or racquet. Similarly, lumps of wax applied to a surfboard increase the roughness of the board's surface, thereby increasing the coefficient of friction between the board and the surfer's feet. The application of a thin, smooth coat of wax to the bottom of cross country skis is designed to decrease the coefficient of friction between the skis and the snow, with different waxes used for various snow conditions.

■ Racquetball and golf gloves are designed to increase the friction between the hand and the racquet or club, as are the grips on the handles of the racquets and clubs themselves.

A widespread misconception about friction is that greater contact surface area generates more friction. Advertisements often imply that wide track automobile tires provide better traction than tires of normal width. However, the only factors known to affect friction are the coefficient of friction and the normal reaction force. Because wide track tires typically weigh more than normal tires, they do increase friction to the extent that they increase R. However, the same

Figure 11-9
From a mechanical perspective, it is easier to pull than to push an object such as a desk, since pulling tends to decrease the magnitude of R and F, whereas pushing tends to increase R and F.

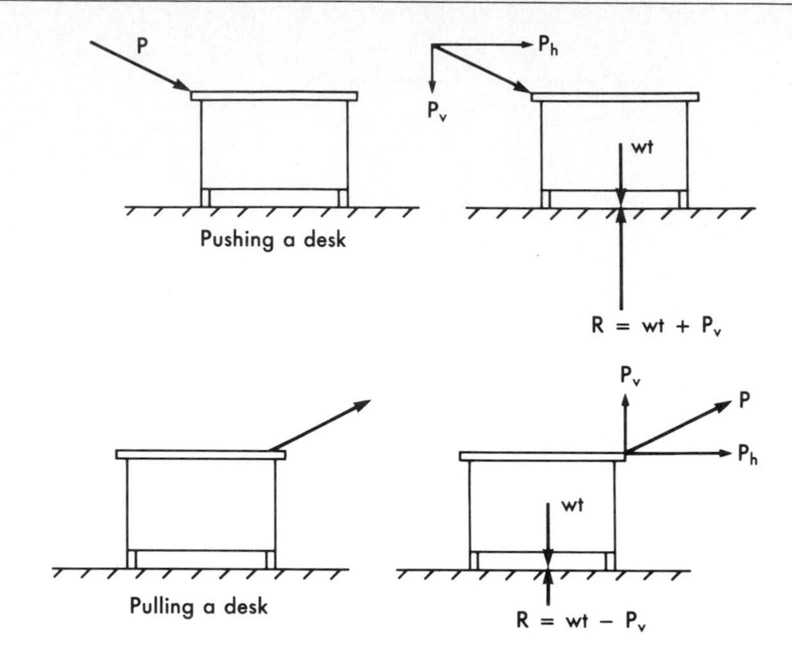

The coefficient of friction between a dancer's shoes and the floor must be small enough to allow freedom of motion but large enough to prevent slippage.

effect can be achieved by carrying bricks or cinder blocks in the trunk of the car, a practice often followed by people who regularly drive on icy roads.

Friction exerts an important influence during many daily activities. Walking depends on a proper coefficient of friction between a person's shoes and the supporting surface. If the coefficient of friction is too low, as when a person with smooth-soled shoes walks on a patch of ice, slippage will occur. The bottom of a wet bathtub or shower stall should provide a coefficient of friction with the soles of bare feet that is sufficiently large to prevent slippage.

The amount of friction present between ballet shoes and the dance studio floor must be controlled so that movements involving some amount of sliding or pivoting—such as *glissades, assembles,* and pirouettes—can be executed smoothly but without slippage. Rosin is often applied to dance floors because it provides a large coefficient of static friction but a significantly smaller coefficient of dynamic friction (11). This helps prevent slippage in static situations and allows desired movements to occur freely.

The amount of friction present during sport situations has engendered heated controversies. The National Football League Players

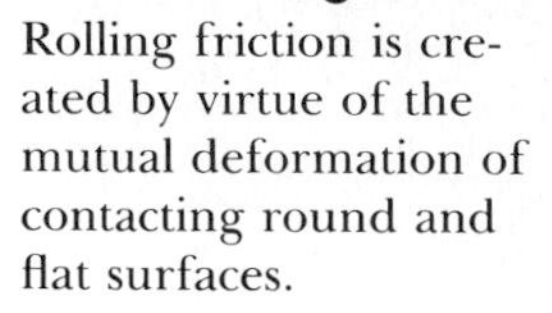

Figure 11-10
Rolling friction is created by virtue of the mutual deformation of contacting round and flat surfaces.

Association has attempted to have artificial turf declared a "hazardous substance" partly because the high coefficient of friction between artificial turf and a football shoe often does not allow rotation of a planted foot. Many knee injuries have been attributed to the immobility of a foot planted on artificial turf when a player is tackled. However, playing on artificial turf continues because the Consumer Products Safety Commission concluded that there is insufficient evidence supporting the claim (16).

Another controversial disagreement occurred between Glenn Allison, a retired professional bowler and member of the American Bowling Congress Hall of Fame, and the American Bowling Congress. The dispute arose over the amount of friction present between Allison's ball and the lanes on which he bowled a perfect score of 300 in three consecutive games. According to the congress, his scores could not be recognized because the lanes he used did not conform to congress standards for the amount of conditioning oil present (10).

■ Rolling friction is influenced by the weight, radius, and deformability of the rolling object, as well as by the coefficient of friction between the two surfaces.

The magnitude of the rolling friction present between a rolling object, such as a bowling ball or an automobile tire, and a flat surface is approximately one hundredth to one thousandth of that present between sliding surfaces. Rolling friction occurs because both the curved and the flat surfaces are slightly deformed during contact (Figure 11-10). The coefficient of friction between the surfaces in contact, the normal reaction force, and the size of the radius of curvature of the rolling body all influence the magnitude of rolling friction. For bicycle tires, rolling friction is inversely proportional to the wheel diameter (18). It decreases with bicycle tire width and increases with reduced tire pressure (13).

Figure 11-11
A horse can pull a cart if the horse's hooves generate more friction than the wheels of the cart.

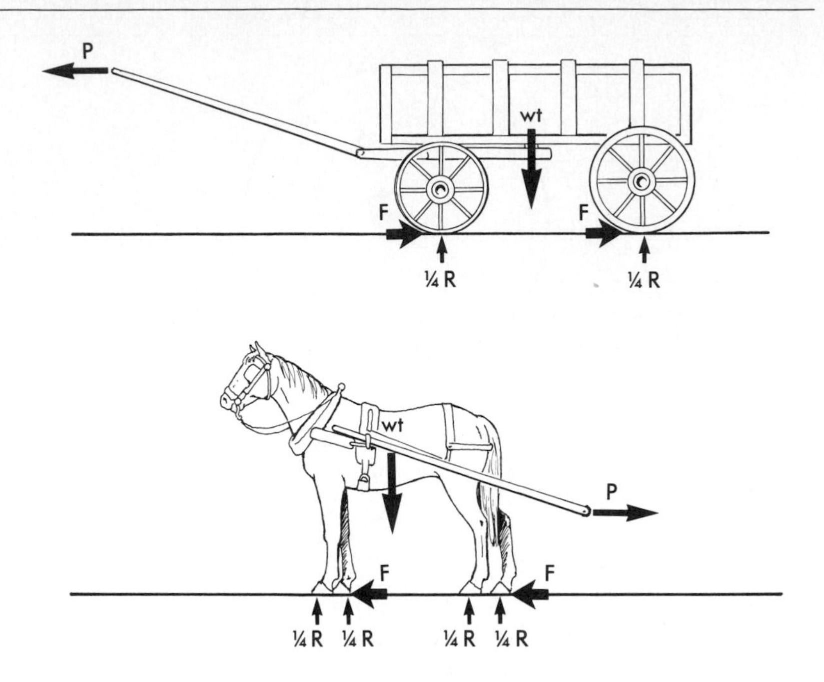

■ The synovial fluid present at many of the joints of the human body greatly reduces the friction between the articulating bones.

The amount of friction present in a sliding or rolling situation is dramatically reduced when a layer of fluid, such as oil or water, intervenes between two surfaces in contact. The amount of friction between a bowling ball and a properly oiled lane is extremely small, and according to the American Bowling Congress, an insufficient amount of oil on the lanes gave Allison the unfair advantage of added ball traction (10).

The force of friction is the determining factor for movement in the horse and cart mechanical system. The system moves forward if the magnitude of the friction force generated by the horse's hooves against the ground exceeds that produced by the wheels of the cart against the ground (Figure 11-11). Because most horses are shod to increase the amount of friction between their hooves and the ground and most cart wheels are round and smooth to minimize the amount of friction they generate, the horse is usually at an advantage. However, if the horse stands on a slippery surface or if the cart rests in deep sand or is heavily loaded, motion may not be possible.

Momentum

momentum
the product of a body's mass and its velocity

Another factor that affects the outcome of interactions between two bodies is **momentum,** a mechanical quantity that is particularly important in situations involving collisions. Momentum may be defined generally as the quantity of motion that an object possesses. More specifically, linear momentum is the product of an object's mass and its velocity:

$$M = mv$$

A static object (with 0 velocity) has no momentum; that is, its momentum equals 0. A change in a body's momentum may be caused by either a change in the body's mass or a change in its velocity. In most human movement situations, changes in momentum result from changes in velocity. Units of momentum are units of mass multiplied by units of velocity, expressed in terms of kg·m/s. Because velocity is a vector quantity, momentum is also a vector quantity and is subject to the rules of vector composition and resolution.

Momentum is a vector quantity.

When a head-on collision between two objects occurs, there is a tendency for both objects to continue moving in the direction of motion originally possessed by the object with the greatest momentum. If a 90 kg hockey player traveling at 6 m/s to the right collides head-on with an 80 kg player traveling at 7 m/s to the left, the momentum of the first player is the following:

$$\begin{aligned} M &= mv \\ &= (90\text{ kg})\,(6\text{ m/s}) \\ &= 540\text{ kg·m/s} \end{aligned}$$

The momentum of the second player is expressed as follows:

$$\begin{aligned} M &= mv \\ &= (80\text{ kg})\,(7\text{ m/s}) \\ &= 560\text{ kg·m/s} \end{aligned}$$

Since the second player's momentum is greater, both players would tend to continue moving in the direction of the second player's original velocity after the collision. Actual collisions are also affected by the extent to which the players become entangled, by whether one or both players remain on their feet, and by the elasticity of the collision.

Neglecting these other factors that may influence the outcome of the collision, it is possible to calculate the magnitude of the combined velocity of the two hockey players after the collision using a modified statement of Newton's first law of motion. Newton's first law may be restated as the *principle of conservation of momentum:*

In the absence of external forces, the total momentum of a given system remains constant.

The principle is expressed in equation format as the following:

$$\begin{aligned} M_1 &= M_2 \\ (mv)_1 &= (mv)_2 \end{aligned}$$

Figure 11-12

SAMPLE PROBLEM 3

A 90 kg hockey player traveling with a velocity of 6 m/s collides head-on with an 80 kg player traveling at 7 m/s. If the two players entangle and continue traveling together as a unit following the collision, what is their combined velocity?

Known

$m_1 = 90$ kg

$v_1 = 6$ m/s

$m_2 = 80$ kg

$v_2 = -7$ m/s

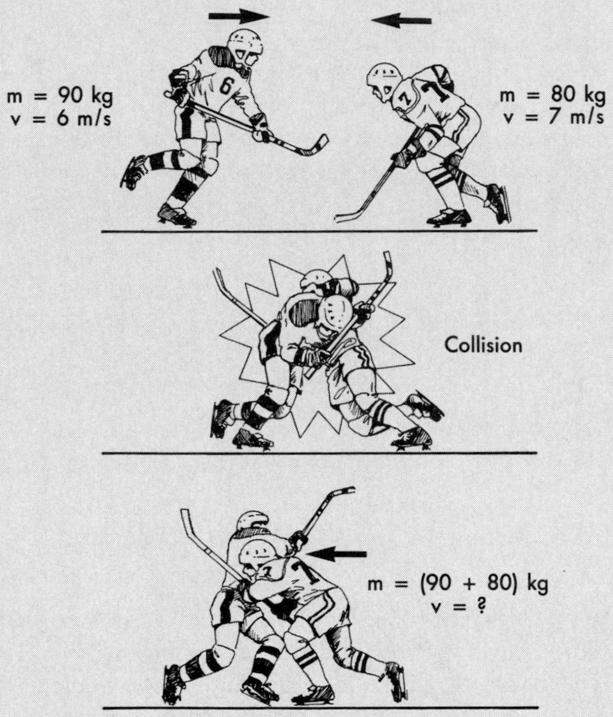

Solution

The law of conservation of momentum may be used to solve the problem, with the two players considered as the total system.

Before collision *After collision*

$$m_1v_1 + m_2v_2 = (m_1 + m_2)(v)$$

$$(90 \text{ kg})(6 \text{ m/s}) + (80 \text{ kg})(-7 \text{ m/s}) = (90 \text{ kg} + 80 \text{ kg})(v)$$

$$540 \text{ kg·m/s} - 560 \text{ kg·m/s} = (170 \text{ kg})(v)$$

$$-20 \text{ kg·m/s} = (170 \text{ kg})(v)$$

$v = 0.12$ m/s in Player B's original direction of travel

Subscript 1 designates an initial point in time and subscript 2 represents a later time.

Applying this principle to the hypothetical example of the colliding hockey players, the vector sum of the two players' momenta before the collision is equal to their single, combined momentum following the collision (Figure 11-12). In reality, friction and air resistance are external forces that typically act to reduce the total amount of momentum present.

Impulse

When external forces do act, they change the momentum present in a system predictably. Changes in momentum depend not only on the magnitude of the acting external forces but also on the length of time that each force acts. The product of force and time is known as **impulse:**

impulse
the product of a force and the time interval over which the force acts

$$\text{Impulse} = Ft$$

When an impulse acts on a system, the result is a change in the system's total momentum. The *impulse-momentum relationship* may be expressed as the following:

$$Ft = \Delta M$$

$$Ft = (mv)_2 - (mv)_1$$

Subscript 1 designates an initial time and subscript 2 represents a later time. An application of this relationship is presented in Figure 11-13.

Significant changes in an object's momentum may result from a small force acting over a large time interval or from a large force acting over a small time interval. A golf ball rolling across a green gradually loses momentum because of the small force of rolling friction exerted while the ball is in motion. The momentum of a baseball struck vigorously by a bat also changes because of the large force exerted by the bat during the fraction of a second it is in contact with the ball.

The amount of impulse generated by the human body is often intentionally manipulated. When a vertical jump is executed, the larger the impulse generated against the floor, the greater the change in the performer's momentum and the higher the resulting jump. Minimizing the landing shock or the maximum ground reaction force generated is advantageous during a landing from a jump. A performer who lands rigidly will experience a relatively large ground reaction force sustained over a relatively short time interval. Alternatively, allowing the hip, knee, and ankle joints to undergo flexion during the landing increases the time interval over which the landing force is absorbed, thereby reducing the magnitude of the

Figure 11-13

SAMPLE PROBLEM 4

A toboggan race begins with the two crew members pushing the toboggan to get it moving as quickly as possible before they climb in. If crew members apply an average force of 100 N in the direction of motion of the 90 kg toboggan for a period of 7 seconds before jumping in, what is the toboggan's speed (neglecting friction) at that point?

Known

F = 100 N

t = 7 s

m = 90 kg

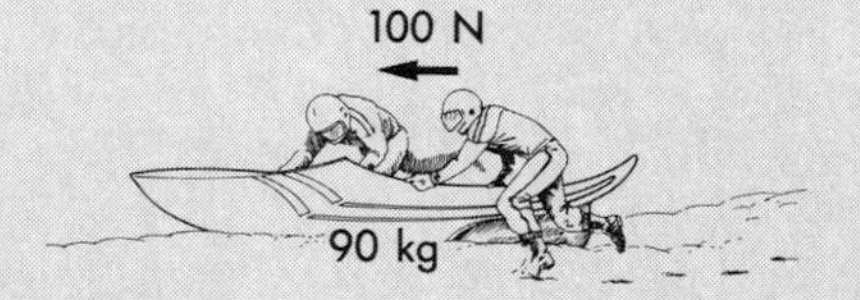

Solution

The crew members are applying an impulse to the toboggan to change the toboggan's momentum from 0 to a maximum amount. The impulse-momentum relationship may be used to solve the problem.

$$Ft = (mv)_2 - (mv)_1$$

$$(100\ N)\ (7\ s) = (90\ kg)\ (v) - (90\ kg)\ (0)$$

v = 7.78 m/s in the direction of force application

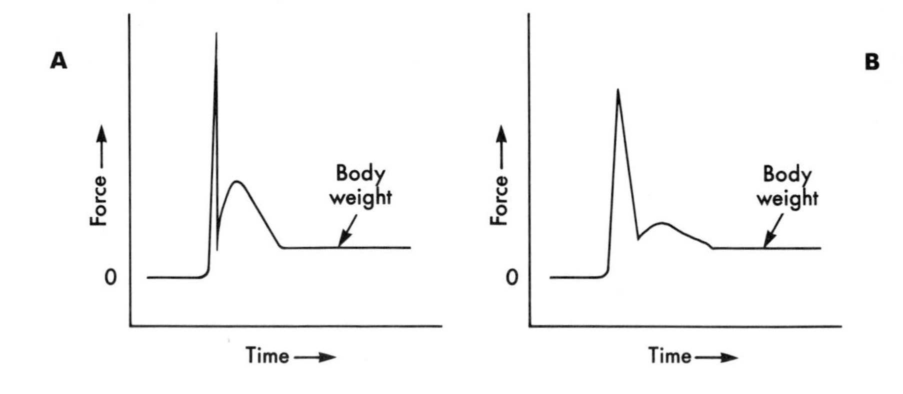

Figure 11-14
Representations of ground reaction forces during vertical jump performances. **A,** A rigid landing. **B,** A landing with hip, knee, and ankle flexion occurring. Note the differences in the magnitudes and times of the landing impulses. (Courtesy Gail Evans, San Jose, California.)

"Giving" with the ball during a catch serves to lessen the magnitude of the impact force sustained.

force sustained. Figure 11-14 illustrates this difference in landing style with force-time histories for vertical jumps performed on a force platform. Takeoff and landing impulses are equal to the areas under their respective portions of the force-time curve.

It is also useful to manipulate impulse when catching a hard-thrown ball. "Giving" with the ball after it initially contacts the hands or the glove before bringing the ball to a complete stop will prevent the force of the ball from causing the hands to sting. The greater the time period between initial hand contact with the ball and bringing the ball to a complete stop, the less the magnitude of the force exerted by the ball against the hand and the less the likelihood of experiencing a sting.

Impact

The type of collision that occurs between a struck baseball and a bat is known as an **impact.** An impact involves the collision of two bodies over an extremely small time interval during which the two bodies exert relatively large forces on each other. The behavior of two objects following an impact depends not only on their collective momentum but also on the nature of the impact.

impact
a collision characterized by the exchange of a large force during a small time interval

For the hypothetical case of a **perfectly elastic impact** the relative velocities of the two bodies after impact are the same as their relative velocities before impact. The impact of a superball with a hard sur-

perfectly elastic impact
an impact during which the velocity of the system is conserved

face approaches perfect elasticity because the ball's speed diminishes little during its collision with the surface. At the other end of the range is the **perfectly plastic impact,** during which at least one of the bodies in contact deforms and does not regain its original shape and the bodies do not separate. This occurs when modeling clay is dropped on a surface.

perfectly plastic impact
an impact resulting in the total loss of system velocity

Most impacts are neither perfectly elastic nor perfectly plastic but are somewhere between the two. The **coefficient of restitution** describes the relative elasticity of an impact. It is a unitless number between 0 and 1. The closer the coefficient of restitution is to 1, the more elastic the impact, and the closer the coefficient is to 0, the more plastic the impact.

coefficient of restitution
a number that serves as an index of the elasticity of a collision

The coefficient of restitution governs the relationship between the relative velocities of two bodies before and after an impact. This relationship, which was originally formulated by Newton, may be stated as follows:

> When two bodies undergo a direct collision, the difference in their velocities immediately after impact is proportional to the difference in their velocities immediately before impact.

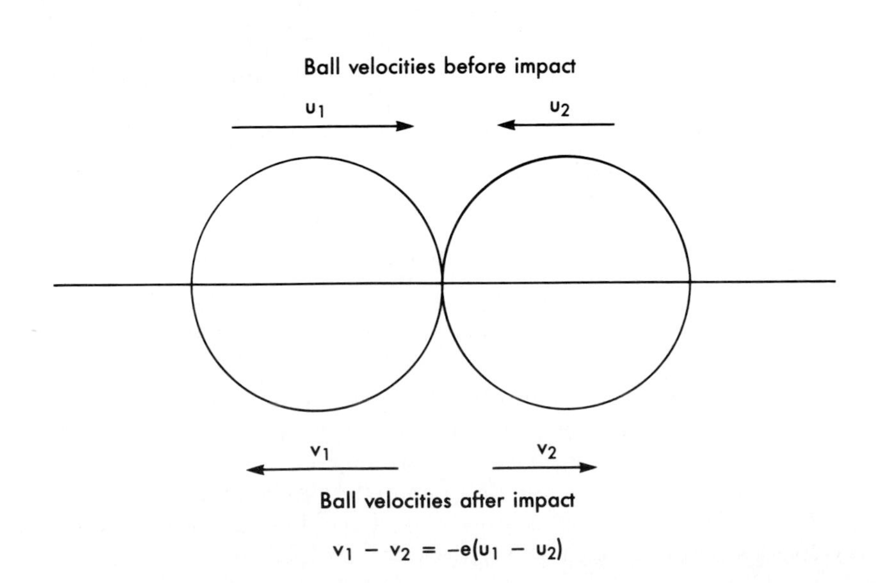

Figure 11-15
The differences in two ball's velocities before impact is proportional to the difference in their velocities after impact.

This relationship can also be expressed algebraically as the following:

$$-e = \frac{\text{relative velocity after impact}}{\text{relative velocity before impact}}$$

$$-e = \frac{v_1 - v_2}{u_1 - u_2}$$

In this formula, e is the coefficient of restitution, u_1 and u_2 are the velocities of the bodies just before impact, and v_1 and v_2 are the velocities of the bodies immediately after impact (Figure 11-15).

In tennis, the nature of the game depends on the type of impacts between ball and racket and between ball and court. Research has shown that racket size, shape, balance, flexibility, string type and tension, and swing kinematics all interact to influence the outcome of the ball-racket impact (7, 8). The condition of the ball is also significant. Tests have shown that rebound from a surface is higher after 800 impacts than when the balls are new because the loss of nap increases the coefficient of restitution between ball and surface and decreases the ball's aerodynamic drag. However, the major factor affecting ball rebound is the amount of time the ball is out of a pressurized can. A loss of rebound height for both used and unused balls occurs after 5 days out of a can (15). The surface of the court also influences ball rebound during play, with differences in the coefficients of restitution and friction between ball and surface making some courts "fast" and others "slow."

Ball/racket and ball/ court impacts determine the nature of the game in tennis.

In the case of an impact between a moving body and a stationary one, Newton's law of impact can be simplified because the velocity of the stationary body remains 0. The coefficient of restitution between a ball and a flat, stationary surface onto which the ball is dropped may be approximated using the following formula:

$$e = \sqrt{\frac{h_b}{h_d}}$$

In this equation, e is the coefficient of restitution, h_d is the height from which the ball is dropped, and h_b is the height to which the ball bounces (Figure 11-16). The coefficient of restitution describes the interaction between two bodies during an impact; it is *not* descriptive of any single object or surface. Dropping a golf ball, racquetball, basketball, and baseball onto several different surfaces demonstrates that some balls bounce higher on certain types of surfaces (Figure 11-17).

Figure 11-16

SAMPLE PROBLEM 5

A basketball is dropped from a height of 2 m onto a gymnasium floor. If the coefficient of restitution between ball and floor is 0.9, how high will the ball bounce?

Known:

$h_d = 2$ m

$e = 0.9$

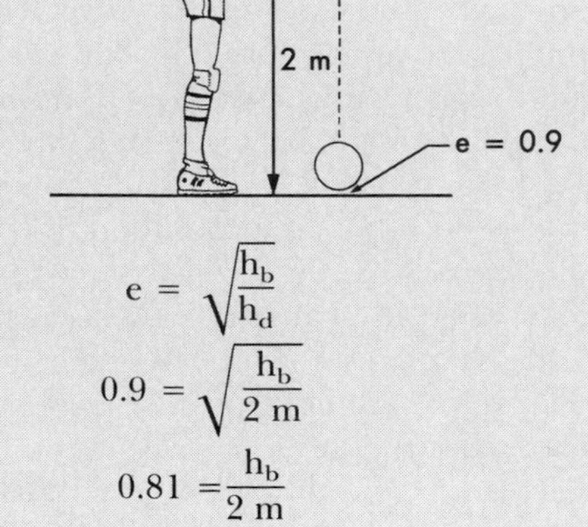

Solution

$$e = \sqrt{\frac{h_b}{h_d}}$$

$$0.9 = \sqrt{\frac{h_b}{2\text{ m}}}$$

$$0.81 = \frac{h_b}{2\text{ m}}$$

$h_b = 1.6$ m

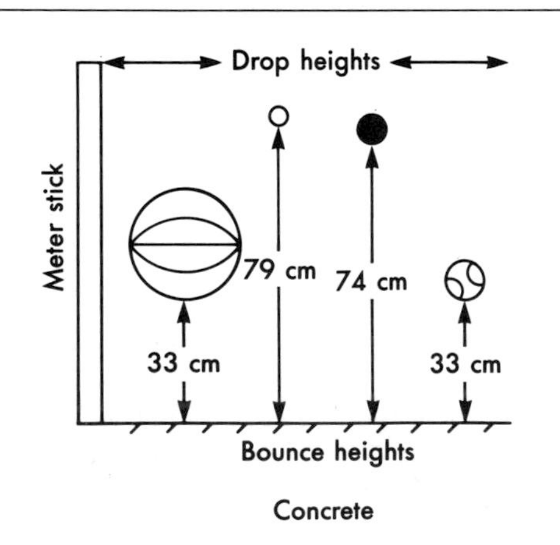

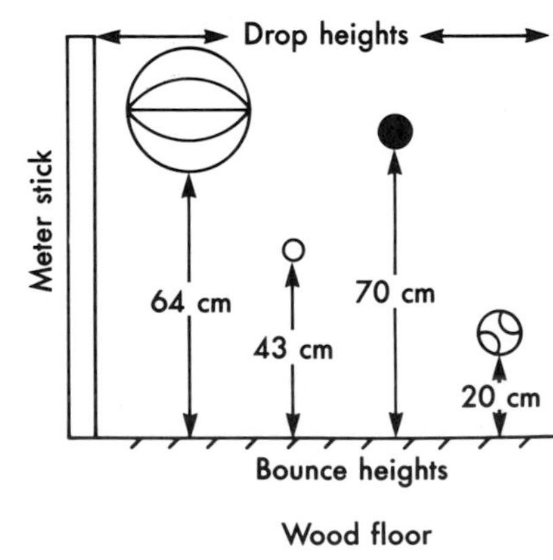

Figure 11-17
Maximum bounce heights of a basketball, golf ball, racquetball, and baseball all dropped onto the same surface from a height of 1 m.

WORK, POWER, AND ENERGY

Work

The word *work* is commonly used in a variety of contexts. A person can speak of "working out" in the weight room, doing "yard work," or "working hard" to prepare for an exam. However, the definition of mechanical **work** differs from the implications of the more common usages. From a mechanical standpoint, work is defined as force applied against a resistance, multiplied by the distance the resistance is moved:

work
the expression of mechanical energy that is calculated as force multiplied by the distance the resistance is moved

■ Mechanical work should not be confused with caloric expenditure.

$$W = Fd$$

When a body is moved a given distance as the result of the action of an applied force, the body has had work performed on it, with the quantity of work equal to the product of the magnitude of the applied force and the distance through which the body was moved. When a force is applied to a body but no net force results because of opposing forces such as friction or the body's own weight, no mechanical work has been done, since there was no movement of the body.

When the muscles of the human body produce tension that results in the motion of a body segment, the mechanical work performed may be characterized as either positive or negative work according to the type of muscle action involved. The production of muscle tension concentrically, such as when a weight lifter elevates a barbell from the floor, is *positive* work. In this case the muscles are performing work on the barbell. Eccentric action of the muscles, as is typically associated with lowering a barbell back to the floor, is *negative* work. When negative work is performed, the barbell (or other moving body) is performing work on the muscles. Performing positive mechanical work typically requires greater caloric expenditure than performing the same amount of negative mechanical work. However, no simple relationship between the caloric energy required for performing equal amounts of positive and negative mechanical work has been discovered. Aura and Komi (1) have reported that when subjects are monitored during performances of positive and negative mechanical work, energy expenditures vary considerably both among subjects and for individual subjects.

Positive mechanical work is performed during elevation of a barbell.

Units of work are units of force multiplied by units of distance. In the metric system, the common unit of force (N) multiplied by a common unit of distance (m) is termed the *joule* (J).

$$1 \text{ J} = 1 \text{ Nm}$$

Figure 11-18

SAMPLE PROBLEM 6

A 580 N person runs up a flight of 30 stairs of riser (height) of 25 cm during a 15 second period. How much mechanical work is done? How much mechanical power is generated?

Known

wt (F) = 580 N

h = 30 × 25 cm

t = 15 s

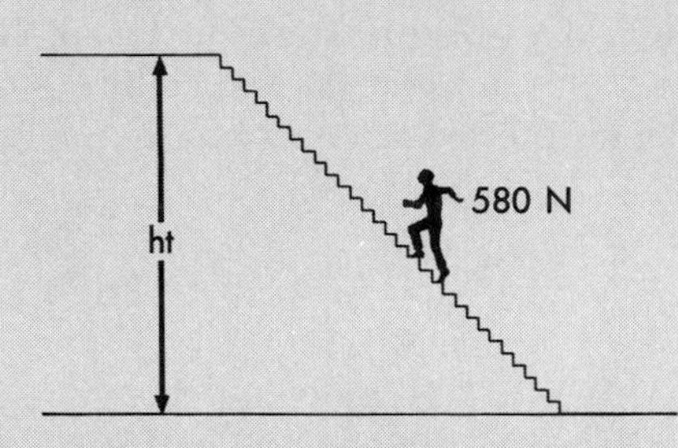

Solution

For mechanical work:

$$W = Fd$$

$$W = (580\ N)\ (30 \times 0.25\ m)$$

$$\boxed{W = 4350\ J}$$

For mechanical power:

$$P = \frac{W}{t}$$

$$P = \frac{4350\ J}{15\ s}$$

$$\boxed{P = 290\ W}$$

Power

power
the rate of work production that is calculated as work divided by the time during which the work was done

Another term used in different contexts is **power.** In mechanics, power refers to the amount of mechanical work performed in a given time:

$$\text{Power} = \frac{\text{Work}}{\text{Change in time}}$$

$$P = \frac{W}{\Delta t}$$

Using the relationships previously described, power can also be defined as the following:

$$\text{Power} = \frac{\text{Force} \times \text{distance}}{\text{Change in time}}$$

$$P = \frac{Fd}{\Delta t}$$

Because velocity equals the directed distance divided by the change in time, the equation can also be expressed as the following:

$$P = Fv$$

Units of power are units of work divided by units of time. In the metric system, joules divided by seconds are termed *watts* (W).

$$1 \text{ W} = 1 \text{ J/s}$$

In sports such as throwing, jumping, and sprinting events and Olympic weightlifting, the athlete's ability to exert mechanical power or the combination of force and velocity is critical to successful performance. A sample problem involving mechanical work and power is shown in Figure 11-18.

■ The ability to produce mechanical power is critically important for athletes competing in explosive track and field events.

Energy

Energy is defined generally as the capacity to do work. Mechanical energy is therefore the capacity to do mechanical work. Units of mechanical energy are the same as units of mechanical work (joules in the metric system). There are two forms of mechanical energy—**kinetic energy** and **potential energy.**

kinetic energy
energy of motion

potential energy
stored energy that is calculated as a body's weight multiplied by its height

Kinetic energy (KE) is the energy of motion. A body possesses kinetic energy only when in motion. Formally, the kinetic energy of linear motion is defined as one half of a body's mass multiplied by the square of its velocity:

$$KE = \tfrac{1}{2} mv^2$$

If a body is motionless (v equals 0), its kinetic energy is also 0. Because velocity is squared in the expression for kinetic energy, increases in a body's velocity create dramatic increases in its kinetic energy. For example, a 2 kg ball rolling with a velocity of 1 m/s has a kinetic energy of 1 J:

$$\begin{aligned} KE &= \tfrac{1}{2} mv^2 \\ &= (0.5)(2 \text{ kg})(1 \text{ m/s})^2 \\ &= (1 \text{ kg})(1 \text{ m}^2/\text{s}^2) \\ &= 1 \text{ J} \end{aligned}$$

If the velocity of the ball is increased to 3 m/s, kinetic energy is significantly increased:

$$
\begin{aligned}
KE &= \tfrac{1}{2}\,mv^2 \\
&= (0.5)\,(2\ kg)\,(3\ m/s)^2 \\
&= (1\ kg)\,(9\ m^2/s^2) \\
&= 9\ J
\end{aligned}
$$

The other major category of mechanical energy is potential energy (PE), which is the energy of position. More specifically, potential energy is a body's weight multiplied by its height above a reference surface:

The bent pole stores strain energy for subsequent release as kinetic energy and heat.

$$PE = (wt)(h)$$

$$PE = ma_gh$$

In the second formula, m represents mass, a_g is the acceleration of gravity, and h is the body's height. The reference surface is usually the floor or the ground, but in special circumstances it may be defined as another surface.

Because in biomechanical applications the weight of a body is typically fixed, changes in potential energy are usually based on changes in the body's height. For example, when a 50 kg bar is elevated to a height of 1 m, its potential energy at that point is 490.5 J:

$$
\begin{aligned}
PE &= ma_gh \\
&= (50\ kg)(9.81\ m/s^2)(1m) \\
&= 490.5\ J
\end{aligned}
$$

Potential energy may also be thought of as stored energy. The term *potential* implies potential to do mechanical work or potential for conversion to kinetic energy. A special form of potential energy is called **strain energy** (SE) or elastic energy. Strain energy may be defined as follows:

strain energy
a form of potential energy stored when a body is deformed

$$SE = \tfrac{1}{2}\,kx$$

In this formula, k is a spring constant, representing a material's relative stiffness or ability to store energy on deformation, and x is the distance over which the material is deformed. When an object is stretched, bent, or otherwise deformed, it stores this particular form of potential energy for later use. For example, when the end of a diving board or a trampoline surface is depressed, strain energy is created. Subsequent conversion of the stored strain energy to kinetic energy enables the surface to return to its original shape and position. The poles used by vaulters store strain energy as they bend and then release kinetic energy as they straighten during the performance of the vault. In 1963 the increase of approximately 23 cm in the world record for the pole vault was attributed largely to the advent of vaulting poles made of fiber glass, a material capable of storing more strain energy than the bamboo, steel, or aluminum of which earlier poles were constructed (9).

Conservation of Mechanical Energy

Consider the changes that occur in the mechanical energy of a ball tossed vertically into the air. As the ball gains height, it also gains potential energy (ma_gh). However, since the ball is losing velocity with increasing height because of gravitational acceleration, it is also losing kinetic energy ($\frac{1}{2}mv^2$). At the apex of the ball's trajectory (the instant between rising and falling), its height and potential energy are at a maximum value and its velocity and kinetic energy are 0. As the ball starts to fall, it progressively gains kinetic energy while losing potential energy.

The correlation between the kinetic and potential energies of the vertically tossed ball illustrate a concept that applies to all bodies when the only external force acting is gravity. The concept is known as the *law of conservation of mechanical energy,* which may be stated as follows:

When gravity is the only acting external force, a body's mechanical energy remains constant.

Since the mechanical energy a body possesses is the sum of its potential and kinetic energies, the relationship may also be expressed as the following:

$$(PE + KE) = C$$

In this formula, C is a constant; that is, it is a number that remains constant throughout the period of time during which gravity is the only external force acting. Figure 11-19 quantitatively illustrates this principle.

The Principle of Work and Energy

There is a special relationship between the quantities of mechanical work and mechanical energy. This relationship is described as the *principle of work and energy,* which may be stated as follows:

The work of a force is equal to the change in energy that it produces in the object acted on.

Algebraically, the principle may be represented as the following:

$$W = \Delta KE + \Delta PE + \Delta TE$$

In this formula, KE is kinetic energy, PE is potential energy, and TE is thermal energy (heat). The algebraic statement of the principle of work and energy indicates that the change in the sum of the forms

Figure 11-19

SAMPLE PROBLEM 7

A 2 kg ball is dropped from a height of 1.5 m. What is its velocity immediately before impact with the floor?

Known

m = 2 kg

h = 1.5 m

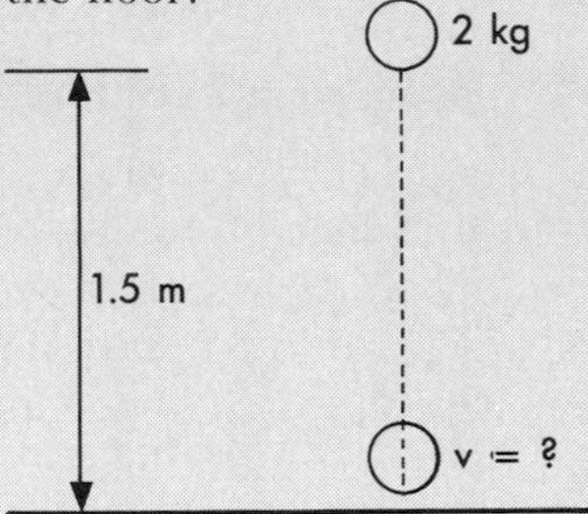

Solution

The principle of the conservation of mechanical energy may be used to solve the problem. The total energy possessed by the ball when it is held at a height of 1.5 m is its potential energy. Immediately before impact, the ball's height (and potential energy) may be assumed to be 0, and 100% of its energy at that point is kinetic.

Total (constant) mechanical energy possessed by the ball:

$$PE + KE = C$$

$$(wt)\,(h) + \tfrac{1}{2}\,mv^2 = C$$

$$(2\ kg)\,(9.81\ m/s^2)\,(1.5\ m) + 0 = C$$

$$29.43\ J = C$$

Velocity of the ball before impact:

$$PE + KE = 29.43\ J$$

$$(wt)\,(h) + \tfrac{1}{2}\,mv^2 = 29.43\ J$$

$$(2\ kg)\,(9.81\ m/s^2)\,(0) + \tfrac{1}{2}\,(2\ kg)\,v^2 = 29.43\ J$$

$$v^2 = 29.43\ J/kg$$

$$\boxed{v = 5.42\ m/s}$$

of energy produced by a force is quantitatively equal to the mechanical work done by that force. When a tennis ball is projected into the air by a ball-throwing machine, the mechanical work performed on the ball by the machine results in changes in both the potential and kinetic energies of the ball (Figure 11-20). In this situation, the change in the ball's thermal energy is negligible.

The work-energy relationship is also evident during movements of the human body. For example, the arches of the feet in runners act as a mechanical spring to store and subsequently return strain energy as they cyclically deform and then regain their resting

SAMPLE PROBLEM 8

Figure 11-20

How much mechanical work is required to catch a 1.3 kg ball traveling at a velocity of 40 m/s?

Known

m = 1.3 kg

v = 40 m/s

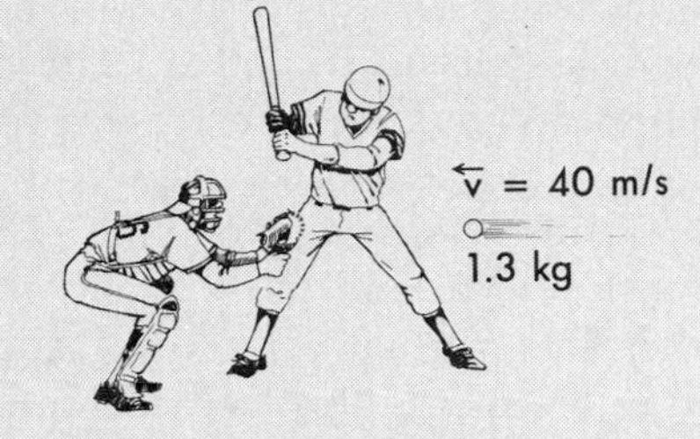

Solution

The principle of work and energy may be used to calculate the mechanical work required to change the ball's kinetic energy to 0. Assume that the potential energy and thermal energy of the ball do not change:

$$W = \Delta KE$$

$$W = (\tfrac{1}{2}mv^2)_2 - (\tfrac{1}{2}mv^2)_1$$

$$W = 0 - (\tfrac{1}{2})(1.3\ kg)(40\ m/s)^2$$

W = 26 J

shapes. For a 70 kg man running at 4.5 m/s, each arch stores approximately 17 J of energy at midstance. Combined with the estimated 35 J stored by each of the Achilles tendons, this equals a storage and partial return of approximately one half of the mechanical energy expended or one half of the mechanical work required of the muscles during the stance phase (2). The ability of the arches to function as a spring reduces the amount of mechanical work that would otherwise be required during running.

It is important not to confuse the production of mechanical energy or mechanical work by the muscles of the human body with the consumption of chemical energy or caloric expenditure. Factors such as concentric versus eccentric muscular contractions, the transfer of energy between body segments, elastic storage and reuse of energy, and limitations in joint ranges of motion complicate direct quantitative calculation of the relationship between mechanical and physiological energy estimates (19). Approximately 25% of the energy consumed by the muscles is converted into work, with the remainder changed to heat or used in the body's chemical processes.

SUMMARY

Linear kinetics is the study of the forces associated with linear motion. The interrelationships among many basic kinetic quantities are identified in the physical laws formulated by Sir Isaac Newton.

Friction is a force generated at the interface of two surfaces in contact when there is motion or a tendency for motion of one surface with respect to the other. The magnitudes of maximum static friction and kinetic friction are determined by the coefficient of friction between the two surfaces and by the normal reaction force pressing the two surfaces together. The direction of friction force always opposes the direction of motion or motion tendency.

Other factors that affect the behavior of two bodies in contact when a collision is involved are momentum and elasticity. Linear momentum is the product of an object's mass and its velocity. The total momentum present in a given system remains constant barring the action of external forces. Changes in momentum result from impulses—external forces acting over a time interval. The elasticity of an impact governs the amount of velocity present in the system following the impact. The relative elasticity of two impacting bodies is represented by the coefficient of restitution.

Mechanical energy has two major forms: kinetic and potential. When gravity is the only acting external force, the sum of the kinetic and potential energies possessed by a given body remains constant. Changes in a body's energy are equal to the mechanical work done by an external force.

INTRODUCTORY PROBLEMS

1. How much force must be applied by a kicker to give a stationary 2.5 kg ball an acceleration of 40 m/s^2? (Answer: 100 N)
2. A high jumper with a body weight of 712 N exerts a force of 3 kN against the ground during takeoff. How much force is exerted by the ground on the high jumper? (Answer: 3 kN)
3. What factors affect the magnitude of friction?
4. If μ_s between a basketball shoe and a court is 0.56 and the normal reaction force acting on the shoe is 350 N, how much horizontal force is required to cause the shoe to slide? (Answer: >196N)
5. A football player pushes a 670 N tackling sled. The coefficient of static friction between sled and grass is 0.73 and the coefficient of dynamic friction between sled and grass is 0.68.
   a. How much force must the player exert to start the sled in motion?
   b. How much force is required to keep the sled in motion?
   c. Answer the same two questions with a 100 kg coach standing on the back of the sled.

   (Answer: a. >489.1 N; b. 455.6 N; c. >1205.2 N, 1122.7 N)

6. Lineman A has a mass of 100 kg and is traveling with a velocity of 4 m/s when he collides head-on with Lineman B, who has a mass of 90 kg and is traveling at 4.5 m/s. If both players remain on their feet, what will happen? (Answer: Lineman B will push Lineman A backward with a velocity of 0.03 m/s)
7. Two skaters gliding on ice run into each other head-on. If the two skaters hold onto each other and continue to move as a unit after the collision, what will be their resultant velocity? Skater A has a velocity of 5 m/s and a mass of 65 kg. Skater B has a velocity of 6 m/s and a mass of 60 kg. (Answer: v = 0.28 m/s in the direction originally possessed by Skater B)
8. A ball dropped on a surface from a 2 m height bounces to a height of 0.98 m. What is the coefficient of restitution between ball and surface? (Answer: 0.7)
9. A set of 20 stairs, each of 20 cm height, is ascended by a 700 N man in a period of 1.25 seconds. Calculate the mechanical work, power, and change in potential energy during the ascension. (Answer: W = 2800 J, P = 2240 W, PE = 2800 J)
10. A pitched ball with a mass of 1 kg reaches a catcher's glove traveling at a velocity of 28 m/s.
    a. How much momentum does the ball have?
    b. How much impulse is required to stop the ball?
    c. If the ball is in contact with the catcher's glove for 0.5 seconds during the catch, how much average force is applied by the glove?

    (Answer: a. 28 kg m/s; b. 28 N s; c. 56 N)

ADDITIONAL PROBLEMS

1. Select one sport or daily activity and identify the ways in which the amount of friction present between surfaces in contact affects performance outcome.
2. A 2 kg block sitting on a horizontal surface is subjected to a horizontal force of 7.5 N. If the resulting acceleration of the block is 3 m/s^2, what is the magnitude of the friction force opposing the motion of the block? (Answer: 1.5 N)
3. A worker must push a 400 N crate across a floor with μ_s = 0.35 and μ_k = 0.30. The worker pushes at an angle that is 30 degrees below the horizontal. How much force must be exerted to initiate motion of the crate? To maintain motion of the crate? (Answer: $>$161.7 N; $>$138.6 N)
4. What is the magnitude of the normal reaction force for a 200 N sled sitting on a 6 degree incline? (Answer: 198.9 N)
5. Explain in what ways mechanical work is and is not related to caloric expenditure. Include in your answer the distinction between positive and negative work and the influence of anthropometric factors.

6. A 108 cm, 0.73 kg golf club is swung for 0.5 seconds with a constant acceleration of 10 rad/s^2. What is the linear momentum of the club head when it impacts the ball? (Answer: 15.8 kg·m/s)
7. A 6.5 N ball is thrown with an initial velocity of 20 m/s at a 35 degree angle from a height of 1.5 m.
   a. What is the velocity of the ball if it is caught at a height of 1.5 m?
   b. If the ball is caught at a height of 1.5 m, how much mechanical work is required?

   (Answer: a. 20 m/s; b. 132.5 J)
8. A 50 kg person performs a maximum vertical jump with an initial velocity of 2 m/s.
   a. What is the performer's maximum kinetic energy during the jump?
   b. What is the performer's maximum potential energy during the jump?
   c. What is the performer's minimum kinetic energy during the jump?
   d. How much is the performer's center of mass elevated during the jump?

   (Answer: a. 100 J; b. 100 J; c. 0; d. 20 cm)
9. Using the principle of conservation of mechanical energy, calculate the maximum height achieved by a 7 N ball tossed vertically upward with an initial velocity of 10 m/s. (Answer: 5.1 m)
10. Select one of the following sport activities and speculate about the changes that take place between kinetic and potential forms of mechanical energy.
    a. A single leg support during running
    b. A tennis serve
    c. A pole vault performance
    d. A springboard dive

REFERENCES

1. Aura O and Komi PV: Mechanical efficiency of pure positive and pure negative work with special reference to the work intensity, Int J Sports Med, 7:44, 1986.
2. Bennett MS et al: Elastic properties of the human foot and their significance for running. In Bennett MS et al: Biomechanics in sport, London, 1988, Mechanical Engineering Publications, Ltd.
3. Burke J: The day the universe changed, Boston, 1985, Little, Brown & Co, Inc.
4. Cavanagh PR and Lafortune MA: Ground reaction forces in distance running, J Biomech 13:397, 1980.
5. Cohen BI: Newton's discovery of gravity, Sci Am 244:166, 1981.
6. Dapena J: Biomechanics of elite high jumpers. In Terauds J et al, eds: Sports biomechanics, Del Mar, Calif, 1984, Academic Publishers.

7. Groppel JL et al: Effects of different string tension patterns and racket motion on tennis racket-ball impact, Int J Sport Biomech 3:142, 1987.
8. Groppel JL et al: The effects of string type and tension on impact in midsized and oversized tennis racquets, Int J Sport Biomech 3:40, 1987.
9. Jerome J: Pole vaulting: biomechanics at the bar. In Schrier EW and Allman WF, eds: Newton at the bat, New York, 1984, Charles Scribner's Sons.
10. Kiefer J: Bowling: the great oil debate. In Schrier EW and Allman WF, eds: Newton at the bat, New York, 1984, Macmillan-Charles Scribner's Sons, 1984.
11. Laws K: The physics of dance, New York, 1984, Schirmer Books.
12. Nigg BM et al: Load sport shoes and playing surfaces. In Frederick EC, ed: Sport shoes and playing surfaces, Champaign, Ill, 1984, Human Kinetics Publishers, Inc.
13. Pons DJ and Vaughan CL: Mechanics of cycling. In Vaughan CL, ed: Biomechanics of sport, Boca Raton, Fla, 1989, CRC Press, Inc.
14. Putnam CA and Kozey JW: Substantive issues in running. In Vaughan CL, ed: Biomechanics of sport, Boca Raton, Fla, 1989, CRC Press, Inc.
15. Rand KT, Hyer MW, and Williams MH: A dynamic test for comparison of rebound characteristics of three brands of tennis balls. In Groppel JL, ed: Proceedings of the national symposium on racquet sports, Champaign, Ill, 1979.
16. Rapoport R: Artificial turf: is the grass greener? In Schrier EW and Allman WF, eds: Newton at the bat, New York, 1984, MacMillan-Charles Scribner's Sons.
17. van Ingen Schenau GJ, DeBoer RW, and DeGroot G: Biomechanics of speed skating. In Vaughan CL, ed: Biomechanics of sport, Boca Raton, Fla, 1989, CRC Press, Inc.
18. Whitt FR and Wilson DG: Bicycling science: ergonomics and mechanics, Cambridge, Mass, 1974, The MIT Press.
19. Williams KR: The relationship between mechanical and physiological energy estimates, Med Sci Sports Exerc 17:317, 1985.

ANNOTATED READINGS

Barham JN: Introduction to kinetics. In Barham JN: Mechanical kinesiology, St Louis, 1978, CV Mosby Co.

Introduces basic kinetic concepts from a mechanical perspective.

McMahon TA and Greene PR: Fast running tracks, Sci Am 239:148, 1978.

Includes a fascinating discussion of the thought processes behind the degree of elasticity engineered into Harvard's indoor track. Using this track, runners can better their best times by nearly 3%.

Townend MS: Mathematics in sport, New York, 1984, John Wiley & Sons, Inc.

The chapter on jumping provides in-depth analyses of changes in quantities such as momentum, mechanical work, and mechanical energy during performance of the straddle and Fosbury flop high jump techniques, the pole vault, the long jump, and the triple jump.

Wiktorin CV and Nordin M: Introduction to problem solving in biomechanics, Philadelphia, 1986, Lea & Febiger.

Provides an introduction to kinetic concepts in the context of clinical applications for physical therapists.

12 MOVEMENT

Equilibrium

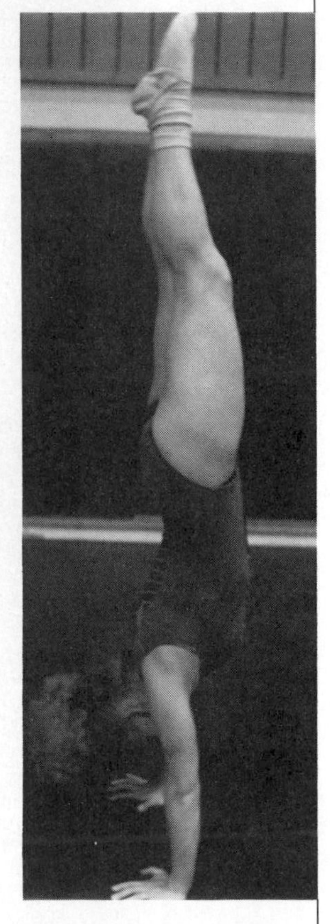

Many athletic skills require mechanical stability.

After reading this chapter, the student will be able to:

Define torque, quantify resultant torques, and identify the factors that affect resultant joint torques.

Identify the mechanical advantages associated with the different classes of levers and explain the ways in which muscles and bones function as levers.

Solve basic quantitative problems using the equations of static equilibrium.

Define center of gravity and explain the significance of center of gravity location in the human body.

Explain the ways in which mechanical factors affect a body's stability.

Why do long jumpers and high jumpers lower their centers of gravity before takeoff? What mechanical factors enable a wheelchair to remain stationary on a graded ramp or Sumo wrestlers to resist the attack of their opponents? A body's mechanical stability is based on its resistance to both linear and angular motion. In this chapter the kinetics of angular motion will be introduced along with the factors that determine overall mechanical stability.

STATIC AND DYNAMIC EQUILIBRIUM

Torque

When an eccentric (off-center) force is applied to a freely moving object, such as a pencil on a desk top, or to an object with a fixed axis of rotation, such as an airplane propeller, the factors that determine the amount of angular motion resulting involve the magnitude of the applied force and the distance from the force's line of action to the object's center of rotation.

■ Application of eccentric force results in rotation of a body with a fixed axis of rotation and in general motion of an unrestrained body.

Consider the swinging door shown in Figure 12-1. To open the door, you must apply a force of a particular magnitude at a certain distance from the door's axis of rotation, which is its hinges. Does it require less force to open the door if force is applied in the center of the door, close to the hinges, or close to the outer end of the door? If the answer is not clear, you can experiment by applying a one-finger force at various distances from a door's hinges.

■ It is easiest to initiate rotation when force is applied perpendicularly and as far away as possible from the axis of rotation.

The rotary effect created by an applied force is known as **torque** or *moment of force.* Torque, which may be thought of as *rotary force,* is the angular equivalent of linear force. Algebraically, torque is the product of force and the perpendicular distance from the force's line of action to the axis of rotation:

torque
the rotary force that produces angular acceleration

$$T = Fd_{\perp}$$

Thus, an increase in the magnitude of the applied force or in the perpendicular distance of the force's line of action to the axis of ro-

Figure 12-1
Which position of force application is best for opening the swinging door? Experience should verify that position **C** is best.

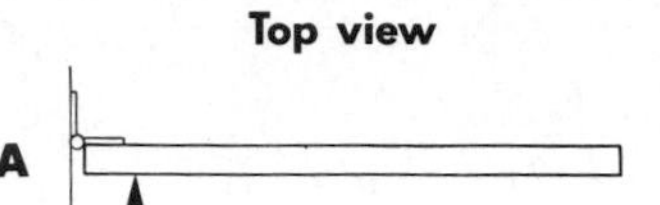

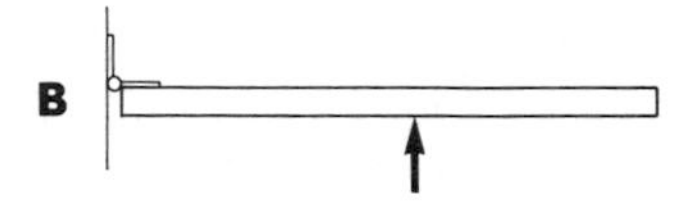

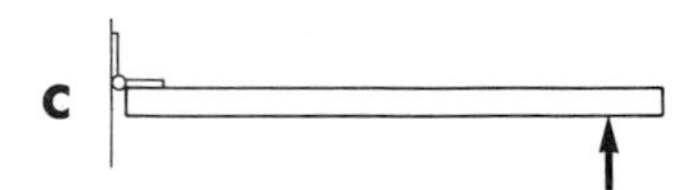

Figure 12-2
The moment arms of the forces are the perpendicular distances from the forces' lines of action to the door hinge, which is the door's axis of rotation.

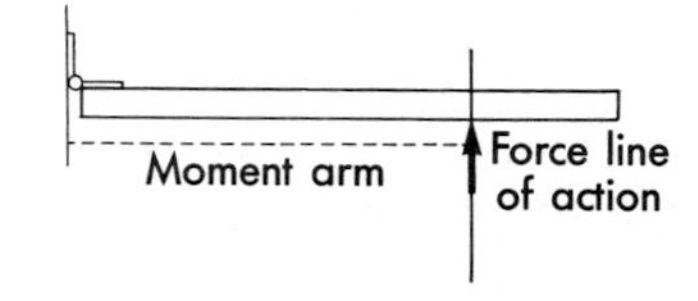

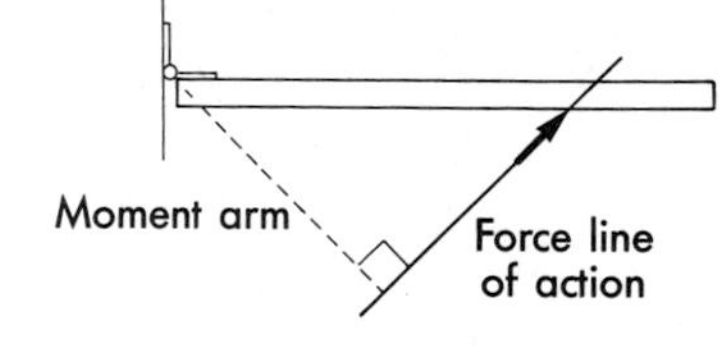

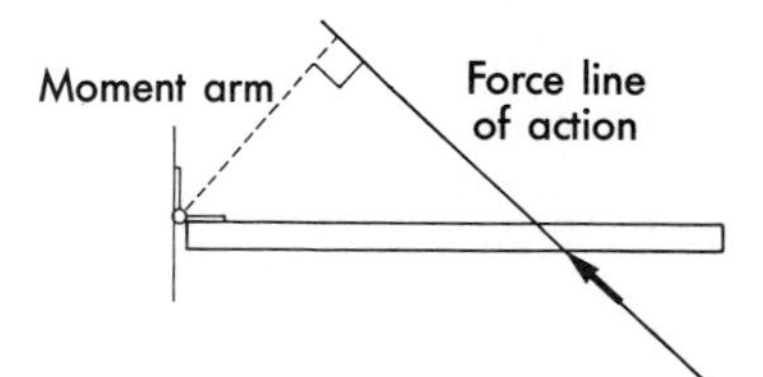

tation results in an increase in the acting torque. The greater the amount of torque acting at the axis of rotation, the greater the tendency for rotation to occur and the greater the angular acceleration of a given body. A force directed through a body's axis of rotation produces no torque because $d_{\perp}$ equals 0.

If a force is applied to a swinging door at other than a 90 degree angle, the perpendicular distance from its line of action to the axis of rotation is changed (Figure 12-2). The perpendicular distance between the force's line of action and the axis of rotation is known as the **moment arm.** The moment arm is also the *shortest* distance between the force's line of action and the axis of rotation. Any change in the orientation of a force's line of action with respect to the axis of rotation results in a change in the moment arm and a subsequent change in the amount of torque generated by the force at the axis of rotation. The torque generated by a force of a given magnitude is always proportional to the length of the moment arm.

moment arm
the shortest (perpendicular) distance between a force's line of action and an axis of rotation

An example of the significance of moment arm length is provided by a dancer's choice of foot placement when preparing to execute a total body rotation around the vertical axis. When a dancer initiates a turn, the torque producing the turn is provided by equal and oppositely directed forces exerted by the feet against the floor. Each such pair of forces is known as a force **couple.** Because the oppositely directed forces in a couple are positioned on opposite sides of the axis of rotation, they produce torque in the same direction. The torque generated by a couple is therefore the sum of the products of each force and its moment arm. Turning from fifth position, with a small distance between the feet, requires greater force production by a dancer than turning at the same rate from fourth position in which the moment arms of the forces in the couple are longer (Figure 12-3). Significantly more force is required when the torque is generated

couple
a pair of equal, oppositely directed forces that act on opposite sides of an axis of rotation to produce torque

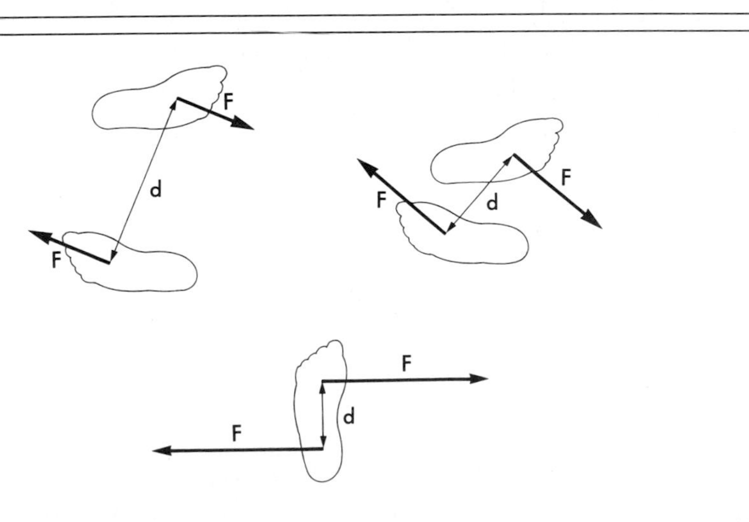

Figure 12-3
The wider a dancer's stance, the greater the moment arm for the force couple generated by the feet when a turn is executed. When rotation is initiated from a single foot stance, the moment arm becomes the distance between the support points of the foot.

by a single support foot for which the moment arm is reduced to the distance between the metatarsals and the calcaneus (16).

Torque is a vector quantity and is therefore characterized by both magnitude and direction. The magnitude of the torque created by a given force is equal to $Fd_{\perp}$, and the direction of a torque may be described as clockwise or counterclockwise. As discussed in Chapter 10, the counterclockwise direction is conventionally referred to as the positive (+) direction, and the clockwise direction is regarded as neg-

Figure 12-4

SAMPLE PROBLEM 1

Two children sit on opposite sides of a playground seesaw. If Joey, weighing 200 N, is 1.5 m from the seesaw's axis of rotation and Susie, weighing 190 N is 1.6 m from the axis of rotation, which end of the seesaw will drop?

Known

Joey: $wt(F_J) = 200\ N$

$d_{\perp J} = 1.5\ m$

Susie: $wt(F_S) = 190\ N$

$d_{\perp S} = 1.6\ m$

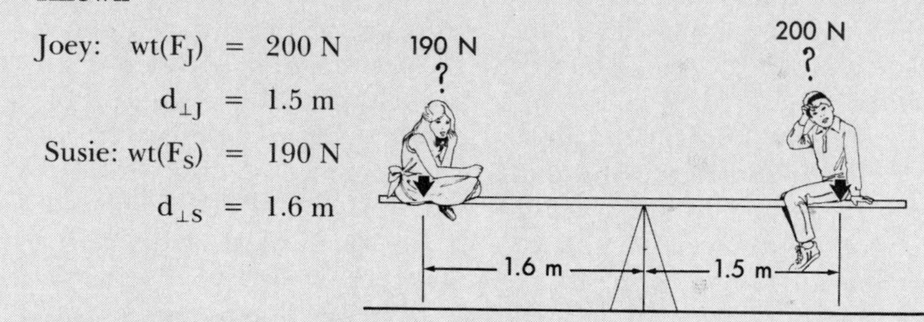

Solution

The seesaw will rotate in the direction of the resultant torque at its axis of rotation. To find the resultant torque, the torques created by both children are summed according to the rules of vector composition. The torque produced by Susie's body weight is in a counterclockwise (positive) direction, and the torque produced by Joey's body weight is in a clockwise (negative) direction.

$$T_a = (F_S)(d_{\perp S}) - (F_J)(d_{\perp J})$$

$$T_a = (190\ N)(1.6\ m) - (200\ N)(1.5\ m)$$

$$= 304\ N\text{-}m - 300\ N\text{-}m$$

$$= 4\ N\text{-}m$$

The resultant torque is in a positive direction, and Susie's end of the seesaw will fall.

ative (−). Two or more torques acting at a given axis of rotation can be added using the rules of vector composition (Figure 12-4).

In the sport of rowing a small lateral oscillation of the stern of the boat results with the traditionally positioned crew comprising four members (21). This oscillation results because the rowers on one side of the boat are positioned farther from the stern than their counterparts on the other side, thus causing a net torque about the stern during rowing. The Italian rig eliminates this lateral oscillation by positioning rowers so that no net torque is produced, assuming that the force produced by each rower with each stroke is nearly the same(Figure 12-5). Italian and German rowers have also developed alternative positionings for the traditional eight-member crew (Figure 12-6).

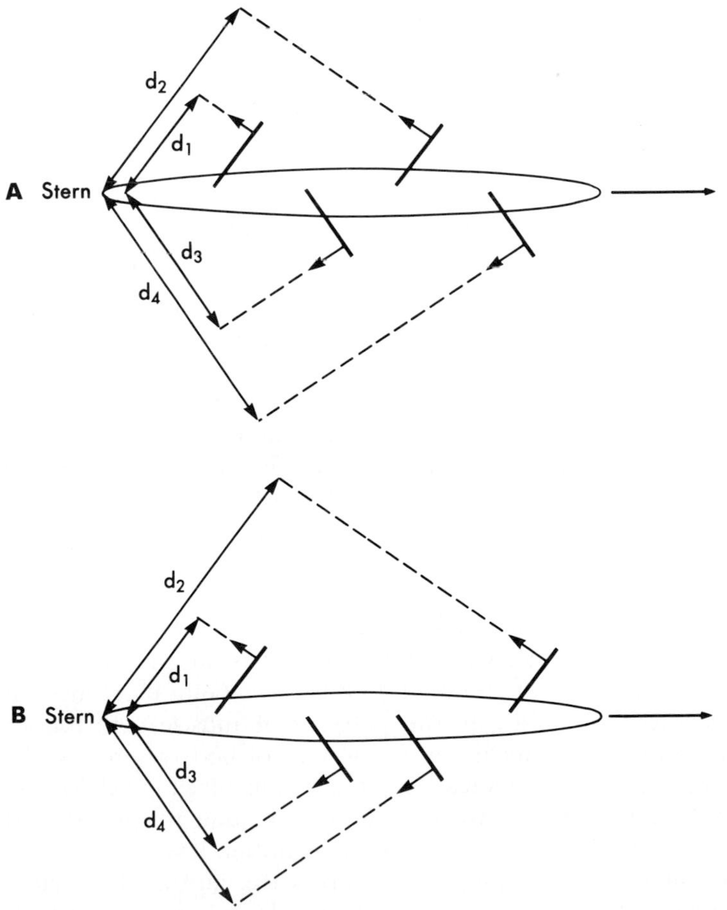

Figure 12-5

A, This crew arrangement creates a net torque about the stern of the boat because the sum of the top side oar moment arms ($d_1 + d_2$) is less than the sum of the bottom side moment arms ($d_3 + d_4$). **B,** This arrangement eliminates the problem, assuming that all rowers stroke simultaneously and produce equal force because ($d_1 + d_2$) = ($d_3 + d_4$).

Figure 12-6
The Italians and Germans have used alternative positionings for eight-member crews. The torques produced by the oar forces with respect to the stern are balanced in arrangements **B** and **C,** but not in the traditional arrangement shown in **A.**

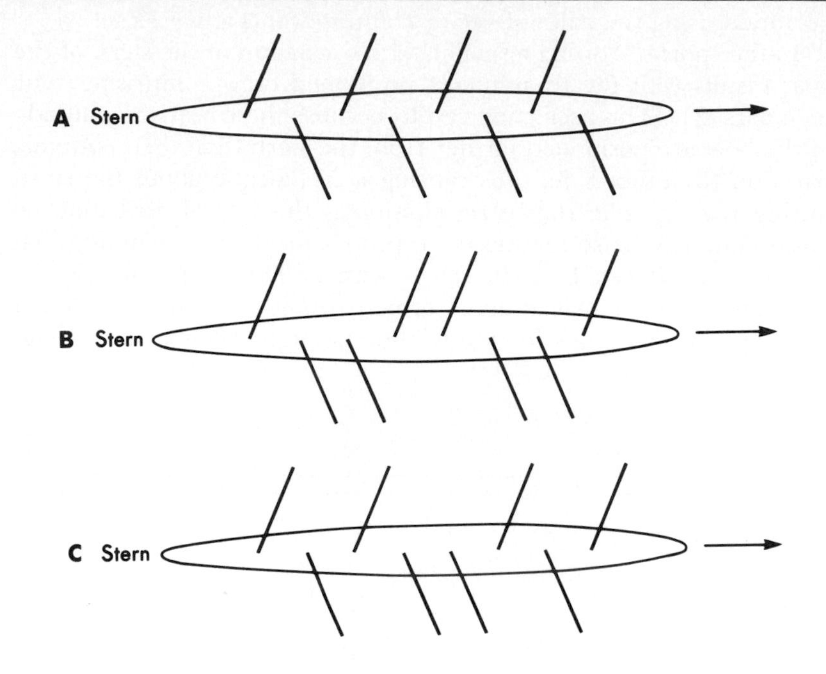

Resultant Joint Torques

The concept of torque is important in the study of human movement because torque produces movement of the body segments. When a muscle crossing a joint develops tension, it produces a force pulling on the bone to which it attaches, thereby creating torque at the joint the muscle crosses. Much human movement involves tension development in the agonist and antagonist muscle groups. The tension in the antagonists controls the velocity of the movement and enhances the stability of the joint at which the movement is occurring. Since antagonist tension development creates torque in the direction opposite that of the torque produced by the agonist, the resulting movement at the joint is a function of the net torque.

Because directly measuring the forces produced by muscles during the execution of most movement skills is not practical, measurements or estimates of resultant joint torques (joint moments), are often studied to investigate the patterns of muscle contributions. A number of factors, including the weight of body segments, the motion of the body segments, and the action of external forces, may contribute to net joint torques. However, joint torques do provide rough estimates of muscle group contribution levels (2).

Resultant joint torques at the hip, knee, and ankle during running have consequently been evaluated by many investigators (19).

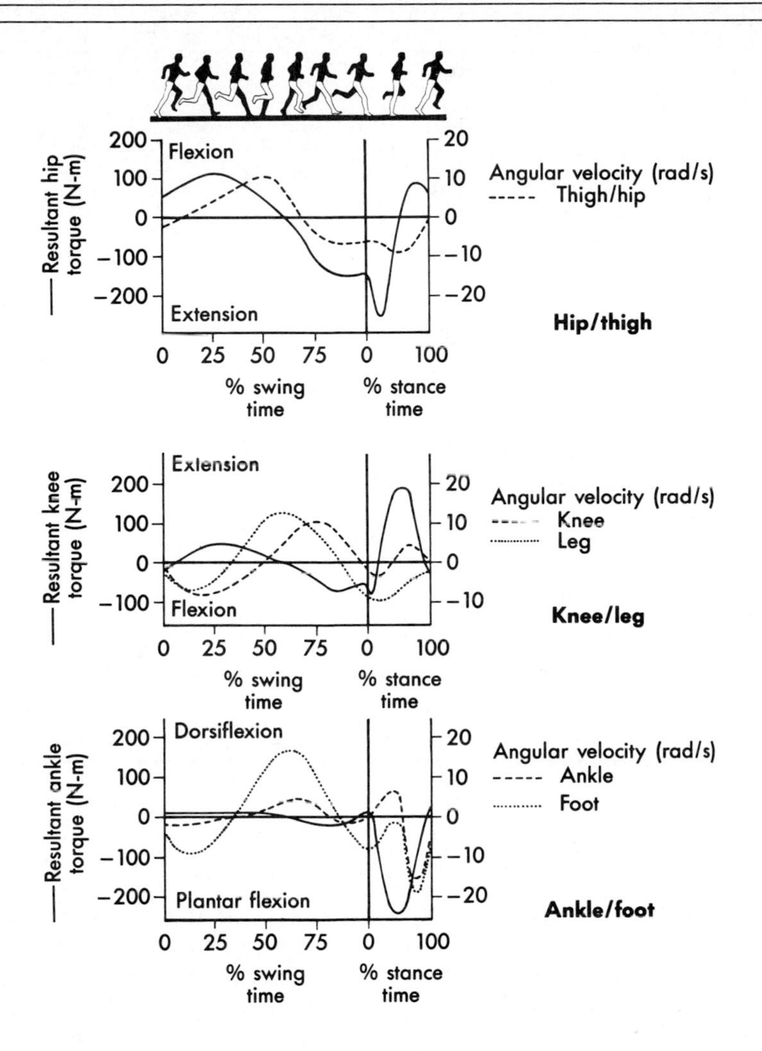

Figure 12-7
Representative resultant joint torque and segment and joint angular velocity curves for hip and thigh, knee and leg, and ankle and foot. (Modified from Putnam CA and Kozey JW: Substantive issues in running. In Vaugn CL, ed: Biomechanics of sport, Boca Raton, Fla, 1989, CRC Press, Inc.)

Figure 12-7 displays representative resultant joint torques for the hip, knee, and ankle, along with joint and segmental velocities during a running stride. During certain phases of the stride, the net torque at each joint is in the same direction as the angular velocity at the joint. During other phases the reverse is true. When net torque and joint movement occur in the same direction, the torque is termed *concentric,* and torque in the direction opposite joint motion is considered to be *eccentric.* Although these terms are generally useful descriptors in analysis of muscular function, their application to the muscles of the lower extremity is complicated by the presence of multi-joint muscles. (19).

Figure 12-8
Absolute average joint torques for the hip, knee, and ankle versus pedaling rate during cycling. (Modified from Redfield R and Hull ML: On the relation between joint moments and pedalling at constant power in bicycling, J Biomech 19:317, 1986.)

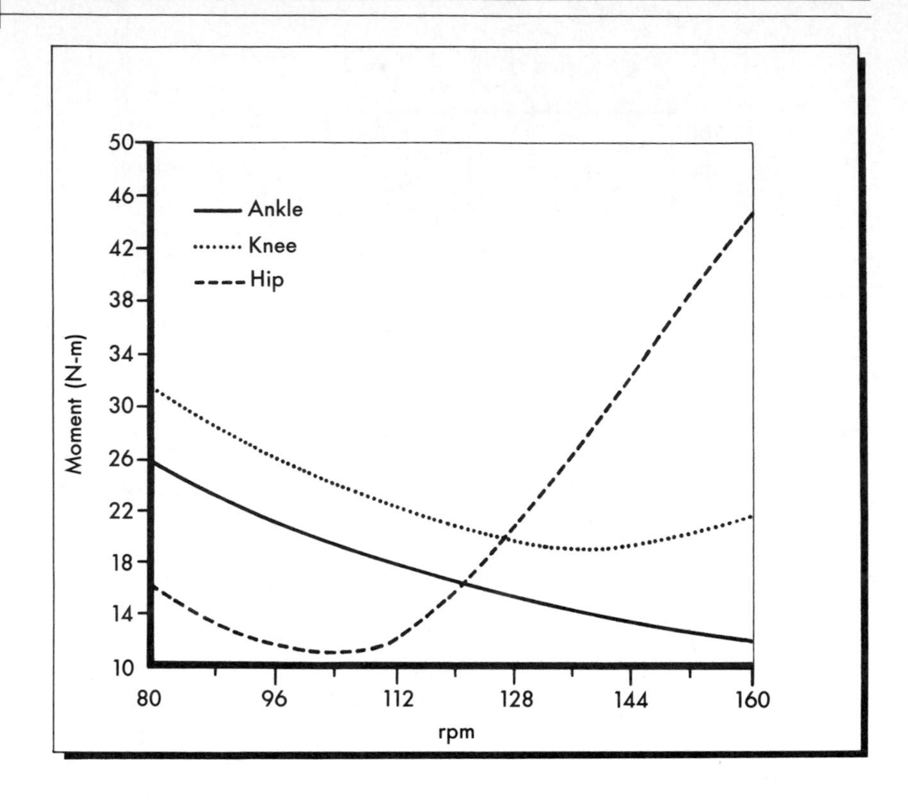

The torques required at the hip, knee, and ankle during cycling at a given power are influenced by different factors.

Lower extremity joint torques during cycling at a given power are affected by pedaling rate, seat height, length of the pedal crank arm, and the distance from the pedal spindle to the ankle joint. Average hip and knee torques during cycling under cruising conditions have been reported to be minimum at approximately 105 rotations per minute (20). Figure 12-8 shows the changes in average resultant torque at the hip, knee, and ankle joints with changes in pedaling rate at a constant power.

It is widely assumed that the muscular force requirements of resistance exercise increase as the amount of resistance increases. However, resultant torques at the hip, knee, and ankle during performance of the squat are proportional to the load lifted only if the kinematics of the exercise remain constant (11). Resultant joint torques during squat exercises have also been shown to increase with movement speed (3, 12). However, movement speed increases during weight training should be undertaken with caution because increased speed increases not only the muscie tension required but also the likelihood of incorrect technique and subsequent injury (7).

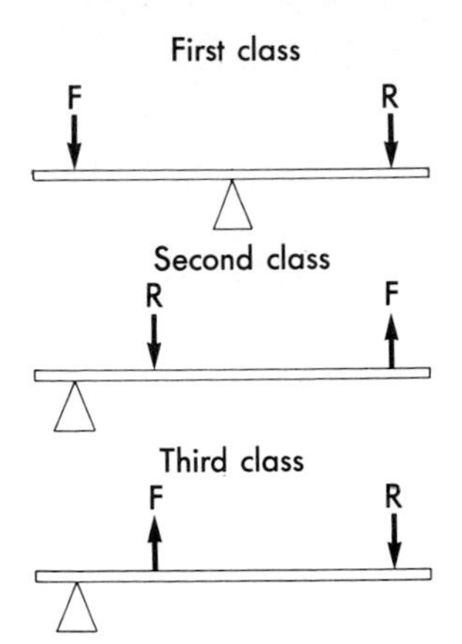

Figure 12-9
Relative locations of the applied force, the resistance, and the fulcrum or axis of rotation determine lever classifications.

Levers

When muscles develop tension, pulling on bones to support or move the resistance created by the weight of the body segment or segments and possibly the weight of an added load, the muscle and bone are functioning mechanically as a **lever.** A lever is a rigid bar that rotates about an axis or **fulcrum.** Force or forces applied to the lever move a resistance. In the human body, the bone acts as the rigid bar, the joint is the axis or fulcrum, and the muscles apply force. The three relative arrangements of the applied force, resistance, and axis of rotation for a lever are shown in Figure 12-9.

With a **first class lever** the applied force and resistance are located on opposite sides of the axis. The playground seesaw is an example of a first class lever, as are a number of commonly used tools, including scissors, pliers, and crowbars. Within the human body the simultaneous action of agonist and antagonist muscle groups on opposite sides of a joint axis is analogous to the functioning of a first class lever, with the agonists providing the applied force and the antagonists supplying a resistance force. With a first class lever, the applied force and resistance may be at equal distances from the axis, or one may be farther away from the axis than the other.

In a **second class lever** the applied force and resistance are on the same side of the axis, with the resistance closer to the axis. A wheelbarrow, lug nut wrench, and nutcracker are examples of second class levers, although there are no completely analogous examples in the human body.

With a **third class lever** the force and resistance are on the same side of the axis, but the applied force is closer to the axis. A canoe

lever
a simple machine consisting of a relatively rigid barlike body that may be made to rotate about an axis

fulcrum
the point of support or axis about which a lever may be made to rotate

first class lever
a lever positioned with the applied force and the resistance on opposite sides of the axis of rotation

second class lever
a lever positioned with the fulcrum located between the applied force and the resistance

third class lever
a lever positioned with the applied force between the fulcrum and the resistance

Figure 12-10

A, First class levers. **B,** Second class levers. **C,** Third class levers. Note that the paddle and shovel function as third class levers only when the top hand does not apply force but serves as a fixed axis of rotation.

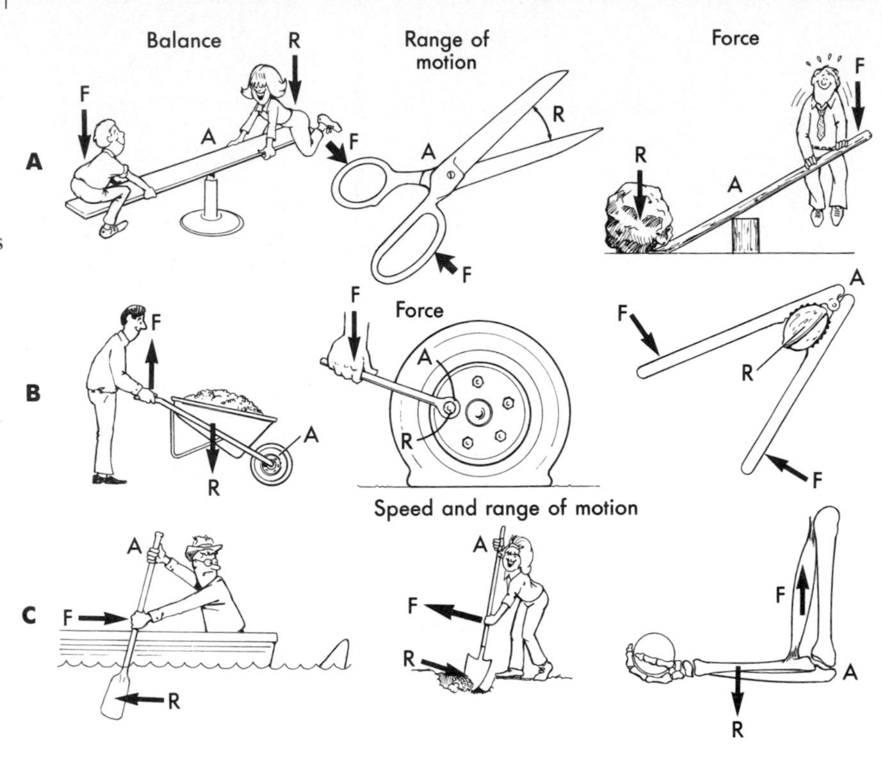

paddle and shovel can serve as third class levers (Figure 12-10). Most muscle-bone lever systems of the human body are also of the third class, with the muscle supplying the applied force and attaching to the bone at a short distance from the joint center compared to the distance at which the resistance supplied by the weight of the body segment or that of a more distal body segment acts.

The moment arm of an applied force can also be referred to as the *force arm,* and the moment arm of a resistance can be referred to as the *resistance arm.*

A lever system can serve one of two purposes (Figure 12-11). Whenever the moment arm of the applied force is greater than the moment arm of the resistance, the magnitude of the applied force needed to move a given resistance is less than the magnitude of the resistance. Whenever the resistance arm is longer than the force arm, the resistance may be moved through a relatively large distance. The mechanical effectiveness of a lever for moving a resistance may be expressed quantitatively as its **mechanical advantage,** which is the ratio of the moment arm of the force to the moment arm of the resistance:

mechanical advantage
the ratio of force arm/resistance arm for a given lever

$$\text{Mechanical advantage} = \frac{\text{Moment arm (force)}}{\text{Moment arm (resistance)}}$$

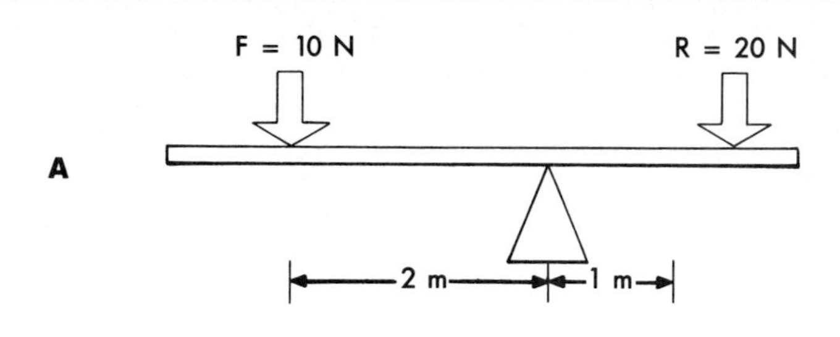

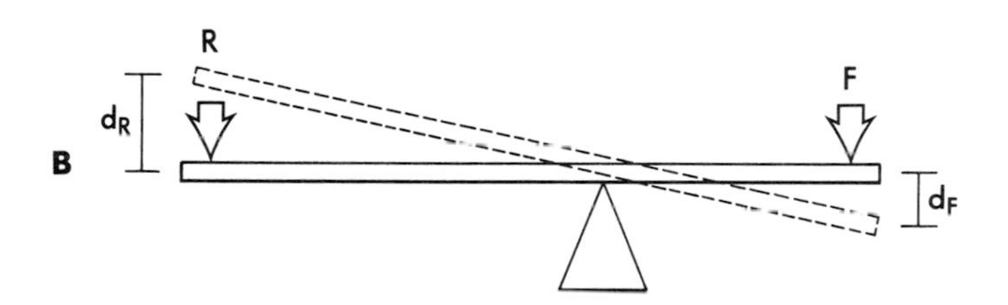

Figure 12-11
A, A force can balance a larger resistance when its moment arm is longer than the moment arm of the resistance. **B,** A force can move a resistance through a larger range of motion when the moment arm of the force is shorter than the moment arm of the resistance.

Whenever the moment arm of the force is longer than the moment arm of the resistance, the mechanical advantage ratio reduces to a number that is greater than one, and the magnitude of the applied force required to move a resistance is less than the magnitude of the resistance. The ability to move a resistance with a force that is smaller than the resistance is mechanically effective.

Alternatively, when the mechanical advantage ratio is less than one, a force that is larger than the resistance must be applied to cause motion of the lever. Although this arrangement is less effective in the sense that more force is required, a small movement of the lever at the point of force application moves the resistance through a larger range of motion (Figure 12-11).

Anatomical Levers

Skilled athletes in many sports intentionally maximize the length of the effective moment arm for force application to maximize the effect of the torque produced by muscles about a joint. During execution of the spike in volleyball, expert hitters not only strike the ball with the arm fully extended but also vigorously rotate the body in the transverse plane during the hit, making the axis of rotation the spine and maximizing the length of the anatomical lever delivering the force. The same strategy is employed by skilled tennis players executing the serve or the smash and by accomplished baseball pitchers.

Skilled pitchers often maximize the length of the moment arm between the ball hand and the total body axis of rotation during the delivery of a pitch to maximize the effect of the torque produced by the muscles.

In the human body, most muscle-bone lever systems are of the third class and therefore have a mechanical advantage of less than one. Although this arrangement promotes range of motion and angular speed of the body segments, the muscle forces generated must be in excess of the resistance force or forces if positive mechanical work is to be done.

The angle at which a muscle pulls on a bone also affects the mechanical effectiveness of the muscle-bone lever system. The force of muscular tension is resolved into two force components—one perpendicular to the attached bone and one parallel to the bone (Figure 12-12). As discussed in Chapter 5, only the component of muscle force acting perpendicular to the bone—the rotary component—actually causes the bone to rotate about the joint center. The component of muscle force is directed parallel to the bone and either pulls the bone away from the joint center (a dislocating component) or toward the joint center (a stabilizing component), depending on whether the angle between the bone and the attached muscle is less than or greater than 90 degrees. The angle of maximum mechanical advantage for any muscle is the angle at which the most rotary force can be produced. At a joint such as the elbow, the relative angle present at the joint is close to the angles of attachment of the elbow flexors. The maximum mechanical advantages for the brachialis, bi-

Figure 12-12

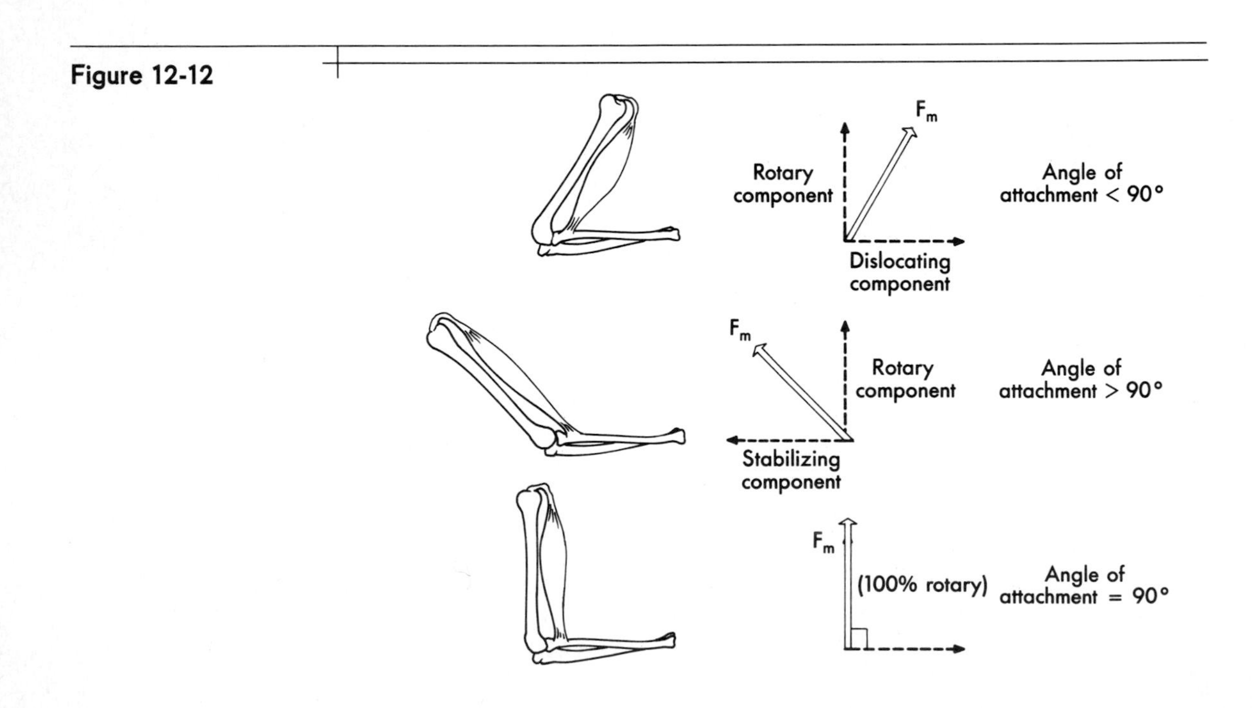

ceps, and brachioradialis occur between angles at the elbow of approximately 75 and 90 degrees (Figure 12-13).

As joint angle and mechanical advantage change, muscle length also changes. Alterations in the lengths of the elbow flexors associated with changes in angle at the elbow are shown in Figure 12-14. These changes affect the amount of tension a muscle can generate, as discussed in Chapter 4. The angle at the elbow at which maximum flexion torque is produced is approximately 80 degrees, with torque capability progressively diminishing as the angle at the elbow changes in either direction (22).

The varying mechanical effectiveness of muscle groups for producing joint rotation with changes in joint angle is the underlying basis for the design of modern variable-resistance strength training devices. These machines are designed to match the changing torque-generating capability of a muscle group throughout the range of motion at a joint. Machines manufactured by Universal (the Centurion) and Nautilus are examples. Although these machines offer more relative resistance through the extremes of joint range of mo-

■ The force-generating capability of a muscle is affected by muscle length and cross-sectional area, the angle and distance from the joint center of the tendon attachment to bone, the velocity of muscle shortening, and the state of neuromuscular training.

■ Variable resistance training devices are designed to match the resistance offered to the torque-generating capability of the muscle group as it varies throughout a range of motion.

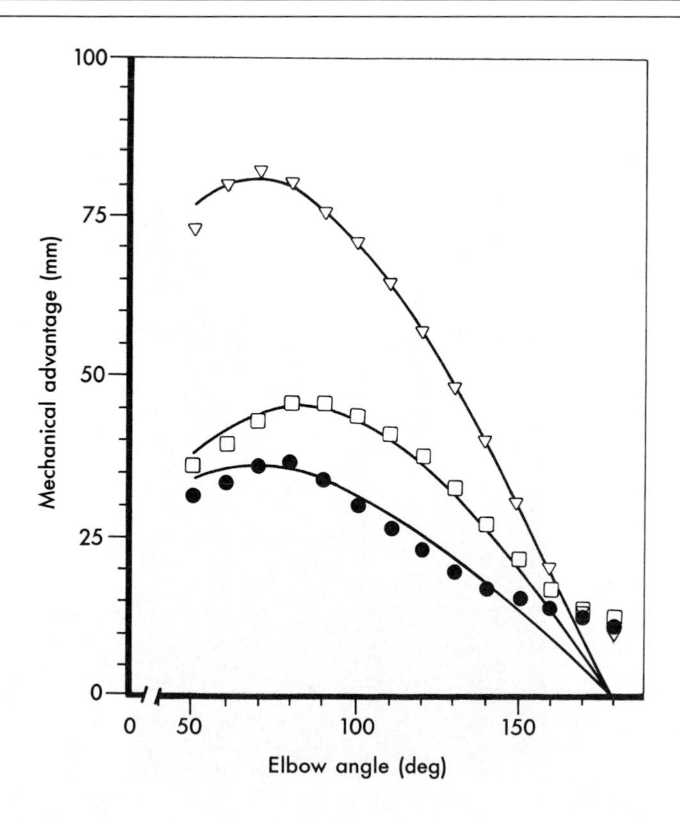

Figure 12-13
Mechanical advantage of the brachialis (●), biceps (□), and brachioradialis (▽) as a function of elbow angle. (Modified from van Zuylen EJ, van Zelzen A, and van der Gon JJD: A biomechanical model for flexion torques of human arm muscles as a function of elbow angle, J Biomech 21:183, 1988.)

Figure 12-14
Contractile length of the brachialis (●), biceps (□), and brachioradialis (▽) as a function of elbow angle. (Modified from van Zuylen EJ, van Zelzen A, and van der Gon JJD: A biomechanical model for flexion torques of human arm muscles as a function of elbow angle, J Biomech 21:183, 1988.)

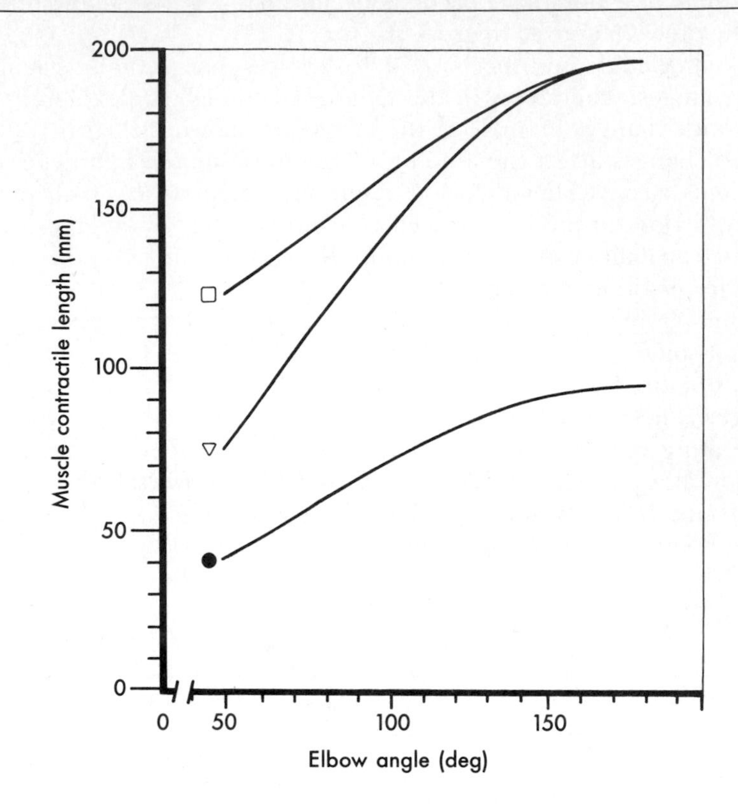

A cam in a variable resistance training machine is designed to match the resistance offered to the mechanical advantage of the muscle.

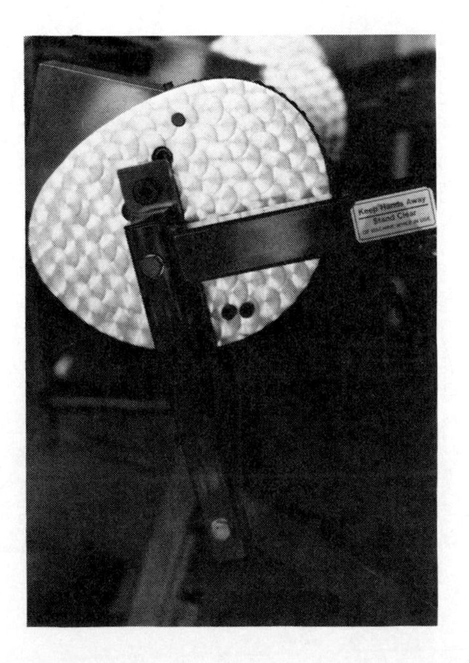

tion than free weights, the resistance patterns incorporated do not match average human strength curves (7).

Isokinetic machines represent another approach to matching torque-generating capability with resistance. These devices are generally designed so that an individual applies force to a lever arm that rotates at a constant angular velocity. If the joint center is aligned with the center of rotation of the lever arm, the body segment rotates with the same (constant) angular velocity of the lever arm. If volitional torque production by the involved muscle group is maximum throughout the range of motion, a maximum matched resistance is theoretically achieved. However, when force is initially applied to the lever arm of isokinetic machines, acceleration occurs and the angular velocity of the arm fluctuates until the set rotational speed is reached (7). It is also possible to use isokinetic resistance machines without applying maximal effort throughout the range of motion, which is one reason some individuals prefer other modes of resistance training.

■ The term *isokinetic* implies constant angular velocity at a joint when applied to exercise machinery.

Equations of Static Equilibrium

Whenever a body is completely motionless, it is in **static equilibrium.** Unless the net vector sum of the forces acting on an object is 0, the object is undergoing acceleration in the direction of the resultant force and is therefore in motion. Conversely, if an object is motionless, it may be inferred that the sum of all forces acting on the object is equal to 0. However, the vector sum of the forces acting on an object may equal 0 so that no translation occurs and yet rotation is produced (Figure 12-15).

static equilibrium
a motionless state characterized by $\Sigma F_v = 0$, $\Sigma F_h = 0$, and $\Sigma T = 0$

■ The application of any unopposed (net) force to a body results in acceleration of the body.

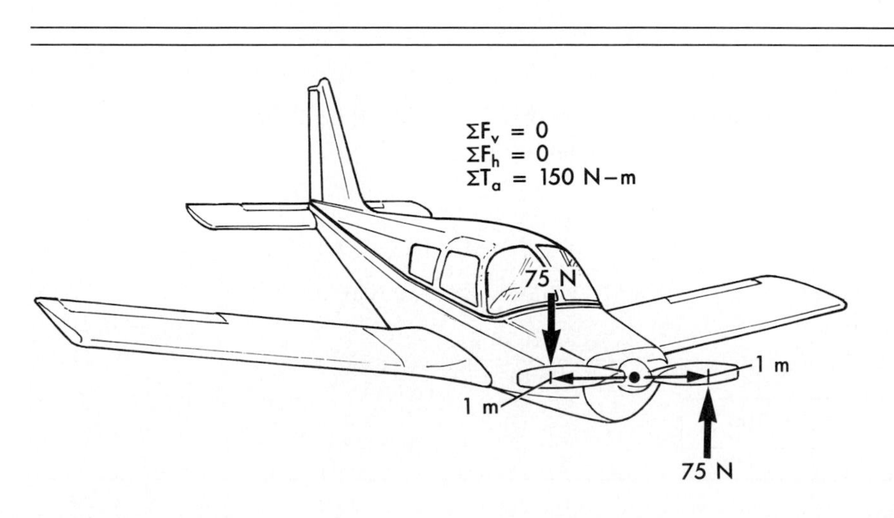

Figure 12-15
Although the sums of the vertical and horizontal forces are both 0, a torque is generated at the propeller's axis of rotation.

Consequently, three conditions must be met for a body to be in a state of static equilibrium: 1) The sum of all vertical forces (or force components) acting on the body must be 0, 2) the sum of all horizontal forces (or force components) acting on the body must be 0, and 3) the sum of all torques must be zero:

$$\Sigma F_v = 0$$

$$\Sigma F_h = 0$$

$$\Sigma T = 0$$

Figure 12-16

SAMPLE PROBLEM 2

How much force must be produced by the biceps brachii, attaching at 90 degrees to the radius at 3 cm from the center of rotation at the elbow joint, to support a weight of 70 N held in the hand at a distance of 30 cm from the elbow joint? (Neglect the weight of the forearm and hand, and neglect any action of other muscles.)

Known

d_m = 3 cm

wt = 70 N

d_{wt} = 30 cm

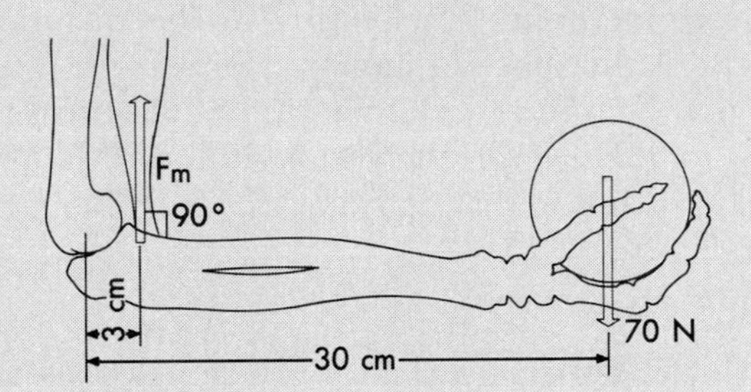

Solution

Since the situation described is static, the sum of the torques acting at the elbow must be equal to 0.

$$\Sigma T_e = 0$$

$$\Sigma T_e = (F_m)(d_m) - (wt)(d_{wt})$$

$$0 = (F_m)(0.03 \text{ m}) - (70 \text{ N})(0.30 \text{ m})$$

$$F_m = \frac{(70 \text{ N})(0.30 \text{ m})}{.03 \text{ m}}$$

$$\boxed{F_m = 700 \text{ N}}$$

The capital Greek letter sigma (Σ) means *the sum of*, F_v represents vertical forces, F_h represents horizontal forces, and T is torque. Whenever an object is in a static state, it is assumed that all three conditions are in effect. The conditions of static equilibrium are valuable tools for solving problems relating to human movement (Figure 12-16 to Figure 12-18).

SAMPLE PROBLEM 3

Figure 12-17

Two individuals apply force to opposite sides of a frictionless swinging door. If A applies a 30 N force at a 40 degree angle 45 cm from the door's hinge and B applies force at a 90 degree angle 38 cm from the door's hinge, what amount of force is applied by B if the door remains in a static position?

Known

$F_A = 30$ N

$d_{\perp A} = (0.45 \text{ m})(\sin 40)$

$d_{\perp B} = 0.38$ m

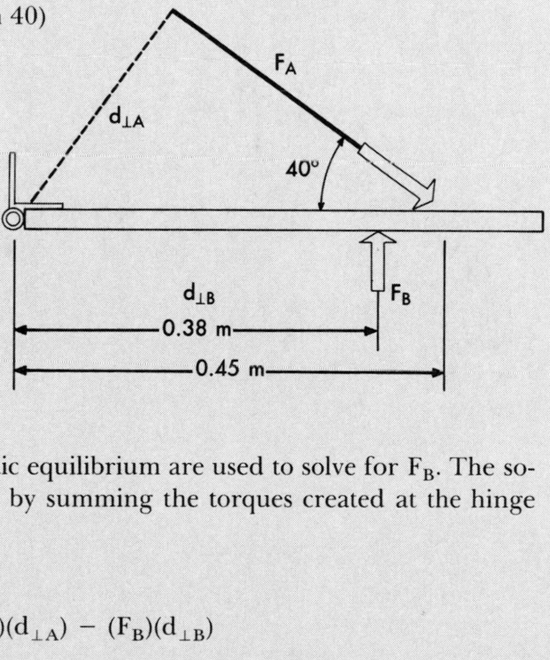

Solution

The equations of static equilibrium are used to solve for F_B. The solution may be found by summing the torques created at the hinge by both forces.

$$\Sigma T_h = 0$$

$$\Sigma T_h = (F_A)(d_{\perp A}) - (F_B)(d_{\perp B})$$

$$0 = (30 \text{ N})(0.45 \text{ m})(\sin 40) - (F_B)(0.38 \text{ m})$$

$$\boxed{F_B = 22.8 \text{ N}}$$

Figure 12-18

SAMPLE PROBLEM 4

The quadriceps tendon attaches to the tibia at a 30 degree angle 4 cm from the joint center at the knee. When an 80 N weight is attached to the ankle 28 cm from the knee joint, how much force is required of the quadriceps to maintain the leg in a horizontal position? What is the magnitude and direction of the reaction force exerted by the femur on the tibia? (Neglect the weight of the leg and the action of other muscles.)

Known

$wt = 80\text{ N}$

$d_{wt} = 0.28\text{ m}$

$d_F = 0.04\text{ m}$

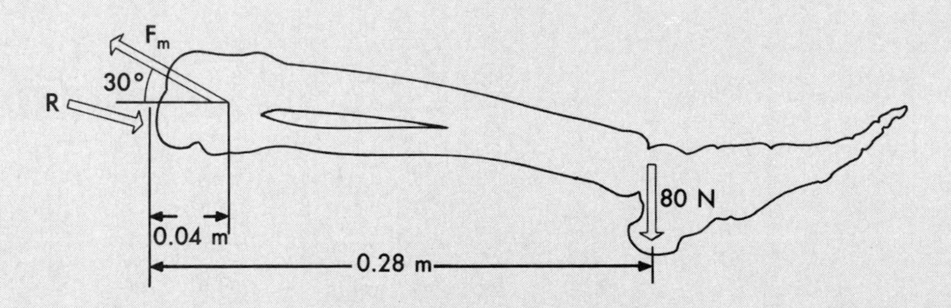

Solution

The equations of static equilibrium can be used to solve for the unknown quantities:

$$\Sigma T_k = 0$$

$$\Sigma T_k = (F_m \sin 30)(d_F) - (wt)(d_{wt})$$

$$0 = (F_m \sin 30)(0.04\text{ m}) - (80\text{ N})(0.28\text{ m})$$

$$\boxed{F_m = 1120\text{ N}}$$

The equations of static equilibrium can be used to solve for the vertical and horizontal components of the reaction force exerted by the femur on the tibia. Summation of vertical forces yields the following:

$$\Sigma F_v = 0$$

$$\Sigma F_v = R_v + (F_m \sin 30) - wt$$

$$0 = R_v + 1120 \sin 30 \text{ N} - 80 \text{ N}$$

$$R_v = -480 \text{ N}$$

Summation of horizontal forces yields the following:

$$\Sigma F_h = 0$$

$$\Sigma F_h = R_h - (F_m \cos 30)$$

$$0 = R_h - 1120 \cos 30 \text{ N}$$

$$R_h = 970 \text{ N}$$

The Pythagorean theorem can now be used to find the magnitude of the resultant reaction force:

$$R = \sqrt{(-480 \text{ N})^2 + (970 \text{ N})^2}$$

$$R = 1082 \text{ N}$$

The tangent relationship can be used to find the angle of orientation of the resultant reaction force:

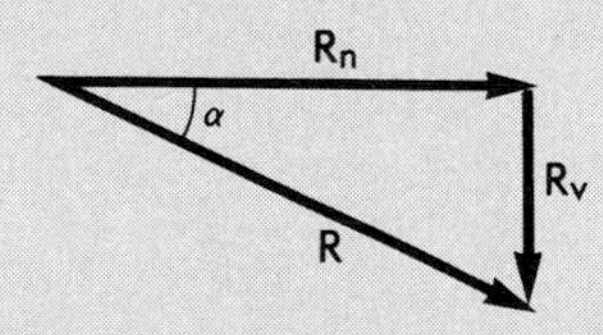

$$\tan\alpha = \frac{480 \text{ N}}{970 \text{ N}}$$

$$\alpha = 26.3$$

R = 1082 N 26.3 degrees

Equations of Dynamic Equilibrium

dynamic equilibrium (D'Alembert's principle)
the concept indicating a balance between applied forces and inertial forces for a body in motion

Bodies in motion are considered to be in a state of **dynamic equilibrium,** with all acting forces resulting in equal and oppositely directed inertial forces. This general concept was first identified by the French mathematician D'Alembert and is known as *D'Alembert's principle.* Modified versions of the equations of static equilibrium, which incorporate factors known as *inertia vectors,* describe the conditions of dynamic equilibrium. The equations of dynamic equilibrium may be stated as follows:

$$\Sigma F_x - m\bar{a}_x = 0$$

$$\Sigma F_y - m\bar{a}_y = 0$$

$$\Sigma T_G - \bar{I}\alpha = 0$$

Figure 12-19

SAMPLE PROBLEM 5

A 580 N skydiver in free fall is accelerating at $-8.8\ m/s^2$ rather than $-9.81\ m/s^2$ because of the force of air resistance. How much drag force is acting on the skydiver?

Known

$$wt = -580\ N$$

$$a = -8.8\ m/s^2$$

$$\text{Mass} = \frac{580\ N}{9.81\ m/s^2} = 59.12\ kg$$

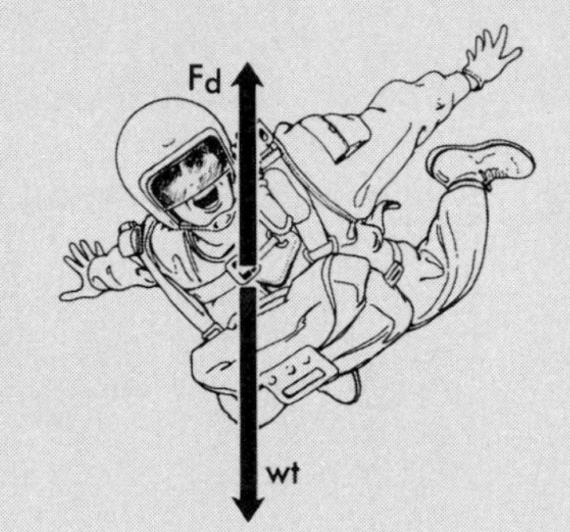

Solution

Since the skydiver is considered to be in dynamic equilibrium, D'Alembert's principle may be used. All identified forces acting are vertical forces, so the equation of dynamic equilibrium summing the vertical forces to zero is used:

$$\Sigma F_y - \bar{m}a_y = 0$$

Given that $\Sigma F_y = -580\ N + F_d$, substitute the known information into the equation:

$$-580\ N + F_d - (59.12\ kg)(-8.8\ m/s^2) = 0$$

$$\boxed{F_d = 59.7\ N}$$

The sums of the horizontal and vertical forces acting on a body are ΣF_x and ΣF_y, $m\bar{a}_x$ and $m\bar{a}_y$ are the products of the body's mass and the horizontal and vertical accelerations of the body's center of mass, ΣT_G is the sum of torques about the body's center of mass, and $\bar{I}\alpha$ is the product of the body's moment of inertia about the center of mass and the body's angular acceleration (Figure 12-19). (The concept of moment of inertia is discussed in Chapter 13.)

A familiar example of the effect of D'Alembert's principle is the change in vertical force experienced when riding in an elevator. As the elevator accelerates upward, an inertial force in the opposite direction is created and body weight as measured on a scale in the elevator increases. As the elevator accelerates downward, an upwardly directed inertial force decreases body weight.

CENTER OF GRAVITY

A body's mass is the matter of which it is composed. A unique point is associated with every body, around which the body's mass is equally distributed in all directions. This point is known as the **center of mass** or the **mass centroid** of the body. In the analysis of bodies subject to gravitational force, the center of mass may also be referred to as the **center of gravity** (CG), the point about which a body's weight is equally balanced in all directions or the point about which the sum of torques produced by the weights of the body segments is equal to 0. This definition does not imply that the weights positioned on opposite sides of the CG are equal but that the torques created by the weights on opposite sides of the CG are equal. As illustrated in Figure 12-20, equal weight and equal torque generation

center of mass
mass centroid
center of gravity
the point around which the mass and weight of a body are balanced in all directions

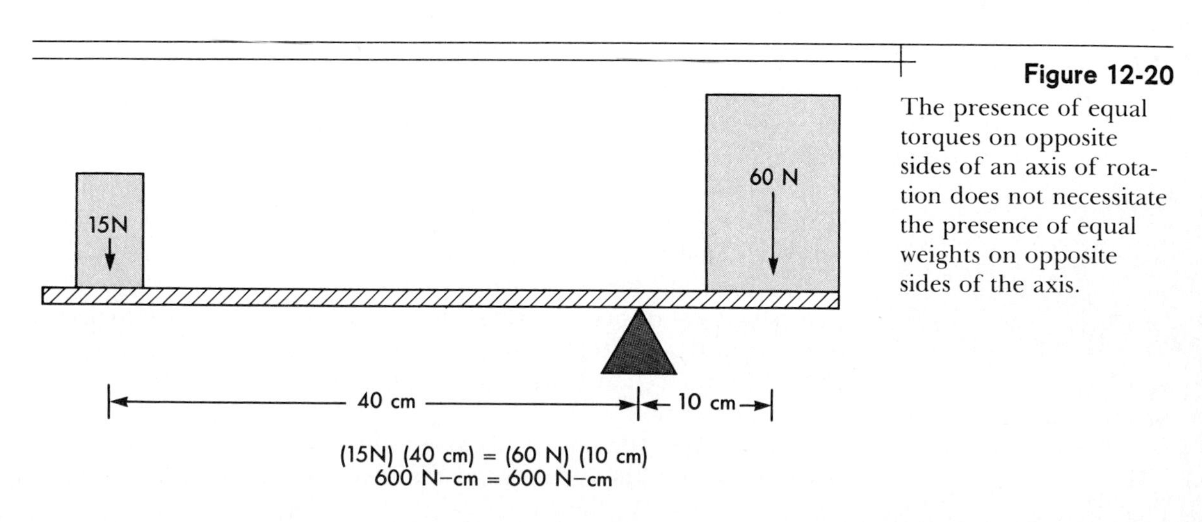

Figure 12-20
The presence of equal torques on opposite sides of an axis of rotation does not necessitate the presence of equal weights on opposite sides of the axis.

Figure 12-21
The center of gravity is the single point associated with a body around which the body's weight is equally balanced in all directions.

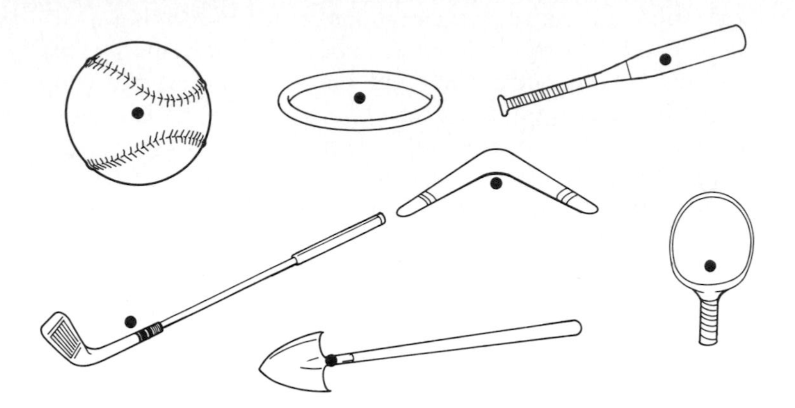

on opposite sides of a point can be quite different. The terms *center of mass* and *center of gravity* are more commonly used for biomechanics applications than *mass centroid,* although all three terms refer to exactly the same point. Because the masses of bodies on the earth are subject to gravitational force, the center of gravity is probably the most accurately descriptive of the three to use for biomechanical applications.

The CG of a perfectly symmetrical object of homogeneous density and therefore homogeneous mass and weight distribution is at the exact center of the object. For example, the CG of a spherical shot or a solid rubber ball is at its geometric center. If the object is a homogeneous ring, the CG is located in the hollow center of the ring. An object's CG is not always located physically inside of the object (Figure 12-21).

The running kinematics of a young child include noticeable vertical oscillations of the CG.

Locating the Center of Gravity

The location of the CG for a one-segment object, such as a baseball bat, a broom, or a shovel, can be approximately determined using two approaches. The first involves the use of a fulcrum to determine the location of a balance point for the object in three different planes. Because the CG is the point around which the mass of a body is equally distributed, it is also the point around which the body is balanced in all directions. Since balancing some objects in certain planes may be difficult, suspending a plumb line from the object in question in three different positions is an alternative procedure. The point of intersection of the three paths of the plumb line approximates the location of the CG (Figure 12-22).

The location of a body's CG is of interest because, mechanically, a body behaves as though all of its mass were concentrated at the CG.

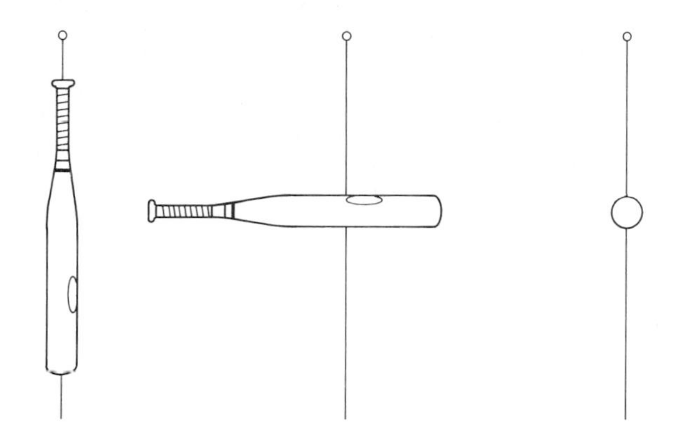

Figure 12-22

The suspension method for determining CG location involves suspending an object from a given point about which it is free to rotate. When the object comes to rest, a plumb line suspended from the same point can be used to mark a vertical line on the object about which it is balanced. When the procedure is repeated with the object aligned in two other planes, the intersection of the three lines approximates the location of the CG.

For example, when the human body acts as a projectile, the body's CG follows a parabolic trajectory, regardless of any changes in the configurations of the body segments while in the air. Another implication is that when a weight vector is drawn for an object displayed in a free body diagram, the weight vector acts at the CG. Because the body's mechanical behavior can be traced by following the path of the total body CG, this factor has been studied as a possible indicator of performance proficiency in several sports.

For example, it has been hypothesized that skilled runners display less vertical oscillation of the CG during performance. Although this has not been well documented for adult runners (24), in children of 4 to 7 years of age, vertical oscillation of the CG decreases with age and concomitant maturation of running gait (18, 23).

The pattern of movement of the CG during takeoff for the high jump is one factor believed to distinguish skilled from less-skilled performances. Using the ground reaction force to convert horizontal CG velocity to upwardly directed vertical CG velocity during takeoff for the high jump is an effective procedure (5). Research indicates that better Fosbury flop style of high jumpers employ both body lean and body flexion (especially of the support leg) just before takeoff to lower the CG and prolong support foot contact time, thus resulting in increased takeoff impulse (1).

CG motion is also believed to be an index of performance outcome during takeoff in the long jump. A skilled athlete's CG is lowered approximately 10% of its normal height during the final strides of the approach (10). This lengthens the vertical path over which the body is accelerated during takeoff, thus facilitating a high vertical velocity at takeoff (Figure 12-23). The speed and angle of takeoff primarily determine the trajectory of the performer's CG during the

The speed and projection angle of an athlete's total body center of mass largely determines performance outcome in the long jump.

Figure 12-23
Height of the athlete's CG during preparation for takeoff in the long jump. (Modified from Nixdorf E and Bruggemann P: Zur Absprungvorbereitung beim Weitsprung—Eine biomechanische Untersuchung zum problem der Korperschwerpunktsenkung, Lehre Leichtathlet p. 1539, 1983.)

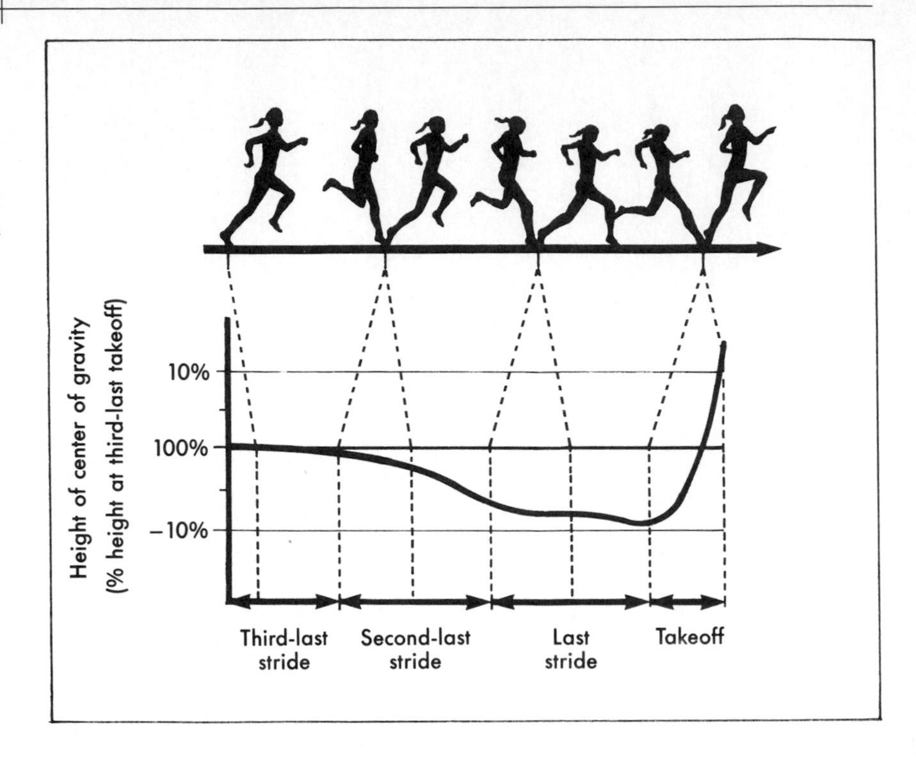

jump. The only other influencing factor is air resistance, which exerts an extremely small effect on long jump performance.

Examination of CG plots is a useful tool for coaches and athletes who want to enhance performance. The United States Olympic Committee's "Elite Athlete Project" allows biomechanists, track and field coaches, and athletes to collectively analyze factors such as CG motion that may distinguish skilled from less skilled performance.

Locating the Human Body Center of Gravity

■ Location of the CG of the human body is complicated by the fact that its constituents (such as bone, muscle, and fat) have different densities and are unequally distributed throughout the body.

Locating the CG for a body containing two or more movable, interconnected segments is more difficult than for a nonsegmented body because every time the body changes configuration, its weight distribution and CG location are changed. Every time an arm, leg, or finger moves, the CG location as a whole is shifted at least slightly in the direction in which the weight is moved.

Some relatively simple procedures exist for determining the location of the CG of the human body. In the seventeenth century, the Italian mathematician Borelli used a simple balancing procedure for

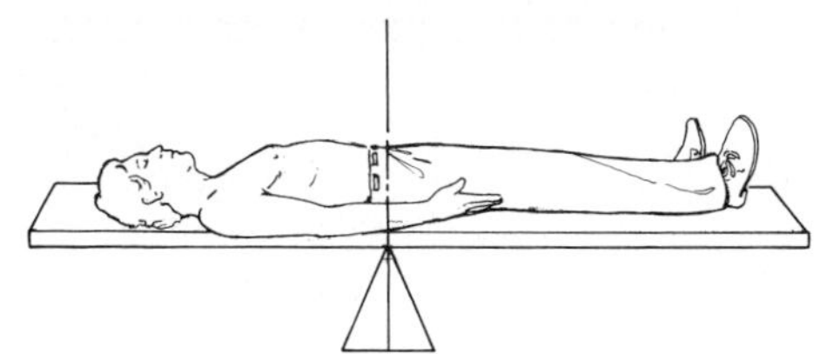

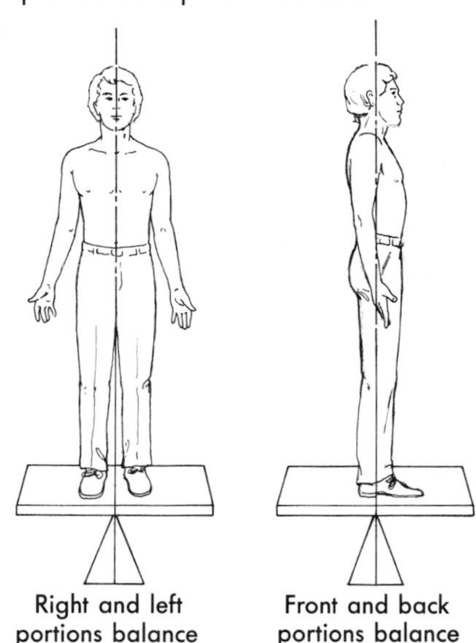

Figure 12-24
The relatively crude procedure devised by seventeenth century mathematician Borelli for approximating the CG location of the human body.

CG location that involved positioning a person on a wooden board (Figure 12-24) (17). A more sophisticated version of this procedure enables calculation of the location of the plane passing through the CG of a person positioned on a **reaction board.** This procedure requires the use of a scale, a platform of the same height as the weighing surface of the scale, and a rigid board with sharp supports on either end (Figure 12-25). The calculation of the location of the plane containing the CG involves the summation of torques acting about the platform support. The subject's body weight, which acts at the CG, creates a torque at the platform support in one direction, and the reaction force of the scale on the platform, which the reading on the scale indicates, creates a torque at the platform support in the opposite direction. Although the platform also exerts a reaction force on the board, it creates no torque because the distance of that force from the platform support is 0. Since the reaction board and

reaction board
a specially constructed board for determining the center of gravity location of a body positioned on top of it

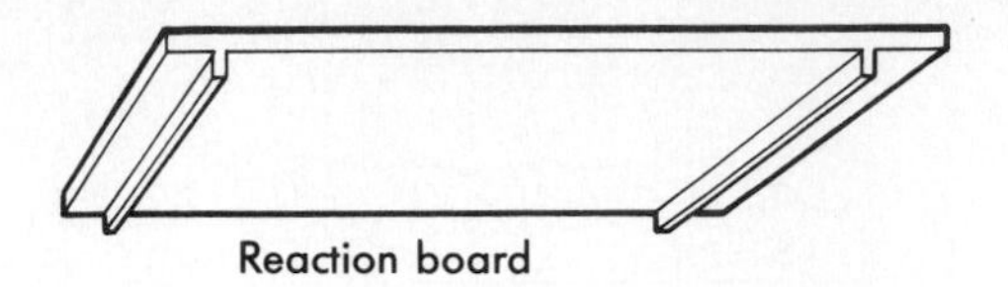

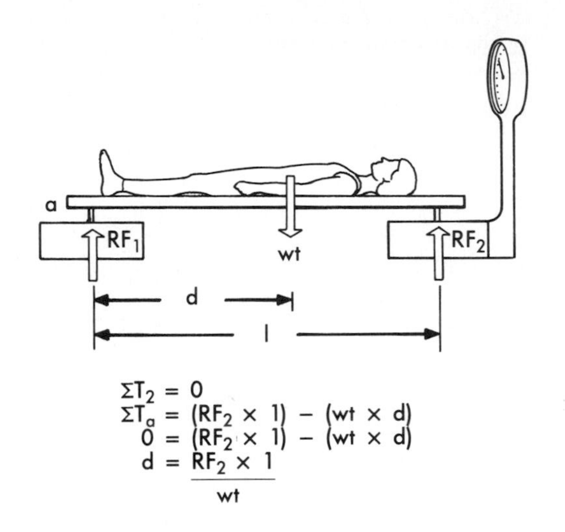

Figure 12-25
By summing torques at point a, d (the distance from a to the subject's CG) may be calculated.

subject are in static equilibrium, the sum of the two torques acting at the platform support must be 0, and the distance of the subject's CG plane to the platform may be calculated (Figure 12-26).

When considered separately, each segment of the human body possesses its own CG. Researchers have proposed different techniques for determining the CG location of the human body and its segments (8, 9, 13, 14).

A commonly used procedure for estimating the location of the total body CG from projected film images of the human body is known as the **segmental method.** This procedure is based on the concept that since the body is composed of individual segments (each with an individual CG), the location of the total body CG is a function of the locations of the respective segmental CGs. Some body segments, however, are much more massive than others and have a larger influence on the location of the total body CG. When the products of each body segment's CG location and its mass are summed and subsequently divided by the sum of all segmental masses (total body mass), the result is the location of the total body CG. The segmental

segmental method
the procedure for determining total body center of mass location based on the masses and center of mass locations of the individual body segments

SAMPLE PROBLEM 6

Figure 12-26

Find the distance from the platform support to the subject's CG, given the following information (with subject positioned, adjusted for weight of board alone as positioned):

Known

$$\text{Mass (subject)} = 73 \text{ kg}$$

$$\text{Scale reading} = 44 \text{ kg}$$

$$l_b = 2 \text{ m}$$

Solution

$$wt_s = (73 \text{ kg})(9.81 \text{ m/s}^2) = 716.13 \text{ N}$$

$$wt_b = (44 \text{ kg})(9.81 \text{ m/s}^2) = 431.64 \text{ N}$$

Use an equation of static equilibrium:

$$\Sigma T_a = 0$$

$$\Sigma T_a = (wt_b)(l_b) - (wt_s)(d)$$

$$0 = (431.64 \text{ N})(2 \text{ m}) - (716.13 \text{ N})(d)$$

$$d - \frac{(431.64 \text{ N})(2 \text{ m})}{(716.1\text{N})}$$

$$\boxed{d = 1.2 \text{ m}}$$

method uses data for average locations of individual body segment CGs as related to a percentage of segment length (4, 6, 9):

$$X_{cg} = \Sigma\ (x_s)\ (m_s)/\Sigma m_s$$

$$Y_{cg} = \Sigma\ (y_s)\ (m_s)/\Sigma m_s$$

In this formula, X_{cg} and Y_{cg} are the coordinates of the total body CG, x_s and y_s are the coordinates of the individual segment CGs, and m_s is individual segment mass. Thus the x coordinate of each segment's CG location is identified and multiplied by the mass of that respective segment. The (x_s) (m_s) products for all of the body segments are then summed and subsequently divided by total body mass to yield the x coordinate of the total body CG location. The same procedure is followed to calculate the y coordinate for total body CG location (Figure 12-27).

■ The location of the CG of a multisegmented object is more influenced by the positions of the heavier segments than by those of the lighter segments.

■ The segmental method is most commonly implemented through a computer program that reads x,y coordinates of joint centers from a file created by a digitizer.

Figure 12-27

SAMPLE PROBLEM 7

The x,y coordinates of the CGs of the upper arm, forearm, and hand segments are provided on the diagram below. Use the segmental method to find the CG for the entire arm, using the data provided for segment masses.

Known

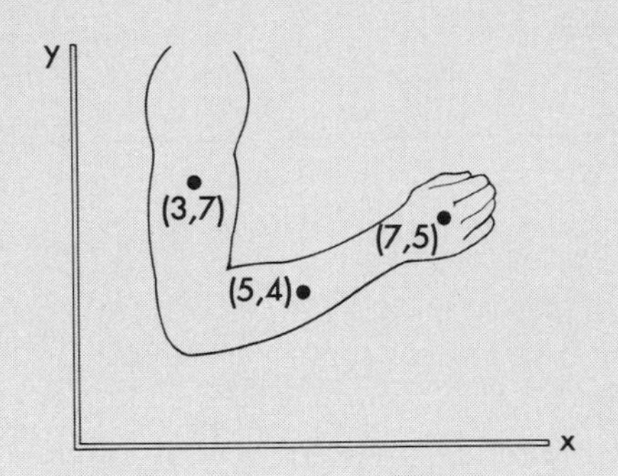

SEGMENT	MASS %	x	(x)(MASS %)	y	(y)(MASS %)
Upper arm	0.45				
Forearm	0.43				
Hand	0.12				
Σ					

Solution

First list the x and y coordinates in their respective columns, and then calculate and insert the product of each coordinate and the mass percentage for each segment into the appropriate columns. Sum the product columns, which yield the x,y coordinates of the total arm CG.

SEGMENT	MASS %	x	(x)(MASS %)	y	(y)(MASS %)
Upper arm	0.45	3	1.35	7	3.15
Forearm	0.43	5	2.15	4	1.72
Hand	0.12	7	0.84	5	0.60
Σ			4.34		5.47

x = 4.34
y = 5.47

STABILITY AND BALANCE

A concept closely related to the principles of equilibrium is **stability.** Stability is defined mechanically as resistance to both linear and angular acceleration or resistance to disruption of equilibrium. In some circumstances, such as a Sumo wresting contest or the defense of a quarterback by an offensive lineman, maximizing stability is desirable. In other situations, an athlete's best strategy is to intentionally minimize stability. Sprinters and swimmers in the preparatory stance before the start of a race intentionally assume a body position allowing them to accelerate quickly and easily at the sound of the starter's pistol. An individual's ability to control equilibrium is known as **balance.**

stability
the resistance to disruption of equilibrium

balance
the ability to control equilibrium

Different mechanical factors affect a body's stability. According to Newton's second law of motion (F = ma), the more massive an object is, the greater the force required to produce a given acceleration. Football linemen who are expected to maintain their positions despite the forces exerted on them by opposing linemen are therefore more mechanically stable if they are more massive. In contrast, gymnasts are at a disadvantage with greater body mass because execution of most gymnastic skills involves disruption of stability.

The greater the amount of friction between an object and the surface or surfaces it contacts, the greater the force requirement for initiating or maintaining motion. Toboggans and racing skates are designed so that the friction they generate against the ice will be minimal, enabling a quick disruption of stability at the beginning of a run or race. However, racquetball gloves and the racquet grip are designed to increase the stability of the player's grip on the racquet.

Another factor affecting stability is the size of the **base of support.** This consists of the area enclosed by the outermost edges of the body in contact with the supporting surface or surfaces (Figure 12-28). When the line of action of a body's weight (directed from the CG) moves outside the base of support, a torque is created that tends to cause angular motion of the body toward the CG, thereby disrupting stability. The larger the base of support is, the less the likelihood that this will occur. Martial artists typically assume a wide stance during defensive situations to increase stability. Alternatively, sprinters in the starting blocks maintain a relatively small base of support so that they can quickly disrupt stability at the start of the race. Maintaining balance during an *arabesque on pointe* requires continual adjustment of CG location through subtle body movements (16).

base of support
the area bound by the outermost regions of contact between a body and support surface or surfaces

The horizontal location of the CG relative to the base of support can also influence stability. The closer the horizontal location of the CG to the boundary of the base of support, the smaller the force required to push it outside the base of support, thereby disrupting equilibrium. Athletes in the starting position for a race consequently assume stances that position the CG close to the forward edge of the

Figure 12-28

The base of support for **A,** a square stance, **B,** an angled stance, **C,** a one foot stance, **D,** a three point stance, and **E,** a four point stance. Areas of contact between body parts and the support surface are darkly shaded. The base of support is the area enclosed by the dashed line.

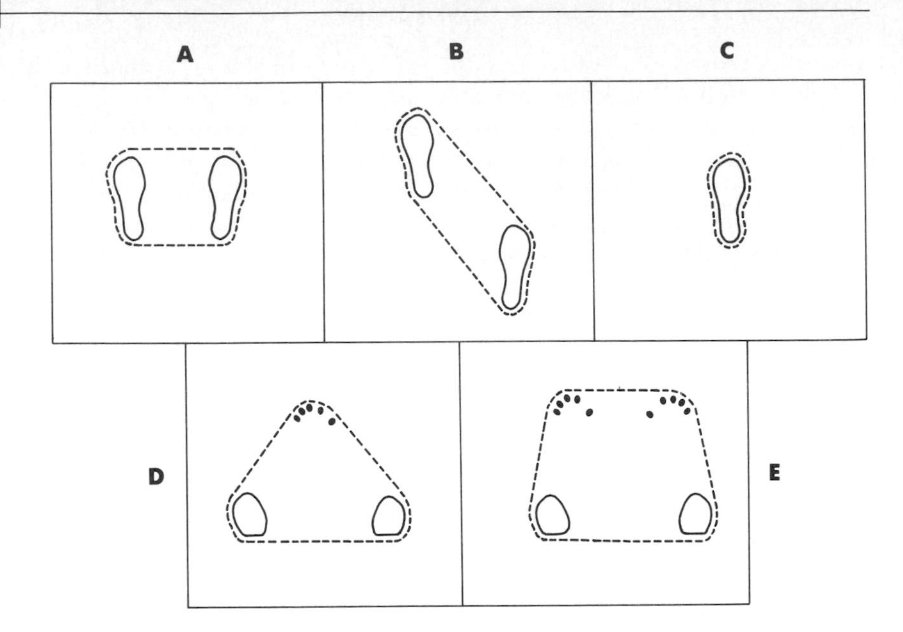

Performing an *arabesque on pointe* requires excellent balance because lateral movement of the dancer's line of gravity outside the small base of support will result in loss of balance.

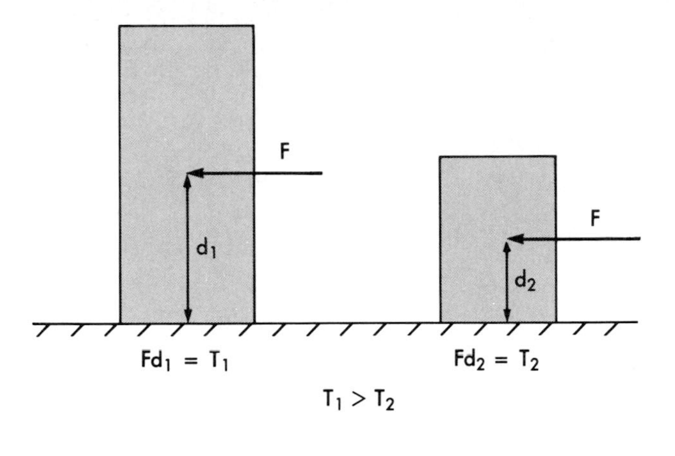

Figure 12-29
The higher the CG location, the greater the amount of torque its motion creates about the intersection of the line of gravity and the support surface.

base of support. Alternatively, if a horizontal force must be sustained, stability is enhanced if the CG is positioned closer to the oncoming force, since the CG can be displaced farther before being moved outside the base of support. Sumo wrestlers lean toward their opponents when being pushed backward.

The height of the CG relative to the base of support can also affect stability. The higher the positioning of the CG, the greater the potentially disruptive torque created if the body undergoes an angular displacement (Figure 12-29). Athletes in many sports crouch when added stability is desirable. The principles of mechanical stability are summarized in the box.

A swimmer on the blocks positions her CG close to the front boundary of her base of support to prepare for forward acceleration.

PRINCIPLES OF MECHANICAL STABILITY

When other factors are held constant, a body's ability to maintain equilibrium is increased by the following:

1. Increasing body mass
2. Increasing friction between the body and the surface or surfaces contacted
3. Increasing the size of the base of support in the direction of the line of action of an external force
4. Horizontally positioning the center of gravity near the edge of the base of support on the side of an oncoming external force
5. Vertically positioning the center of gravity as low as possible

Although the principles of stability are generally true, their application to the human body should be made only with the recognition that neuromuscular factors are also influential. Changes in foot position have been found to affect two measures of standing balance—the location of the line of gravity and postural sway. Research has shown that frontal plane sway decreases when the base of support is widened to 15 cm but that increasing the width of the base of support beyond 15 cm does not further reduce body sway in the frontal plane. Researchers have also found that when people stand with one foot 30 cm in front of the other, frontal plane sway increases but a surprising increase in sagittal plane sway occurs as well. The angle of foot position during normal stance does not affect balance; however, an extreme toe-in position causes more body sway (15). More research is needed to clarify the application of the principles of stability to human balance.

SUMMARY

Rotary motion is caused by torque, a vector quantity with magnitude and direction. When a muscle develops tension, it produces torque at the joint or joints that it crosses. Rotation of the body segment occurs in the direction of the resultant joint torque.

Mechanically, muscles and bones function as levers. Most joints function as third class lever systems, well structured for maximizing range of motion and movement speed but requiring muscle force of greater magnitude than that of the resistance to be overcome. The angle at which a muscle pulls on a bone also affects its mechanical effectiveness because only the rotary component of muscle force produces joint torque.

When a body is motionless, it is in static equilibrium. The three conditions of static equilibrium are $\Sigma F_v = 0$, $\Sigma F_h = 0$, and $\Sigma T = 0$. A body in motion is in dynamic equilibrium when inertial factors are considered.

The mechanical behavior of a body subject to force or forces is greatly influenced by the location of its center of gravity—the point around which the body's weight is equally balanced in all directions. Different procedures are available for determining center of gravity location.

A body's mechanical stability is its resistance to both linear and angular acceleration. A number of factors influence a body's stability, including mass, friction, center of gravity location, and base of support.

INTRODUCTORY PROBLEMS

1. Why does a force directed through an axis of rotation not cause rotation at the axis?

2. Why does the orientation of a force acting on a body affect the amount of torque it generates at an axis of rotation within the body?
3. A 23 kg boy sits 1.5 m from the axis of rotation of a seesaw. At what distance from the axis of rotation must a 21 kg boy be positioned on the other side of the axis to balance the seesaw? (Answer: 1.6 m)
4. How much force must be produced by the biceps brachii at a perpendicular distance of 3 cm from the axis of rotation at the elbow to support a weight of 200 N at a perpendicular distance of 25 cm from the elbow? (Answer: 1667 N)
5. Two people push on opposite sides of a swinging door. If A exerts a force of 40 N at a perpendicular distance of 20 cm from the hinge and B exerts a force of 30 N at a perpendicular distance of 25 cm from the hinge, what is the resultant torque acting at the hinge and which way will the door swing? (Answer: T_h = 0.5 N-m; in the direction that A pushes)
6. Which lever classes do a golf club, swinging door, and broom belong to? Explain your answers, using diagrams if necessary.
7. Is the mechanical advantage of a first class lever greater than, less than, or equal to one? Explain.
8. Using a diagram, identify the magnitudes of the rotary and stabilizing components of a 100 N muscle force that acts at an angle of 20 degrees to a bone. (Answer: rotary component = 34 N, stabilizing component = 94 N)
9. A 10 kg block sits motionless on a table in spite of an applied horizontal force of 2 N. What are the magnitudes of the reaction force and friction force acting on the block? (Answer: R = 98.1 N, F = 2 N)
10. Given the following data for the reaction board procedure, calculate the distance from the platform support to the subject's CG: RF_2 = 400 N, 1 = 2.5 m, wt = 600 N. (Answer: 1.67 m)

ADDITIONAL PROBLEMS

1. Draw two free body diagrams—one of a person performing a push-up with the elbows fully extended and one of a person performing a push-up with the body in its lowest position. Is one of these positions easier to maintain? Why or why not?
2. Select another exercise of your choice that can be performed with two or more variations. Construct free body diagrams of two comparable but different body positions during these exercises and explain which exercise is more difficult.
3. A 35 N hand and forearm are held at a 45 degree angle to the vertically oriented humerus. The CG of the forearm and hand is located at a distance of 15 cm from the joint center at

the elbow, and the elbow flexor muscles attach at an average distance of 3 cm from the joint center. (Assume that the muscles attach at an angle of 45 degrees to the bones.)

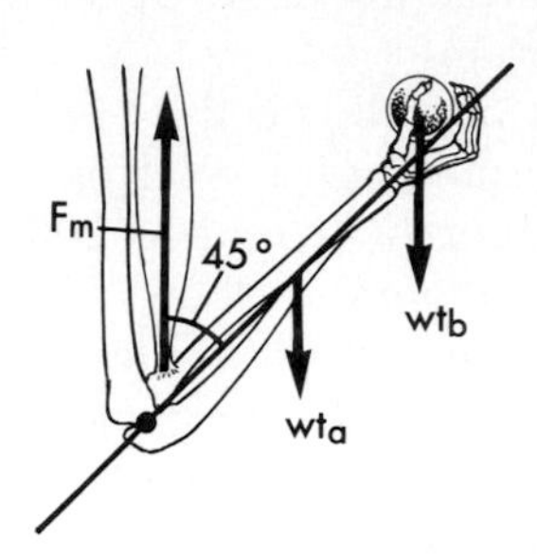

a. How much force must be exerted by the forearm flexors to maintain this position?
b. How much force must the forearm flexors exert if a 50 N weight is held in the hand at a distance along the arm of 25 cm?

(Answer: a. 175 N; b. 591.7 N)

4. A hand exerts a force of 90 N on a scale at 32 cm from the joint center at the elbow. If the triceps attach to the ulna at a 90 degree angle and at a distance of 3 cm from the elbow joint center and if the weight of the forearm and hand is 40 N with the hand/forearm CG located 17 cm from the elbow joint center, how much force is being exerted by the triceps?

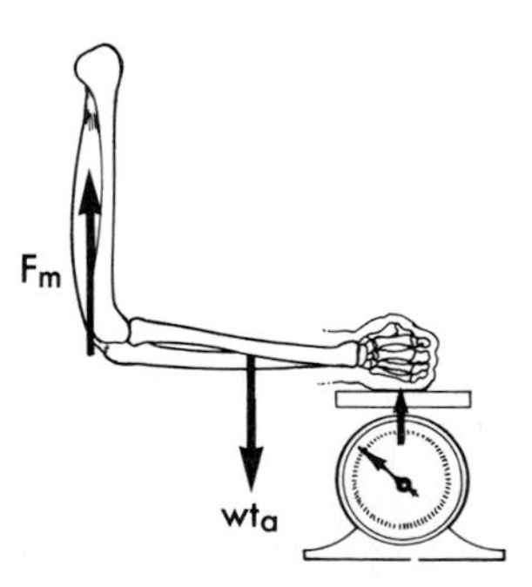

(Answer: 733.3 N)

5. A patient rehabilitating a knee injury performs knee extension exercises wearing a 15 N weight boot. Calculate the amount of torque generated at the knee by the weight boot

for the four positions shown, given a distance of 0.4 m between the weight boot's CG and the joint center at the knee.

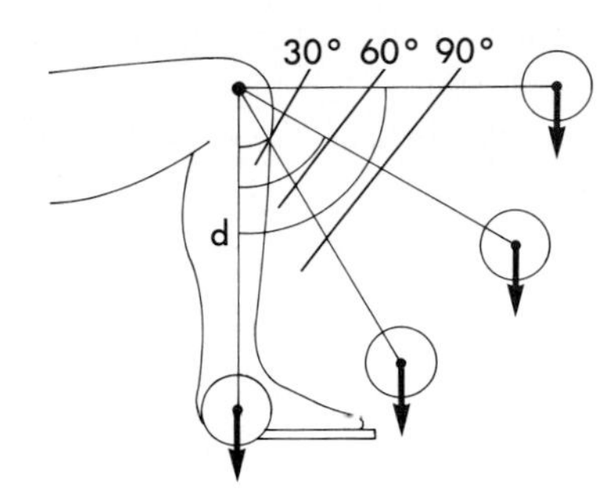

(Answer: a. 0; b. 3N-m; c. 5.2 N-m; d. 6 N-m)

6. A 600 N person picks up a 180 N suitcase positioned so that the suitcase's CG is 20 cm lateral to the location of the person's CG before picking up the suitcase. If the person does not lean compensate for the added load in any way, where is the combined CG location for the person and suitcase with respect to the person's original CG location? (Answer: Shifted 4.6 cm toward the suitcase)

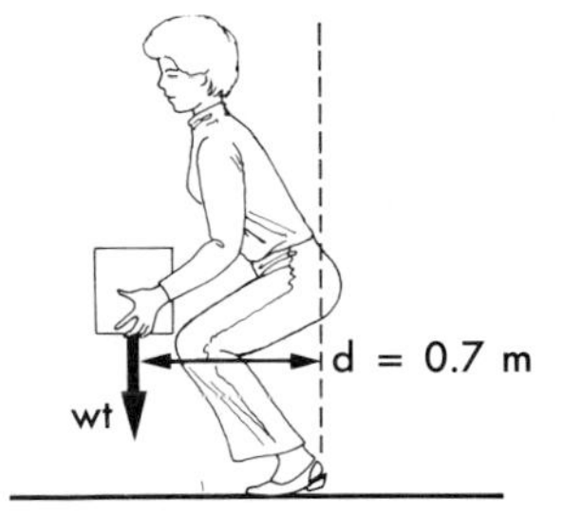

7. A worker leans over and picks up a 90 N box at a distance of 0.7 m from her low back muscles. Neglecting the effect of body weight, how much added force is required of the low back muscles with an average moment arm of 1.5 cm to stabilize the box in the position shown? (Answer: 4200 N)

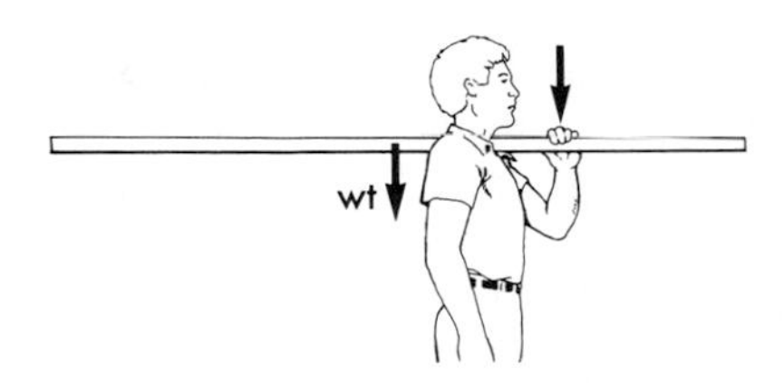

8. A man carries a 3 m, 32 N board over his shoulder. If the board extends 1.8 m behind the shoulder and 1.2 m in front of the shoulder, how much force must the man apply vertically downward with his hand that rests on the board 0.2 m in front of the shoulder to stabilize the board in this position? (Assume that the weight of the board is evenly distributed throughout its length). (Answer: 48 N)

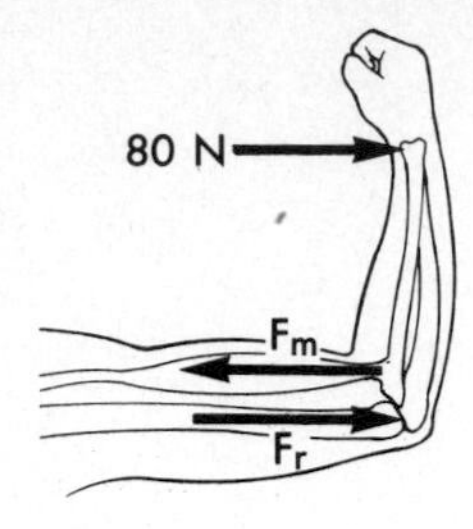

9. A therapist applies a lateral force of 80 N to the forearm at a distance of 25 cm from the axis of rotation at the elbow. The biceps attaches to the radius at a 90 degree angle and at a distance of 3 cm from the elbow joint center.
   a. How much force is required of the biceps to stabilize the arm in this position?
   b. What is the magnitude of the reaction force exerted by the humerus on the ulna?

   (Answer: a. 666.7 N; b. 586.7 N)

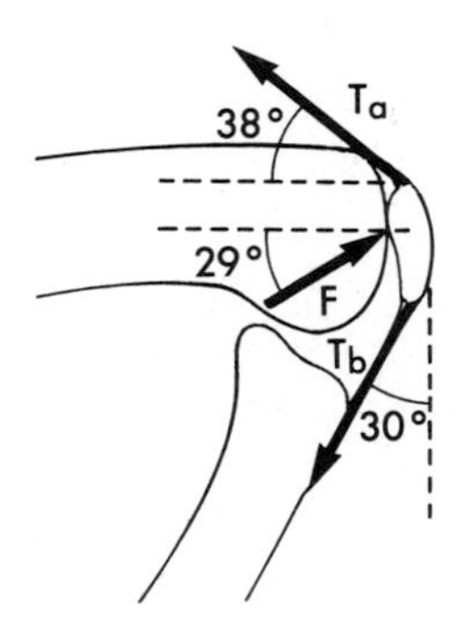

10. Tendon forces T_a and T_b are exerted on the patella. The femur exerts force F on the patella. If the magnitude of T_b is 80 N, what are the magnitudes of T_a and F, if no motion is occurring at the joint? (Answer: T_a = 44.8 N, F = 86.9 N)

REFERENCES

1. Ae M et al: Biomechanical analysis of the preparatory motion for takeoff in the Fosbury flop, Int J Sport Biomech, 2:66, 1986.
2. Andrews JG: On the relationship between resultant joint torques and muscular activity, Med Sci Sports Exerc 14:361, 1982.
3. Cappozzo A et al: Lumbar spine loading during half-squat exercises, Med Sci Sports Exerc 17:613, 1985.
4. Clauser CE, McConville JT, and Young JW: Weight, volume, and center of mass of segments of the human body, AMRL Tech Rep, 1969, Wright-Patterson Air Force Base.
5. Dapena J: Biomechanics of elite high jumpers. In Terauds J et al, eds: Sports biomechanics, Del Mar, Calif, 1984, Academic Publishers.
6. Dempster WT: Space requirements of the seated operator, WADC Tech Rep 55-159, 1955, Wright-Patterson Air Force Base.
7. Garhammer J: Weight lifting and training. In Vaughan CL, ed: Biomechanics of sport, 1989.
8. Hall SJ and DePauw KP: A photogrammetrically based model for predicting total body mass centroid location, Res Q Exerc Sport 53:37, 1982.
9. Hanavan EP: A mathematical model of the human body, AMRL Tech Rep 64-102, 1964, Wright-Patterson Air Force Base.

10. Hay JG: The biomechanics of the long jump, Exerc Sport Sci Rev 14:401, 1986.
11. Hay JG, Andrews JG, and Vaughan CL: The influence of external load on joint torques exerted in a squat exercise, Proceedings of the biomechanics symposium, Bloomington, Ind, 1980.
12. Hay JG et al: Load, speed, and equipment effects in strength training exercises. In Matsui H and Kobayashi K, eds: Biomechanics VIII-B, Champaign, Ill, 1983, Human Kinetic Publishers, Inc. Press.
13. Jensen RK: Estimation of the biomechanical properties of three body types using a photogrammetric method, J Biomech 11:349, 1978.
14. Katch V, Weltman A, and Gold E: Validity of anthropometric measurements and the segment-zone method for estimating segmental and total body volume, Med Sci Sports 6:271, 1974.
15. Kirby RL, Price NA, and MacLeod DA: The influence of foot position on standing balance, J Biomech 20:423, 1987.
16. Laws K: The physics of dance, New York, 1984, Schirmer Books.
17. LeVeau B: Williams and Lissner: biomechanics of human motion, ed 2, Philadelphia, 1977, WB Saunders Co.
18. Miyamaru M et al: Path of the whole body center of gravity for young children in running. In Jonsson B, ed: Biomechanics X-B, Champaign, Ill, 1987, Human Kinetics Publishers, Inc.
19. Putnam CA and Kozey JW: Substantive issues in running. In Vaughan CL, ed: Biomechanics of sport, Boca Raton, Fla, 1989, CRC Press, Inc.
20. Redfield R and Hull ML: On the relation between joint moments and pedalling rates at constant power in bicycling, J Biomech 19:317, 1986.
21. Townend MS: Mathematics in sport, New York, 1984, John Wiley & Sons, Inc.
22. van Zuylen EJ, van Zelzen A, and van der Gon JJD: A biomechanical model for flexion torques of human arm muscles as a function of elbow angle, J Biomech 21:183, 1988.
23. Wickstrom RL: Fundamental motor patterns, ed 3, Philadelphia, 1983, Lea & Febiger.
24. Williams KR: Biomechanics of running, Exerc Sport Sci Rev 13:389, 1985.

ADDITIONAL READINGS

Garhammer J: Weight lifting and training. In Vaughan CL, ed: Biomechanics of sport, Boca Raton, Fla, 1989, CRC Press, Inc.

Critically reviews the scientific literature on weight lifts and resistance training.

Hay JG: The biomechanics of the long jump, Exerc Sports Sci Rev 14:401, 1986.

Reviews the scientific literature on factors related to performance in the long jump, with particular attention paid to the positioning of the athlete's center of gravity.

Laws K: The physics of dance, New York, 1984, Schirmer Books.

Relates principles of torque production and balance to the performance of various ballet movements.

Wiktorin CV and Nordin M: Introduction to problem solving in biomechanics, Philadelphia, 1982, Lea & Febiger.

Contains numerous sample problems with solutions provided.

13 MOVEMENT

Angular Kinetics

After reading this chapter, the student will be able to:

Identify the angular analogues of mass, force, and momentum.

Explain why changes in the configuration of a rotating airborne body can produce changes in the body's angular velocity.

Identify and provide examples of the angular analogues of Newton's laws of motion.

Define centripetal force and centrifugal force and explain where and how they act.

Solve quantitative problems relating to the factors that cause or modify angular motion.

Why do sprinters run with more swing phase flexion at the knee than distance runners? Why do dancers and ice skaters spin more rapidly when their arms are brought in close to the body? Why do cats always land on their feet? In this chapter more concepts pertaining to angular kinetics are explored from the perspective of the similarities and differences between linear and angular kinetic quantities.

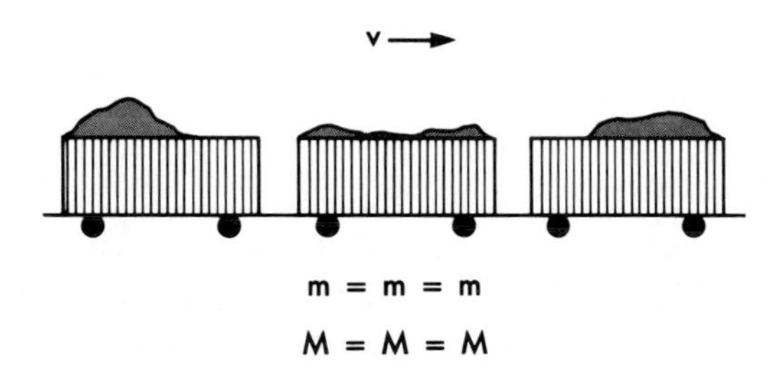

Figure 13-1
The distribution of mass in a system does not affect its linear momentum.

RESISTANCE TO ANGULAR ACCELERATION
Moment of Inertia

Inertia is a body's tendency to resist acceleration (see Chapter 2). Although inertia itself is a concept rather than a quantity that can be measured in units, a body's inertia is directly proportional to its mass (Figure 13-1). According to Newton's second law, the greater a body's mass, the greater its resistance to linear acceleration. Therefore, mass is a body's inertial characteristic for considerations relative to linear motion.

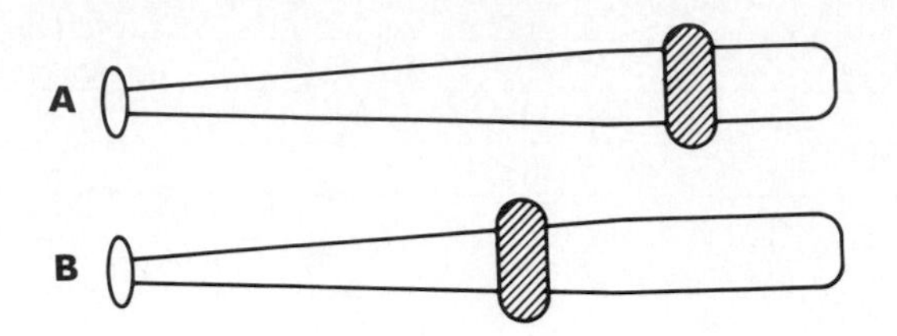

Figure 13-2
Although both bats have the same mass, Bat A is harder to swing than Bat B because the weight ring on it is positioned farther from the axis of rotation.

■ The more closely mass is distributed to the axis of rotation, the easier it is to initiate or stop angular motion.

Resistance to angular acceleration is also a function of a body's mass. The greater the mass, the greater the resistance to angular acceleration. However, the relative ease or difficulty of initiating or halting angular motion depends on an additional factor—the distribution of mass with respect to the axis of rotation.

Consider the baseball bat shown in Figure 13-2. Suppose a player warming up in the batter's box adds a weight ring to the bat he is swinging. Will the relative ease of swinging the bat be greater with the weight positioned near the striking end of the bat or near the bat's grip? Is it easier to swing a bat held with the hands placed on the grip of the bat (the normal position) or with the bat turned around and the hands placed on the barrel of the bat?

Experimentation with a baseball bat or some similar object makes it apparent that the more closely concentrated the mass to the axis of rotation, the easier it is to swing the object. Conversely, the more mass positioned away from the axis of rotation, the more difficult it is to initiate (or stop) angular motion. Resistance to angular acceleration therefore depends not only on the amount of mass possessed by an object but also on the distribution of that mass with respect to the axis of rotation. The inertial property for angular motion must incorporate both factors.

moment of inertia
the inertial property for rotating bodies that increases with both mass and the distance the mass is distributed from the axis of rotation

The inertial property for angular motion is known as **moment of inertia,** represented as *I*. Because every body is composed of molecules (particles of mass), each with its own particular distance from a given axis of rotation, the moment of inertia of a single particle of mass may be represented as the following:

$$I = mr^2$$

In this formula, m is the particle's mass and r is the particle's radius of rotation. The moment of inertia of an entire body is the sum of the moments of inertia of all the mass particles the object contains (Figure 13-3):

$$I = \Sigma mr^2$$

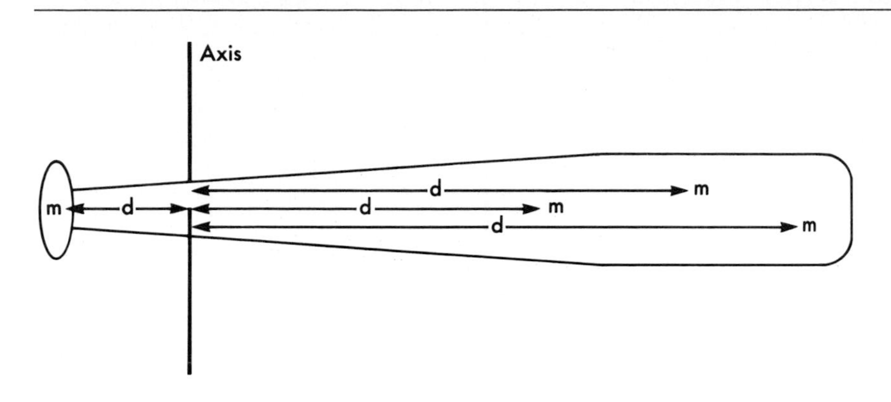

Figure 13-3
Moment of inertia is the sum of the products of the mass of the particles of which an object is composed and the square of each particle's radius of rotation.

The distribution of mass with respect to the axis of rotation is more significant than the total amount of body mass in determining resistance to angular acceleration because r is squared. Values of r change as the axis of rotation changes. Thus, when a player grips a baseball bat, "choking up" on the bat reduces the bat's moment of inertia with respect to the axis of rotation at the player's wrists and concomitantly increases the relative ease of swinging the bat. Little League baseball players often unknowingly make use of this concept when swinging bats that are longer and heavier than they can properly handle.

Changes in movement kinematics can affect changes in the moments of inertia of body limbs, thereby affecting the relative ease or difficulty of moving the limbs. When a multisegmented limb such as the arm or leg undergoes rotation, the joint angles present at the knee and at the elbow significantly influence the limb's moment of inertia. For example, the primary axis of rotation for a leg passes through the hip joint during running. The distribution of a given leg's mass and therefore its moment of inertia with respect to the hip depends on the angle present at the knee (Figure 13-4). Maximum flexion at the knee during the swing phase of gait increases with running speed (15). In sprinting, maximum angular acceleration of the legs is desired, and considerably more flexion at the knee is present during the swing phase than during running at slower speeds. This action greatly reduces the moment of inertia of the leg with respect to the hip joint, thus reducing resistance to hip flexion. During activities such as walking in which minimal angular acceleration of the legs is required, the flexion occurring at the knee during the swing phase remains relatively small, and the leg's moment of inertia with respect to the hip is relatively large. This tends to enhance stability because a larger torque at the hip is required to move the leg.

During sprinting, extreme flexion at the knee reduces the moment of inertia of the swinging leg.

Figure 13-4
The moments of inertia of the leg segments about the hip change as the angle at the knee changes because of alterations in the positions of the leg segment masses with respect to the hip.

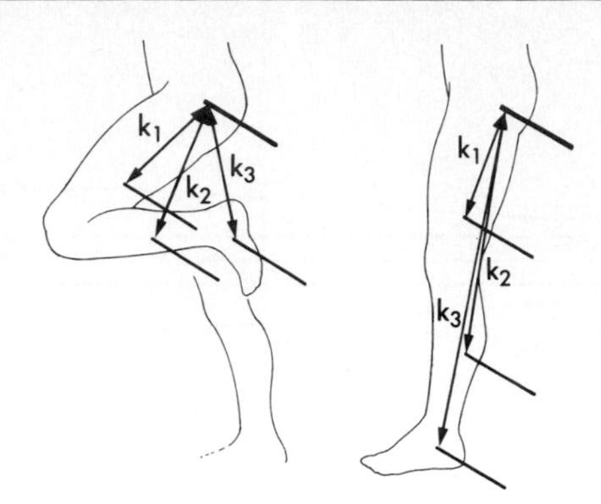

Determining Moment of Inertia

The fact that bone, muscle, and fat have different densities and are distributed dissimilarly in individuals complicates efforts to calculate human body segment moments of inertia.

Because there are formulas available for calculating the moment of inertia of regularly shaped solids, some investigators have modeled the human body as a composite of various geometric shapes.

Assessing moment of inertia for a body with respect to an axis by measuring the distance of each particle of body mass from an axis of rotation and then applying the formula is obviously impractical. In practice, mathematical procedures are used to calculate moment of inertia for bodies of regular geometric shapes and known dimensions. Because the human body is composed of segments that are of irregular shapes and heterogenous mass distributions, either experimental procedures or mathematical models are used to approximate moment of inertia values for individual body segments and for the body as a whole in different positions. Different approaches to moment of inertia assessment for the human body and its segments have been proposed. These procedures include the use of average values from cadaver studies, measurement of the acceleration of a swinging limb, photogrammetric methods, and mathematical modeling (7).

Once moment of inertia for a body of known mass has been assessed, the value may be characterized using the following formula:

$$I = mk^2$$

radius of gyration
the distance from the axis of rotation to a point where the body's mass could be concentrated without altering its rotational characteristics

In this formula, I is moment of inertia with respect to an axis, m is total body mass, and k is a distance known as the **radius of gyration.** The radius of gyration represents the object's mass distribution with respect to a given axis of rotation. It is the distance from the axis of rotation to a point at which the mass of the body can theoretically be concentrated without altering the inertial characteristics of the rotating body. The radius of gyration is a useful index of moment of inertia when a given body's resistance to rotation with respect to different axes is discussed. Units of moment of inertia parallel the formula definition of the quantity and therefore consist of units of mass multiplied by units of length squared (kg·m^2).

Human Body Moment of Inertia

Moment of inertia can only be defined with respect to a specific axis of rotation. The axis of rotation for a body segment in sagittal and frontal plane motions is typically an axis passing through the center of a body segment's proximal joint. When a segment rotates around its own longitudinal axis, its moment of inertia is quite different than its moment of inertia during flexion and extension or abduction and adduction because its mass distribution and therefore its moment of inertia are markedly different with respect to this axis of rotation. Figure 13-5 illustrates the difference in the lengths of the radii of gyration for the forearm with respect to the transverse and longitudinal axes of rotation.

The moment of inertia of the human body as a whole is also different with respect to different axes. When the entire human body rotates free of support, it moves around three **principal axes**—the transverse, the anteroposterior, or the longitudinal axis, each of which passes through the total body center of gravity. Moment of inertia with respect to one of these axes is known as a **principal moment of inertia.** Figure 13-6 shows quantitative estimates of principal moments of inertia for the human body in several different positions. When the body assumes a tucked position during a somersault, its principal moment of inertia (and resistance to angular motion) about the transverse axis is clearly less than when the body

▪ The ratio of muscular strength (the ability of a muscle group to produce torque about a joint) to segmental moments of inertia (resistance to rotation at a joint) is an important contributor to performance capability in gymnastic events.

principal axes
the three mutually perpendicular axes passing through the center of gravity that are referred to respectively as the transverse, anteroposterior, and longitudinal axes

principal moment of inertia
the total body moment of inertia relative to one of the principal axes

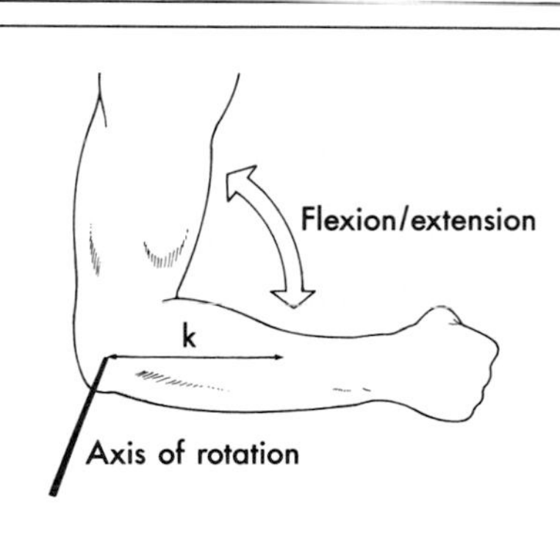

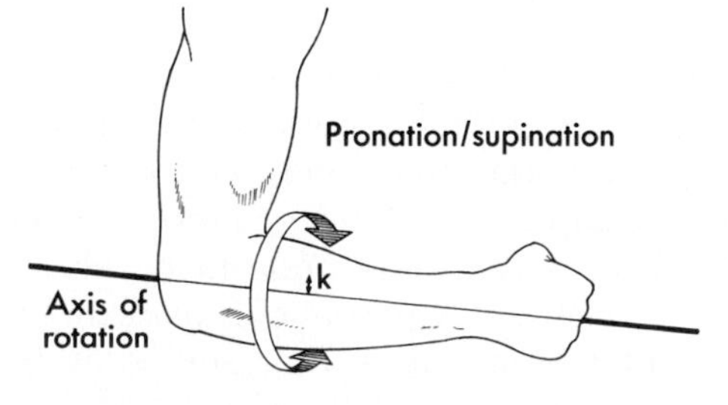

Figure 13-5
The radius of gyration varies with the distribution of mass relative to a given axis of rotation. The radius of gyration of the forearm for flexion/extension movements is much larger than that for pronation/supination.

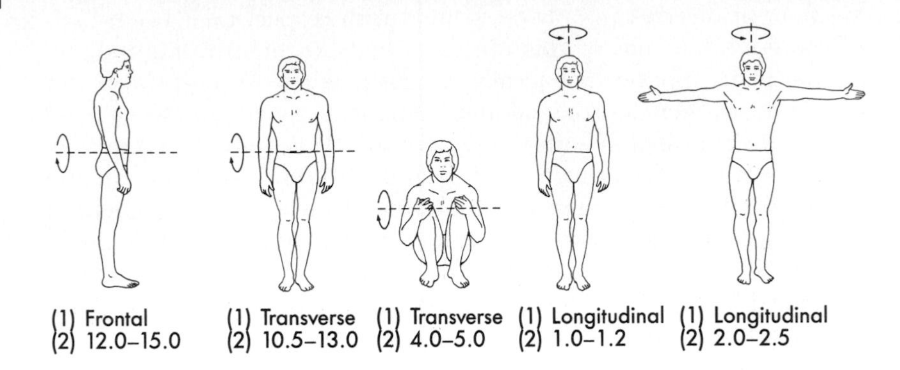

Figure 13-6
Principal moments of inertia of the human body in different positions with respect to different principal axes. *(1)*, Principal axis. *(2)*, Moment of inertia ($kg{\cdot}m^2$). (Modified from Hochmuth G: Biomechanik sportlicher bewegungen, Frankfurt, Germany, 1967, Wilhelm Limpert, Verlag.)

is in anatomical position. Divers performing a somersaulting dive undergo changes in principal moment of inertia about the transverse axis on the order of 15 $kg{\cdot}m^2$ to 6.5 $kg{\cdot}m^2$ as the body goes from a layout position to a pike position (4).

The adolescent growth spurt results in major changes in moments of inertia of the body segments.

As children grow from childhood through adolescence and into adulthood, developmental changes result in changing proportions of body segment lengths, masses, and radii of gyration, all affecting segment moments of inertia (6). Segment moments of inertia affect resistance to angular rotation and therefore performance capability in sports such as gymnastics and diving. Several prominent female gymnasts who achieved world-class status during early adolescence have faded from the public view before reaching the age of 20 because of declines in their performance capabilities generally attributed to changes in body proportions with growth. Substantial changes in principal moments of inertia for the body segments occur with age, and large interindividual differences in the growth patterns of these principal moments of inertia exist (7). According to Jenson (6), the best predictor of moment of inertia values among children is the product of body mass and body height squared rather than age.

ANGULAR MOMENTUM

angular momentum
the quantity of angular motion possessed by a body that is equal to the product of moment of inertia and angular velocity

In its role as the index of inertia for rotary movement, moment of inertia is an important component of other angular kinetic quantities. The quantity of motion that an object possesses is referred to as its *momentum* (see Chapter 11). Linear momentum is the product of the linear inertial property (mass) and linear velocity. The quantity of angular motion that a body possesses is likewise known as **angular momentum.** Angular momentum, represented as H, is the product of the angular inertial property, moment of inertia, and angular velocity:

$$\text{For linear motion: } M = mv$$

$$\text{For angular motion: } H = I\omega$$

$$\text{Or: } H = mk^2\omega$$

Three factors affect the magnitude of a body's angular momentum—its mass (m), the distribution of that mass with respect to the axis of rotation (k), and the angular velocity of the body (ω). If a body has no angular velocity, it has no angular momentum. As mass or angular velocity increases, angular momentum increases proportionally. The factor that most dramatically influences angular momentum is the distribution of mass with respect to the axis of rotation because angular momentum is proportional to the square of the radius of gyration (Figure 13-7). Units of angular momentum result from multiplying units of mass, units of length squared, and units of angular velocity, which yields $kg \cdot m^2/s$.

Because angular velocity is a vector quantity, angular momentum is also a vector quantity. A description of angular momentum must therefore include not only magnitude but also direction, either counterclockwise (+) or clockwise (−). Two or more angular momentum vectors can be added using the rules of vector composition.

The arm swing during takeoff contributes significantly to the diver's angular momentum.

During takeoff from a springboard or platform, a competitive diver must attain sufficient linear momentum to reach the necessary height (and safe distance from the board or platform) and sufficient angular momentum to perform the required number of rotations. For multiple rotation, nontwisting platform dives, the angular momentum generated at takeoff increases as the rotational requirements of the dive increase (3). Angular momentum values as high as 66 $kg \cdot m^2/s$ and 70 $kg \cdot m^2/s$ have been reported for multiple gold medalist Greg Louganis during his back two and a half and forward three and a half springboard dives, respectively (10). When a twist is also incorporated into a somersaulting dive, the angular momentum required is further increased. Inclusion of a twist during forward one and a half springboard dives is associated with increased angular momentum at takeoff of 6% to 19% (13).

The angular momentum required for execution of somersaulting dives is derived from the interaction between the feet and the platform or springboard during takeoff. Rotation of the body segments and body orientation with respect to the platform or springboard govern the forces exerted by the diver through the feet against the support surface. The rotation of the arms at takeoff generally contributes more to angular momentum than the motion of any other segment during both platform and springboard dives (3, 11). Highly skilled divers perform the arm swing with the arms fully extended,

Figure 13-7

SAMPLE PROBLEM 1

Consider a rotating 10 kg body for which $k = 0.2$ m and $\omega = 3$ rad/s. What is the effect on the body's angular momentum if the mass doubles? The radius of gyration doubles? The angular velocity doubles?

Solution

The body's original angular momentum is the following:

$$H = mk^2\omega$$

$$H = (10\ \text{kg})\ (0.2\ \text{m})^2(3\ \text{rad/s})$$

$$H = 1.2\ \text{kg}\cdot\text{m}^2/\text{s}$$

With mass doubled:

$$H = mk^2\omega$$

$$H = (20\ \text{kg})\ (0.2\ \text{m})^2(3\ \text{rad/s})$$

$$H = 2.4\ \text{kg}\cdot\text{m}^2/\text{s}$$

H is doubled.

With k doubled:

$$H = mk^2\omega$$

$$H = (10\ \text{kg})\ (0.4\ \text{m})^2(3\ \text{rad/s})$$

$$H = 4.8\ \text{kg}\cdot\text{m}^2/\text{s}$$

H is quadrupled.

With ω doubled:

$$H = mk^2\omega$$

$$H = (10\ \text{kg})\ (0.2\ \text{m})^2(6\ \text{rad/s})$$

$$H = 2.4\ \text{kg}\cdot\text{m}^2/\text{s}$$

H is doubled.

thus maximizing the moment of inertia of the arms and the angular momentum generated. Less-skilled divers must often use flexion at the elbow to reduce the moment of inertia of the arms about the shoulders so that arm swing can be completed during the time available (11).

During performances of other skills the motions of the body segments *counteract* the tendency for total body rotation resulting from

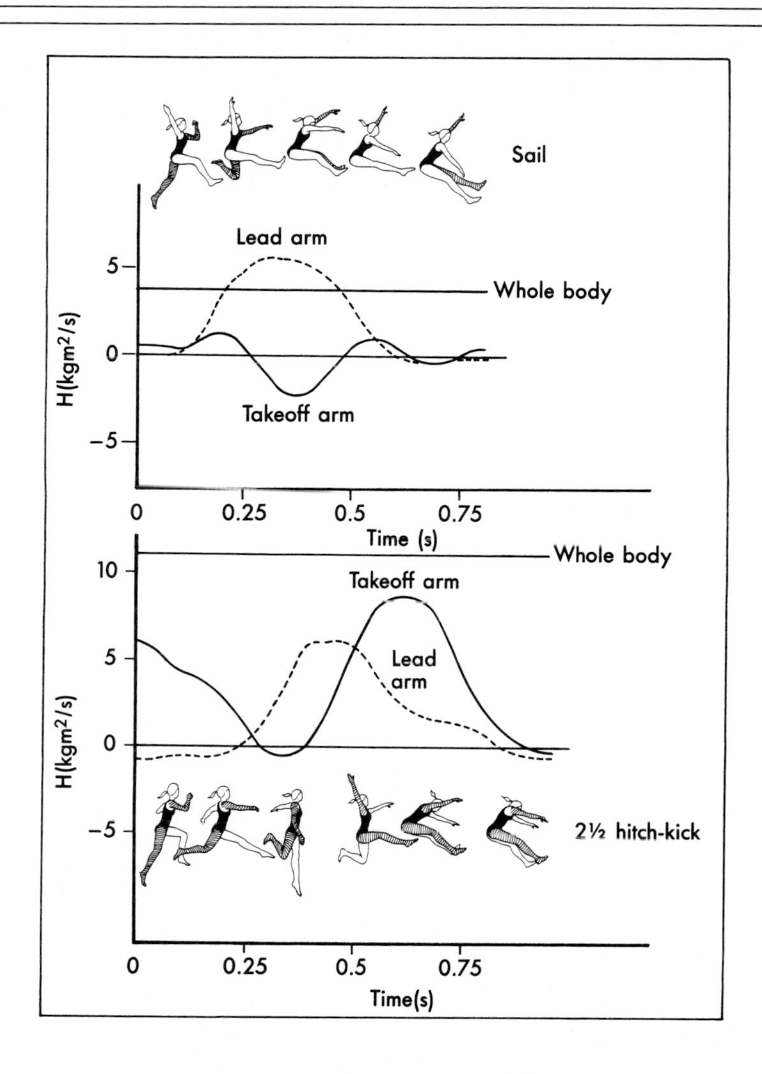

Figure 13-8

Angular momentum versus time during the sail and two and a half hitch-kick styles of long jump showing the amounts of whole body angular momentum counteracted by the arms. (Modified from Herzog W: Maintenance of body orientation in the flight phase of long jumping, Med Sci Sports Exerc 18:231, 1986.)

the angular momentum generated at takeoff. In the long jump, rotation of the arms and legs during the airborne phase offsets the tendency for forward rotation of the body around the transverse principal axis produced during takeoff and enables maintenance of the trunk in a nearly upright position. The athlete's principal moment of inertia with respect to the transverse axis during the long jump is approximately 11.05 $kg \cdot m^2/s$ during the two and a half hitch-kick style of long jump and 4.17 $kg \cdot m^2/s$ during the sail long jump. During the hitch-kick jump the actions of the takeoff leg and takeoff arm primarily counteract total body angular momentum. During the sail the major contributions are from the takeoff leg and the lead arm (Figures 13-8 to 13-9) (5).

One approach to dealing with the angular momentum generated during the long jump takeoff has been to incorporate a flip into the airborne phase.

Figure 13-9
Angular momentum versus time during the sail and two and a half hitch-kick styles of long jump showing the amounts of whole body angular momentum counteracted by the legs. (Modified from Herzog W: Maintenance of body orientation in the flight phase of long jumping, Med Sci Sports Exerc 18:231, 1986.)

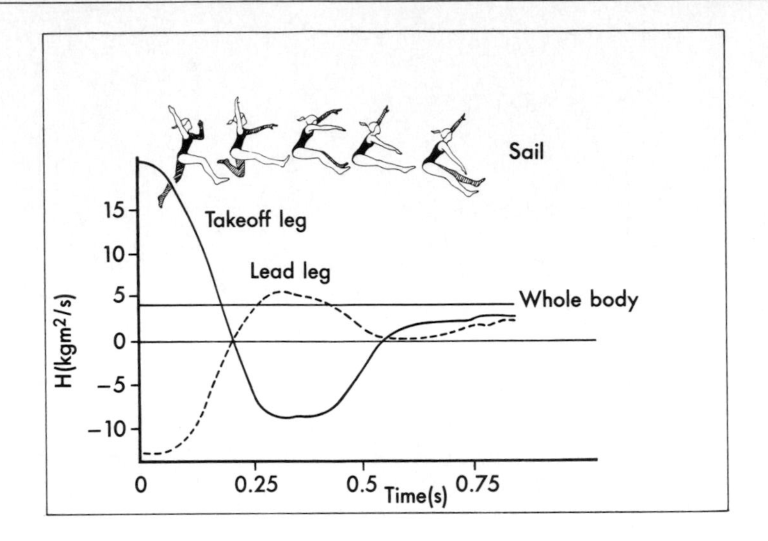

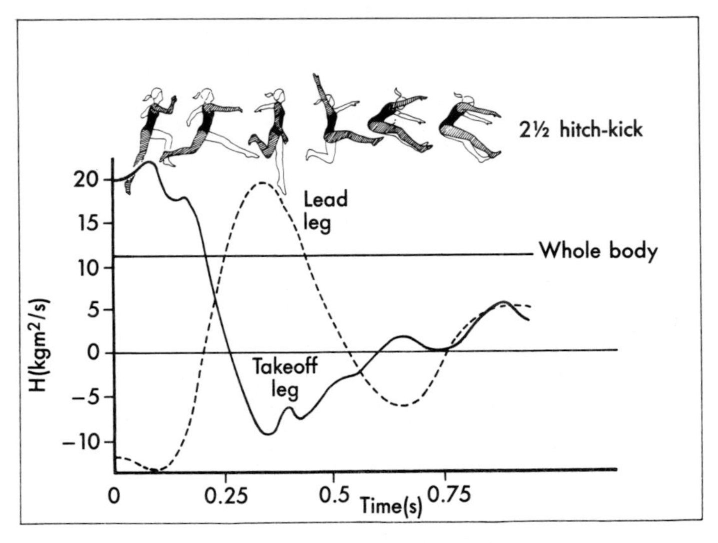

Conservation of Angular Momentum

Whenever gravity is the only acting external force, angular momentum is conserved. For angular motion, the principle of conservation of momentum may be stated as follows:

The total angular momentum of a given system remains constant in the absence of external torques.

Gravitational force acting at a body's center of gravity produces no torque because $d_\perp$ equals 0 and therefore creates no change in angular momentum.

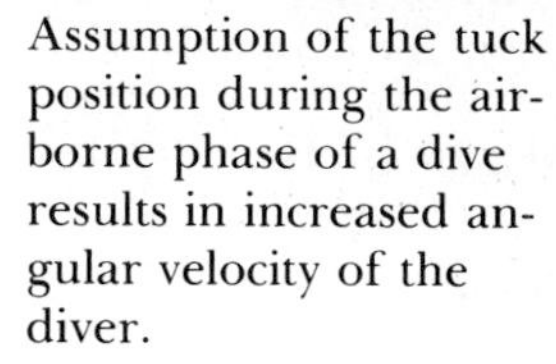

Figure 13-10
Assumption of the tuck position during the airborne phase of a dive results in increased angular velocity of the diver.

The principle of conservation of angular momentum is particularly useful in the mechanical analysis of diving, trampolining, and gymnastics events in which the human body undergoes controlled rotations while airborne. In a one and a half front somersault dive the diver leaves the springboard with a fixed amount of angular momentum produced largely by the forceful rotation of the body segments during the recoil action of the board at takeoff (Figure 13-10) (11). According to the principle of conservation of angular momentum, the amount of angular momentum present at the instant of takeoff remains constant throughout the dive. As the diver goes from an extended layout position into a tuck, the radius of gyration is decreased, thus reducing the body's principal moment of inertia about the transverse axis. Because angular momentum remains constant, a compensatory increase in angular velocity must accompany the decrease in moment of inertia. The tighter the diver's tuck, the greater the angular velocity. Once the somersault is completed, the diver extends to a full layout position, thereby increasing total body moment of inertia with respect to the axis of rotation. Again, because angular momentum remains constant, an equivalent decrease in angular velocity occurs. For the diver to appear to enter the water perfectly vertically, minimal angular velocity is desirable. The sample problem in Figure 13-11 quantitatively illustrates this example.

■ The magnitude and direction of the angular momentum vector for an airborne performer are established at the instant of takeoff.

Because the mass of the human body does not change appreciably during a given movement, changes in angular velocity when external torques are not present must compensate for changes in body segment configuration that result in changes in moment of inertia. Novice gymnasts performing a handspring with insufficient angular mo-

Figure 13-11

SAMPLE PROBLEM 2

A 60 kg diver is positioned so that his radius of gyration is 0.5 m as he leaves the board with an angular velocity of 4 rad/s. What is the diver's angular velocity when he assumes a tuck position, altering his radius of gyration to 0.25 m?

Known

$m = 60$ kg

$k = 0.5$ m

$\omega = 4$ rad/s

$m = 60$ kg

$k = 0.25$ m

Solution

To find ω, calculate the amount of angular momentum that the diver possesses when he leaves the board, since angular momentum remains constant during the airborne phase of the dive:

Position 1:

$$H = mk^2\omega$$

$$H = (60 \text{ kg})\ (0.5 \text{ m})^2(4 \text{ rad/s})$$

$$H = 60 \text{ kg}\cdot\text{m}^2/\text{s}$$

Use this constant value for angular momentum to determine ω when $k = 0.25$ m:

$$\text{Position 2:} \quad H = mk^2\omega$$

$$60 \text{ kg}\cdot\text{m}^2/\text{s} = (60 \text{ kg})\ (0.25 \text{ m})^2(\omega)$$

$$\boxed{\omega = 16 \text{ rad/s}}$$

mentum may need to maintain a tucked position all the way through landing to undergo enough angular displacement to avoid landing on their backs.

Other examples of conservation of angular momentum occur when an airborne performer has a total body angular momentum of 0 and a forceful movement, such as a jump pass or volleyball spike, is executed. When a volleyball player performs a spike, moving the hitting arm with a high angular velocity and a large angular momentum, there is a compensatory rotation of the lower body producing an equal amount of angular momentum in the opposite direction (Figure 13-12). Since the moment of inertia of the two legs with respect to the hip is much greater than that of the spiking arm with respect to the shoulder, the angular velocity of the legs generated to counter the angular momentum of the swinging arm is much less than the angular velocity of the spiking arm.

Figure 13-12
During the airborne execution of a spike in volleyball, total body angular momentum is conserved.

Transfer of Angular Momentum

Although angular momentum remains constant in the absence of external torques, transferring angular velocity at least partially from one principal axis of rotation to another is possible. This occurs when a diver changes from a primarily somersaulting rotation to one that is primarily twisting and vice versa. An airborne performer's angular velocity vector does not necessarily occur in the same direction as the angular momentum vector. It is possible for a body's somersaulting angular momentum and its twisting angular momentum to be altered in midair, though the vector sum of the two, the total angular momentum, remains constant in magnitude and direction.

Researchers have observed several procedures for changing the total body axis of rotation. Asymmetrical arm movements and rotation of the hips (termed *hula movement*) can tilt the axis of rotation out of the original plane of motion (Figure 13-13) (14, 17). Liu and Nelson (9) have identified some principles for optimizing arm swing movements when tilting the axis of rotation is desired. These include initiating the arm swing at the point during which somersaulting velocity is highest, employing a high velocity of the swinging arm along the longitudinal axis of the body, and maximizing the mass moved with the swinging arm by including shoulder elevation with the arm movement.

Figure 13-13
Asymmetrical positioning of the arms with respect to the transverse principal axis can shift rotation partially to the longitudinal axis.

Airborne human performers employ several twisting techniques. The methods of twist production used in freestyle aerial skiing have been studied by Yeadon (16). He found that some skiers left the ground with all angular momentum around the principal transverse axis and initiated twisting through asymmetrical arm and hip movements. Other skiers left the ground with angular momentum around both transverse and vertical principal axes. Among the latter group, some also employed asymmetrical arm and hip movements to increase the angles of rotational tilt (16).

■ Asymmetrical movement of the arms is used more frequently to tilt the axis of rotation during airborne performances than rotational movement of the hips.

Even when total body angular momentum is 0, generating a twist in midair is possible using skillful manipulation of a body composed of at least two segments. Prompted by the observation that a domestic cat seems to always land on its feet no matter what the position from which it falls, scientists have studied this apparent contradiction of the principle of conservation of angular momentum (2). Gymnasts and divers can use this procedure, referred to as *cat rotation,* without violating the conservation of angular momentum.

Cat rotation is basically a two phase process (Figure 13-14). It is accomplished most effectively when the two body segments are in a 90 degree pike position so that the radius of gyration of one segment is maximum with respect to the longitudinal axis of the other segment. The first phase consists of the internally generated rotation of Segment 1 around its longitudinal axis. Because angular momentum is conserved, there is a compensatory rotation of Segment 2 in the opposite direction around the longitudinal axis of Segment 1. However, the resulting rotation is of a relatively small velocity because k for Segment 2 is relatively large with respect to Axis 1. The

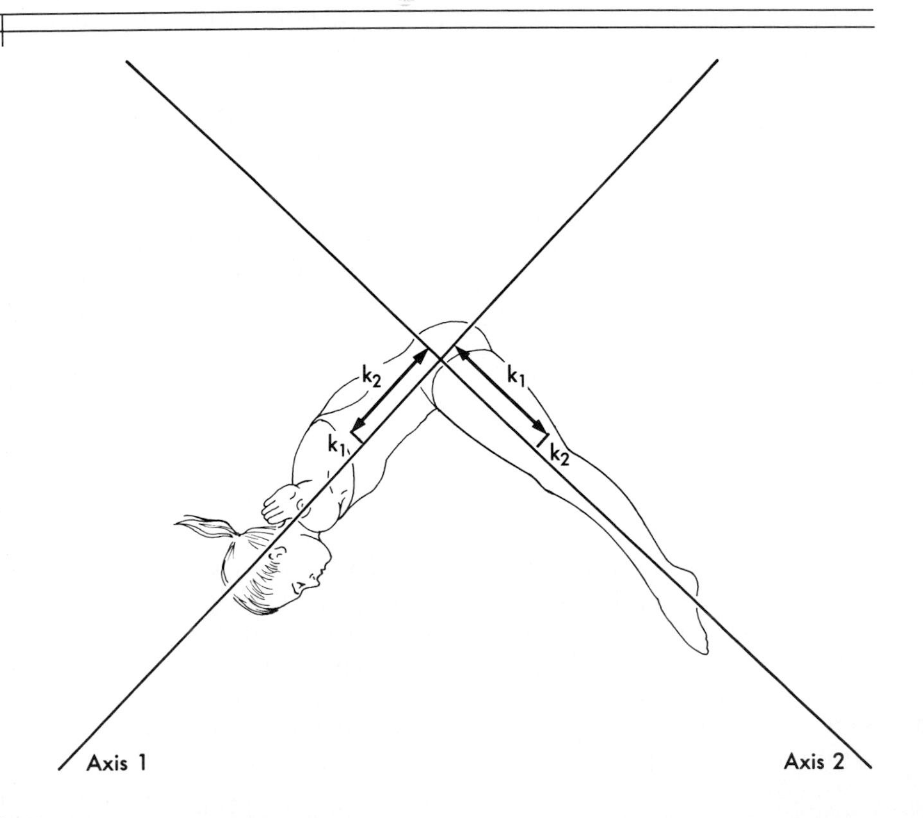

Figure 13-14
Cat rotation.

second phase of the process consists of rotation of Segment 2 around its longitudinal axis in the same direction originally taken by Segment 1. Accompanying this motion is a compensatory rotation of Segment 1 in the opposite direction around Axis 2. Again, angular velocity is relatively small because k for Segment 1 is relatively large with respect to Axis 2. Using this procedure, a skilled diver can initiate a twist in midair and turn through as much as 450 degrees (4). Cat rotation is performed around the longitudinal axes of the two major body segments. It is easier to initiate rotation about the longitudinal principal axis than either the transverse or anteroposterior principal axes, because total body moment of inertia with respect to the longitudinal axis is the smallest of the three.

Angular momentum can also be transferred from one body segment to another to increase the effectiveness of a skill involving maximum velocity of a distal segment. During the execution of a sling style underhand softball pitch, angular momentum can be progressively transferred from proximal to distal body segments (Figure 13-15). The pitch begins as the pitcher draws the ball hand back behind the body through extension at the shoulder and elbow of the pitching arm. As this occurs, the total body center of gravity is shifted forward and the forward step with the foot opposite the throwing arm starts at approximately the same time as flexion of the throwing arm at the shoulder. While the throwing arm is in flexion at the shoulder, the forward movement of the total body center of gravity stops, thereby transferring the momentum of the moving body entirely to the throwing arm. Because the throwing arm has a fixed mass, it accommodates this added momentum with dramatically increased angular velocity. As the throwing arm reaches an approximately vertical position during its forward progression, the movement of the upper arm stops, causing a transfer of angular momentum to the forearm and ball hand with a consequent increase in

■ Transfer of angular momentum from proximal to distal segments occurs during the execution of both underhand and overhand pitching motions among skilled performers.

Figure 13-15
A sling style of softball pitch.

the forearm's angular velocity. The final transfer of angular momentum occurs just before ball release when the forearm stops, resulting in a large angular velocity at the wrist at the time of ball release. The throwing arm functions as a whip, with transfer of angular momentum resulting in greater angular velocities of distal segments as the proximal segments are stopped.

Because momentum is not conserved during a pitch, its transfer among body segments is not complete. The extent to which momentum is transferred depends on the kinematics of the movement, which are influenced by agonist and antagonist muscle group tensions and joint ranges of motion. It is possible to perform a pitch so that no transfer of momentum among body segments occurs. However, skillful pitching is characterized by some transfer of momentum.

Change in Angular Momentum

When an external torque does act, it changes the amount of angular momentum present in a system predictably. Just as with changes in linear momentum, changes in angular momentum depend not only on the magnitude and direction of acting external torques but also on the length of the time interval over which each torque acts:

$$\text{Linear impulse} = Ft$$

$$\text{Angular impulse} = Tt$$

angular impulse
this action produces a change in angular momentum that is equal to the product of torque and the time interval over which the torque acts

When an **angular impulse** acts on a system, the result is a change in the total angular momentum of the system. The impulse-momentum relationship for angular quantities may be expressed as the following:

$$Tt = \Delta H$$

$$Tt = (I\omega)_2 - (I\omega)_1$$

As before, the symbols T, t, H, I, and ω represent torque, time, angular momentum, moment of inertia, and angular velocity respectively, and subscripts 1 and 2 denote initial and second or final points in time. Because angular impulse is the product of torque and time, significant changes in an object's angular momentum may result from the action of a large torque over a small time interval or from the action of a small torque over a large time interval. Since torque is the product of a force's magnitude and the perpendicular distance to the axis of rotation, both of these factors affect angular impulse. The effect of angular impulse on angular momentum is shown in Figure 13-16.

In the throwing events in track and field, the object is to maximize the angular impulse exerted on an implement before release to maximize its momentum and the ultimate horizontal displacement fol-

SAMPLE PROBLEM 3

Figure 13-16

What average amount of force must be applied by the shoulder flexors inserting at an average perpendicular distance of 1.5 cm from the axis of rotation at the shoulder over a period of 0.3 seconds to stop the motion of the 3.5 kg arm swinging with an angular velocity of 5 rad/s when k = 20 cm?

Known

$d = 0.015$ m

$t = 0.3$ s

$m = 3.5$ kg

$k = 0.20$ m

$\omega = 5$ rad/s

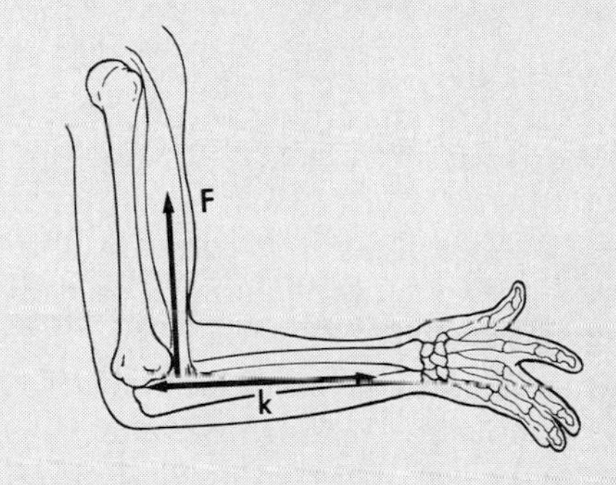

Solution

The impulse-momentum relationship for angular motion can be used:

$$Tt = \Delta H$$

$$Fdt = (mk^2\omega)_2 - (mk^2\omega)_1$$

$$F(0.015 \text{ m})(0.3 \text{ s}) = 0 - (3.5 \text{ kg})(0.20 \text{ m})^2(5 \text{ rad/s})$$

$$F = 155.56 \text{ N}$$

lowing release. As discussed in Chapter 10, linear velocity is directly related to angular velocity, with the radius of rotation serving as the factor of proportionality. As long as the moment of inertia (mk^2) of a rotating body remains constant, increased angular momentum translates directly to increased linear momentum when the body is projected. This concept is particularly evident in the hammer throw in which the athlete first swings the hammer two or three times around the body with the feet planted followed by the next three or four whole body turns executed with the athlete facing the hammer before release. Some hammer throwers perform the first one or two of the whole body turns with the trunk in slight flexion (called *countering with the hips*), thereby enabling a farther reach with the hands

Figure 13-17
A hammer thrower must counter the centrifugal force of the hammer to avoid being pulled out of the throwing ring. **A,** Countering with the shoulders results in a smaller radius of rotation for the hammer than **B,** countering with the hips.

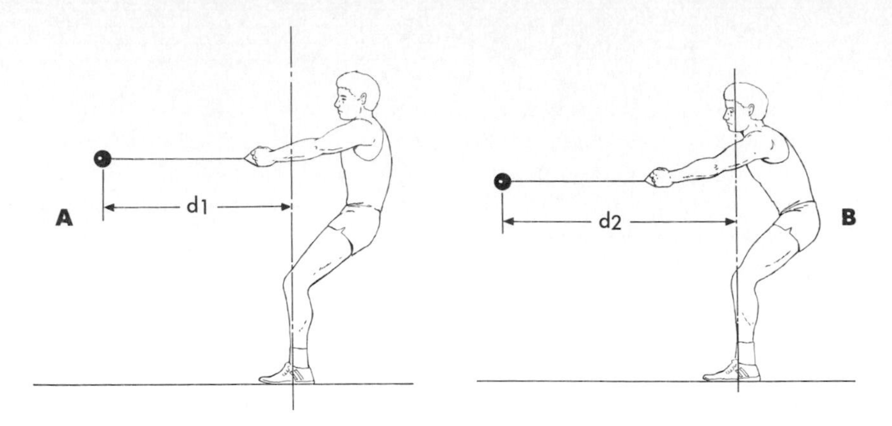

(Figure 13-17). This tactic increases the radius of rotation and thus the moment of inertia of the hammer with respect to the axis of rotation so that if angular velocity is not reduced, the angular momentum of the thrower/hammer system is increased. For this strategy the final turns are completed with the entire body leaning away from the hammer or *countering with the shoulders.* According to Dapena and McDonald (1), although the ability to lean forward throughout the turns should increase the angular momentum imparted to the hammer, factors such as a natural tendency to protect against excessive spinal stresses or shoulder strength limitations may prevent the thrower from accomplishing this technique modification.

The surface reaction force is used by the dancer to generate angular momentum during the takeoff of the *tour jeté.*

The angular momentum necessary for the total body rotations executed during aerial skills is primarily generated by the angular impulse created by the reaction force of the support surface during takeoff. During back dives performed from a platform, the major angular impulse is produced during the final weighting of the platform when the diver comes out of a crouched position through extension at the hip, knee, and ankle joints and executes a vigorous arm swing simultaneously (12). The vertical component of the platform reaction force, acting in front of the diver's center of gravity, primarily creates the backward angular momentum required (Figure 13-18).

Angular impulse produced through the support surface reaction force is also essential for performance of the *tour jeté,* a dance movement that consists of a jump accompanied by a 180 degree turn, with the dancer landing on the foot opposite the takeoff foot. When the movement is performed properly, the dancer appears to rise straight up and then rotate about the principal vertical axis in the air. In reality, the jump must be executed so that a reaction torque around the dancer's vertical axis is generated by the floor. The extended leg at the initiation of the jump creates a relatively large moment of in-

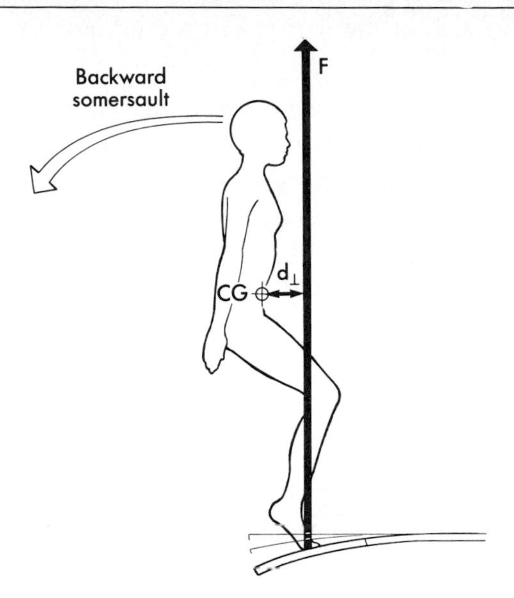

Figure 13-18

The product of the vertical reaction force (F) and its moment arm with respect to the diver's center of gravity ($d_{\perp}$) creates a torque, which generates the angular impulse that produces the diver's angular momentum at takeoff.

ertia relative to the axis of rotation, thereby resulting in a relatively low total body angular velocity. At the peak of the jump the dancer's legs simultaneously cross the axis of rotation and the arms simultaneously come together overhead, close to the axis of rotation. These movements dramatically reduce moment of inertia, thus increasing angular velocity (8).

▪ When a support surface reaction force is directed through the performer's center of gravity, linear but not angular impulse is generated.

ANGULAR ANALOGUES OF NEWTON'S LAWS OF MOTION

Table 13-1 presents linear and angular kinetic quantities in a parallel format. With the many parallels between linear and angular motion, it is not surprising that Newton's laws of motion may also be stated in terms of angular motion. It is necessary to remember that torque and moment of inertia are the angular equivalents of force and mass in substituting terms.

Table 13-1

LINEAR AND ANGULAR KINETIC QUANTITIES

LINEAR	ANGULAR
Mass (m)	Moment of inertia (I)
Force (F)	Torque (T)
Momentum (M)	Angular momentum (H)
Impulse (Ft)	Angular impulse (Tt)

Newton's First Law

The angular version of the first law of motion may be stated as follows:

> A rotating body will maintain a state of constant angular motion unless acted on by an external torque.

In the analysis of human movement in which mass remains constant throughout, this angular analogue forms the underlying basis for the principle of conservation of angular momentum. Because angular velocity may change to compensate for changes in moment of inertia resulting from alterations in the radius of gyration, the quantity that remains constant in the absence of external torque is angular momentum.

Figure 13-19

SAMPLE PROBLEM 4

The knee extensors insert on the tibia at an angle of 30 degrees at a distance of 3 cm from the axis of rotation at the knee. How much force must the knee extensors exert to produce an angular acceleration at the knee of 1 rad/s^2, given a mass of the lower leg and foot of 4.5 kg and k = 23 cm?

Known

$d = 0.03$ m

$\alpha = 1$ rad/s^2

$m = 4.5$ kg

$k = 0.23$ m

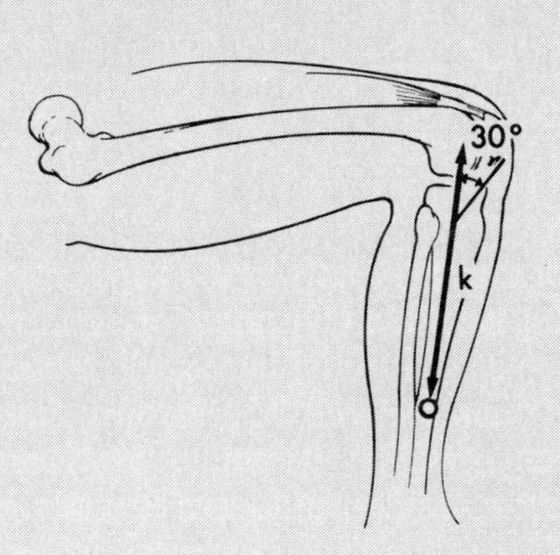

Solution

The angular analogue of Newton's second law of motion may be used to solve the problem:

$$T = I\alpha$$

$$Fd = mk^2\alpha$$

$$(F \sin 30\ N)\ (0.03\ m) = (4.5\ kg)\ (0.23\ m)^2(1\ rad/s^2)$$

$$\boxed{F = 15.9\ N}$$

Newton's Second Law

In angular terms, Newton's second law may be stated algebraically and in words as the following:

$$T = I\alpha$$

An external torque produces angular acceleration of a body that is directly proportional to the magnitude of the torque, in the same direction as the torque, and inversely proportional to the body's moment of inertia.

When the forearm flexors develop tension, the resulting torque at the elbow produces forearm flexion. In accordance with Newton's second law for angular motion, the angular acceleration of the forearm is directly proportional to the magnitude of the net torque at the elbow and in the direction (flexion) of the net torque at the elbow. The greater the moment of inertia with respect to the axis of rotation at the elbow, the smaller the angular acceleration occurring (Figure 13-19).

Newton's Third Law

The law of reaction may be stated in angular form as the following:

For every torque exerted by one body on another, there is an equal and opposite torque exerted by the second body on the first.

When a baseball player forcefully swings a bat, rotating the mass of the upper body, a torque is created around the player's longitudinal axis. If the batter's feet are not firmly planted, the lower body tends to rotate around the longitudinal axis in the opposite direction. However, since the feet usually are planted, the torque generated by the upper body is translated to the ground, where the earth generates a torque of equal magnitude and opposite direction on the cleats of the batter's shoes.

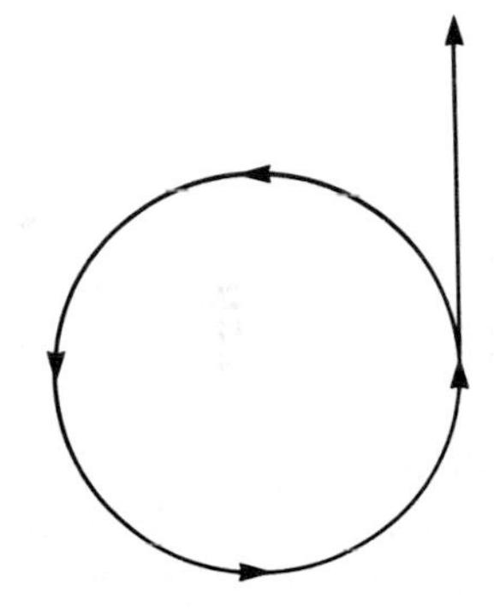

Figure 13-20
An object swung in a circle and then released will follow a linear path tangential to the curve at the point of release.

CENTRIPETAL AND CENTRIFUGAL FORCES

Bodies undergoing rotary motion around a fixed axis are also subject to a linear force. When an object attached to a line is whirled around in a circular path and then released, the object flies off on a path that forms a tangent to the circular path it was following at the point at which it was released (Figure 13-20). **Centripetal force** prevents the rotating body from leaving its circular path while rotation occurs around a fixed axis. The direction of a centripetal force is always toward the center of rotation, which is the reason it is also known as *center-seeking force.* Centripetal force produces the radial component of the acceleration of a body traveling on a curved path (see Chapter 10). The following formula quantifies the magnitude of a centripetal force in terms of the tangential linear velocity of the rotating body:

$$F_c = \frac{mv^2}{r}$$

centripetal force
the force directed toward the center of rotation of any rotating body

In this formula, F_c is centripetal force, m is mass, v is the tangential linear velocity of the rotating body at a given point in time, and r is the radius of rotation. Centripetal force may also be defined in terms of angular velocity:

$$F_c = mr\omega^2$$

As is evident from both equations, the speed of rotation is the most influential factor on the magnitude of centripetal force because centripetal force is proportional to the square of velocity or angular velocity.

centrifugal force
the reaction force equal in magnitude and opposite in direction to centripetal force

In accordance with Newton's third law, there is a force of equal magnitude and opposite direction that is created as a reaction to centripetal force. This reaction force is termed **centrifugal force.** Because centrifugal force exists only in the form of a reaction force, centripetal and centrifugal forces always act on different bodies. When a ball attached to a line is swung in a circle, the line exerts a centripetal force on the ball, causing the ball to remain on its circular path, and the ball exerts a centrifugal force on the line, keeping the line taut (Figure 13-21). If the line breaks and the ball flies off on a tangent to its circular path, both centripetal and centrifugal forces cease to exist.

■ Because centrifugal force exists only as a reaction to centripetal force, many physics texts do not identify centrifugal force by name.

The influence of centripetal and centrifugal forces on sport-related activities are numerous. When a player swings a tennis racquet through the air along an angular path, the hand gripping the racquet supplies the centripetal force that keeps the racquet traveling on its angular path, and the racquet exerts a centrifugal force on the hand. When the implement being swung is more massive, as in the hammer throw, the amounts of centripetal and centrifugal force are proportionally increased so that the athlete must lean away from the

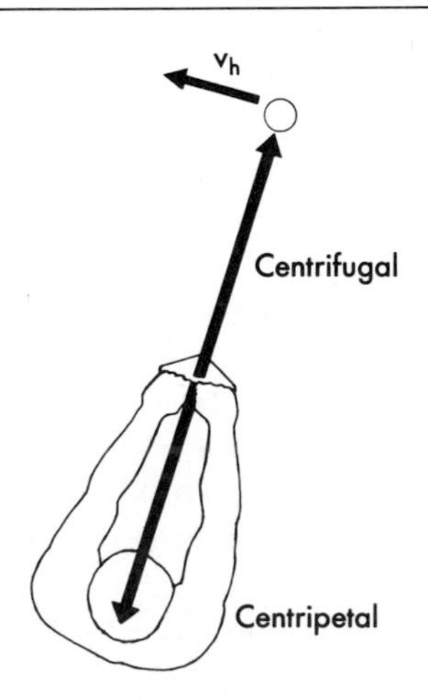

Figure 13-21
When a ball on a string is rotated, the thrower exerts a centripetal force on the ball and the ball exerts an equal and opposite centrifugal force on the thrower.

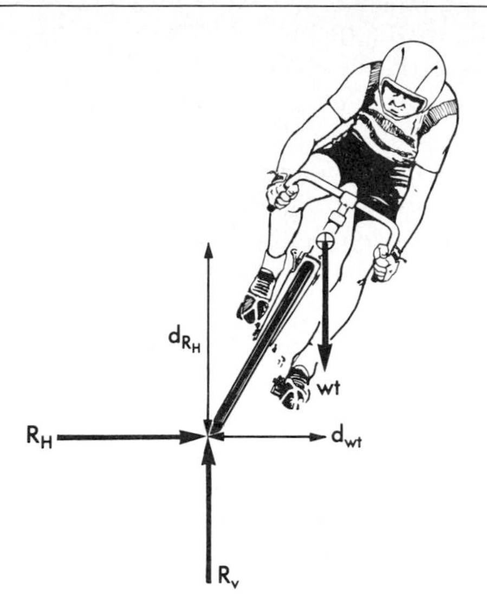

Figure 13-22
Free body diagram of a cyclist on a curve. R_H is centripetal force. When the cyclist is balanced, $(wt)(d_{wt}) = (R_H)(D_{RH})$.

rotating hammer to avoid being pulled off balance by the centrifugal force. Similarly, the heavier the gymnast performing a giant swing on a bar, the more the grip strength required to counter the centrifugal force the rotating body exerts on the bar.

When a cyclist rounds a curve, the ground exerts centripetal force on the tires of the cycle. The forces acting on the cycle/cyclist system are weight and the ground reaction force (Figure 13-22). The horizontal component of the ground reaction force (centripetal force) creates a torque about the cycle/cyclist center of gravity. To prevent rotation toward the outside of the curve, the cyclist must lean to the inside of the curve so that the moment arm of the system's weight relative to the contact point with the ground is large enough to produce an oppositely directed torque of equal magnitude.

Cyclists and runners lean into a curve to offset the torque created by centripetal force acting on the base of support.

SUMMARY

Whereas as a body's resistance to linear acceleration is proportional to its mass, resistance to angular acceleration is also related to the distribution of body mass with respect to the axis of rotation. Resistance to angular acceleration is known as moment of inertia, a quantity that incorporates both the amount of mass and its distribution relative to the center of rotation.

Just as linear momentum is the product of the linear inertial property, mass, and linear velocity, angular momentum is the product of moment of inertia and angular velocity. In the absence of external torques, angular momentum is conserved. An airborne human performer can alter total body angular velocity by manipulating moment of inertia through changes in body configuration relative to the principal axis around which rotation is occurring. Skilled performers can also alter the axis of rotation and initiate rotation when no angular momentum is present while airborne. The principle of conservation of angular momentum is based on the angular version of Newton's first law of motion. The second and third laws of motion may also be expressed in angular terms by substituting moment of inertia for mass, torque for force, and angular acceleration for linear acceleration.

A linear force that acts on all rotating bodies is centripetal or center-seeking force, which is always directed toward the center of rotation. The magnitude of centripetal force depends on the mass, speed, and radius of rotation of the rotating body. Centrifugal force, which exists only in the form of a reaction force, is equal in magnitude and opposite in direction to centripetal force.

INTRODUCTORY PROBLEMS

1. Select three sport or daily living implements and explain the ways in which you can modify each implement's moment of inertia with respect to the axis of rotation.
2. Construct a table displaying common units of measure for both linear and angular quantities of the inertial property, momentum, and impulse.
3. Skilled performance of a number of sport skills is characterized by "follow through." Explain the value of "follow through" in terms of the concepts discussed in this chapter.
4. Explain the reason the product of body mass and body height squared is a good predictor of body moment of inertia.
5. A 1.1 kg racquet has a moment of inertia about a grip axis of rotation of 0.4 $kg \cdot m^2$. What is its radius of gyration? (Answer: 0.6 m)
6. How much angular impulse must be supplied by the hamstrings to bring a leg swinging at 8 rad/s to a stop, given that the leg's moment of inertia is 0.7 $kg \cdot m^2$? (Answer: 5.6 $kg \cdot m^2/s$)
7. Given the following principal transverse axis moments of inertia and angular velocities, calculate the angular momentum of each of the following gymnasts. What body configurations do these moments of inertia represent?

	I_{cg} ($kg \cdot m^2$)	(rad/s)
A	3.5	20.00
B	7.0	10.00
C	15.0	4.67

(Answer: A = 70 $kg \cdot m^2$; B = 70 $kg \cdot m^2$; C = 70 $kg \cdot m^2$)

8. A volleyball player's 3.7 kg arm moves at an average angular velocity of 15 rad/s during execution of a spike. If the average moment of inertia of the extending arm is 0.45 kg·m^2, what is the average radius of gyration for the arm during the spike? (Answer: 0.35 m)
9. A 50 kg diver in a full layout position, with a total body radius of gyration with respect to her transverse principal axis equal to 0.45 m, leaves a springboard with an angular velocity of 6 rad/s. What is the diver's angular velocity when she assumes a tuck position, reducing her radius of gyration to 0.25 m? (Answer: 19.4 rad/s)
10. If the centripetal force exerted on a swinging tennis racket by a player's hand is 40 N, how much centrifugal force is exerted on the player by the racket? (Answer: 40 N)

ADDITIONAL PROBLEMS

1. The radius of gyration of the thigh with respect to the transverse axis at the hip is 54% of the segment length. The mass of the thigh is 10.5% of total body mass, and the length of the thigh is 23.2% of total body height. What is the moment of inertia of the thigh with respect to the hip for males of the following body masses and heights?

	MASS (kg)	HEIGHT (m)
A	60	1.6
B	60	1.8
C	70	1.6
D	70	1.8

(Answer: A = 0.25 kg·m^2, B = 0.32 kg·m^2, C = 0.30 kg·m^2, D = 0.37 kg·m^2)

2. If you had to design a model of the human body composed entirely of regular geometric solids, which solid shapes would you choose? Using a straight edge, sketch a model of the human body that incorporates the solid shapes you have selected.
3. A 0.68 kg tennis ball is given an angular momentum of 2.72×10^{-3} kg·m^2/s when struck by a racquet. If its radius of gyration is 2 cm, what is its angular velocity? (Answer: 10 rad/s)
4. A 7.27 kg shot makes seven complete revolutions during its 2.5 second flight. If its radius of gyration is 2.54 cm, what is its angular momentum? (Answer: 0.0817 kg·m^2/s)
5. What is the resulting angular acceleration of a 1.7 kg forearm and hand when the forearm flexors, attaching 3 cm from the center of rotation at the elbow, produce 10 N of tension, given

a 90 degree angle at the elbow and a forearm and hand radius of gyration of 20 cm? (Answer: 4.41 rad/s^2)

6. The patellar tendon attaches to the tibia at a 20 degree angle 3 cm from the axis of rotation at the knee. If the tension in the tendon is 400 N, what is the resulting acceleration of the 4.2 kg lower leg and foot given a radius of gyration of 25 cm for the lower leg/foot with respect to the axis of rotation at the knee? (Answer: 15.6 rad/s^2)
7. A cavewoman swings a 0.75 m sling of negligible weight around her head with a centripetal force of 220 N. What is the initial velocity of a 9 N stone released from the sling? (Answer: 13.4 m/s)
8. A 7.27 kg hammer on a 1 m wire is released with a linear velocity of 28 m/s. What centrifugal force is exerted on the thrower by the hammer at the instant before release? (Answer: 5.7 kN)
9. Discuss the effect of banking a curve on a racetrack. Construct a free body diagram to assist with your analysis.
10. Using the data in Appendix D, calculate the locations of the radii of gyration of all body segments with respect to the proximal joint center for a 1.7 m tall woman.

REFERENCES

1. Dapena J and McDonald C: A three-dimensional analysis of angular momentum in the hammer throw, Med Sci Sports Exerc 21:206, 1989.
2. Frohlich C: The physics of somersaulting and twisting, Sci Am 242:154, 1980.
3. Hamill J, Ricard MD, and Golden DM: Angular momentum in multiple rotation nontwisting platform dives, Int J Sport Biomech 2:78, 1986.
4. Hay JG: The biomechanics of sports techniques, ed 3, Englewood Cliffs, NJ, 1985, Prentice-Hall, Inc.
5. Herzog W: Maintenance of body orientation in the flight phase of long jumping, Med Sci Sports Exerc 18:231, 1986.
6. Jensen RK: The growth of children's moment of inertia, Med Sci Sports Exerc 18:440, 1987.
7. Jensen RK and Nassas G: Growth of segment principal moments of inertia between four and twenty years, Med Sci Sports Exerc 20:594, 1988.
8. Laws K: The physics of dance, New York, 1984, Schirmer Books.
9. Liu ZC and Nelson RC: Analysis of twisting somersault dives using computer diagnostics. In Winter DA et al, eds: Biomechanics IX-B, Champaign, Ill, 1985, Human Kinetics Publishers, Inc.
10. Miller DI and Munro CF: Body segment contributions to height achieved during the flight of a springboard dive, Med Sci Sports Exerc 16:234, 1984.

11. Miller DI, Jones IC, and Pizzimenti MA: Taking off: Greg Louganis' diving style, Soma 2:20, 1988.
12. Miller DI et al: Kinetic and kinematic characteristics of 10-m platform performances of elite divers: I. back takeoffs, Int J Sport Biomech 5:60, 1989.
13. Sanders RH and Wilson BD: Angular momentum requirements of the twisting and nontwisting forward 1½ somersault dive, Int J Sport Biomech 3:47, 1900.
14. Van Gheluwe B: A biomechanical simulation model for airborne twist in backward somersaults, J Hum Movement Studies 7:1, 1981.
15. Williams KR: Biomechanics of running, Exerc Sport Sci Rev 13:389, 1985.
16. Yeadon MR: Twisting techniques used in freestyle aerial skiing, Int J Sport Biomech 5:275, 1989.
17. Yeadon MR and Atha J: The production of a sustained aerial twist during a somersault without the use of asymmetrical arm action. In Winter DA et al, eds: Biomechanics IX-B, Champaign, Ill, 1985, Human Kinetics Publishers, Inc.

ANNOTATED READINGS

Frohlich C: The physics of somersaulting and twisting, Sci Am 242:154, 1980.

Describes the way in which divers, gymnasts, astronauts, and cats perform rotational maneuvers in midair that seem to violate the conservation of angular momentum.

Miller DI, Jones IC, and Pizzimenti MA: Taking off: Greg Louganis' diving style, Soma 2:20, 1988.

Discusses the linear and angular momentum requirements for performing total body rotations at the world class level and describes the methods by which these factors may be studied.

Miller DI et al: Kinetic and kinematic characteristics of 10-m platform performances of elite divers: I. back takeoffs, Int J Sport Biomech 5:60, 1989.

Miller DI et al: Kinetic and kinematic characteristics of 10-m platform performances of elite divers: II. reverse takeoffs, Int J Sport Biomech 6:283, 1990.

Presents a detailed description of a film, video, and force platform study of participants in the Fifth World Diving Championships.

Townend MS: Mathematics in sport, New York, 1984, John Wiley & Sons, Inc.

A chapter on running, a chapter on jumping, and sections on gymnastics and high-board diving discuss the principle of conservation of angular momentum with respect to various applications in the identified sports.

14 THE IMPORTANCE OF FLUIDS

Introduction to Fluid Mechanics

After reading this chapter, the student will be able to:

Explain the ways in which the composition and flow characteristics of a fluid affect fluid forces.

Define buoyancy and explain the variables that determine whether a human body will float.

Define drag, identify the components of drag, and identify the factors that affect the magnitude of each component.

Define lift and explain the ways in which it can be generated.

Discuss the three theories regarding propulsion of the human body in swimming.

Why does a javelin travel farther in the air than in a vacuum? Why are there dimples in a golf ball? What makes a boomerang come back to the thrower? Why are some people able to float while others cannot? Why are cyclists, swimmers, snow skiers, and ice skaters concerned with streamlining their bodies during competition?

Both air and water are fluid mediums that exert forces on bodies moving through them. Some of these forces slow the progress of a moving body; others provide support or propulsion. A general understanding of the actions of fluid forces on human movement activities is an important component of the study of the biomechanics of human movement. This chapter is designed to provide an introduction to the effects of fluid forces on both human and projectile motion.

The ability to control the action of fluid forces differentiates elite from average swimmers.

THE NATURE OF FLUIDS

Although in general conversation the term *fluid* is often used interchangeably with the term *liquid,* from a mechanical perspective, a **fluid** is any substance that tends to flow or continuously deform when acted on by a shear force (15). Both gases and liquids are fluids with similar mechanical behaviors.

fluid
a substance that flows when subjected to a shear stress

Relative Motion

Because a fluid is a medium capable of flow, the influence of the fluid on an object moving through it depends not only on the object's velocity but also on the velocity of the fluid. Consider the case

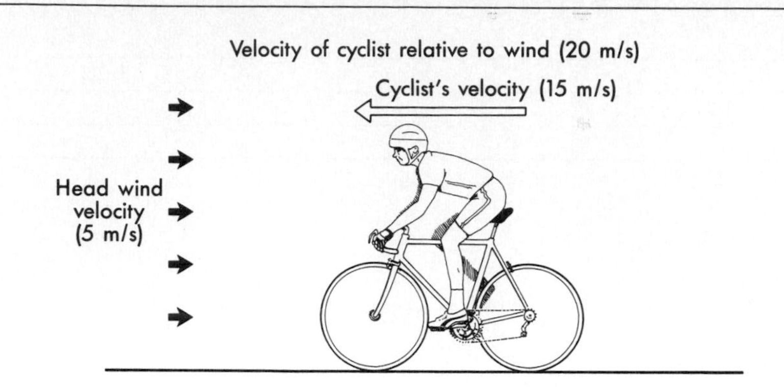

Figure 14-1
The relative velocity of a moving body with respect to a fluid is equal to the velocity of the body minus the absolute velocity of the wind.

of waders standing in the shallow portion of a river with a moderately strong current. If they stand still, they feel the force of the current against their legs. If they walk upstream against the current, the current's force against their legs is even stronger. If they walk downstream, the current's force is reduced and perhaps even imperceptible.

relative velocity
the velocity of a body with respect to the velocity of something else, such as the surrounding fluid

When a body moves through a fluid, the **relative velocity** of the body with respect to the velocity of the fluid influences the acting forces. Whenever the direction of an object's motion directly opposes the direction of the fluid flow, the magnitude of the object's relative velocity is the algebraic sum of the magnitude of its own velocity and that of the fluid (Figure 14-1). When the direction of an object's motion is the same as that of the surrounding fluid, the magnitude of the relative velocity is the difference in the magnitudes of the velocities of the object and the fluid. Figure 14-2 provides a quantitative illustration of the concept of relative velocity.

SAMPLE PROBLEM 1

Figure 14-2

A sailboat is traveling at an absolute speed of 3 m/s against a 0.5 m/s current and with a 6 m/s tail wind. What is the velocity of the current with respect to the boat? What is the velocity of the wind with respect to the boat?

Known

$v_b = 3$ m/s →

$v_c = 0.5$ m/s ←

$v_w = 6$ m/s →

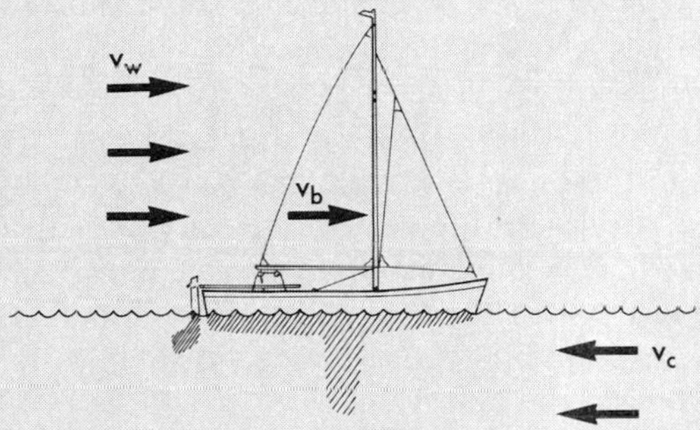

Solution

The absolute velocity of the current is equal to the vector sum of the absolute velocity of the boat and the velocity of the current with respect to the boat.

$$v_c = v_b + v_{c/b}$$

$$(0.5\text{m/s} \leftarrow) = (3\text{m/s} \rightarrow) + v_{c/b}$$

$$v_{c/b} = (3.5\text{m/s} \leftarrow)$$

The velocity of the current with respect to the boat is 3.5 m/s in the direction opposite that of the boat.

The absolute velocity of the wind is equal to the vector sum of the absolute velocity of the boat and the velocity of the wind with respect to the boat.

$$v_w = v_b + v_{w/b}$$

$$(6\text{m/s} \rightarrow) = (3\text{m/s} \rightarrow) + v_{w/b}$$

$$v_{w/b} = (3\text{m/s} \rightarrow)$$

The velocity of the wind with respect to the boat is 3 m/s in the direction in which the boat is sailing.

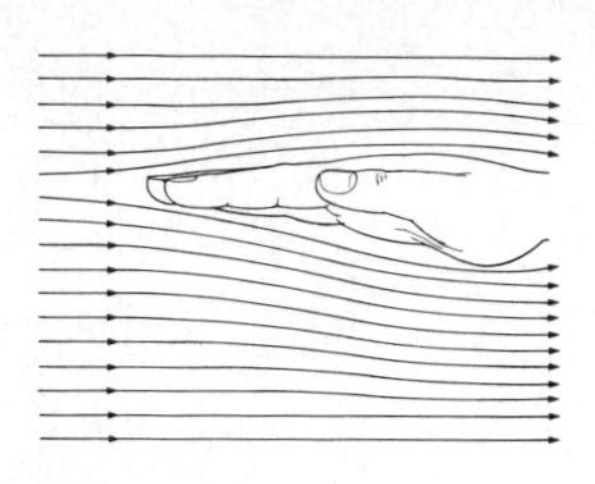

Figure 14-3
Laminar flow.

laminar flow
a flow characterized by smooth, parallel layers of fluid

turbulent flow
a flow characterized by mixing of adjacent fluid layers

Laminar Versus Turbulent Flow

When an object such as a human hand or a canoe paddle moves through water, there is little apparent disturbance of the immediately surrounding water if the relative velocity of the object with respect to the surrounding fluid is low. However, if the relative velocity of motion through the water is sufficiently high, waves and eddies appear.

When an object moves with sufficiently low velocity relative to any fluid medium, the flow of the adjacent fluid is termed **laminar flow.** It is characterized by smooth layers of fluid molecules flowing parallel to one another (Figure 14-3).

When an object moves with sufficiently high velocity relative to a surrounding fluid, the layers of fluid near the surface of the object mix and the flow is termed *turbulent.* The rougher the surface of the body, the lower the relative velocity at which turbulence is caused. Laminar flow and **turbulent flow** are distinct categories. If any turbulence is present, the flow is nonlaminar. The nature of the fluid flow surrounding an object can dramatically affect the fluid forces exerted on the object.

Fluid Properties

Other factors that influence the magnitude of the forces a fluid generates are the fluid's density, specific weight, and viscosity. As discussed in Chapter 2, density (ρ), is defined as mass/volume, and the ratio of weight to volume is known as specific weight (γ). The denser and heavier the fluid medium surrounding a body, the greater the magnitude of the forces the fluid exerts on the body. The property of fluid viscosity involves the resistance of a fluid to flow. The greater the extent to which a fluid resists flow under an applied force, the more viscous the fluid is. A thick molasses, for example, is

Table 14-1
APPROXIMATE PHYSICAL PROPERTIES OF COMMON FLUIDS

FLUID	DENSITY (kg/m^3)	SPECIFIC WEIGHT (N/m^3)	VISCOSITY ($Nsec/m^2$)
Air	1.20	11.8	0.000018
Water	998	9790	0.0010
Sea water*	1026	10070	0.0014
Ethyl alcohol	799	7850	0.0012
Mercury	13550	133000	0.0015

Fluids are measured at 20° C and standard atmospheric pressure.

*At 10° C, 3.3% salinity.

more viscous than a liquid honey, which is more viscous than water. Increased fluid viscosity results in increased forces exerted on bodies exposed to the fluid.

Atmospheric pressure and temperature influence a fluid's density, specific weight, and viscosity, with more mass concentrated in a given unit of fluid volume at higher atmospheric pressures and lower temperatures. Because molecular motion in gases increases with temperature, the viscosity of gases also increases. The viscosity of liquids decreases with increased temperature because of a reduction in the cohesive forces among the molecules. The densities, specific weights, and viscosities of common fluids are shown in Table 14-1.

BUOYANCY

Characteristics of the Buoyant Force

Buoyancy is a fluid force that always acts vertically upward. The factors that determine the magnitude of the buoyant force were originally explained by the ancient Greek mathematician Archimedes. **Archimedes' principle** states that the magnitude of the buoyant force acting on a given body is equal to the weight of the fluid displaced by the body. The latter factor is calculated by multiplying the specific weight of the fluid by the volume of the portion of the body that is surrounded by the fluid. Buoyancy (F_b) is calculated as the product of the displaced volume of fluid (V_d) and the fluid's specific weight (γ):

Archimedes principle
the buoyant force acting on a body is equal to the weight of the fluid displaced by the body

$$F_b = V_d \gamma$$

For example, if a water polo ball with a volume of 0.2 m^3 is completely submerged in water at 20° C, the buoyant force acting on the ball is equal to the ball's volume multiplied by the specific weight of water at 20° C:

$$F_b = V_d \gamma$$

$$F_b = (0.2\ m^3)\ (9790\ N/m^3)$$

$$F_b = 1958\ N$$

The more dense the surrounding fluid, the greater the magnitude of the buoyant force. Since sea water is more dense than distilled water, a given object's buoyancy is greater in sea water than in distilled water. Because the magnitude of the buoyant force is directly related to the volume of the submerged object, the point at which the buoyant force acts is the object's **center of volume,** which is also known as the *center of buoyancy*. The center of volume is the point around which a body's volume is equally distributed in all directions.

center of volume
the point around which a body's volume is equally balanced and at which the buoyant force acts

Flotation

The ability of a body to float in a fluid medium depends on the relationship between the body's buoyancy and its weight. When weight and the buoyant force are the only two forces acting on a body and their magnitudes are equal, the body floats in a motionless state, in accordance with the principles of static equilibrium. If the magnitude of the weight is greater than that of the buoyant force, the body sinks, moving downward in the direction of the net force.

Most objects float statically in a partially submerged position. The volume of the object needed to generate a buoyant force equal to the object's weight is the volume that is submerged.

Flotation of the Human Body

In the study of biomechanics, buoyancy is most commonly of interest relative to the flotation of the human body in water. Some individuals cannot float in a motionless position, and others float with little effort. This difference in floatability is a function of body density. Since the density of bone and muscle is greater than the density of fat, individuals who are extremely muscular and have little body fat have higher average body densities than individuals with less muscle, less dense bones, or more body fat. If two individuals have an identical body volume, the one with the higher body density weighs more. Alternatively, if two people have the same body weight, the person with the higher body density has a smaller body volume. For flotation to occur, the body volume must be large enough to create a buoyant force greater than or equal to body weight (Figure 14-4). Many individuals can float only when holding a large volume of inspired air in the lungs, a tactic that increases body volume without practically altering body weight.

■ People who cannot float in swimming pools may float in Utah's Great Salt Lake, in which the density of the water surpasses even that of sea water.

The orientation of the human body as it floats in water is determined by the relative position of the total body center of gravity relative to the total body center of volume. The exact locations of the center of gravity and center of volume vary with anthropometric dimensions and body composition. Typically, the center of gravity is inferior to the center of volume due to the relatively large volume and relatively small weight of the lungs. Because weight acts at the center of gravity and buoyancy acts at the center of volume, a torque is created that rotates the body until it is positioned so that these two acting forces are vertically aligned and the torque ceases to exist (Figure 14-5).

When beginning swimmers try to float on their backs, they typically assume a horizontal body position. Once the swimmer relaxes, the lower end of the body sinks because of the acting torque. An experienced teacher instructs beginning swimmers to assume a more diagonal position in the water before relaxing into the back float. This position minimizes torque and the concomitant sinking of the lower extremity.

SAMPLE PROBLEM 2

Figure 14-4

When holding a big breath of air in her lungs, a 22 kg girl has a body volume of 0.025 m^3. Can she float in fresh water if γ equals 9810 N/m^3? Given her body volume, how much could she weigh and still be able to float?

Known

$m = 22$ kg

$V = 0.025\ m^3$

$\gamma = 9810\ N/m^3$

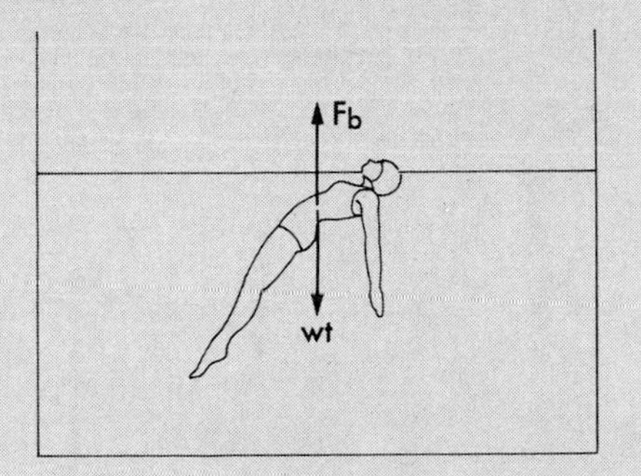

Solution

The free body diagram shows that two forces are acting on the girl—her weight and the buoyant force. According to the conditions of static equilibrium, the sum of the vertical forces must be equal to zero for the girl to float in a motionless position. If the buoyant force is less than her weight, she will sink, and if the buoyant force is equal to her weight, she will float completely submerged. If the buoyant force is greater than her weight, she will float partly submerged. The magnitude of the buoyant force acting on her total body volume is the product of the volume of displaced fluid (her body volume) and the specific weight of the fluid:

$$F_b = V\gamma$$

$$F_b = (0.025\ m^3)(9810\ N/m^3)$$

$$F_b = 245.25\ N$$

Her body weight is equal to her body mass multiplied by the acceleration of gravity:

$$wt = (22\ kg)(9.81\ m/s^2)$$

$$wt = 215.82\ N$$

Since the buoyant force is greater than her body weight, the girl will float partly submerged in fresh water.

Yes, she will float.

To calculate the maximum weight that the girl's body volume can support in fresh water, multiply the body volume by the specific weight of water:

$$wt_{max} = (0.025\ m^3)(9810\ N/m^3)$$

$wt_{max} = 245.25\ N$

Figure 14-5

A, A torque is created on a swimmer by body weight (acting at the center of gravity) and the buoyant force (acting at the center of volume). **B,** When the center of gravity and the center of volume are vertically aligned, this torque is eliminated.

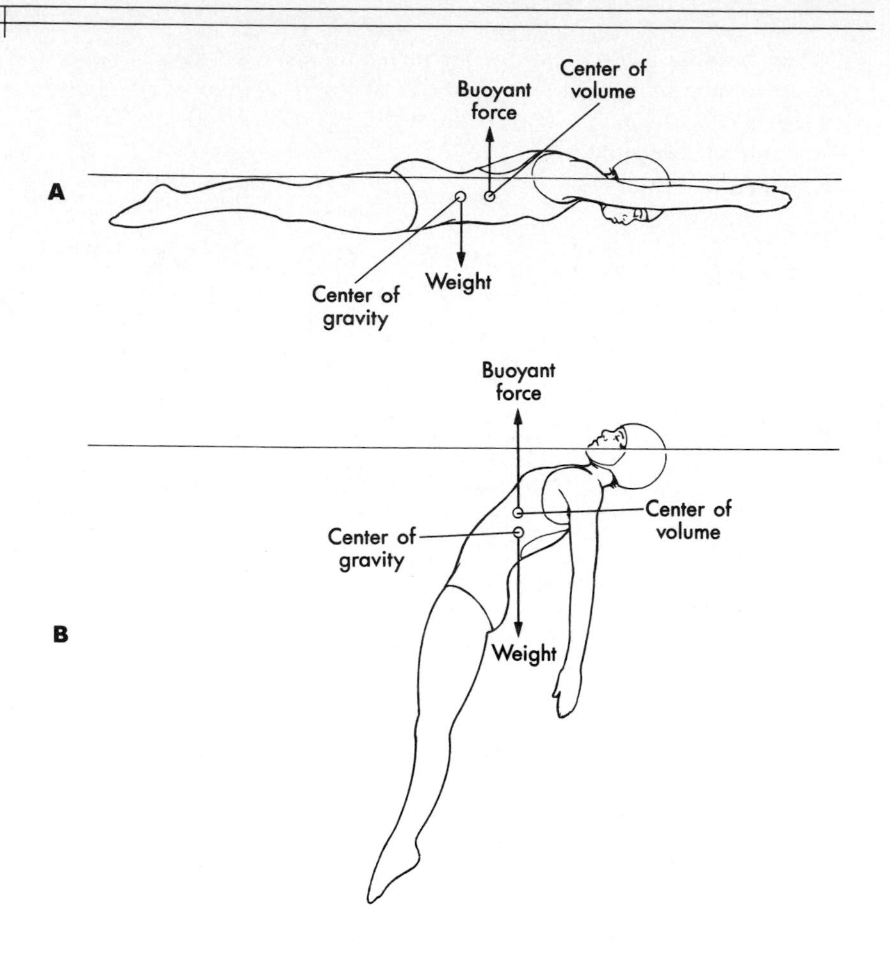

DRAG

Drag is a force caused by the dynamic action of a fluid that acts in the direction of the freestream fluid flow. Generally, drag is a *resistance* force—a force that slows the motion of a body moving through a fluid. The drag force acting on a body in relative motion with respect to a fluid is defined by the following formula:

$$F_D = \frac{1}{2} C_D \rho A_p v^2$$

coefficient of drag
a unitless number that is an index of a body's ability to generate drag

In this formula, F_D is drag force, C_D is the **coefficient of drag,** ρ is the fluid density, A_p is the projected area of the body or the surface area of the body oriented perpendicular to the fluid flow, and v is the relative velocity of the body with respect to the fluid. The coefficient of drag is an index of the amount of drag an object can generate. Its size depends on the shape and orientation of a body relative

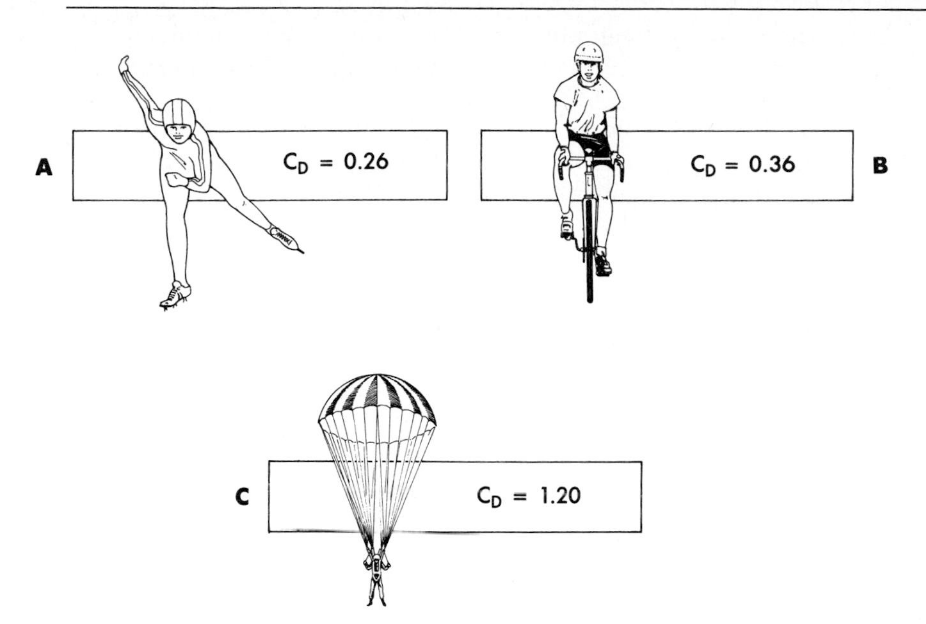

Figure 14-6
Approximate coefficient of drag of the human body. **A,** Frontal drag on a speed skater. **B,** Frontal drag on a cyclist in touring position. **C,** Vertical drag on a parachutist falling with the parachute fully opened. (Modified from Roberson JA and Crowe CT: Engineering fluid mechanics, ed 2, Boston, 1980, Houghton Mifflin Co.)

to the fluid flow, with long, streamlined bodies generally having lower coefficients of drag than blunt or irregularly shaped objects. Approximate coefficients of drag for the human body in positions assumed during several sports are shown in Figure 14-6.

The formula for the total drag force demonstrates the exact way in which each of the identified factors affects drag. If the coefficient of drag, the fluid density, and the projected area of the body remain constant, drag increases with the square of the relative velocity of motion. This relationship is referred to as the **theoretical square law.** According to this law, if cyclists double their speed and other factors remain constant, the drag force opposing them increases fourfold. The effect of drag is more consequential when a body is moving with a high velocity, which occurs in sports such as sprinting, cycling, skiing, the bobsled, and the luge.

theoretical square law
drag increases approximately with the square of velocity when relative velocity is low

Increase or decrease in the fluid density also results in a proportional change in the drag force. Because air density decreases with increasing altitude, many world records set at the 1968 Olympic Games in Mexico City where the elevation is 2250 m may have been partially attributable to the reduced air resistance acting on the competitors. Mathematical model-based estimates indicate that the reduction in drag attributable to the lesser air density in Mexico City accounts for 0.08 seconds of performance time in the 100 m sprint and 0.16 seconds of race time in the 200 m event (11). Bob Beamon's noteworthy long jump performance of 8.9 m during the games was 2.4 cm longer than if the same jump had been performed at sea level (24).

Three forms of resistance contribute to the total drag force. The component of resistance that predominates depends on the nature of the fluid flow immediately adjacent to the body.

Skin Friction

skin friction
surface drag
viscous drag
the drag derived from friction between adjacent layers of fluid near a body moving through the fluid

One component of the total drag is known as **skin friction, surface drag,** or **viscous drag.** This drag is similar to the friction force described in Chapter 11. Skin friction is derived from the sliding contacts between successive layers of fluid close to the surface of a moving body (Figure 14-7). The layer of fluid particles immediately adjacent to the moving body is slowed because of the shear stress the body exerts on the fluid. The next adjacent layer of fluid particles moves with slightly less speed because of friction between the adjacent molecules, and the next layer moves with an even lesser speed. The number of layers of affected fluid becomes progressively larger as the flow moves in the downstream direction along the body. The entire region within which fluid velocity is diminished because of the shearing resistance caused by the boundary of the moving body is the **boundary layer.** The forwardly directed force the body exerts on the fluid in creating the boundary layer results in a reaction force exerted by the fluid on the body. This backwardly directed reaction force is known as *skin friction.*

boundary layer
the layer of fluid immediately adjacent to a body.

Several factors affect the magnitude of skin friction drag. It increases proportionally with increases in the relative velocity of fluid

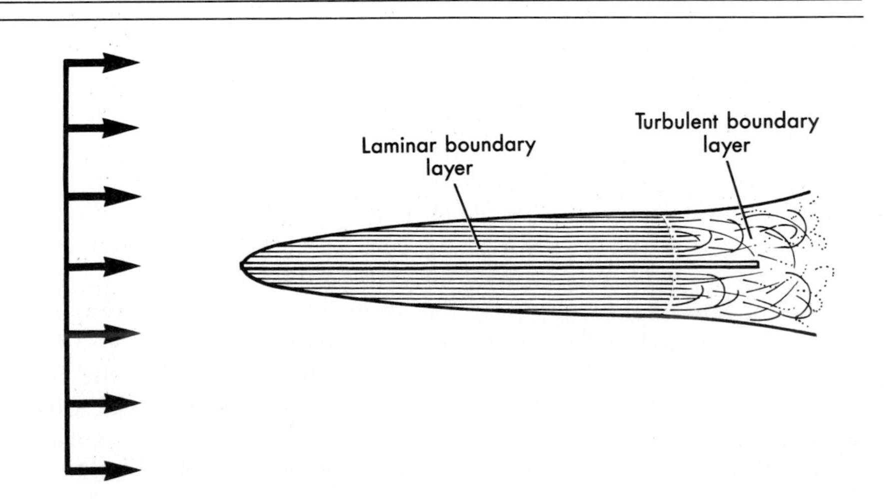

Figure 14-7
The boundary layer for a thin, flat plate, shown from the side view. The laminar boundary layer gradually becomes thicker as flow progresses along the plate.

Modern cycling attire and equipment is designed to reduce form drag and skin friction.

flow, the surface area of the body over which the flow occurs, the roughness of the body surface, and the viscosity of the fluid.

Among these factors, the one that a competitive athlete can readily alter is the relative roughness of the body surface. Athletes can wear tight-fitting clothing composed of a smooth fabric rather than loose-fitting clothing or clothing made of a rough fabric. A 10% reduction of drag occurs when a speed skater wears a smooth, spandex suit as opposed to the traditional wool outfit (23). A 6% decrease in air resistance results from cyclists using appropriate clothing, including sleeves, tights, and smooth covers over the laces of the shoes (10). The long cotton socks and loose shorts and singlets commonly worn by runners are particularly aerodynamically inefficient. Smoothing the running shoe and laces and either covering or shaving the body hair could reduce drag on runners by as much as 10% (11). Competitive male swimmers often shave their body hair to reduce drag.

The other factor affecting skin friction that athletes can alter in some circumstances is the amount of surface area in contact with the fluid. Carrying an extra passenger such as a cox in a rowing event results in a larger wetted surface area of the hull because of the added weight and, as a result, skin friction drag is increased.

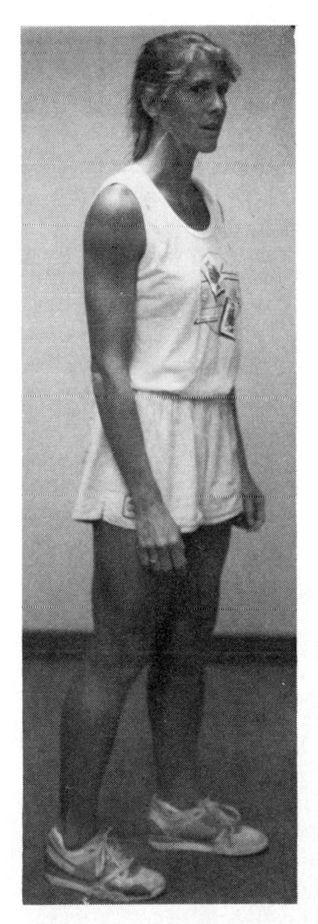

Common running attire is aerodynamically inefficient.

Figure 14-8
Form drag results from the suctionlike force created between the positive pressure zone on a body's leading edge and the negative pressure zone on the trailing edge when turbulence is present.

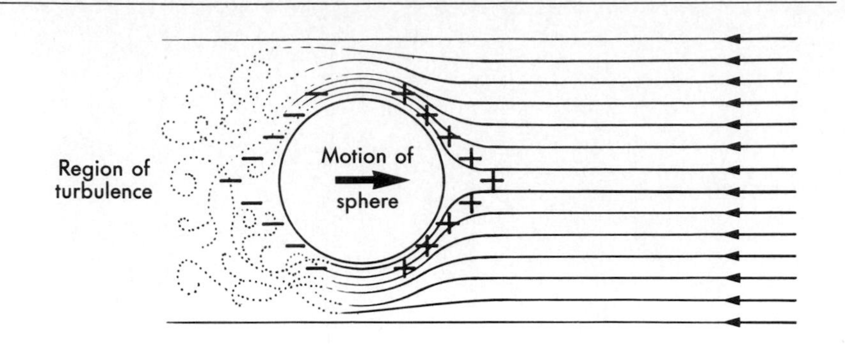

form drag
profile drag
pressure drag
the drag created by a pressure differential between the lead and rear sides of a body moving through a fluid

Form Drag

A second component of the total drag acting on a body moving through a fluid is **form drag,** which is also known as **profile drag** or **pressure drag.** Form drag is the major contributor to overall drag during most human and projectile motion.

When a body moves through a fluid medium with sufficient velocity to create a pocket of turbulence behind the body, an imbalance in the pressure surrounding the body—a *pressure differential*—is created (Figure 14-8). At the upstream end of the body where fluid particles meet the body head-on, a zone of relative high pressure is formed. At the downstream end of the body where turbulence is present, a zone of relative low pressure is created. Whenever a pressure differential exists, a force is directed from the region of high pressure to the region of low pressure. For example, a vacuum cleaner creates a suction force because a region of relative low pressure (the vacuum) exists inside the machine housing. This force, directed from front to rear of the body in relative motion through a fluid, constitutes form drag.

Hay and Thayer (9) have employed an array of tufts attached to a swimmer's suit and body to visualize flow patterns in studying the types of drag forces acting on a swimmer (Figure 14-9). Because form drag is characterized by the separation of flow from the boundary and the formation of eddies downstream, identifying when and where form drag acts during a particular stroke may be possible using this technique. For example, the use of tufts has shown that form drag on the legs during the breaststroke kick is less when swimmers use a dolphin style rather than a flat style of kick.

Several factors affect the magnitude of form drag, including the relative velocity of the body with respect to the fluid, the magnitude of the pressure gradient between the front and rear ends of the body, and the size of the surface area that is aligned perpendicular to the flow. Both the size of the pressure gradient and the amount of

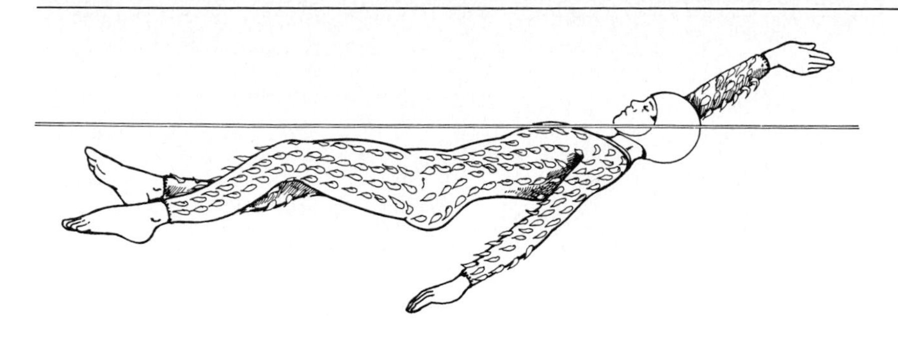

Figure 14-9
Tufts are used to study the patterns of fluid flow associated with swimming strokes. (Modified from Hay JG and Thayer AM: Flow visualization of competitive swimming techniques: the tufts method, J Biomech 22:11, 1989.)

surface area perpendicular to the fluid flow can be reduced to minimize the effect of form drag on the human body. For example, streamlining the overall shape of the body reduces the magnitude of the pressure gradient. Streamlining minimizes the amount of turbulence created and hence minimizes the negative pressure that is created at the object's rear (Figure 14-10). Assuming a more crouched body position also reduces the body's projected surface area oriented perpendicular to the fluid flow.

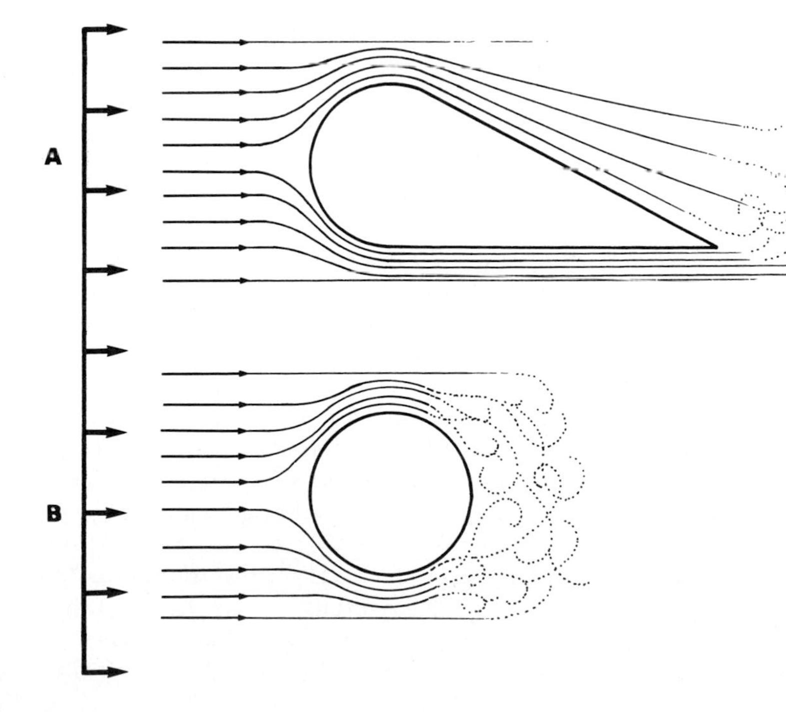

Figure 14-10
The effect of streamlining is a reduction in the turbulence created at the trailing edge of a body in a fluid. **A,** A streamlined shape. **B,** A sphere.

A streamlined cycling helmet.

Competitive cyclists, skaters, and skier's assume a streamlined body position with the smallest possible area of the body oriented perpendicular to the oncoming airstream. Similarly, race cars, yacht hulls, and some cycling helmets are designed with streamlined shapes. Many competitive cyclists use solid wheels that create less air turbulence than spoked wheels. Using a triathlon wet suit can reduce the drag on a competitor swimming at a typical triathlon race pace of 1.25 m/sec by as much as 14%. Researchers have attributed this effect to the buoyant effect of the wetsuit resulting in reduced form drag on the swimmer (19). For this reason, wet suits are typically banned in triathlon competitions.

The nature of the boundary layer at the surface of a body moving through a fluid can also influence form drag by affecting the pressure gradient between the front and rear ends of the body. When the boundary layer is primarily laminar, the fluid separates from the boundary close to the front end of the body, creating a large turbulent pocket with a large negative pressure and thereby a large form drag (Figure 14-11). In contrast, when the boundary layer is turbulent, the point of flow separation is closer to the rear end of the body, the turbulent pocket created is smaller, and the resulting form drag is smaller.

The nature of the boundary layer depends on the roughness of the body's surface and the body's velocity relative to the flow. As the relative velocity of motion for an object such as a golf ball increases, changes in the acting drag occur (Figure 14-12). As relative velocity

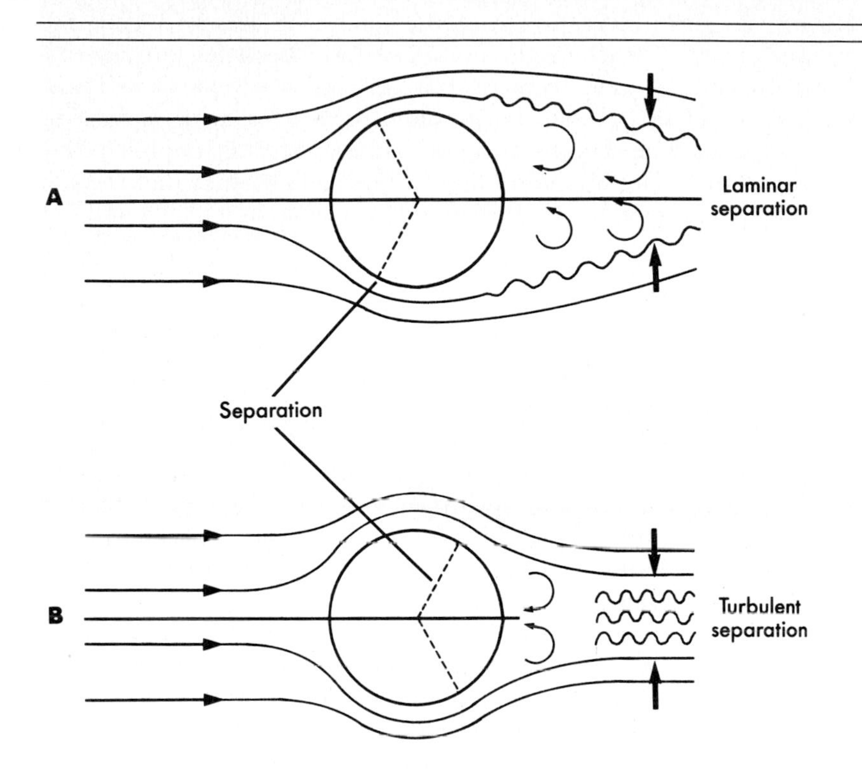

Figure 14-11

A, Laminar flow results in an early separation of flow from the boundary and a larger drag producing wake as compared to **B,** turbulent boundary flow.

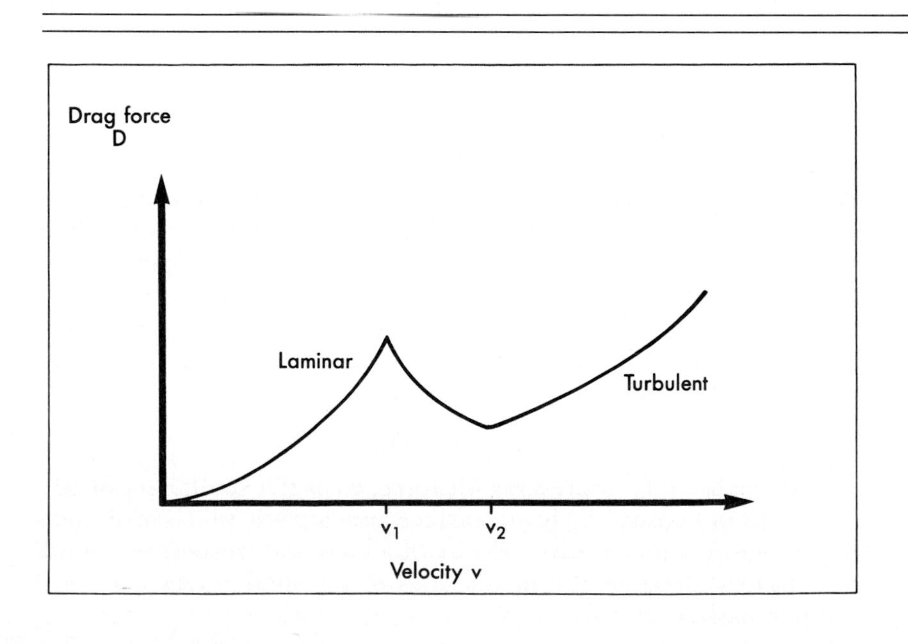

Figure 14-12

Drag increases with the square of velocity until there is sufficient relative velocity (v_1) to generate a turbulent boundary layer. As velocity increases beyond this point, form drag decreases. After a second critical point in the relative velocity increase (v_2), the drag again increases.

increases to a certain critical point, the theoretical square law is in effect, with drag increasing with the square of velocity. After this critical velocity is reached, the boundary layer becomes more turbulent than laminar, and form drag is reduced. As velocity increases further, the effects of skin friction and form drag grow, increasing the total drag. The dimples in a golf ball are carefully engineered to produce a turbulent boundary layer at the ball's surface that reduces form drag on the ball over the range of velocities at which a golf ball travels.

Wave Drag

wave drag
the drag created by the generation of waves at the interface between two different fluids, such as air and water

The third type of drag acts at the interface of two different fluids, for example, at the interface between water and air. Although bodies that are completely submerged in a fluid are not affected by **wave drag,** this form of drag can be a major contributor to the overall drag acting on a human swimmer, particularly when the swim is done in open water. When a swimmer moves a body segment along, near, or across the air and water interface, a wave is created in the more dense fluid (the water). The reaction force the water exerts on the swimmer constitutes wave drag.

The magnitude of wave drag increases with greater up-and-down motion of the body and increased swimming speed. The height of the bow wave generated in front of a swimmer increases proportionally with swimming velocity, although at a given velocity, skilled swimmers produce smaller waves than less-skilled swimmers (18). At fast swimming speeds, wave drag is generally the largest component of the total drag acting on the swimmer. In most swimming pools the lane lines are designed to minimize wave action by dissipating moving surface water.

LIFT FORCE

lift
the force acting on a body in a fluid in a direction perpendicular to the fluid flow

While drag forces act in the direction of the freestream fluid flow, another force, known as **lift,** is generated perpendicular to the fluid flow. Although the name *lift* suggests that this force is directed vertically upward, it may assume any direction as determined by the direction of the fluid flow and the orientation of the body. The factors affecting the magnitude of lift are basically the same factors that affect the magnitude of drag:

$$F_L = \frac{1}{2} C_L \rho A_p v^2$$

coefficient of lift
a unitless number that is an index of a body's ability to generate lift

In this equation, F_L represents lift force, C_L is the **coefficient of lift,** ρ is the fluid density, A_p is the surface area against which lift is generated, and v is the relative velocity of a body with respect to a fluid. The factors affecting the magnitudes of the fluid forces discussed are summarized in Table 14-2.

The bow wave generated by a competitive swimmer.

The lane lines in modern swimming pools are designed to minimize wave action, enabling faster racing times.

Table 14-2
FACTORS AFFECTING THE MAGNITUDES OF FLUID FORCES

FORCE	FACTORS
Buoyant force	1. Specific weight of the fluid 2. Volume of fluid displaced
Skin friction	1. Relative velocity of the fluid 2. Amount of body surface area exposed to the flow 3. Roughness of the body surface 4. Fluid viscosity
Form drag	1. Relative velocity of the fluid 2. Pressure differential between leading and rear edges of the body 3. Amount of body surface area perpendicular to the flow
Wave drag	1. Relative velocity of the wave 2. Amount of surface area perpendicular to the wave 3. Fluid viscosity
Lift force	1. Relative velocity of the fluid 2. Fluid density 3. Size, shape, and orientation of the body

Foil Shape

foil
a shape capable of generating lift in the presence of a fluid flow

Bernoulli principle
an expression of the inverse relationship between relative velocity and relative pressure in a fluid flow

One way in which lift force may be created is for the shape of the moving body to resemble that of a **foil** (Figure 14-13). When the fluid stream encounters a foil, the fluid separates, with some flowing over the curved surface and some flowing straight back along the flat surface on the opposite side. The fluid that flows over the curved surface reaches the rear edge of the foil at the same time as the fluid flowing over the flat side of the foil. Since the fluid on the curved side has a longer distance to travel, it must move with a higher velocity. This difference in the velocity of the fluid flow creates a pressure difference in the fluid, in accordance with a relationship derived by the Italian scientist Bernoulli. According to the **Bernoulli principle,** regions of relative high velocity fluid flow are associated with regions of relative low pressure, and regions of relative low velocity flow are associated with regions of relative high pressure. When these regions of relative low and high pressure are created on opposite sides of the foil, the result is a lift force directed perpendicular to the foil from the high pressure zone toward the low pressure zone.

Different factors affect the magnitude of the lift force acting on a foil. The greater the velocity of the foil relative to the fluid, the greater the pressure differential and the lift force generated. Other contributing factors are the fluid density and the surface area of the flat side of the foil. As both of these variables increase, lift is greater. An additional factor of influence is the *coefficient of lift,* which indicates a body's ability to generate lift based on its shape.

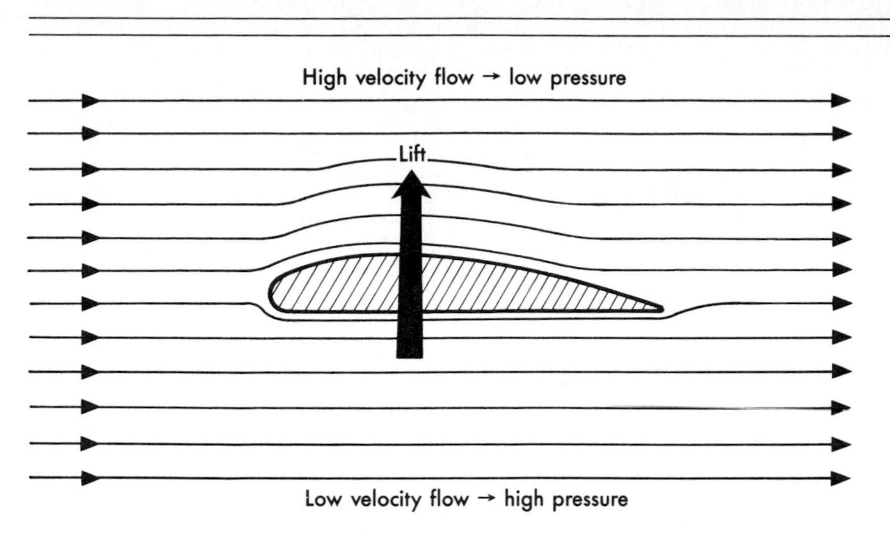

Figure 14-13
Lift force generated by a foil shape is directed from the region of relative high pressure on the flat side of the foil toward the region of relative low pressure on the curved side of the foil.

A lateral view of the human hand shows that it resembles a foil shape. When a swimmer slices a hand through the water, it generates lift force directed perpendicular to the palm. Synchronized swimmers rapidly slice their hands back and forth (sculling) to maneuver their bodies through various positions in the water. The lift force generated by rapid sculling motions enables elite synchronized swimmers to support their bodies in an inverted position with both legs extended out of the water (2).

The semifoil shapes of projectiles such as the discus, javelin, football, boomerang, and frisbee generate some lift force when oriented at appropriate angles with respect to the direction of fluid flow. Spherical projectiles such as a shot or a ball, however, do not sufficiently resemble a foil and cannot generate lift by virtue of shape.

The angle of orientation of the projectile with respect to the fluid flow—the **angle of attack**—is an important factor in launching a lift-producing projectile for maximum range (horizontal displacement). A positive angle of attack is necessary to generate a lift force (Figure 14-14). As the angle of attack increases, the amount of surface area exposed perpendicularly to the fluid flow also increases, thereby increasing the amount of form drag acting. With too steep of an attack angle, the fluid cannot flow along the curved side of the foil or create lift. Airplanes that assume too steep an ascent can stall and lose altitude until pilots reduce the orientation of the wings to enable lift.

angle of attack
the angle between the longitudinal axis of a body and the direction of the fluid flow

It is advantageous to maximize lift and minimize drag to maximize the flight distance of a projectile such as the discus or javelin. Form drag, however, is minimum at an angle of attack of 0, which is

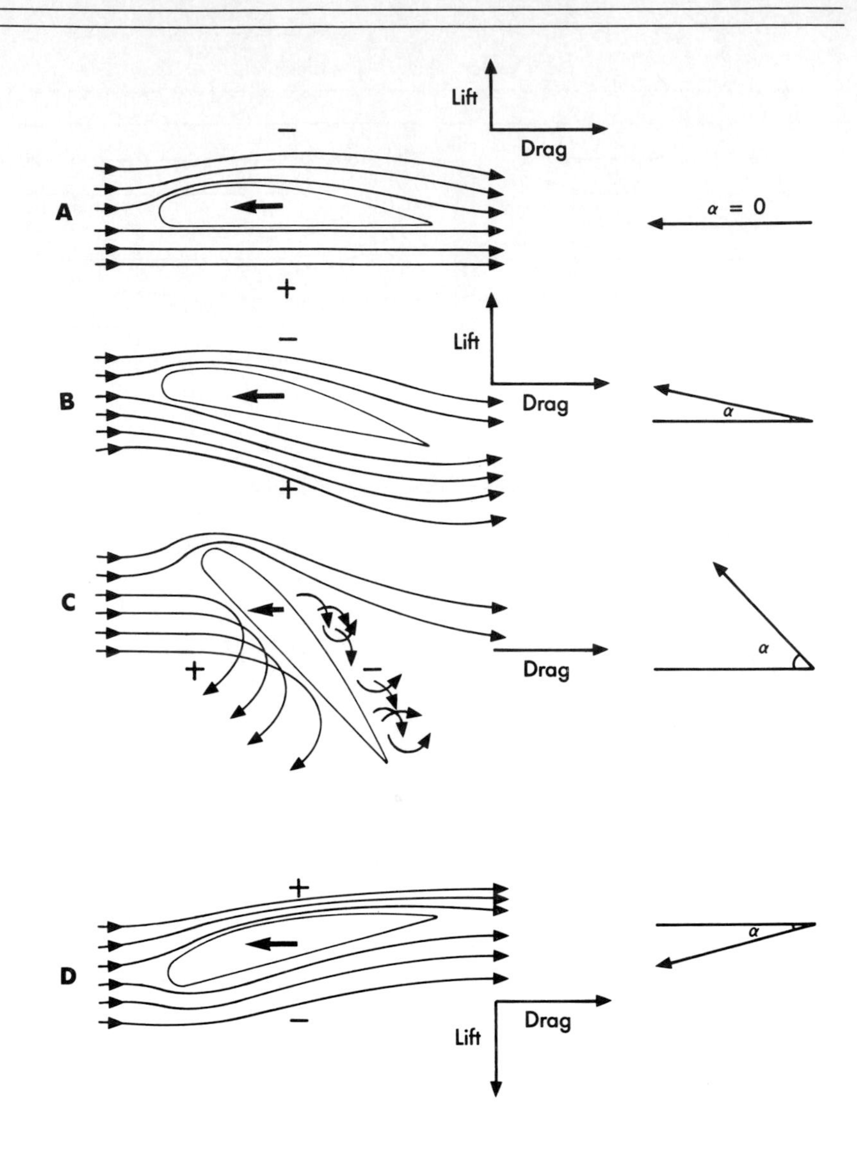

Figure 14-14
A, Drag and lift are small because the angle of attack (α) does not create a sufficiently high pressure differential across the top and bottom surfaces of the foil. **B,** An angle of attack that promotes lift. **C,** When the angle of attack is too large, the fluid cannot flow over the curved surface of the foil and no lift is generated. **D,** When the angle of attack is below the horizontal, lift is created in a downward direction. (Modified from Maglischo E: Swimming faster: a comprehensive guide to the science of swimming, Palo Alto, Calif, 1982, Mayfield Publishing Co.)

lift/drag ratio
the magnitude of the lift force divided by the magnitude of the total drag force acting on a body at a given time

a poor angle for generating lift. The optimum angle of attack for maximizing range is the angle at which the **lift/drag ratio** is maximum. The largest lift/drag ratio for a discus traveling at a relative velocity of 24 m/s is generated at an angle of attack of 10 degrees (7).

When the projectile is the human body during the performance of a jump, maximizing the effects of lift while minimizing the effects of drag is more complicated. Because of the relatively long period of time during which the body is airborne, the lift/drag ratio for the human body is particularly important for optimizing performance in the ski jump.

After studying factors affecting ski jump performance, researchers suggested the following: Ski jumpers should have a flattened body with a large frontal area (for generating lift) and a small body weight (for enabling greater acceleration) during takeoff. During the first part of the flight, jumpers should assume a small angle of attack to minimize drag (Figure 14-15). During the latter part of the flight, they should increase attack angle up to that of maximum lift. On smaller jumping hills where takeoff velocities are less, jumpers should assume the attack angle for maximum lift earlier in the flight because the effect of drag is not as great (14).

Other researchers investigating the ski jump have noted that the maximum length of the jump occurs neither at the body angle of attack for maximum lift nor at the attack angle for which the lift/drag angle is maximum but somewhere between the two (Figure 14-16) (5). The ski jump consists of takeoff, flight, and landing, with body position during each phase influencing body position during the subsequent phase or phases. The effects of body position during all phases of the jump must be considered for the modeling of an optimal performance.

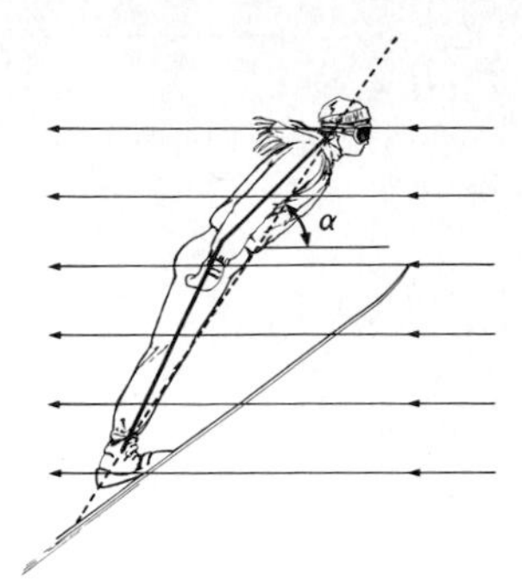

Figure 14-15

The angle of attack is the angle formed between the primary axis of a body and the direction of the fluid flow.

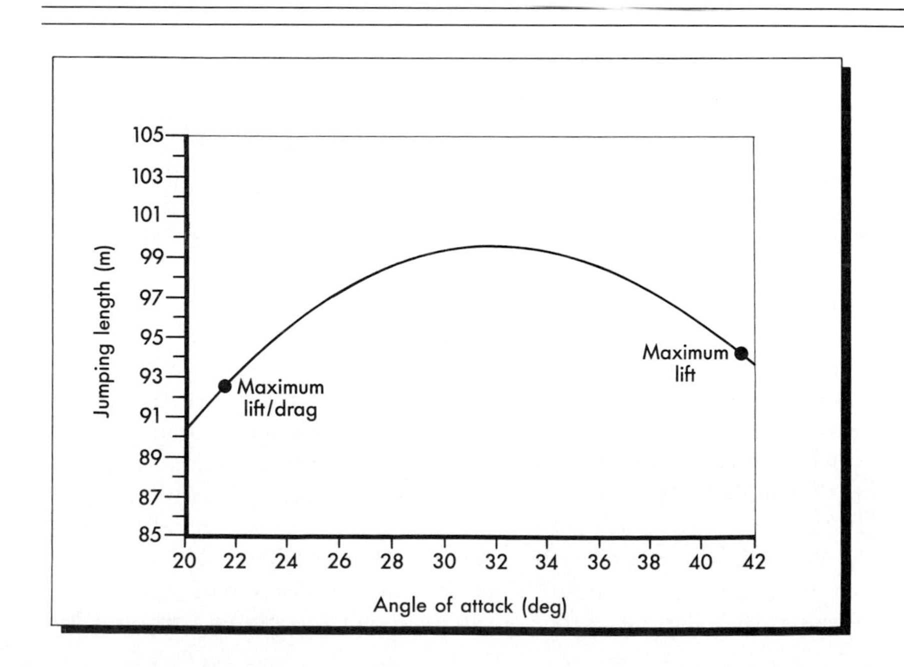

Figure 14-16

The relationship between ski jump length and the performer's angle of attack. (Modified from Denoth J, Luethi SM, and Gasser HH: Methodological problems in optimization of the flight phase in ski jumping, Int J Sport Biomech 3:404, 1987.)

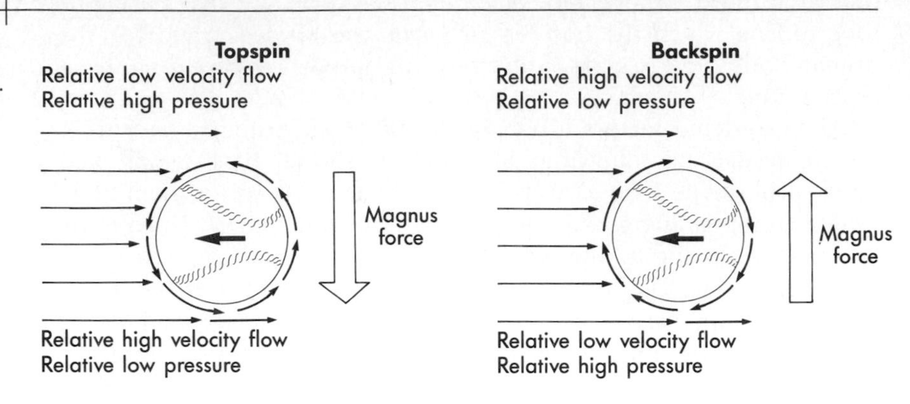

Figure 14-17
The Magnus force created by a spinning body.

Magnus Effect

Spinning objects also generate lift. When an object in a fluid medium spins, the boundary layer of fluid molecules adjacent to the object spins with it. When this happens, the fluid molecules on one side of the spinning body collide head-on with the molecules in the fluid freestream (Figure 14-17). This creates a region of relative low velocity and high pressure. On the opposite side of the spinning object the boundary layer moves in the same direction as the fluid flow, thereby creating a zone of relative high velocity and low pressure. The pressure differential creates what is called the **Magnus force,** a lift force directed from the high pressure region to the low pressure region.

Magnus force
the lift force created by spin

Magnus force affects the flight path of a spinning projectile as it travels through the air, causing the path to deviate progressively in the direction of the spin, a deviation known as the **Magnus effect** (Figure 14-18). When a tennis ball or table tennis ball is hit with topspin, the ball drops more rapidly than it would without spin and the ball tends to rebound low and fast, often making it more difficult for the opponent to return the shot. The nap on a tennis ball traps a relatively large boundary layer of air with it as it spins, thereby accentuating the Magnus effect. The Magnus effect can also result from sidespin, as when a pitcher throws a curve ball. The modern-day version of the curve ball is a ball that is intentionally pitched with spin, causing it to follow a curved path in the direction of the spin throughout its flight path. Curve balls thrown by major league pitchers spin as quickly as 27 revolutions per second and deviate horizontally as much as 40 cm over the pitcher to batter distance (20).

Magnus effect
the deviation in the trajectory of a spinning object toward the direction of spin resulting from the Magnus force

In 1982 a controversy between professional baseball players and scientists arose over the behavior of a pitched curve ball. The players claimed that the path taken by a curve ball was along a straight line

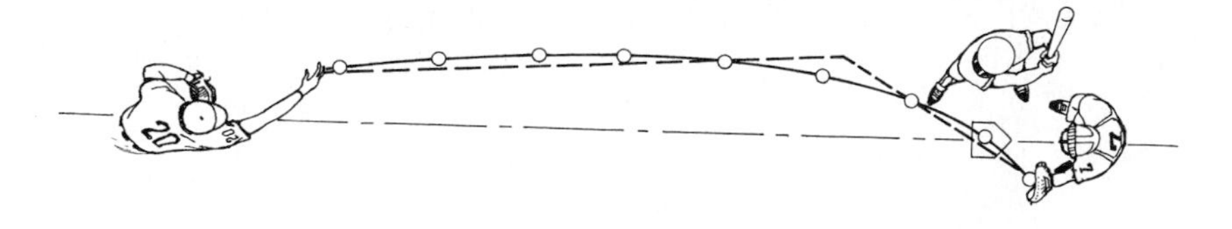

Figure 14-18
The trajectory of a ball thrown with sidespin is curved due to the Magnus effect. The dashed line shows the illusion seen by the players on the field.

until a certain critical point at which the ball "broke" and suddenly curved. This effect is enhanced when topspin is imparted to the ball, since the Magnus effect of topspin accentuates the effect of gravity. However, the actual path of a curve ball is a smooth arc, which has been documented with high-speed movie film (1).

Soccer players can use the Magnus effect when it is advantageous for a kicked ball to follow a curved path, such as when a player executing a free kick attempts to score. The "banana shot" consists of a kick executed so that the kicker places a lateral spin on the ball, curving it around the wall of defensive players in front of the goal (Figure 14-19).

The Magnus effect is maximal when the axis of spin is perpendicular to the direction of relative fluid velocity. Golf clubs are designed

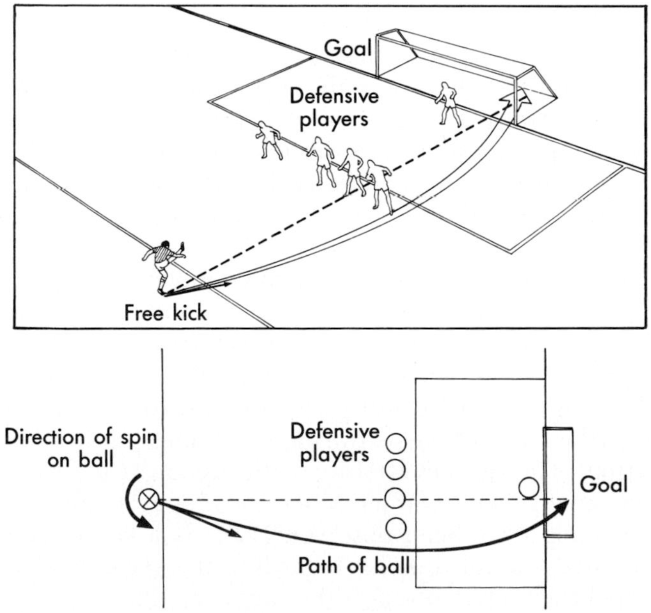

Goal
Defensive players
Free kick
Direction of spin on ball
Defensive players
Goal
Path of ball

Figure 14-19
A banana shot in soccer.

Figure 14-20
The loft on a golf club is designed to produce backspin on the ball. A properly hit ball rises because of the Magnus effect.

to impart some backspin to the struck ball, thereby creating an upwardly directed Magnus force that increases flight time and flight distance (Figure 14-20). When a golf ball is hit laterally off-center, a spin about a vertical axis is also produced, causing a laterally deviated Magnus force that causes the ball to deviate from a straight path. When backspin and sidespin have been imparted to the ball, the resultant effect of the Magnus force on the path of the ball depends on the orientation of the ball's resultant axis of rotation to the airstream and on the velocity with which the ball was struck. A golf ball struck laterally off-center may therefore display a slight or a more severe deviation to one side.

PROPULSION IN A FLUID MEDIUM

Whereas a headwind slows a runner or cyclist by increasing the acting drag force, a tailwind can actually contribute to forward propulsion. Theoretical calculations indicate that a tailwind of 2 m/s improves running time during a 100 m sprint by approximately 0.18 seconds (25). A tailwind affects the relative velocity of a body with respect to the air, thereby modifying the resistive drag acting on the body. Thus, a tailwind of a velocity greater than the velocity of the moving body produces a drag force in the direction of motion (Figure 14-21). This force has been termed **propulsive drag.**

propulsive drag
the drag acting in the direction of a body's motion

Analyzing the fluid forces acting on a swimmer is more complicated. Resistive drag acts on a swimmer, yet the propulsive forces exerted by the water in reaction to the swimmer's movements are responsible for the swimmer's forward motion through the water. The motions of the body segments during swimming produce a complex combination of drag and lift forces throughout each stroke cycle, and even among elite swimmers a wide range of kinetic patterns during stroking have been observed (17). As a result, researchers have proposed several theories regarding the ways in which swimmers propel themselves through the water (8, 12).

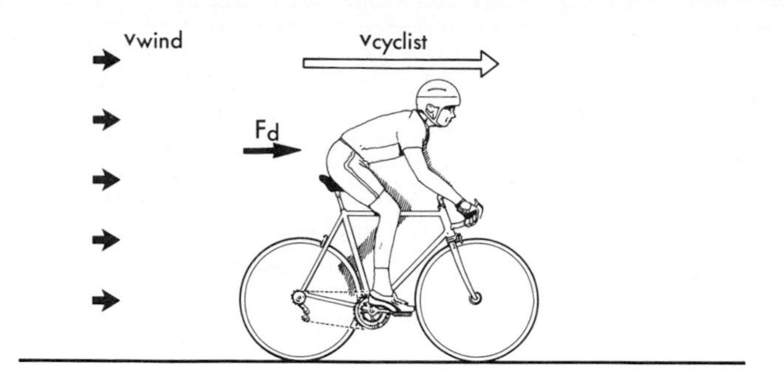

Figure 14-21
Drag force acting in the same direction as the body's motion may be thought of as propulsive drag because it contributes to the forward velocity of the body.

Propulsive Drag Theory

propulsive drag theory
a theory attributing propulsion in swimming to propulsive drag on the swimmer

The oldest theory of swimming propulsion is the **propulsive drag theory,** which was proposed by Counsilman and Silvia (4) and is based on Newton's third law of motion. According to this theory, as a swimmer's hands and arms move backward through the water, the forwardly directed reaction force generated by the water produces propulsion. The theory also suggests that the horizontal components of the downward and backward motion of the foot and the upward and backward motion of the opposite foot generate a forwardly directed reaction force from the water.

When high-speed movie films of skilled swimmers revealed that swimmer's hands and feet followed a zigzag rather than a straight-back path through the water, the theory was modified. It was suggested that this type of movement pattern enabled the body segments to push against still or slowly moving water instead of water already accelerated backward, thereby creating more propulsive drag. However, propulsive drag may not be the major contributor to swimming propulsion (8).

Propulsive Lift Theory

propulsive lift theory
a theory attributing propulsion in swimming at least partially to lift acting on the swimmer

The **propulsive lift theory** was proposed by Counsilman (3) in 1971. It states that swimmers use the foil-like shape of the hand by employing rapid lateral movements through the water to generate lift. The lift is resisted by downward movement of the hand and by stabilization of the shoulder joint, which translates the forwardly directed force to the body, propelling it past the hand. The theory was modified by Firby (6) in 1975 with the suggestion that swimmers use their hands and feet as propellers, constantly changing the pitches of the body segments to use the most effective angle of attack (Figure 14-22).

A number of investigators have since studied the forces generated by the body segments during swimming (16, 22, 26). It has been

Figure 14-22
Use of the hand in swimming may generate lift and drag for propulsion by acting as a propeller. (Modified from Maglischo E: Swimming faster: a comprehensive guide to the science of swimming, Palo Alto, Calif, 1982, Mayfield Publishing Co.)

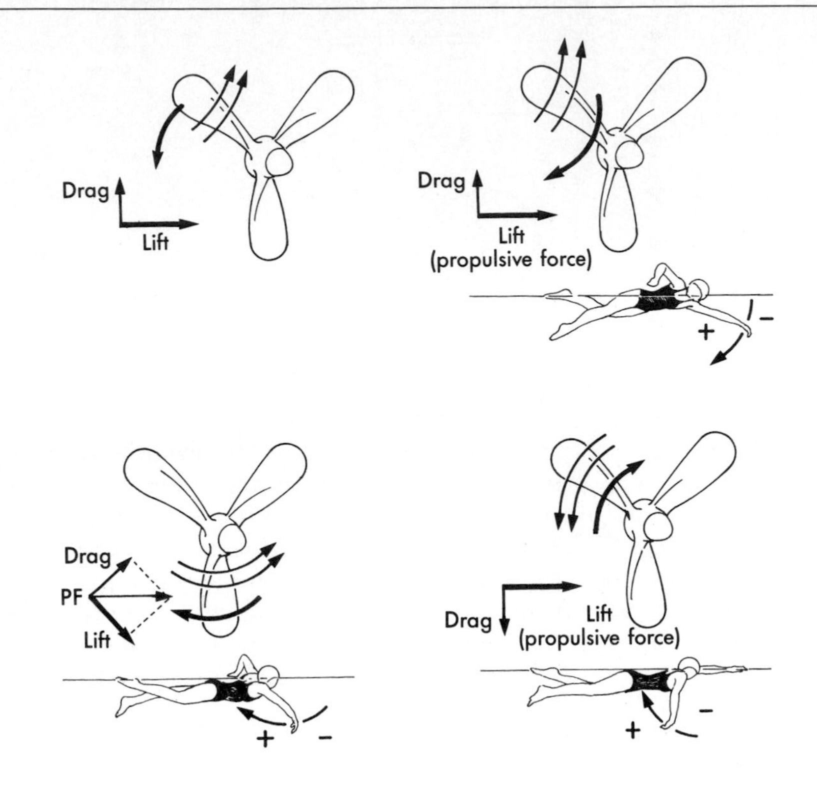

shown that lift does contribute to propulsion and that a combination of lift and drag forces act throughout a stroke cycle. The relative contributions of lift and drag vary with the stroke performed, the phase within the stroke, and the individual swimmer. For example, lift is the primary force acting during the breaststroke, whereas lift and drag contribute differently to various phases of the front crawl stroke (16).

Vortex Theory

vortex theory
a theory attributing propulsion in swimming to the creation of bound vortices in the water by the swimmer

The most recently proposed theory on propulsion in swimming is known as the **vortex theory.** After observing the water flow patterns surrounding a swimmer, researchers (21) proposed that propulsion is derived from the creation of a bound vortex or rotational fluid flow by propelling body segment. A bound vortex consists of rotating water. A swimmer performing the dolphin kick leaves behind a series of bound vortices (Figure 14-23). The formation of these vortices is proposed to cause propulsion of the swimmer. Better ankle

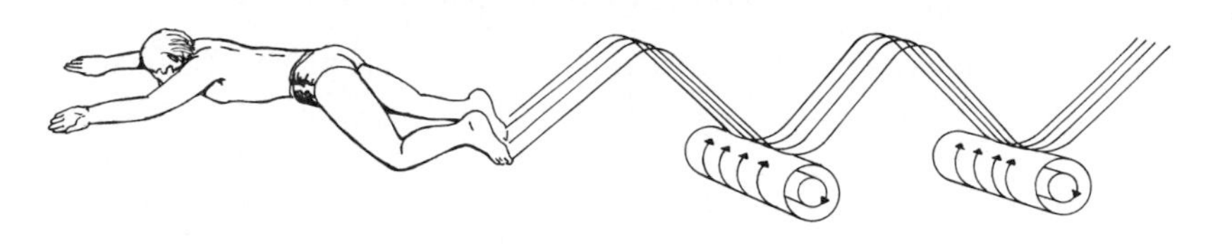

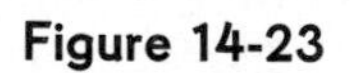

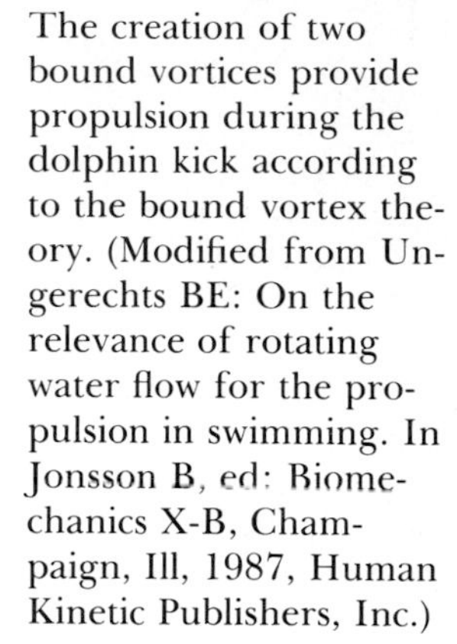

Figure 14-23
The creation of two bound vortices provide propulsion during the dolphin kick according to the bound vortex theory. (Modified from Ungerechts BE: On the relevance of rotating water flow for the propulsion in swimming. In Jonsson B, ed: Biomechanics X-B, Champaign, Ill, 1987, Human Kinetic Publishers, Inc.)

flexibility enhances a swimmer's ability to form bound vortices during the kick (21).

More research is needed to clarify the mechanical factors that contribute to swimming efficiency. As with performance in many sports, different techniques are likely to be more effective for athletes of different anthropometric characteristics.

SUMMARY

The relative velocity of a body with respect to a fluid and the density, specific weight, and viscosity of the fluid affect the magnitudes of fluid forces. The fluid force that enables flotation is buoyancy. The buoyant force acts vertically upward, its point of application is the body's center of volume, and its magnitude is equal to the product of the volume of the displaced fluid and the specific gravity of the fluid. A body floats in a static position only when the magnitude of the buoyant force and body weight are equal and when the center of volume and the center of gravity are vertically aligned.

Drag is a type of fluid force that acts in the direction of the freestream fluid flow. Skin friction is the major form of drag present when the flow is primarily laminar. When the boundary layer of fluid molecules next to the surface of the moving body is primarily turbulent, form drag predominates. Wave drag is created at the interface between two different fluids, such as water and air. It may be the major resistive force acting on swimmers moving with high relative velocities.

Lift force can be generated perpendicular to the freestream fluid flow. Lift is created by a pressure differential in the fluid on opposite sides of a body that results from differences in the velocity of the fluid flow. The lift generated by spin is known as the Magnus force. Propulsion in swimming may result from a complex interplay of propulsive drag and lift forces and possibly from the creation of bound vortices in the water by the swimmer.

INTRODUCTORY PROBLEMS

For all problems, assume that the specific weight of fresh water equals 9810 N/m^3 and the specific weight of sea (salt) water equals 10,070 N/m^3.

1. A boy is swimming with an absolute speed of 1.5 m/s in a river where the speed of the current is 0.5 m/s. What is the velocity of the swimmer with respect to the current when the boy swims directly upstream? Directly downstream? (Answer: 2 m/s in the upstream direction; 1 m/s in the downstream direction)
2. A cyclist is riding at a speed of 14 km/hr into a 16 km/hr headwind. What is the wind velocity relative to the cyclist? What is the cyclist's velocity with respect to the wind? (Answer: 30 km/hr in the direction of the wind; 30 km/hr in the direction of the cyclist)
3. A skier traveling at 5 m/s has a speed of 5.7 m/s relative to a headwind. What is the absolute wind speed? (Answer: 0.7 m/s)
4. A 700 N man has a body volume of 0.08 m^3. If submerged in fresh water, will he float? Given his body volume, how much could he weigh and still float? (Answer: Yes; 784.8 N)
5. A racing shell has a volume of 0.38 m^3. When floating in fresh water, how many 700 N people can it support? (Answer: 5)
6. How much body volume must a 60 kg person have to float in fresh water? (Answer: 0.06 m^3)
7. Explain the implications for flotation due to the difference between the specific weight of fresh water and the specific weight of sea water.
8. What strategy can people use to improve their chances of floating in water? Explain your answer.
9. What types of individuals may have a difficult time floating in water? Explain your answer.
10. A beach ball weighing 1 N and with a volume of 0.03 m^3 is held submerged in sea water. How much force must be exerted vertically downward to hold the ball completely submerged? To hold the ball one-half submerged? (Answer: 301.1 N; 150.05 N)

ADDITIONAL PROBLEMS

1. A cyclist riding against a 12 km/hr headwind has a velocity of 28 km/hr with respect to the wind. What is the cyclist's absolute velocity? (Answer: 16 km/hr)

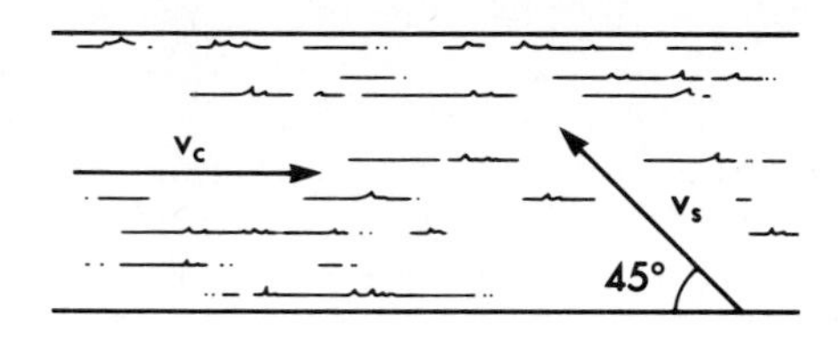

2. A swimmer crossing a river proceeds at an absolute speed of 1.5 m/s on a course oriented at a 45 degree angle to the 1 m/s current. Given that the absolute velocity of the swimmer is equal to the vector sum of the velocity of the current and the velocity of the swimmer with respect to the current, what is the magnitude and direction of the velocity of the swimmer with respect to the current? (Answer: 2.08 m/s at an angle of 31 degrees to the current)
3. What maximum average density can a body possess if it is to float in fresh water? Sea water?
4. A scuba diver carries camera equipment in a cylindrical container that is 45 cm long, 20 cm in diameter, and 22 N in weight. For optimal maneuverability of the container under water, how much should its contents weigh? (Answer: 120.36 N)
5. A 50 kg person with a body volume of 0.055 m^3 floats in a motionless position. How much body volume is above the surface if the water is fresh? If the water is salt? (Answer: 0.005 m^3; 0.0063 m^3)
6. A 670 N swimmer oriented horizontally in fresh water has a body volume of 0.07 m^3 and a center of volume located 3 cm superior to the center of gravity.
   a. How much torque does the swimmer's weight generate?
   b. How much torque does the buoyant force acting on the swimmer generate?
   c. What can the swimmer do to counteract the torque and maintain a horizontal position?
   (Answer: 0; 20.6 N-m)
7. Based on your knowledge of the action of fluid forces, speculate as to why a properly thrown boomerang returns to the thrower.
8. Explain the aerodynamic benefits of drafting on a bicycle or in an automobile.
9. What is the practical effect of streamlining? How does streamlining alter the fluid forces acting on a moving body?
10. Explain why a curve ball curves. Include a discussion of the aerodynamic role of the seams on the ball.

REFERENCES

1. Allman WF: Pitching rainbows: the untold physics of the curve ball. In Schrier EW and Allman WF, eds: Newton at the bat, New York, 1981, Macmillan-Charles Scribner's Sons.
2. Atwater AE and Francis PR: Cinematographic analysis of the mechanics of support sculling. In Dillman CJ and Bauer SJ, eds: Abstracts of biomechanical research, Colorado Springs, Colo, 1983, United States Olympic Committee.
3. Brown RM and Counsilman JE: The role of lift in propelling swimmers. In Cooper JM, ed: Biomechanics, Chicago, 1971, Athletic Institute.
4. Counsilman JE: Science of swimming, Englewood Cliffs, NJ, 1968, Prentice-Hall, Inc.
5. Denoth J, Luethi SM, and Gasser HH: Methodological problems in optimization of the flight phase in ski jumping, Int J Sport Biomech 3:404, 1987.
6. Firby H: Howard Firby on swimming, London, 1975, Pelham Books, Ltd.
7. Ganslen RV: Aerodynamic factors which influence discus flight. In Hay JG: The biomechanics of sports techniques, ed 3, Englewood Cliffs, NJ, 1985, Prentice-Hall, Inc.
8. Hay JG: Swimming biomechanics: a brief review, Swimming Technique p 15, Nov 1986-Jan 1987.
9. Hay JG and Thayer AM: Flow visualization of competitive swimming techniques: the tufts method, J Biomec 22: 11, 1989.
10. Kyle CR and Burke E: Improving the racing bicycle, Mechanical Engineering. 106:34, 1984.
11. Kyle CR and Caiozzo VJ: The effect of athletic clothing aerodynamics upon running speed, Med Sci Sports Exerc 18:509, 1986.
12. Maglischo C et al: The swimmer: a study of propulsion and drag, Soma 2:40, 1987.
13. Maglischo E: Swimming faster: a comprehensive guide to the science of swimming, Palo Alto, Calif, 1982, Mayfield Publishing Co.
14. Remizov LP: Biomechanics of optimal flight in ski jumping, J Biomech 17:167, 1984.
15. Roberson JA and Crowe CT: Engineering fluid mechanics, ed 2, Boston, 1980, Houghton Mifflin Co.
16. Schleihauf RE: A hydrodynamic analysis of swimming propulsion. In Terauds J and Bedingfield E, eds: Swimming III, Baltimore, 1979, University Park Press.
17. Schleihauf RE et al: Models of aquatic skill sprint front crawlstroke, N Z J Sports Med p 6, Mar 1986.

18. Takamoto M, Ohmichi H, and Miyashita M: Wave height in relation to swimming velocity and proficiency in front crawl stroke. In Winter D et al, eds: Biomechanics IX-B, Champaign, Ill, 1985, Human Kinetics Publishers, Inc.
19. Toussaint HM et al: Effect of a triathlon wet suit on drag during swimming, Med Sci Sports Exerc 21:325, 1989.
20. Townend MS: Mathematics in sport, New York, 1984, John Wiley & Sons, Inc.
21. Ungerechts BE: On the relevance of rotating water flow for the propulsion in swimming. In Jonsson B, ed: Biomechanics X-B, Champaign, Ill, 1987, Human Kinetics Publishers, Inc.
22. Valiant GA, Holt LE, and Alexander AB: The contributions of lift and drag force components of the hand/forearm to a swimmer's propulsion. In Terauds J, ed: Biomechanics in sports, Del Mar, Calif, 1983, Academic Publishers.
23. van Ingen Schenau GJ: The influence of air friction in speed skating, J Biomech 15:449, 1982.
24. Ward-Smith AJ: The influence of aerodynamic and biomechanical factors on long jump performance, J Biomech 16:655, 1983.
25. Ward-Smith AJ: A mathematical analysis of the influence of adverse and favourable winds on sprinting, J Biomech 18:351, 1985.
26. Wood TC: A fluid dynamics analysis of the propulsive potential of the hand and forearm in swimming. In Terauds J and Bedingfield E, eds: Swimming III, Baltimore, 1979, University Park Press.

ANNOTATED READINGS

Hay JG: Swimming biomechanics: a brief review, Swimming Technique. p 15, Nov 1986-Jan 1987.

Reviews the research on factors contributing to performance in competitive swimming.

Kyle CR: Athletic clothing, Sci Am 254:104, 1986.

Presents an overview of the aerodynamic improvements that have been made in clothing for several different sports based on wind tunnel research on drag.

Scrier EW and Allman WF, eds: Newton at the bat, New York, 1981, Macmillan-Charles Scribner's Sons.

Chapters on baseball pitching, golf, boomerangs, and frisbee include interesting and entertaining discussions of the actions of fluid forces.

Townend MS: Mathematics in sport, New York, 1984, John Wiley & Sons, Inc.

The chapter on sailing provides a technical analysis of the aerodynamics and hydrodynamics of sailing and windsurfing.

15 ASSESSMENT

Movement Analysis

After reading this chapter, the student will be able to:

Explain why thorough understanding of a movement is prerequisite to analysis of its performance.

Demonstrate an understanding of the general protocol used for effective human movement analyses.

Identify the strategies that can be employed to maximize the value of observational data collected on a human movement of interest.

Identify the procedures used to maximize the thoroughness of the analysis of observational data collected.

Describe the organizational framework that can be useful for communicating the results of a human movement analysis.

Identify and describe the uses of the instrumentation commonly used by biomechanic researchers.

Many jobs require conducting qualitative analyses daily.

What teaching cues can help eliminate a golfer's slice? How can effective coaching tips be formulated for the development of a volleyball player's spike? What gait anomalies can be ameliorated through the use of orthotics? How can an athlete's performance be improved? How does the clinician, coach, or teacher of physical activities arrive at answers to questions such as these? How do people analyze human motion to correct or improve it? In this chapter the problem-solving approach is adapted to provide a template for an approach to solving problems associated with the analysis of human movement.

QUALITATIVE VERSUS QUANTITATIVE ANALYSIS

As introduced in Chapter 1, analysis of human movement may be either qualitative or quantitative. The word *quantitative* implies that numbers are involved, and *qualitative* refers to a description of quality without the use of numbers. Qualitative assessment includes both knowledge of the movement characteristics desired and the ability to observe and analyze whether a given performance incorporates these characteristics (7). Clinicians, coaches, and teachers of physical activities regularly conduct mental analyses of movement quality in the process of formulating advice for their patients, athletes, and students. Biomechanic researchers rely heavily on computerized quantitative analyses in their studies of humans and other organisms. Although these two analysis approaches sound different, there are actually a number of common elements.

REQUIRED SKILLS AND KNOWLEDGE

The process of visual observation is the most commonly used methodological approach for analyzing the mechanics of human movement. Based on information gained from watching an athlete perform a skill, a patient walk down a ramp, or a student attempt a novel task, coaches, clinicians, and teachers make judgements and recommendations on a daily basis.

To the untrained observer, there may be no differences in the forms displayed by an elite hurdler and a novice hurdler or in the functioning of a normal knee and an injured, partially rehabilitated knee. What skills are necessary and what procedures are used for effective analysis of human movement mechanics?

Baseline knowledge of human anatomy and mechanical principles provide essential background for the serious human movement ana-

Biomechanic researchers use computers for quantitative analyses of human motion.

Figure 15-1

Movement analysts operate from background knowledge of the general anatomical and mechanical considerations affecting human movement. They acquire knowledge of the skill to be analyzed and of the appropriate analysis protocol.

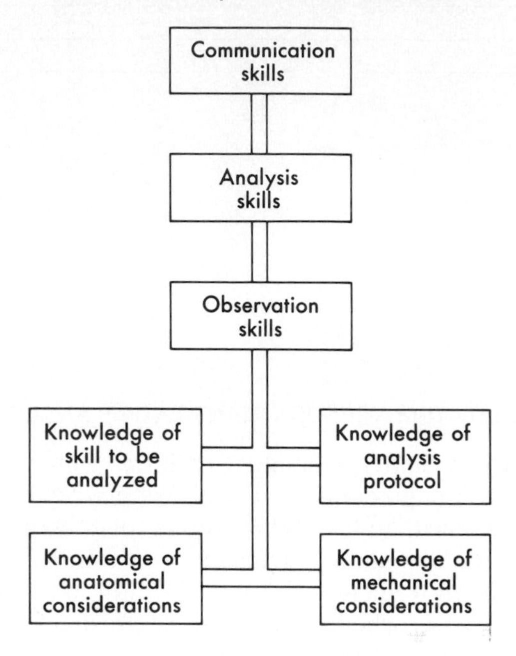

lyst (Figure 15-1). A thorough understanding of the motor skill, technique, or movement pattern to be analyzed is also desirable. Of equal importance is knowledge of the appropriate *protocol* or procedural format for conducting the movement analysis. Conducting the actual analysis requires the ability to effectively observe and record information about the movement of interest. Once observations have been made, analysts use their analytical skills and knowledge. However, for an analysis to be meaningful for anyone beyond the analyst, communication skills are required.

Knowledge of the Movement to be Analyzed

The ability to effectively analyze a motor skill requires an understanding of the nature and the purpose of the skill of interest. Without a sound understanding of the skill, analysts may have difficulty in identifying the factors that contribute to performance and may misinterpret the observational data collected. Some skills, such as archery, pistol shooting, and rifle shooting, are accorded success based on accuracy. Other skills, including the throwing and jumping events, require the maximum displacement of a projectile. For golf, shots should be both in a particular direction and of a particular distance. Running, swimming, and cycling events are predicated on maximum speed as well as strategy. Although the desired outcomes of these activities are generally well known, the specific kinematic and kinetic patterns contributing to success are often less apparent. Successful analysts must not only understand the purpose of the movement being analyzed but also recognize the factors contributing to skilled performance of the movement.

A skilled performer is usually better equipped to analyze the quality of a performance than a person who is unfamiliar with the skill. Experienced tennis and softball players are more proficient than novices at detecting errors in the forehand tennis stroke and in batting (2, 8). Experts may focus their attention differently than novices when analyzing performance (3). Personal background experiences have also been found to affect the focus of attention during movement analyses performed by physical education student teachers. (1).

Although direct experience in the performance of a sport skill can contribute to the analyst's understanding, direct experience is not the only or necessarily the best way to become familiar with a motor skill. The best athletes do not always become the best coaches, and highly successful coaches may have had little to no participatory experience in the sports that they coach.

The conscientious coach, teacher, and clinician typically use several avenues to pursue information about a given movement and the factors that affect it. One is to read related articles from the biomechanics research literature and sport and coaching journals, although not all movement patterns and skills have been researched and some analysis literature is so esoteric that advanced training in biomechanics is required to understand it. It is important to distinguish between those articles that are supported by research findings and those based primarily on opinion. Some well-respected coaches have erred in their common sense approaches to skill analyses. There are also opportunities to interact directly with individuals who have expert knowledge of particular skills at conferences and symposia.

Analysts may also benefit from an understanding of the nature of human movement patterns and skills in general. For example, distinctions are made between **discrete skills** and **continuous skills** and between **open skills** and **closed skills.** Cyclic skills such as running, swimming, rowing, paddling and cycling are continuous skills because they involve repetition of movement patterns. Skills with definite beginning and ending points are discrete skills, which include volleyball serves, spikes, and passes, football or soccer kicks, and basketball and hockey shots and passes. Open and closed skills, on the other hand are distinguished based on the surrounding environment. When the environment is unpredictable, such as when a football player carries a ball down the field or a basketball player throws

discrete skills
purposeful movement patterns with identifiable beginnings and endings

continuous skills
purposeful movement patterns that are cyclic and repetitive

open skills
skills executed in an unpredictable setting that may require quick adaptations of the performer

closed skills
skills executed in a predictable setting

a ball in from the sideline, the skill is open. A closed skill is performed in a situation that is predictably structured. Examples are serves in volleyball, gymnastic routines, and training runs by marathoners along a familiar route. Both open and closed skills may be either continuous or discrete.

Performer characteristics also affect a movement performance. These include the performer's age, gender, and anthropometry, the developmental and skill levels at which the performer is operating, and any special considerations, such as other physical or personality traits that may impact performance. In many cases, providing a novice, preschool-aged performer with cues for a skilled, mature performance may be counterproductive because children are not scaled-down adults (9). Although training can ameliorate loss of muscular strength and joint range of motion once thought to be inevitably associated with aging, human movement analysts need increased knowledge of and sensitivity to the special needs of senior citizens who wish to develop specialized motor skills. Analysts also need to be

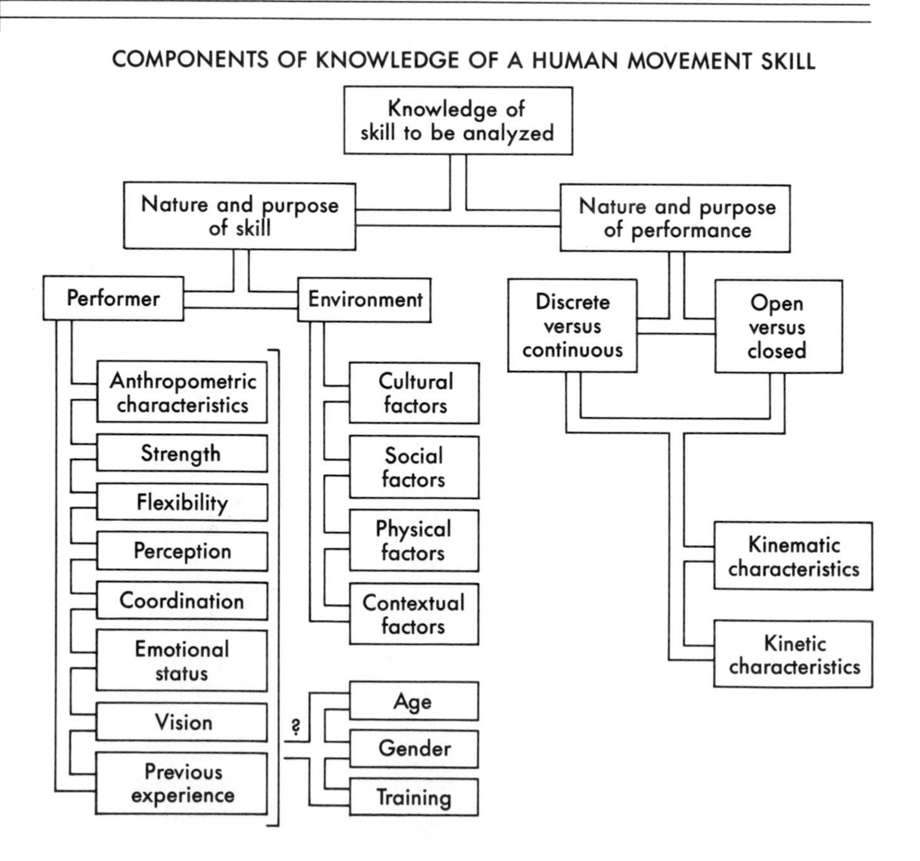

Figure 15-2
A number of factors contribute to a thorough understanding of the performance of a human motor skill.

aware that although differences in performance capability throughout the lifespan have traditionally been attributed to gender, before puberty most gender-associated performance differences are probably culturally derived rather than biologically determined (12). Young girls are usually not expected to be as skilled or even as active as young boys. Analysts of elementary school-aged performers should not reinforce this cultural misunderstanding by lowering their expectations of girls based on gender. Analysts should also be sensitive to the other factors that can influence performance. Has the performer experienced a recent emotional upset? Is the sun in his eyes? Is she tired? A summary of the considerations relative to knowledge of a human movement skill is shown in Figure 15-2.

Knowledge of Analysis Protocol

The number of procedural factors that analysts should consider when planning and conducting a biomechanical analysis increases with the complexity of the movement and the degree of sophistication desired. Even the simplest analysis may yield inadequate or faulty information if approached improperly.

Analysts should first determine whether the movement is primarily planar or nonplanar. If the major movements of interest are largely confined to a single plane, such as with the legs during cycling or the pitching arm during an underhand softball pitch, a single viewing perspective (for example, a side view or a rear view) may be sufficient. If the movement occurs in more than one plane, as with the motions of the arms and legs during the breaststroke or the arm motion during a baseball batter's swing, the observer may need to view the movement from more than one perspective to see all critical aspects. For example, a rear view, a side view, and a top view of a martial artist's kick yield different information about the movement. Figure 15-3 contrasts primarily planar and multiplanar skills.

Even when a movement is primarily planar, using more than a single viewing perspective of the performer may be necessary. In the analysis of some skills, a view enabling observation of the performer's eye movements can yield valuable information. This occurs when analyzing fencing, since focusing the eyes on the rapidly moving tip of an opponent's blade can be a drastic performance error, causing the fencer to be easily misled by the opponent's feints (13). A common error among beginner tennis players is looking across the net at the area where they expect to hit the ball rather than visually tracking enough of the oncoming ball's trajectory to establish racket contact with it.

A tennis player's eyes should follow the oncoming ball trajectory long enough to enable establishing contact of the ball with the racket.

The analyst's viewing distance from the performer should also be selected thoughtfully. If the analyst wishes to observe pronation and supination patterns of a subject walking on a treadmill, a close-up rear view of the lower legs and feet is necessary. Analyzing where a

Figure 15-3

While skills that are primarily planar may require only one viewing perspective, the movement analyst should view multiplanar skills from more than one direction.

Figure 15-4

The observation distance between analyst and performer should be selected based on the nature and focus of the questions of interest.

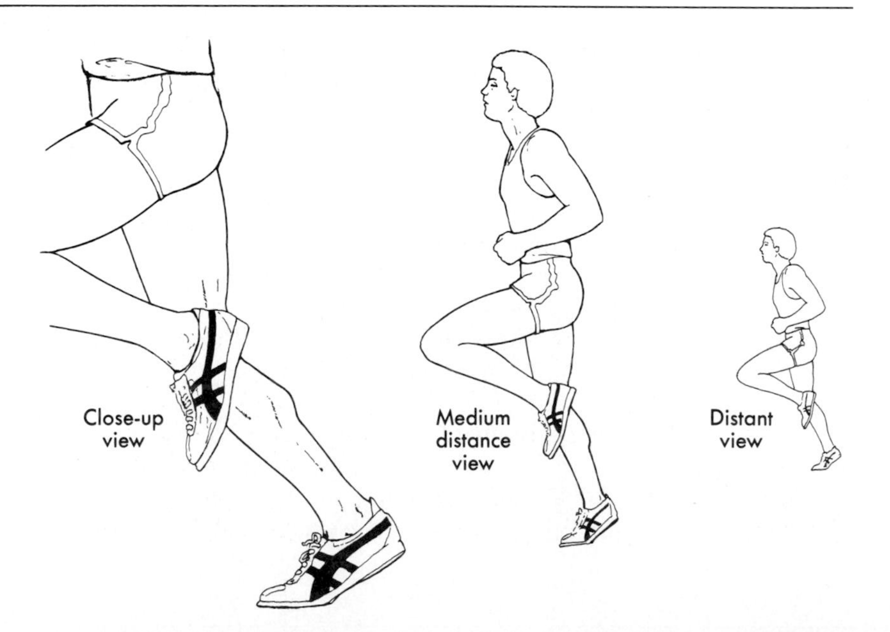

particular volleyball player moves on the court during a series of plays under rapidly changing game conditions, is best accomplished from a reasonably distant view. The effect of viewing distance is illustrated in Figure 15-4.

A second element of consideration relevant to any biomechanical analysis is the number of trials or executions of the movement that should be observed before an analysis is formulated. An elite athlete may display movement kinematics that deviate only slightly across performances, but a child learning to run may take no two steps alike. Basing an analysis on observation of a single performance is usually unwise. The greater the inconsistency in the performer's technique, the larger the number of observations that should be made.

When biomechanic researchers undertake a study of a particular movement, the subjects performing the movement are often minimally attired so that the movements of the body segments will not be obscured by loose-fitting apparel. Although there are many situations, such as instructional classes, competitive events, and team practices, for which this may not be practical, analysts should be aware that loose clothing can obscure subtle motions occurring at the performer's joints. Performing the analysis at a location where there is adequate lighting and where the color of the background contrasts with the color of the subject's skin and garments also improves the visibility of the observed movement. Figure 15-5 provides a summary of the considerations related to knowledge of analysis protocol.

When subjects are minimally attired, analysts are better able to observe the movement kinematics.

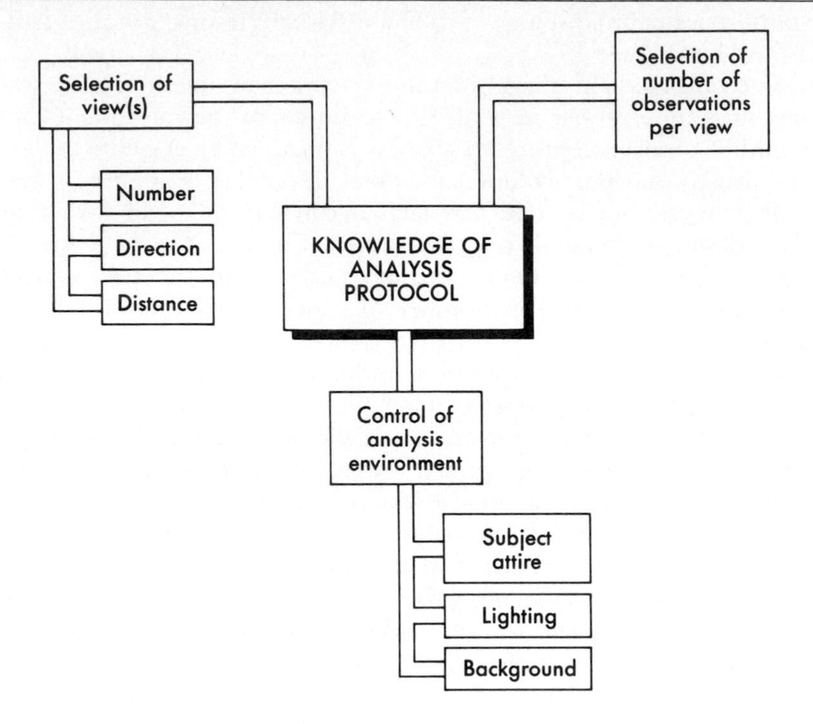

Figure 15-5
The types of decisions that are consciously made in preparation for an effective human movement analysis.

Observation Skills

Observation skills include the abilities to view the characteristics of interest and to focus attention on the critical aspects of the event being observed. A high level of familiarity with the nature of the skill or movement being performed is extremely helpful in identifying what these critical, attention-requiring elements of performance may be. Although visual acuity and selection of the proper view or views contribute to the ability to observe a performance, several techniques also increase the amount of information that a movement analysis may provide.

Analyzing a dynamic event is usually more difficult than analyzing a static pose. As movement occurs, the difficulty of adequately observing an event increases. As movement speed increases, it becomes progressively more difficult to see all of the action of interest. The human eye cannot resolve events that occur in less than approximately one fourth of a second (5). Consequently, even the most careful observer may miss the rapid movements of the hands, arms, legs, or the entire body.

Fortunately, several tools are available to assist observers of human movement in performing qualitative and quantitative analyses. A camera, such as a 35 mm, instamatic, or polaroid, captures a static pose or movement at a critical instant during a dynamic event so that analysts can study a print or projection of the captured image in detail. However, a single picture is typically a poor representation of an entire movement.

With videotape and movie film, observers can replay a dynamic event repeatedly. Better quality playback units also enable slow motion viewing and single picture advance so that the analyst can isolate the critical aspects of a movement. An advantage of videotape over film is that the tape can be erased and used over again once the recorded movement has been analyzed. Movie cameras typically capture more pictures per second than standard video cameras, and a much clearer image is obtained in each separate picture. Advances in video technology are rapidly improving the capabilities of video cameras and recording units.

Although using a video or movie camera usually significantly enhances the analyst's ability to make detailed visual observations of a movement, there is a potential drawback. The subject's awareness of the presence and use of a biomechanical measurement system such as a camera sometimes results in changes in performance (10). Researchers found performance decrements among unskilled subjects and performance increments among skilled subjects performing basketball free throws when a camera was introduced to record motion. Movement analysts should be aware that subjects may be distracted or unconsciously modify their techniques when a recording device is used.

Nonvisual forms of information can also sometimes be useful to the movement analyst. For example, the sounds associated with the execution of some movements provide clues about the way in which the player executed the movement. Proper contact of a golf club with a ball sounds distinctly different than when a golfer "tops" the ball. Similarly, the crack of a baseball bat hitting a ball indicates that the contact was direct rather than glancing. The sound of a double contact of a volleyball player's arms with the ball may identify an illegal hit (Figure 15-6). Listening to the sounds of a knee-injured patient's gait usually reveals whether a limp is present.

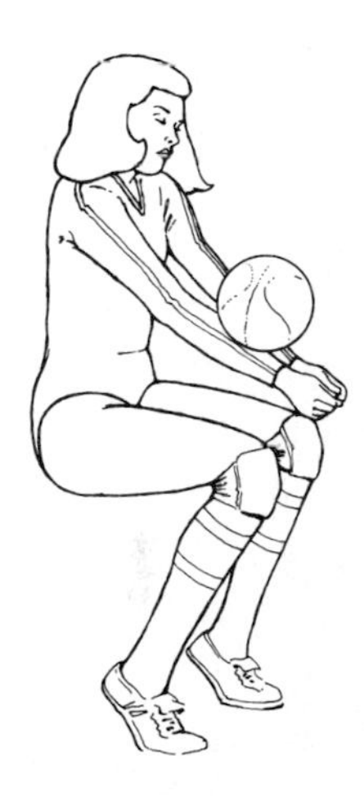

Figure 15-6
Auditory information, such as the sound of a double contact, can be useful to the movement analyst.

Another potential source of information is feedback from the performer being analyzed. A performer who is experienced enough to recognize the way a particular movement feels as compared to the way a slight modification of the same movement feels is a useful source of information. Not all performers are sufficiently kinesthetically attuned to provide meaningful subjective feedback of this nature. The performer being analyzed may also assist in other ways. As Hoffman (6) has pointed out, performance deficiencies may result from errors in technique, perception, or decision making. Identification of perceptual and decision-making errors by the performer often requires more than visual observation of the performance. In

Figure 15-7
The components of observation skills.

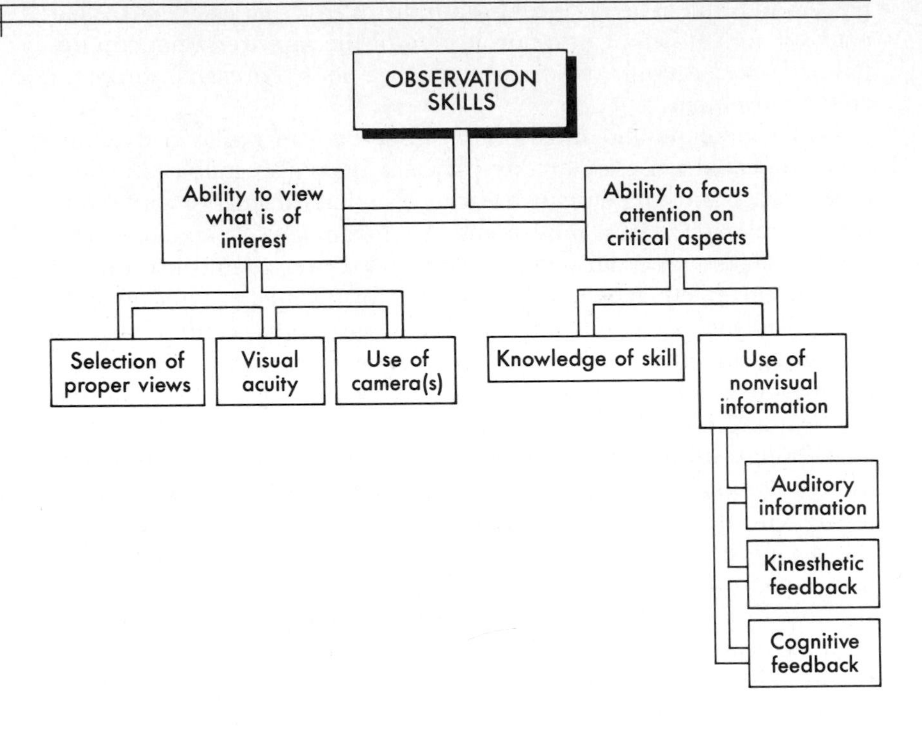

these cases, meaningful questions directed from the analyst to the performer may provide the best clues regarding the major source or sources of difficulty. However, the analyst should consider subjective input from the performer in conjunction with more objective observations.

Analysts must remember that the skill of observation improves with practice. Novice analysts should take every opportunity to practice movement analysis in carefully planned and structured settings. The components of observation skills related to human movement analysis are presented in Figure 15-7.

Analytical Skills

An individual's inherent analytical skills are partially determined by genetic endowment. The analysis of observations of human movement that have been recorded by a camera, written on a notepad, or recorded as mental images, however, can benefit from the adoption of a structured approach.

The first step in most analyses is to identify the major question or questions of interest. Often these questions have already been formulated by the analyst, or they serve as the original purpose for the investigation. For example, how is propulsion generated during performance of the butterfly stroke? What kinematic features distinguish elite sprinters from their less-successful counterparts? What can a particular shot putter do to improve technique? Occasionally, additional questions emerge during the course of collecting observations. For example, what technique changes are occurring over the 30 to 40 m range during performances of the 100 m sprint by a particular runner? What may be contributing to the evident inconsistencies in a golfer's swing? Analysts need to be firmly focused on one or more particular questions or problems to address. Familiarization with the research and coaching literature related to the movement being analyzed is usually of substantial assistance to the analyst in formulating appropriate questions relative to the analysis. The observational data itself may suggest questions of interest and importance that were not previously identified. A thorough and careful analysis is not strictly preprogrammed but is structured (Figure 15-8). A structured analysis typically involves progressive problem solving and often results in the identification of new questions or problems to solve.

The second step is to study the questions of interest and hypothesize as to the ways in which the interplay of mechanical principles, anatomical constraints, and practical considerations may affect the answers to each question. For example, optimal performance in the long jump would require a 45 degree takeoff angle from a mechanical perspective. As discussed in Chapter 9, however, if a long jumper's takeoff angle approaches 45 degrees, horizontal velocity and performance outcome are sacrificed. Analysis of a long jumper's takeoff may benefit from awareness of the range of takeoff angles and speeds actually demonstrated by successful long jumpers of the same age, gender, and approximate skill level as the performer. Regardless of the movement skill being analyzed, maintaining a practical perspective regarding the extent to which mechanical principles can be applied to the human execution of that movement and the degree to which improved skill performance is realistic is important.

Figure 15-8
The back scratch position of a skilled tennis player preparing to serve creates a small moment of inertia for the arm racket unit with respect to the shoulder, enabling a large generation of angular velocity. This position may be one of the checkpoints during an analysis of a tennis serve.

Although completely thorough and sophisticated analyses of human movement require advanced instrumentation and include attention to kinetic and kinematic patterns associated with performance, analysts can obtain useful information by strictly visual means. Hudson (9) has identified six *visual dimensions of movement* an analyst may focus on to infer kinetic and kinematic information from visual observation of a performance. These include the following:

1. The number and nature of segments involved in movement (because more force is typically generated by the movement of more or larger segments)
2. Balance (observed from the approximate position of the line of gravity with respect to the base of support)

Figure 15-9
The analysis skills used in an effective human movement analysis.

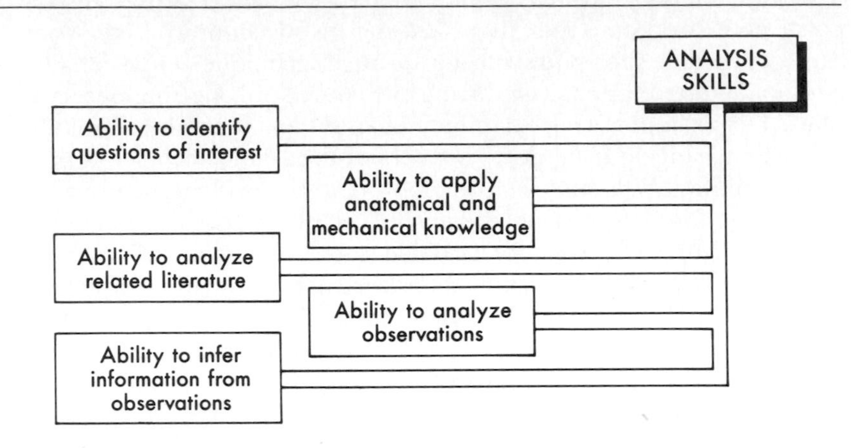

3. Range of motion (with greater force production associated with greater range of motion)
4. The amount of segment extension at the instant of projectile release (with better technique often associated with full extension occurring at the time of release)
5. Limb moment of inertia (based on the positions of rotating segments)
6. Coordination of segments (the sequencing and timing of segment motions)

A thorough analysis is one in which information that may legitimately be inferred from the observational data taken is not overlooked.

Another way to enhance the thoroughness of most analyses is to involve more than one analyst. This reduces the likelihood of overlooking some element of the movement performance being analyzed and may also benefit the accessory analysts. Students in the process of learning a new motor skill may benefit from teaming up to analyze each other's performances under appropriate teacher direction.

The initial analysis may not reveal final answers to the questions that have been formulated. Sometimes an analysis will indicate the need for the collection of additional observations—perhaps from a different view or perspective. Movement modifications may be continuously occurring as learning takes place, especially when the performer is unskilled. The teacher or coach is often involved in a continuous process of formulating an analysis, collecting additional observations, and formulating an updated analysis. This process may also involve the use of communication skills as the teacher or coach relays analytical information back to the student or athlete. The abilities that contribute to human movement analysis skills are summarized in in Figure 15-9.

Communication skills are an important part of the analysis process.

Communication Skills

Communication skills consist of the ability to relay information clearly, accurately, and thoroughly to another individual or a group. The most sophisticated analysis is of little value if it is not properly communicated to the subject or other analysts. Although some individuals have better inherent abilities to organize their thoughts and to clearly speak or write those thoughts, communication skills can be learned with appropriate practice.

The first step in formulating a communication of a movement analysis is to identify what should be communicated. Depending on the target audience, this may include all information derived from the analysis or a small portion of it. Knowledgeable teachers and coaches are aware that novice performers may be overwhelmed and confused by too many cues given at the same time. Similarly, knowledgeable clinicians may wish for a patient to focus attention on only one or two aspects of a rehabilitative program at a given time. If the analyst is communicating with other analysts in either oral or written form, the information presented is usually more comprehensive.

The target audience also largely dictates the type of language that should be employed in the communication of information. Technical terminology, although precise and accurate, may not be understood by a student, athlete, or patient. The words chosen by the analyst must be meaningful to the audience for real communication to occur. If the communication is formal, such as an oral presentation to a group or a written paper, analysts must also consider which tables, graphs, and figures best illustrate the communication.

The next step is to organize the material to be presented. If the intent is to impart a single, simple piece of information to a student learning a novel skill, conscious organization may not be necessary. The more inclusive and complex the material to be presented, the greater the need for organization. Generally, both qualitative and quantitative analyses can benefit from the application of the standard organizational format used for scientific papers. This consists of an **introduction** explaining the rationale for the analysis, an explanation of the observational and analytical **methods** used, a description of the **results** of the analysis, **discussion** of the results, and identification of **conclusions** that have been formulated. An example of this format is shown in the box on p. 462 for a volleyball coach talking to a hitter.

Although a relatively simple interaction may be organized differently without loss of communication, the scientific organizational format is logical, and the more complicated the intended communication, the greater the probability that its clarity will benefit from use of this format.

After the intended communication has been organized, the final step involves its delivery in oral or written form. The better trained

introduction
a section that develops the rationale or purpose for an analysis or investigation

methods section
a section that describes the procedures employed in conducting an analysis

results section
a section that identifies what was observed and calculated during an analysis

discussion section
a section that includes explanatory and interpretive comments on the results of an analysis

conclusions
concise statements logically derived from the results of an analysis

ORGANIZATION OF COMMUNICATION

1. Introduction: "Barb, I've been wondering what we can do to increase the ball speed of your cross-court hits."
2. Methods: "I've been watching you execute cross-court hits from both the back and the side."
3. Results: "It looks to me like you're not getting full extension of the shoulder of your hitting arm in preparation for the hit when you plan to go cross-court."
4. Discussion: "When your arm isn't fully extended at the shoulder before the hit it reduces the range of motion you swing through before contacting the ball. That tends to reduce the velocity you impart to the ball."
5. Conclusion: "Try to be aware of this and to get your arm as far back as possible before executing cross-court hits. I'll help you with this during the next drill."

and the more practiced the analyst is with these forms of communication, the more polished the final product will be.

The skills and knowledge that contribute to effective analysis of human movement can be enhanced through experience with the analysis process. As analysts gain experience, the analysis process becomes more natural and the analyses conducted are likely to become more effective and informational. The components of communication skills are summarized in Figure 15-10.

Figure 15-10
The skills used for effective communication of an analysis of human movement.

COMMUNICATION SKILLS
Ability to identify what should be communicated
Ability to identify how the information should be communicated
Speaking/writing abilities
Choice of terminology
Choice of quantity of information
Choice of organization
Choice of visual aids

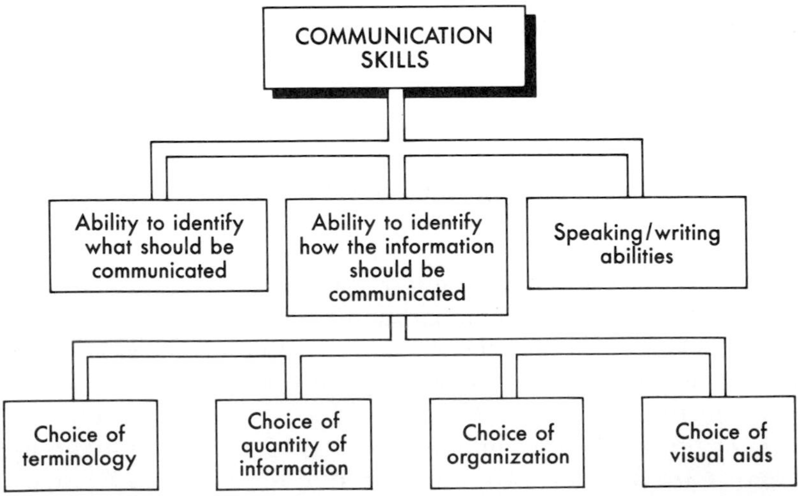

BIOMECHANICS RESEARCH INSTRUMENTATION
Cinematography and Videography

Photographers began employing cameras in the study of human and animal movement during the late nineteenth century. One famous early photographer was Eadweard Muybridge, a British landscape photographer and a rather colorful character who frequently published essays praising his own work (11).

Muybridge used electronically controlled still cameras aligned in sequence with an electromagnetic tripping device to capture serial shots of trotting and galloping horses, thereby resolving the controversy about whether all four hooves are ever airborne simultaneously (they are).

The modern-day movement analyst has many camera types from which to choose. Those most commonly used for documenting human movement sequences are video cameras and 8 and 16 mm movie cameras. The basic video camera captures pictures at 60 Hz (60 fields per second). Depending on the quality of the video playback unit, fewer than 60 pictures per second are typically resolvable during viewing of the tape. More sophisticated and more expensive video cameras are also available that can capture more pictures per second and that provide much better picture resolution through the incorporation of variable speed shuttering. Both 8 mm (super 8) and 16 mm movie cameras provide films with better picture resolution than is available through conventional video systems. Super 8 cameras typically operate at up to 64 frames per second, whereas the most commonly used 16 mm cameras operate at up to 500 frames per second.

Control unit for a video-playback system with single picture advance capability.

The type of movement and the requirements of the analysis largely determine the camera and analysis system of choice. The most widely available and least expensive alternative is conventional video. This medium is adequate for the qualitative analysis of slow movements or for the global tracking of a movement pattern such as a player's path during a game play. However, rapid movements are reduced to a blur on conventional video. Most human movements can be adequately studied by film or video taken with a shuttered camera at rates of 200 Hz, although an extremely fast movement, such as a golf swing, may require 500 or more pictures per second.

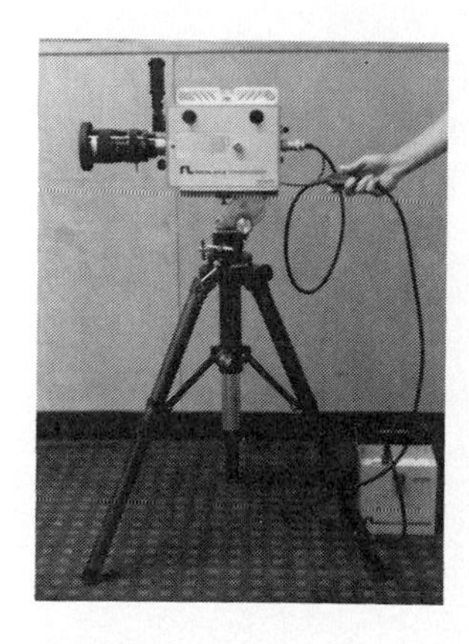

A high-speed movie camera is commonly used by biomechanists to document movement kinematics.

A quantitative film or video analysis is usually performed with computer-linked equipment that enables the calculation of estimates of kinematic and kinetic quantities of interest for each picture. The traditional procedure for analyzing a film or video picture involves a process called *digitizing*. This involves the activation of a hand-held pen, cursor, or mouse over subject joint centers or other points of interest, with the x,y coordinates of each point subsequently stored in a computer data file. Recent technology enables automated digitizing of the film or video by computer software.

A digitizer is an instrument that identifies x,y position coordinates for joint centers and other points of interest on a photograph or film projection.

Optoelectronic Movement Monitoring Systems

Another approach to quantitative analysis of human movement that eliminates the hand-digitizing process involves the attachment of tiny electric lights known as light-emitting diodes (LEDs) or highly reflective markers over the body joint centers. Computer-linked cameras track these special lights or markers, enabling on-line calculation of the quantities of interest. These systems, particularly those using the LEDs, often restrict the subject's maneuverability, rendering them less than optimal for the analysis of movements that cannot be performed naturally in a constrained laboratory setting.

Dynamography

Scientists have devised several types of platforms and portable systems for the measurement of forces and pressure on the plantar surface of the foot. These systems have been employed primarily in gait research but have also been used to study phenomena such as starts, takeoffs, landings, baseball and golf swings, and balance.

Both commercially available and homemade *force platforms* and *pressure platforms* are typically built rigidly into a floor flush with the surface and are interfaced to a computer that calculates kinetic quantities of interest. Force platforms are usually designed to transduce ground reaction forces in vertical, lateral, and anteroposterior directions with respect to the platform itself; pressure platforms provide graphical or digital maps of pressures across the plantar surfaces of the feet. The force platform is a relatively sophisticated instrument, but its limitations include the restrictions of a laboratory setting and potential difficulties associated with the subject consciously targeting the platform.

Portable systems for measuring plantar forces and pressures are also available in commercial and homemade models as instrumented shoes, shoe inserts, and thin transducers that adhere to the plantar surfaces of the feet. These systems provide the advantage of data collection opportunities outside the laboratory but lack the precision of the built-in platforms.

Electromyography

It was reported by the Italian scientist Galvani in the late 1700s that skeletal muscles develop tension when electrically stimulated and produce a detectable current or voltage when they develop tension (4). The latter discovery was of little practical value until the twentieth century when technology became available for the detection and recording of extremely small electrical charges. The technique of recording electrical activity from muscle, or **myoelectric activity,** is known today as *electromyography* (EMG).

myoelectric activity
electric current or voltage produced by a muscle developing tension

Electromyography is used to study neuromuscular function, including identification of which muscles develop tension throughout a movement and which movements elicit more or less tension from a particular muscle or muscle group. It is also used clinically to assess nerve conduction velocities and muscle response in conjunction with the diagnosis and tracking of pathological conditions of the neuromuscular system. Scientists also employ electromyographic techniques to study the ways in which individual motor units respond to central nervous system commands.

The process of electromyography involves the use of **transducers** known as *electrodes* that sense the level of myoelectric activity present at a particular site over time. Depending on the questions of interest, either surface electrodes or fine wire electrodes are used. Surface electrodes, consisting of small discs of conductive material, are positioned on the surface of the skin over a muscle or muscle group to pick up global myoelectric activity. When more localized pick up is desired, in-dwelling, fine wire electrodes are injected directly into a muscle. Output from the electrodes is amplified and graphically displayed or mathematically processed and stored by a computer.

transducers
devices that detect a signal

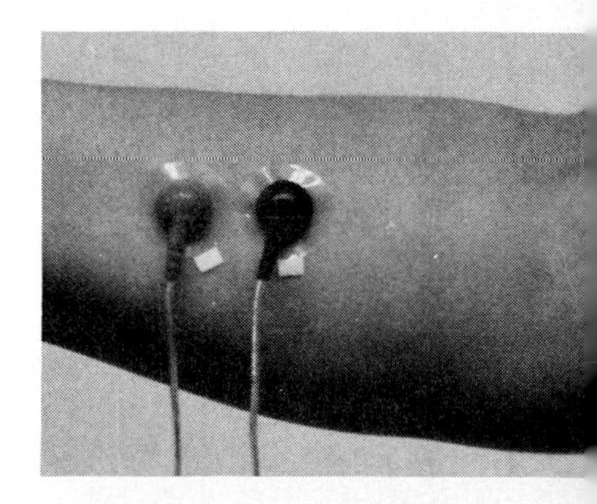

Surface EMG electrodes are small discs that attach directly to the skin over a muscle or muscle group of interest.

Other Assessment Tools

Other tools used by biomechanic researchers are available in commercial and homemade versions. The hand-operated goniometer used for assessing the angle present at a joint is also available in an electronic version known as an *electrogoniometer* or elgon (see Chapter 5). The center of the elgon is positioned over the center of rotation of the joint to be monitored, with the arms of the elgon aligned and firmly attached over the longitudinal axes of the adjacent body segments. When motion occurs at the joint, the electrical output provides a continuous record of the angle present at the joint.

A computerized electromyography system.

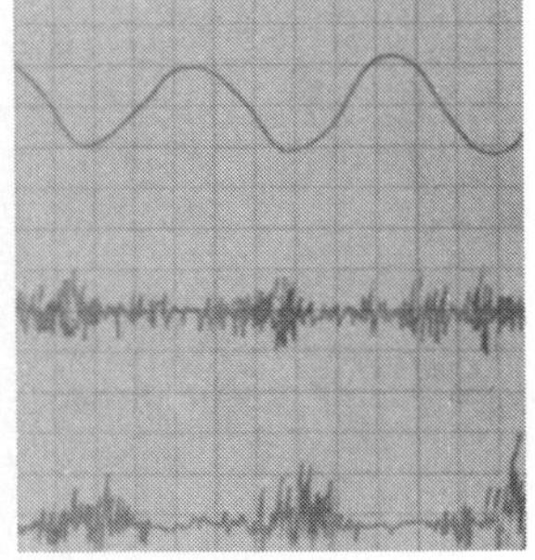

A graphic print out showing changes in joint range of motion and myoelectric activity recorded from two muscle groups.

Systems combining photocells, light beams, and timers can be used to directly measure movement velocity. The system is usually configured so that light beams intercept photocells at two or more carefully measured positions (Figure 15-11). The photocells are electrically connected to a timer so that the time interval between interruption of the light beams by a moving body segment or an object such as a thrown ball can be precisely recorded. The velocity of the moving body is calculated as the measured distance between the photocells divided by the recorded time.

An *accelerometer* is a transducer used for the direct measurement of acceleration. The accelerometer is attached as rigidly as possible

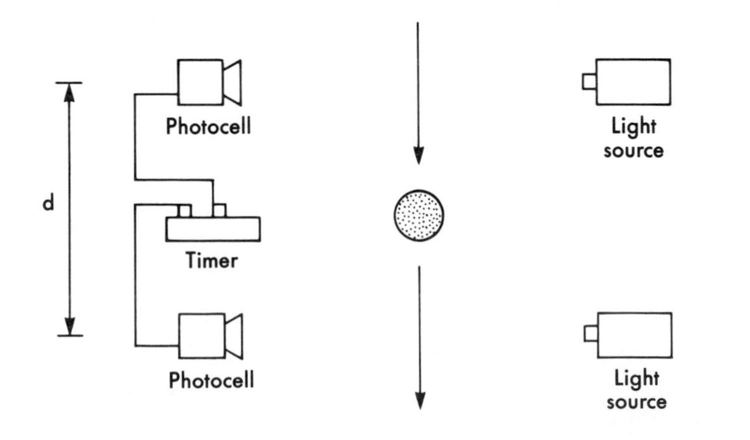

Figure 15-11
A light, photocell, and timer set up for measuring movement velocity. As a ball travels through the apparatus zone, the light beams focused on the photocells are interrupted, sending a signal to the timer. The ball's velocity is calculated as the measured distance between photocells divided by the measured time interval.

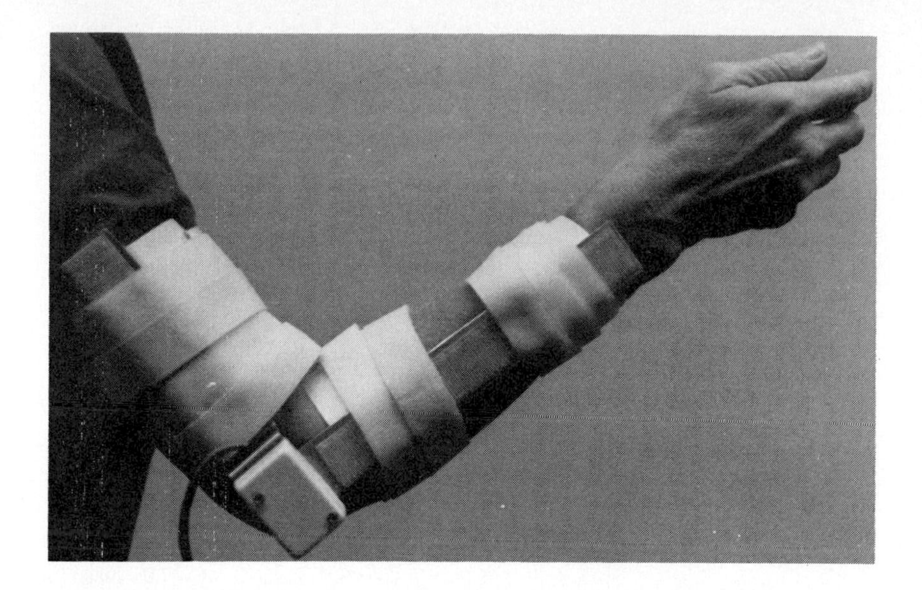

An electrogoniometer transducing the angle present at the elbow.

to the body segment or other object of interest, with electrical output channelled to a recording device. Two or more accelerometers must be used in combination to monitor acceleration during a nonlinear movement.

SUMMARY

Teachers of physical activities, coaches, and clinicians all perform informal analyses to correct or improve human movements. Underlying knowledge of the mechanical and anatomical factors that affect human movement is the foundation for performing these analyses accurately and meaningfully. The ability to effectively carry out such analyses requires other skills and knowledge and can be enhanced with a structured approach.

Before undertaking a human movement analysis, the analyst should have a thorough understanding of the nature and purpose of the movement to be analyzed and of the appropriate protocol or procedural format to use. Conducting the analysis requires observational and analytical skills and knowledge of strategies that maximize the usefulness of the observational data collected. Finally, good communication skills are necessary to share the information gained from the analysis.

INTRODUCTORY PROBLEMS

1. Why is a thorough understanding of a human movement to be analyzed of value to the analyst?

2. Distinguish between movement skills, patterns, and techniques.
3. Select a familiar movement and list the factors that contribute to skilled versus unskilled performance of that movement.
4. Construct an outline of the steps an analyst should take in performing a thorough analysis of a human movement.

5. Test your observation skills by carefully observing the two drawings shown. List the differences that you are able to identify between these two drawings.
6. Observe a single performer executing a movement at both slow and fast speeds. Write a paragraph comparing the kinematics of the two movements. Explain in what ways your ability to observe the two movements may have been different.

7. Choose a familiar sport and list the skills used in that sport that are best observed from close-up, 2 to 3 meters away, and reasonably far away. Write a brief explanation for your choices.
8. Choose a familiar sport and list the skills used in that sport that are best observed from the side view, front view, rear view, and top view. Write a brief explanation for your choices.
9. List the analytical strategies that an analyst of human movement can employ to maximize the information yielded by a set of observational data.
10. Choose one of the instrumentation systems described and write a short paragraph explaining the way in which it might be used to study a question related to analysis of a human movement of interest to you.

ADDITIONAL PROBLEMS

1. Select a familiar movement and identify the ways in which performance of that movement is affected by strength, flexibility, and coordination.
2. List three human movement patterns or skills that might best be observed from a side view, a front or rear view, and a top view.
3. Observe and analyze a single performer executing two similar but different versions of a particular movement (for example, two pitching styles or two gait styles). Explain what viewing perspectives and distances you selected for collecting observational data on each movement. Write a paragraph comparing the kinematics of the two movements.
4. Select a nonplanar movement of interest and list the protocol you would employ in analyzing that movement.
5. Select a partner and plan and carry out an observational analysis of a movement of interest. Write a composite summary analysis of the movement performance. Write a paragraph identifying in what ways the analysis process was changed by the inclusion of a partner.
6. Plan and carry out a videotape session of a slow movement of interest as performed by two different subjects. Write a comparative analysis of the subjects' performances, using the organizational format for scientific papers presented in the chapter.
7. What special expectations, if any, should the analyst have of movement performances if the performer is a senior citizen?

An elementary school-aged girl? A novice? An obese high school-aged boy?

8. Select a partner and choose a novel skill that you both can practice each day over the course of a week (for example, juggling two tennis balls with one hand). Meet briefly with your partner every other day to record observational data on each other's performances. At the same time, record observational data on your own performance. At the end of the week, compare each subjective performance log with the corresponding objective performance log. Write a short description of the ways in which the logs do or do not correspond.
9. Locate an article in the biomechanics research literature that focuses on a movement of interest to you. What instrumentation was used by the researchers? What viewing distances and perspectives were used? How might the analysis described have been improved?
10. Select a movement at which you are reasonably skilled. Plan and carry out observations of a less-skilled individual performing the movement and provide verbal learning cues for that individual if appropriate. Write a short description of the cues provided with a rationale for each cue.

REFERENCES

1. Allison PC: What and how preservice physical education teachers observe during an early field experience, Res Q Exerc Sport 58:242, 1987.
2. Armstrong CW and Hoffman SJ: Effect of teaching experience, knowledge of performer competence, and knowledge of performance outcome on performance error identification, Res Q 50:318, 1979.
3. Bard C et al: Analysis of gymnastics judges' visual search, Res Q Exerc Sport 51:267, 1980.
4. Basmajian JV: Muscles alive, ed 4, Baltimore, 1978, Williams & Wilkins.
5. Eastman Kodak Company: High speed photography, Standard Book No 0-87985-165-1, 1979.
6. Hoffman SJ: The contributions of biomechanics to clinical competence: a view from the gymnasium. In Shapiro R and Marett JR, eds: Proceedings of the second national symposium on teaching kinesiology and biomechanics in sports, Colorado Springs, Colo, 1984, US Olympic Committee.

7. Hoffman SJ: Toward a pedagogical kinesiology, Quest 28:38, 1977.
8. Hoffman SJ and Sembiante JL: Experience and imagery in movement analysis. In Alderson GJK and Tyldesley DA, eds: British proceedings of sport psychology, Salford, England, 1975, British Society of Sports Psychology.
9. Hudson JL: The value of visual variables in biomechanical analysis. In Kreighbaum E and McNeill A, eds: Proceedings of the 1988 symposium of the International Society of Biomechanics in Sports, Bozeman, Mont, 1990, Montana State University.
10. Hudson JL, Lee EJ, and Disch JG: The influence of biomechanical measurement systems on performance. In Adrian M and Deutsch H: Biomechanics: the 1984 Olympic Scientific Congress proceedings, Eugene, Ore, 1986, Norman Ross-Microform Academic Publishers.
11. Mozley AM: Introduction to the dover edition. In Muybridge's complete human and animal locomotion, New York, 1979, Dover Publications, Inc.
12. Thomas JR and Thomas KT: Development of gender differences in physical activity, Quest 40:219, 1988.
13. Williams D and Bradford B: Fighting eyes, Strategies 2:21, 1989.

ANNOTATED READINGS

Dainty DA and Norman RW: Standardizing biomechanical testing in sport, Champaign, Ill, 1987, Human Kinetics Publishers, Inc.

Includes expanded discussion of the characteristics, advantages, and disadvantages of instrumentation systems commonly employed in biomechanics research.

Hay JG: The development of deterministic models for qualitative analysis, Proceedings of the second national symposium on teaching kinesiology and biomechanics in sports, Colorado Springs, Colo, 1984, US Olympic Committee.

Presents a procedure for constructing and using models of variables affecting performance in analyzing human movements.

Hudson JL: The value of visual variables in biomechanical analysis, Proceedings of the 1988 symposium of the International Society of Biomechanics in Sports, (in press).

Describes biomechanical variables that can be assessed visually and discusses their relative value for practitioners.

APPENDIX A

Basic Mathematics and Related Skills

NEGATIVE NUMBERS

Negative numbers are preceded by a minus sign. Although the physical quantities used in biomechanics do not have values that are less than zero in magnitude, the minus sign is often used to indicate the direction opposite the direction regarded as positive. Therefore, it is important to recall the following rules regarding arithmetic operations involving negative numbers:

1. Addition of a negative number yields the same results as subtraction of a positive number of the same magnitude:

$$6 + (-4) = 2$$
$$10 + (-3) = 7$$
$$6 + (-8) = -2$$
$$10 + (-23) = -13$$
$$(-6) + (-3) = -9$$
$$(-10) + (-7) = -17$$

2. Subtraction of a negative number yields the same result as addition of a positive number of the same magnitude:

$$5 - (-7) = 12$$
$$8 - (-6) = 14$$
$$-5 - (-3) = -2$$
$$-8 - (-4) = -4$$
$$-5 - (-12) = 7$$
$$-8 - (-10) = 2$$

3. Multiplication or division of a number by a number of the opposite sign yields a negative result:

$$2 \times (-3) = -6$$
$$(-4) \times 5 = -20$$
$$9 \div (-3) = -3$$
$$(-10) \div 2 = -5$$

4. Multiplication or division of a number by a number of the same sign (positive or negative) yields a positive result:

$$3 \times 4 = 12$$
$$(-3) \times (-2) = 6$$
$$10 \div 5 = 2$$
$$(-15) \div (-3) = 5$$

EXPONENTS

Exponents are superscripted numbers that immediately follow a base number, indicating the number of times that number is to be self-multiplied to yield the result:

$$\begin{aligned} 5^2 &= 5 \times 5 \\ &= 25 \end{aligned}$$

$$\begin{aligned} 3^2 &= 3 \times 3 \\ &= 9 \end{aligned}$$

$$5^3 = 5 \times 5 \times 5$$
$$= 125$$
$$3^3 = 3 \times 3 \times 3$$
$$= 27$$

SQUARE ROOTS

Taking the square root of a number is the inverse operation of squaring a number (multiplying a number by itself). The square root of a number is the number that yields the original number when multiplied by itself. The square root of 25 is 5, and the square root of 9 is 3. Using mathematics notation these relationships are expressed as the following:

$$\sqrt{25} = 5$$

$$\sqrt{9} = 3$$

Because -5 multiplied by itself also equals 25, -5 is also a square root of 25. The following notation is sometimes used to indicate that square roots may be either positive or negative:

$$\sqrt{25} = \pm 5$$
$$\sqrt{9} = \pm 3$$

ORDER OF OPERATIONS

When a computation involves more than a single operation, a set of rules must be used to arrive at the correct result. These rules may be summarized as follows:

1. Addition and subtration are of equal precedence; these operations are carried out from left to right as they occur in an equation:

$$7 - 3 + 5 = 4 + 5$$
$$= 9$$

$$5 + 2 - 1 + 10 = 7 - 1 + 10$$
$$= 6 + 10$$
$$= 16$$

2. Multiplication and division are of equal precedence; these operations are carried out from left to right as they occur in an equation:

$$10 \div 5 \times 4 = 2 \times 4$$
$$= 8$$

$$20 \div 4 \times 3 \div 5 = 5 \times 3 \div 5$$
$$= 15 \div 5$$
$$= 3$$

3. Multiplication and division take precedence over addition and subtraction. In computations involving some combination of operations not of the same level of precedence, multiplication and division are carried out before addition and subtraction are carried out:

$$3 + 18 \div 6 = 3 + 3$$
$$= 6$$

$$9 - 2 \times 3 + 7 = 9 - 6 + 7$$
$$= 3 + 7$$
$$= 10$$

$$8 \div 4 + 5 - 2 \times 2 = 2 + 5 - 2 \times 2$$
$$= 2 + 5 - 4$$
$$= 7 - 4$$
$$= 3$$

4. When parentheses ( ), brackets [ ], or braces { } are used, the operations enclosed are performed first, before the other rules of precedence are applied:

$$2 \times 7 + (10 - 5) = 2 \times 7 + 5$$
$$= 14 + 5$$
$$= 19$$

$$20 \div (2 + 2) - 3 \times 4 = 20 \div 4 - 3 \times 4$$
$$= 5 - 3 \times 4$$
$$= 5 - 12$$
$$= -7$$

USE OF A CALCULATOR

Simple computations in biomechanics problems are often performed quickly and easily with a hand-held calculator. However, the correct result can only be obtained on a calculator when the computation is set up properly and the rules for ordering of operations are followed. Most calculators come with an instruction manual that contains sample calculations. It is worthwhile to completely familiarize yourself with your calculator's capabilities, particularly use of the memory, before using it for solving problems.

PERCENTAGES

A percentage is a part of 100. Thus, 37% represents 37 parts of 100. To find 37% of 80, multiply the number 80 by 0.37:

$$80 \times 0.37 = 29.6$$

The number 29.6 is 37% of 80. If you want to determine the percentage of the number 55 that equals 42, multiply the fraction by 100%:

$$\frac{42}{55} \times 100\% = 76.4\%$$

The number 42 is 76.4% of 55.

SIMPLE ALGEBRA

The solution of many problems involves setting up an equation containing one or more unknown quantities represented as variables such as *x*. An equation is a statement of equality implying that the quantities expressed on the left side of the equal sign are equal to the quantities expressed on the right side of the equal sign. Solving a problem typically requires calculation of the unknown quantity or quantities contained in the equation.

The general procedure for calculating the value of a variable in an equation is to isolate the variable on one side of the equal sign and then to carry out the operations among the numbers expressed on the other side of the equal sign. The process of isolating a variable usually involves performing a series of operations on both sides of the equal sign. As long as the same operation is carried out on both sides of the equal sign, equality is preserved and the equation remains valid:

$$x + 7 = 10$$

Subtract 7 from both sides of the equation:

$$x + 7 - 7 = 10 - 7$$
$$x + 0 = 10 - 7$$
$$x = 3$$

$$y - 3 = 12$$

Add 3 to both sides of the equation:

$$y - 3 + 3 = 12 + 3$$
$$y - 0 = 12 + 3$$
$$y = 15$$

$$z \times 3 = 18$$

Divide both sides of the equation by 3:

$$z \times 3 \div 3 = \frac{18}{3}$$

$$z \times 1 = \frac{18}{3}$$

$$z = 6$$

$$q \div 4 = 2$$

Multiply both sides of the equation by 4:

$$q \div 4 \times 4 = 2 \times 4$$
$$q \times 1 = 2 \times 4$$
$$q = 8$$

$$x \div 3 + 5 = 8$$

Subtract 5 from both sides of the equation:

$$x \div 3 + 5 - 5 = 8 - 5$$
$$x \div 3 = 3$$

Multiply both sides of the equation by 3:

$$x \div 3 \times 3 = 3 \times 3$$
$$x = 9$$

$$y \div 4 - 7 = -2$$

Add 7 to both sides of the equation:

$$y \div 4 - 7 + 7 = -2 + 7$$
$$y \div 4 = 5$$

Multiply both sides of the equation by 4:

$$y \div 4 \times 4 = 5 \times 4$$
$$y = 20$$

$$z^2 = 36$$

Take the square root of both sides of the equation:

$$z = 6$$

MEASURING ANGLES

The following procedure is used for measuring an angle with a protractor:

1. Place the center of the protractor on the vertex of the angle.
2. Align the zero line on the protractor with one of the sides of the angle.
3. The size of the angle is indicated on the protractor scale where the other side of the angle intersects the scale. (Be sure to read from the correct scale on the protractor. Is the angle greater or less than 90 degrees?)

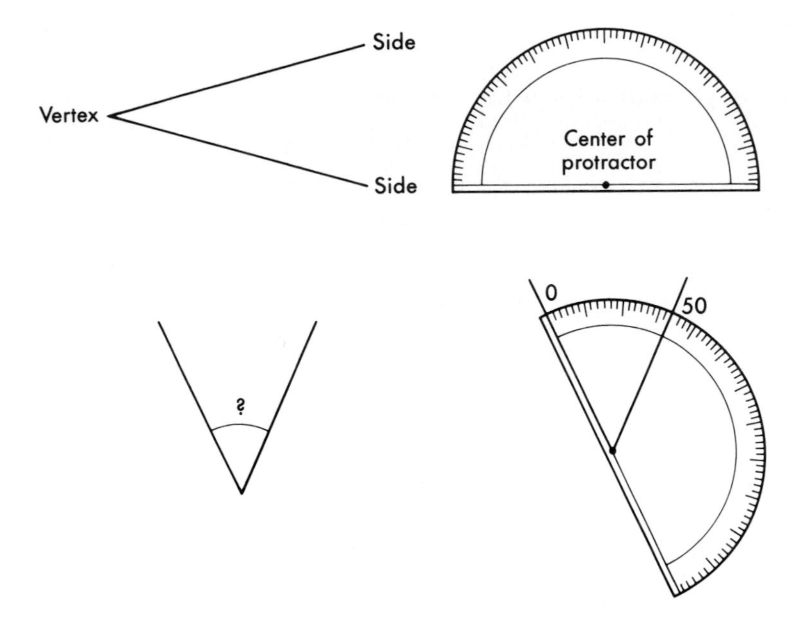

If you are unfamiliar with the use of a protractor, check yourself by verifying the sizes of the following three angles:

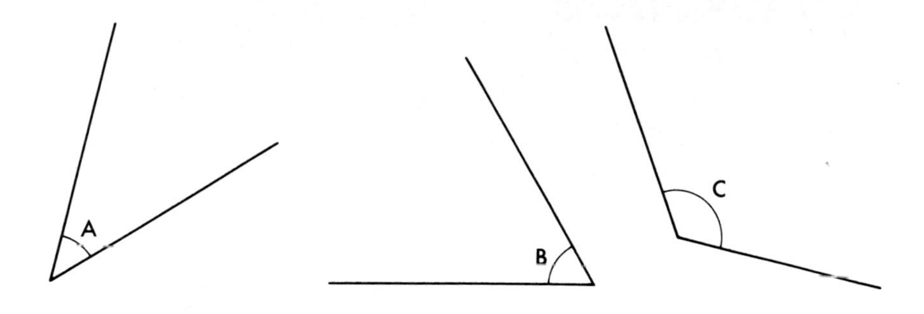

(Answer: A = 45°, B = 60°, C = 123°)

APPENDIX B

Trigonometric Functions

Trigonometric functions are based on relationships present between the sides and angles of triangles. Many functions are derived from a right triangle—a triangle containing a right (90°) angle. Consider the right triangle below with sides A, B, and C, and angles α, β, and γ:

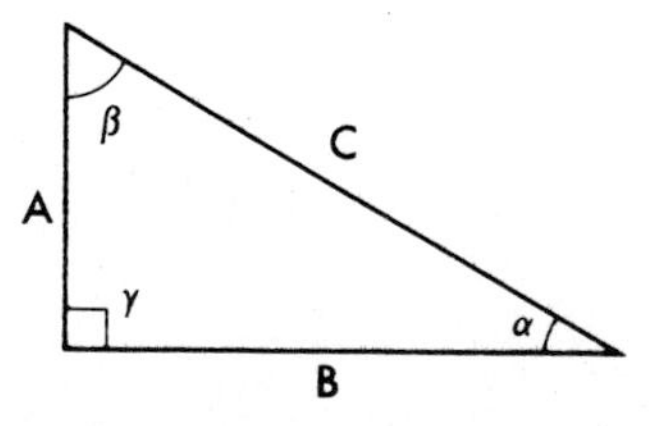

Side C, which is the longest side and the side opposite the right angle, is known as the *hypotenuse of the triangle.*

A commonly used trigonometric relationship for right triangles is the *Pythagorean theorem.* The Pythagorean theorem is an expression of the relationship between the hypotenuse and the other two sides of a right triangle:

The sum of the squares of the lengths of the two sides of a right triangle is equal to the square of the length of the hypotenuse.

Using the sides of the labeled triangle yields the following:

$$A^2 + B^2 = C^2$$

Suppose that sides A and B are 3 and 4 units long, respectively. The Pythagorean Theorem can be used to solve for the length of side C:

$$\begin{aligned} C^2 &= A^2 + B^2 \\ &= 3^2 + 4^2 \\ &= 9 + 16 \\ &= 25 \\ C &= 5 \end{aligned}$$

Three trigonometric relationships are based on the ratios of the lengths of the sides of a right triangle. The sine (abbreviated *sin*) of an angle is defined as the ratio of the length of the side of the triangle opposite the angle to the length of the hypotenuse. Using the labeled triangle yields the following:

$$\sin \alpha = \frac{\text{opposite}}{\text{hypotenuse}} = \frac{A}{C}$$

$$\sin \beta = \frac{\text{opposite}}{\text{hypotenuse}} = \frac{B}{C}$$

With A = 3, B = 4, and C = 5:

$$\sin \alpha = \frac{A}{C} = \frac{3}{5} = 0.6$$

$$\sin \beta = \frac{B}{C} = \frac{4}{5} = 0.8$$

The cosine (abbreviated *cos*) of an angle is defined as the ratio of the length of the side of the triangle adjacent to the angle to the length of the hypotenuse. Using the labeled triangle yields the following:

$$\cos \alpha = \frac{\text{adjacent}}{\text{hypotenuse}} = \frac{B}{C}$$

$$\cos \beta = \frac{\text{adjacent}}{\text{hypotenuse}} = \frac{A}{C}$$

With A = 3, B = 4, and C = 5:

$$\cos \alpha = \frac{B}{C} = \frac{4}{5} = 0.8$$

$$\cos \beta = \frac{A}{C} = \frac{3}{5} = 0.6$$

The third function, the tangent (abbreviated *tan*) of an angle, is defined as the ratio of the length of the side of the triangle opposite the angle to that of the side adjacent to the angle. Using the labeled triangle yields the following:

$$\tan \alpha = \frac{\text{opposite}}{\text{adjacent}} = \frac{A}{B}$$

$$\tan \beta = \frac{\text{opposite}}{\text{adjacent}} = \frac{B}{A}$$

With $A = 3$, $B = 4$, and $C = 5$:

$$\tan \alpha = \frac{A}{B} = \frac{3}{4} = 0.75$$

$$\tan \beta = \frac{B}{A} = \frac{4}{3} = 1.33$$

Two useful trigonometric relationships are applicable to *all* triangles. The first is known as the Law of Sines:

The ratio between the length of any side of a triangle and the angle opposite that side is equal to the ratio between the length of any other side of the triangle and the angle opposite that side.

With respect to the labeled triangle, this may be stated as the following:

$$\frac{A}{\sin \alpha} = \frac{B}{\sin \beta} = \frac{C}{\sin \gamma}$$

A second trigonometric relationships applicable to *all* triangles is the law of cosines:

The square of the length of any side of a triangle is equal to the sum of the squares of the lengths of the other two sides of the triangle minus two times the product of the lengths of the other two sides and the cosine of the angle opposite the original side.

This relationship yields the following for each of the sides of the labeled triangle:

$$A^2 = B^2 + C^2 - 2BC \cos \alpha$$
$$B^2 = A^2 + C^2 - 2AC \cos \beta$$
$$C^2 = A^2 + B^2 - 2AB \cos \gamma$$

APPENDIX C

Common Units of Measurement

This appendix contains factors for converting between metric units commonly used in biomechanics and their English system equivalents. In each case a value expressed in a metric unit can be divided by the conversion factor given to yield the approximate equivalent in an English unit, or a value expressed in and English unit can be multiplied by the conversion factor to find the metric unit equivalent. For example, to convert 100 Newtons to pounds, do the following:

$$\frac{100 \text{ N}}{4.45 \text{ N/lb}} = 22.5 \text{ lb}$$

To convert 100 pounds to Newtons, do the following:

$$(100 \text{ lb})\,(4.45 \text{ N/lb}) = 445 \text{ N}$$

VARIABLE	METRIC UNIT	← MULTIPLY BY DIVIDE BY →	ENGLISH UNIT
Distance	Centimeters	2.54	Inches
	Meters	0.3048	Feet
	Kilometers	1.609	Miles
Speed	Meters/second	0.447	Miles/hour
Mass	Kilograms	14.59	Slugs
Force	Newtons	4.448	Pounds
Work	Joules	1.355	Foot-pounds
Power	Watts	745.63	Horsepower
Energy	Joules	1.355	Foot-pounds
Linear momentum	Kilogram-meters/second	4.448	Slug-feet/second
Impulse	Newton-seconds	4.448	Pound-seconds
Angular momentum	Kilogram-meters^2/second	1.355	Slug-feet^2/second
Moment of inertia	Kilogram-meters^2	1.355	Slug-feet^2
Torque	Newton-meters	1.355	Foot-pounds

APPENDIX D

Anthropometric Parameters for the Human Body*

SEGMENT LENGTHS

SEGMENT	MALES	FEMALES
Head and neck	10.75	10.75
Trunk	30.00	29.00
Upper arm	17.20	17.30
Forearm	15.70	16.00
Hand	5.75	5.75
Thigh	23.20	24.90
Lower leg	24.70	25.70
Foot	4.25	4.25

Segment lengths expressed in percentages of total body height.

SEGMENT WEIGHTS

SEGMENT	MALES	FEMALES
Head	8.26	8.20
Trunk	55.10	53.20
Upper arm	3.25	2.90
Forearm	1.87	1.57
Hand	0.65	0.50
Thigh	10.50	11.75
Lower leg	4.75	5.35
Foot	1.43	1.33

Segment weights expressed in percentages of total body weight.

From Plagenhoef S, Evans FG, and Abdelnour T: Anatomical data for analyzing human motion, Res Q Exerc Sport 54:169, 1983.
*The values reported in these tables represent mean values for limited numbers of subjects as reported in the scientific literature.

SEGMENTAL CENTER OF GRAVITY LOCATIONS

SEGMENT	MALES	FEMALES
Head and neck	55.0	55.0
Trunk	63.0	56.9
Upper arm	43.6	45.8
Forearm	43.0	43.4
Hand	46.8	46.8
Thigh	43.3	42.8
Lower leg	43.4	41.9
Foot	50.0	50.0

Segmental center of gravity locations expressed in percentages of segment lengths; measured from the proximal ends of segments.

SEGMENTAL RADII OF GYRATION MEASURED FROM PROXIMAL AND DISTAL SEGMENT ENDS

	MALES		FEMALES	
SEGMENT	PROXIMAL	DISTAL	PROXIMAL	DISTAL
Upper arm	54.2	64.5	56.4	62.3
Forearm	52.6	54.7	53.0	64.3
Hand	54.9	54.9	54.9	54.9
Thigh	54.0	65.3	53.5	65.8
Lower leg	52.9	64.2	51.4	65.7
Foot	69.0	69.0	69.0	69.0

Segmental radii of gyration expressed in percentages of segment lengths.

GLOSSARY

acceleration	the rate of change in velocity
active stretching	the stretching of muscles, tendons, and ligaments produced by active development of tension in the antagonist muscles.
agonist	the role played by a muscle acting to cause a movement
anatomical reference position	the erect standing position with all body parts, including the palms of the hands, facing forward; used as the starting position for body segment movements
angle of attack	the angle between the longitudinal axis of a body and the direction of the fluid flow
angle of projection	the angle at which a body is projected with respect to the horizontal
angular motion	a motion involving rotation around a central line or point
anisotropic	exhibiting different mechanical properties in response to loads from different directions
antagonist	the role played by a muscle generating torque opposing that generated by the agonists at a joint
anthropometric	pertaining to the dimensions and weights of body segments
apex	the highest point in the trajectory of a projectile
appendicular skeleton	bones of the body appendages
Archimedes Principle	the buoyant force acting on a body is equal in magnitude to the weight of the fluid displaced by the body
articular cartilage	a protective layer of cartilage tissue that covers the ends of articulating bones at diarthrodial joints
articular fibrocartilage	discs or menisci composed of fibrocartilage that intervenes between articulating bones at amphiarthrodial joints
axial	directed along the longitudinal axis of a body
axial skeleton	includes the skull, the vertebrae, the sternum, and the ribs
axis of rotation	an imaginary line oriented perpendicular to the plane of rotation and passing through the center of rotation
balance	the ability to control equilibrium
base of support	an area bound by the outermost regions of contact between a body and support surface

bending	an asymmetric loading that produces tension on one side of a body's longitudinal axis and compression on the other side
Bernoulli principle	the expression of the inverse relationship between relative velocity and relative pressure in a fluid flow
biomechanics	the application of mechanical principles in the study of living organisms
boundary layer	the layer of fluid immediately adjacent to a body
bursae	sacs located around joints that secrete synovial fluid to lessen friction between soft tissues
cancellous bone	bone tissue with high porosity; found in the ends of long bones and in the vertebrae
cardinal plane	one of the three imaginary reference planes that divides the body in half by mass or weight
center of gravity	(center of mass, mass centroid) point around which the mass and weight of a body is balanced in all directions; the point at which gravitational force acts
center of volume	the point around which a body's volume is equally balanced; the point at which the buoyant force acts
centrifugal force	the reaction force equal in magnitude and opposite in direction to centripetal force
centripetal force	a force directed toward the center of rotation of any rotating body; equal in magnitude to mv^2/r
close-packed position	a joint position in which the contact between the articulating bone surfaces is maximal
closed skills	skills executed in a predictable setting
coefficient of drag	a unitless number that serves as an index of a body's ability to generate drag
coefficient of friction	a unitless number that serves as an index of the interaction between two surfaces in contact
coefficient of lift	a unitless number that serves as an index of a body's ability to generate lift
coefficient of restitution	a unitless number that serves as an index of the elasticity of a collision
combined loading	the simultaneous action of more than one of the pure forms of loading
compression	a pressing or squeezing force directed axially through a body
concentric	involving shortening of a muscle

continuous skills	purposeful movement patterns of a cyclic, repetitive nature
cortical bone	compact bone tissue with low porosity; found in the shafts of long bones
couple	a pair of equal, oppositely directed forces that act on opposite sides of an axis of rotation to produce torque
curvilinear	along a curved line
deformation	change in original shape
density	mass per unit of volume
diagonal planes	planes of movement oriented diagonally to the traditionally recognized planes of movement
discrete skills	purposeful movement patterns with identifiable beginnings and endings
displacement	a change in position
drag	the force created by the relative motion of a body in a fluid
dynamic equilibrium	(D'Alembert's Principle) the concept indicating a balance between applied forces and inertial forces for a body in motion
dynamics	the branch of mechanics that deals with systems subject to acceleration
eccentric	involving lengthening of a muscle
elastic limit	the critical point beyond which a stretched material remains stretched and is incapable of regaining its original length
elasticity	the ability to regain original size and shape after a load is removed
English system	the system of weights and measures originally developed in England and used in the United States today
epiphysis	the growth center of a bone that produces new bone tissue as part of the normal growth process until it closes during adolescence or early adulthood
flexibility	a qualitative term used to represent the ranges of motion present at a joint in different directions
fluid	a substance that flows when subjected to a shear stress
force	a push or pull that causes or tends to cause motion; the product of a body's mass and its acceleration
form drag	(profile drag, pressure drag) the drag created by a pressure differential between the lead and rear sides of a body moving through a fluid
free body diagram	a sketch that shows a given body in isolation with vector representations of all the forces acting on the body
frontal plane	a plane in which lateral movements toward and away from the midline of the body occur; the frontal cardinal plane divides the body into front and back halves
friction	a force acting at the area of contact between two surfaces in the direction opposite that of motion or motion tendency
fulcrum	the point of support or axis about which a lever may be made to rotate
general motion	motion involving translation and rotation simultaneously
impact	a collision characterized by the exchange of a large force during a small time interval
impulse	the product of a force and the time interval over which the force acts

inertia	the tendency of a body to maintain a motionless state or a state of constant velocity
initial velocity	the vector quantity incorporating both angle and speed of projection
instant center	the precisely located center of rotation at a joint at a given time
isometric	involving no change in muscle length
joint mobility	a term indicating the relative degree of motion allowed at a joint
joint stability	the ability of a joint to resist abnormal displacement of the articulating bones
kinematics	the form, pattern, or sequencing of movement with respect to time
kinesiology	the study of human movement from the perspectives of art and science
kinetic energy	the energy of motion; calculated as $\frac{1}{2}mv^2$
kinetic friction	the constant friction generated between two surfaces in contact during motion
kinetics	the study of the forces causing or resulting from motion
kyphosis	an extreme curvature in the thoracic region of the spine
laminar flow	a flow characterized by smooth, parallel layers of fluid
lever	a simple machine consisting of a relatively rigid barlike body that may be made to rotate about an axis
lift	the force acting on a body in a fluid in a direction perpendicular to the fluid flow
lift/drag ratio	the magnitude of the lift force divided by the magnitude of the total drag force acting on a body at a given time
linear	following a path along a line that may be straight or curved, with all parts of the body moving in the same direction at the same speed
longitudinal axis	an imaginary line around which transverse plane rotations occur
loose-packed position	all joint positions in which the contact between the articulating bone surfaces is other than maximum
lordosis	an extreme curvature in the lumbar region of the spine
Magnus effect	the deviation in the trajectory of a spinning object toward the direction of spin resulting from the Magnus force
Magnus force	the lift force created by spin
mass	the quantity of matter contained in an object
maximum static friction	the maximum amount of friction that can be generated between two static surfaces
mechanical advantage	the ratio of force arm/resistance arm for a given lever; an MA greater than 1 indicates a force advantage, and an MA less than 1 indicates a range of motion and speed advantage
mechanics	the branch of physics that deals with the actions of forces on particles and on mechanical systems
meter	the most common international unit of length on which the metric system is based; equal in English units to 3.281 feet
metric system	a system of weights and measures used in scientific applications and adopted by every major country in the world except the United States

moment arm	the shortest (perpendicular) distance between a force's line of action and an axis of rotation
moment of inertia	the inertial property for rotating bodies; it increases with both mass and the distance that the mass is distributed from the axis of rotation
momentum	the product of a body's mass and its velocity
motion segment	two adjacent vertebrae and the intervening soft tissues of the joint; the functional unit of the spine
motor unit	a single motor neuron and all of the fibers it innervates
myoelectric activity	the electric current or voltage produced by a muscle developing tension
net force	the resultant force derived from the composition of two or more forces
neutralizer	the role played by a muscle acting to eliminate an unwanted action produced by an agonist
open skills	skills executed in an unpredictable setting that may require quick adaptations of the performer
osteoblasts	bone cells that function to build new bone tissue during bone growth and in response to increased mechanical stress
osteoclasts	bone cells that resorb bone tissue during bone maturation and in response to reduced mechanical stress
osteoporosis	a pathological condition of decreased bone mass and strength
passive stretching	the stretching of muscles, tendons, and ligaments produced by a stretching force other than tension in the antagonist muscles
potential energy	stored energy; calculated as a body's weight multiplied by its height
power	the rate of work production; calculated as work divided by the time during which the work was done
pressure	force per unit of area
primary spinal curves	the thoracic and sacral curves; curves that are present at birth
principal axes	the three mutually perpendicular axes passing through the center of gravity; referred to respectively as the transverse, anteroposterior, and longitudinal axes
principal moment of inertia	total body moment of inertia relative to one of the principal axes
projectile	a body in free fall that is subject only to the forces of gravity and air resistance
projection speed	the magnitude of projection velocity
proprioceptive neuromuscular facilitation	(PNF) an effective stretching procedure involving use of a partner to provide resistance and assist with stretching
propulsive drag	drag acting in the direction of a body's motion
qualitative	pertaining to quality; nonnumerical
quantitative	involving the use of numbers
radial acceleration	the component of angular acceleration directed toward the center or curvature; indicates change in direction
radian	a unit of angular measure used in angular-linear kinematic quantity conversions; equal to 57.3 degrees

radius of gyration	the distance from the axis of rotation to a point where the body's mass could be concentrated without altering its rotational characteristics
radius of rotation	the distance from the axis of rotation to a point of interest on a rotating body
range	the horizontal displacement of a projectile at landing
reaction board	a specially constructed board for determining the center of gravity location of a body positioned on top of it
reciprocal inhibition	the inhibition of the antagonist muscles resulting from activation of muscle spindles
rectilinear	along a straight line
relative projection height	the difference between projection height and landing height
relative velocity	the velocity of a body with respect to the velocity of something else, such as the surrounding fluid
repetitive loading	the application of a nontraumatic load, usually of relatively low magnitude, on a repeated basis
resultant	the single vector that results from vector composition
right hand rule	the procedure for identifying the direction of an angular motion vector
rotator cuff	four muscles (subscapularis, supraspinatus, infraspinatus, and teres minor) that have tendinous attachments to the capsule of the glenohumeral joint
sagittal plane	a plane in which forward and backward movements of the body and body segments occur; the sagittal cardinal plane divides the body into right and left halves
scalar quantity	a physical quantity that is completely described when its magnitude is known
scoliosis	a lateral spinal curvature
secondary spinal curves	the cervical and lumbar curves, which do not develop until the weight of the body begins to be supported in sitting and standing positions
shear	a force directed parallel to a surface
skin friction	(surface drag, viscous drag) the drag derived from friction between adjacent layers of fluid near a body moving through the fluid
specific weight	(weight density) weight per unit of volume
stability	the resistance to disruption of equilibrium
stabilizer	the role played by a muscle acting to stabilize a body part against some other force
static equilibrium	a motionless state characterized by $\Sigma F_v = 0$, $\Sigma F_h = 0$, and $\Sigma T = 0$
statics	the branch of mechanics that deals with systems not subject to acceleration
stiffness	the ratio of stress to strain in a loaded material; the stress divided by the relative amount of change in the structure's shape
strain	the amount of deformation divided by the original length of the structure or the original angular orientation of the structure
strain energy	a form of potential energy stored when a body is deformed
stress	the distribution of force within a body; quantified as force divided by the area over which the force acts
stretch reflex	the monosynaptic reflex initiated by stretching of muscle spindles and resulting in immediate development of muscle tension

tangential acceleration the component of angular acceleration directed along a tangent to the path of motion; indicates change in linear speed

tension pulling or stretching force directed axially through a body

torque (moment of force) a rotary force that produces angular acceleration

trajectory the flight path of a projectile

transducers devices that detect a signal of some sort

translation synonymous with linear motion

transverse plane a plane in which body movements parallel to the ground occur when the body is in an erect standing position; the transverse cardinal plane divides the body into top and bottom halves

turbulent flow a flow characterized by mixing of adjacent fluid layers

valgus the condition of outward deviation in alignment from the proximal to the distal end of a body segment

varus the condition of inward deviation in alignment from the proximal to the distal end of a body segment

vector composition	**(vector addition) the process of determining a single vector from two or more vectors**
vector quantity	**a physical quantity that possesses both magnitude and direction**
vector resolution	**an operation that replaces a single vector with two perpendicular vectors so that the vector composition of the two perpendicular vectors yields the original vector**
velocity	**a change in position with respect to time**
volume	**the space occupied by a body**
wave drag	**the drag created by the generation of waves at the interface between two different fluids, such as air and water**
weight	**the attractive force that the earth exerts on a body**
work	**the expression of mechanical energy; calculated as force multiplied by the distance the resistance is moved**

INDEX

B

C

D

E

F

G

J

K

L

M

Q

R

S

T

U

V

W